Nanoscale Energy Transport and Conversion

Nanoscale Energy Transport and Conversion
A Parallel Treatment of Electrons, Molecules, Phonons, and Photons

Gang Chen

Massachusetts Institute of Technology

2005

OXFORD
UNIVERSITY PRESS

Oxford University Press, Inc., publishes works that further
Oxford University's objective of excellence
in research, scholarship, and education.

Oxford   New York
Auckland   Cape Town   Dar es Salaam   Hong Kong   Karachi
Kuala Lumpur   Madrid   Melbourne   Mexico City   Nairobi
New Delhi   Shanghai   Taipei   Toronto

With offices in
Argentina   Austria   Brazil   Chile   Czech Republic   France   Greece
Guatemala   Hungary   Italy   Japan   Poland   Portugal   Singapore
South Korea   Switzerland   Thailand   Turkey   Ukraine   Vietnam

Copyright © 2005 by Oxford University Press, Inc.

Published by Oxford University Press, Inc.
198 Madison Avenue, New York, New York 10016

www.oup.com

Oxford is a registered trademark of Oxford University Press

All rights reserved. No part of this publication may be reproduced,
stored in a retrieval system, or transmitted, in any form or by any means,
electronic, mechanical, photocopying, recording, or otherwise,
without the prior permission of Oxford University Press.

Library of Congress Cataloging-in-Publication Data
Chen, Gang PhD
Nanoscale energy transport and conversion: a parallel treatment of electrons,
molecules, phonons, and photons/Gang Chen.
    p. cm.
Includes bibliographical references and index.
ISBN-13 978-0-19-515942-4

  1. Thermodynamics. 2. Heat—Transmission. I. Title

TJ265.C497 2004
621.4021—dc22   2004052066

Printed in the United States of America
on acid-free paper

In Memory of

Chang-Lin Tien

A Pioneer
A Source of Inspiration

Foreword

As the size of devices and structures has become smaller and smaller and entered the nanoscale, the physical principles governing their operation are changing dramatically. We can thus expect the experts in heat transfer and energy conversion at the nanoscale will be in large demand in the coming years, as the electronics, biotech, and aeronautics industries develop smaller and smaller devices with increased heat dissipation density, and increased flows through narrow channels, and the energy industry takes advantage of nanotechnology to improve energy conversion efficiency. This new type of engineer will require more knowledge of the fundamentals of heat transfer and energy exchange between electrons, phonons, photons, and molecules.

These changes in the world around us are profoundly changing educational curricula and programs as we prepare students to work in a future era that is difficult to envision based on what we know today. Students now enrolled in degree programs will face many new demands on their knowledge base – demands that are much broader in scope, more rapidly changing in time, and less poorly defined in ultimate content than anything we have yet experienced in educating students for their lifetime careers.

To address these concerns, mechanical engineering departments nationwide have been developing new courses. It is in this context that this textbook *Nanoscale Energy Transport and Conversion* by Professor Gang Chen, sponsored by the MIT-Pappalardo Series in Mechanical Engineering, makes such a valuable contribution. Furthermore, the American Society for Mechanical Engineers (ASME) recently formed the Nanotechnology Institute with a charge to provide an effective knowledge base in nanotechnology to address future problem-solving needs of practicing mechanical engineers and mechanical engineering students who will become the next generation of practitioners. This book clearly contributes strongly to implementing the vision of the ASME Nanotechnology Institute.

The book itself provides a parallel treatment of electrons, molecules, phonons, and photons, providing many examples, references, and exercises for the student or practicing engineer to further advance their skills and follow up on their interests or needs. The topics are selected to prepare students both to solve problems in heat transfer/energy conversion at the nanoscale and to work with engineers and scientists from other fields who are creating the problems that need solution. Nanoscience and nanotechnology characteristically couple many traditional disciplines, ranging from electrical engineering aerospace engineering, physics, chemistry, biology, and neuroscience, and nanoscale-based problems generally require a strong interdisciplinary approach. Not only does this book prepare students to solve problems arising from other disciplines, but also the applications presented in the book help the student to acquire the multidisciplinary perspective that will be needed to work effectively in the interdisciplinary environment necessary to address the grand challenges of the 21st century and beyond. As it has been used successfully at both UCLA and MIT for several years, I am confident that this book can provide an effective mechanism to give the necessary knowledge base in nanoscience and nanotechnology that will be needed by future mechanical engineers as well as current practitioners now attempting to navigate these waters.

MILDRED S. DRESSELHAUS
MIT

Preface

This book aims to provide microscopic pictures of thermal energy transport, and energy conversion processes from nanoscale continuously to macroscale. Energy conversion and transport are ubiquitous to natural processes, as well as to engineered devices and systems. The macroscopic description of heat transfer is based on phenomenological laws, such as the Fourier law for heat conduction, Newton's law of shear stress and Newton's law of cooling for convection, and the Stefan–Boltzmann law for thermal radiation. These laws, combined with the first and second laws of thermodynamics, lead to equations for determining energy conversion efficiency, heat transfer rates, and temperature fields in a system. The traditional engineering approach pays little attention to the microscopic processes that govern macroscopic energy conversion and heat transport phenomena. While this engineering approach was adequate for most engineering applications in the past, it becomes increasingly insufficient for other evolving and emerging technologies, particularly nanotechnology, direct energy conversion technologies, biotechnology, microelectronics, and photonics.

The fundamentals of heat transfer and energy conversion are embedded in diverse subjects in different disciplines, such as quantum mechanics, solid-state physics, statistical mechanics, kinetics, and electrodynamics. The closest siblings of this book are those covering kinetic theory, which mostly deal with gases. However, emerging technologies call for the understanding of heat transferred by other energy carriers: electrons, phonons, photons, and liquid molecules. The daunting tasks I face are to integrate diverse materials into one book that can be understood by both students and researchers. I took an approach to treat these energy carriers in a parallel fashion wherever appropriate.

The parallel treatment serves several purposes. First, pedagogically, some readers may be more familiar with the energy transport of one type of carriers and can use analogy to understand other types of energy carriers. For example, mechanical engineers may be more familiar with radiation heat transfer by photons, heat conduction by gas molecules, and acoustic wave propagation, while readers with physics or electrical engineering backgrounds may be more familiar with electrons and electromagnetic waves. I assume most readers have not been extensively exposed to modern physics other than a course in undergraduate physics, as is the situation in a typical mechanical engineering curriculum. Thus, readers who already have taken courses in quantum mechanics and solid-state physics will have an easier time going through the contents. Second, the parallel treatment reflects my conviction that we should provide students with interdisciplinary training for their future multidisciplinary working environment. Through the parallel treatment, readers will see that the diverse energy carriers share many common grounds and are often described by the same equations. It will be revealing for some readers to see that the Fourier law of heat conduction, the Newton law of shear stress, and the Ohm law for electrical current flow can all be derived from the Boltzmann equation, and that heat conduction at nanoscale can be similar to thermal radiation at macroscale. Through these examples and many others in the book, I hope that the readers will see that the discipline boundaries are easy to cross. As the traditional disciplines have been extensively plowed in the last century, it is in the boundary areas that more room exists for future discoveries and innovations. Nanotechnology is just one example that intersects many different disciplines. In fact, there is no single nanotechnology, but there are many potential nano-based technologies that await the readers to invent and to develop.

One purpose of the book is to serve as a senior or graduate level textbook in nanoscale heat transfer and energy conversion; the other goal is to provide researchers, veterans as well as starters, with a reference. The contents and mathematical details of this book are a delicate balance between conflicting requirements on breadth versus depth, and instruction versus research. I tried to set the book at such a level that a serious reader with a mechanical engineering background can go through the derivations and use the book as a starting point for his or her own research, but anticipate that researchers will need to dig out original papers and other reference books for details. The references cited emphasize original papers and are not intended to be inclusive of publications in the area of nano- and micro-scale heat transfer, simply because of the wide range of topics covered. The choice of contents of this book is strongly influenced by my own learning and research experience.

Because of the wide range of topics covered in this book, one challenge for the readers is to embrace different terminologies. This can indeed be sometimes daunting. I have taught part of the contents of the book at both the University of California at Los Angeles and Massachusetts Institute of Technology; the students included graduates, seniors, and even several juniors, most of them having mechanical engineering backgrounds. Feedback from students is overwhelmingly positive but it is also clear that the contents can be "shocking" to some students. My attitude is: no pain, no gain. The gap that needs to be filled is large but the reward is also large. As the reading progresses, I advise students often to skim the most recent issues in journals such as *Science*, *Nature*, *Applied Physics Letters*, and the many new journals in nanotechnology, in addition to reading nanoscale heat transfer papers, and hope that they will enjoy the pleasure of starting to understand what are being discussed in some of the papers. Of course, the true pleasure will come from applying what they have learnt from the book to their own work, and I will be happy to share with readers such joys, as well as criticism of the book.

GANG CHEN
Carlisle, Massachusetts

Acknowledgments

I used to think that the acknowledgments are just a routine duty of the author until I got near to finishing this book. It is a personal feeling best expressed in a public form. It actually became the section that I wanted to write most. Seldom can anyone write a book without building on the work of others, and this book is no exception. Some of the contributors are acknowledged in the form of references while contributions of others may have been treated as common knowledge. At the personal level, I have many friends to thank, starting from my college studies at Huazhong University of Science and Technology in China, the University of California at Berkeley, and my colleagues at Duke University, the University of California at Los Angeles, and now at Massachusetts Institute of Technology. I will not list all of them because of the length of the list and the possibility of missing some of them. But there are two persons who are even more special that I must acknowledge. I would not have been able to make this contribution if it were not for my Ph.D. thesis advisor, the late Professor Chang-Lin Tien, who hand-picked me while traveling in China and continued to supervise me even during his tenure as the chancellor at UC Berkeley. His vision and pioneering work led to the rapid development of the field of micro- and nanoscale heat transfer. Although he left us at an early age, his spirit will continue to inspire many of us who had the fortune to associate with him personally or through his work. I am also deeply grateful to Professor Mildred S. Dresselhaus, who serves as my other role model with her dedication and energy. She also read and commented on the first few chapters of the book. Students who have taken my course on the subject at UCLA and MIT provided many inputs to the book and I thank them without mentioning their names. Also to be thanked are students from RPI in Professor T. Borca-Tasciuc's class and from U. Kentucky in Professor M.P. Menguc's class, who provided written feedback on some chapters. Several students in my current group helped greatly in proofreading and figure drawing: D. Borca-Tasciuc, Z. Chen, J. Cybulski, C. Dames, T. Harris, A. Henry, L. Hu, A. Narayanaswamy, A. Schmidt, A. Shah, and, particularly, R.G. Yang, for summarizing students' comments. Several former students and post-docs are also acknowledged for stimulating me and contributing to my knowledge in the areas covered, including Drs T. Borca-Tasciuc, W.L. Liu, D. Song, S.G. Volz, B. Yang, and T.F. Zeng. Professors J.B. Freund (UIUC), P.M. Norris (U. Virginia), and M. Kaviany (U. Michigan) have provided constructive feedback on chapters of the book. I, of course, take responsibility for any mistakes. I would also acknowledge the K.C. Wong Education Foundation in Hong Kong and the Simon-Guggenheim Foundation for supporting my career at various stages. The inclusion of this book in the MIT–Papallardo Series in Mechanical Engineering would not have been possible without the generous support of Neil and Jane Papallardo and the series editors, Professors Rohan Abeyarantne and Nam Suh.

I would like to thank several funding agencies, including DOE, DARPA, JPL, NASA, NSF, and ONR, that provided support for my research in this area, among them particularly the NSF Young Investigator Award that supported my early career and the DOD/ONR Multidisciplinary University Research Initiative that provided me an opportunity to work across disciplines. This book is in many ways an integration of education and research. Several papers that I have written in the past few years were stimulated from making analogies between different energy carriers.

My editor at Oxford University Press, Danielle Christensen, put gentle pressure on me to finish the seemingly never-ending manuscript. Lisa Stallings and Barbara Brown at OUP were of great help in the production phase of the book. Sue Nicholls and Ian Guy at Keyword Publishing Services Ltd. carried out careful editing of the manuscript. My thanks go to them for turning the manuscript into final print.

I would like to thank my parents and my wife's parents for their moral support and prayers. Most of all my thanks go to my dear wife Tracy and our son and daughter, Andrew and Karen. I once read from the preface of a book in which the author (whom I know) claimed that on average every printed page of his book took three hours. I did not believe him until I came close to finishing this book. I started writing this book about five years ago, when we were expecting our daughter Karen and I changed my working habit to have a few hours in the early morning. I thought that the book could be finished as a gift at the birth of my daughter. That goal turned out to be a gross underestimation of the time needed to complete a book. Karen is now four and Andrew is ten. They have often been checking which chapter I was writing while waiting patiently with Tracy. A big portion of the roughly 1500 hours that I spent on the book was stolen from them. I hope the delivery of this manuscript will mean that I will now find more time to spend with them.

Figure Credits

Fig. 1.1(a) and (b): Courtesy of International Business Machines Corporation. Unauthorized use not permitted.

Fig. 1.1(c): Reprinted with permission from Steven Pei

Fig. 1.1(d): Reprinted with permission from Steven Pei. Photo by Michael Weimer.

Fig. 1.2(c): Reprinted with permission from Venkatasubramanian, R. et al., 2001, "Thin-Film Thermoelectric Devices with High Room-Temperature Figures of Merit," *Nature*, vol. 413, pp. 597–602. Copyright 2001 Nature.

Fig. 1.4(c): Reprinted with permission from Ren et al., 1998, *Science*, vol. 282, pp. 1105–1107. Copyright 1998 AAAS.

Fig. 2.6(a): Reprinted from Amnon Yariv, 1989, *Quantum Electronics*, 3rd ed., Wiley, New York, with permission from John Wiley and Sons, Inc.

Fig. 5.8: Reprinted with permission from Swartz, E.T. and Pohl, R.O., 1989, "Thermal Boundary Resistance," *Reviews of Modern Physics*, vol. 61, pp. 605–668. Copyright 1989 by the American Physical Society.

Fig. 5.15(b): Reprinted with permission from Odom, T.W., Huang, J., Kim, P. and Lieber, C.M., 1998, "Atomic Structure and Electronic Properties of Single-Walled Carbon Nanotubes," *Nature*, vol. 391, pp. 62–64. Copyright 1998 Nature.

Fig. 5.17(a): Reprinted from Domoto G.A., Boehm, R.F. and Tien, C.L., 1970, "Experimental Investigation of Radiative Transfer between Metallic Surfaces at Cryogenic Temperatures," *Journal of Heat Transfer*, vol. 92, pp. 412–417, with permission from ASME.

Fig. 5.17(b): Reprinted from Hargreaves, C.M., 1969, "Anomalous Radiative Transfer between Closely-Spaced Bodies," *Physics Letters*, vol. 30A, pp. 491–492, with permission from Elsevier.

Fig. 8.6(a) and (b): Reprinted from Qiu, T.Q. and Tien, C.L., 1993, "Heat Transfer Mechanisms during Short-Pulse Laser Heating of Metals," *Journal of Heat Transfer*, vol. 115, pp. 835–841, with permission from ASME.

Fig. 8.12(c): From Fraas L.M., Avery J.E., Huang H.X. and Martinelli, R.U., 2003, "Thermophotovoltaic System Configurations and Spectral Control," *Semiconductor Science Technology*, vol. 18, pp. S165–S173.

Fig. 9.18: Reprinted with permission from Buffat, Ph. and Borel, J.-P., 1976, "Size Effect on the Melting Temperature of Gold Particles," *Physical Review A.*, vol. 13, pp. 2287–2298. Copyright 1976 by the American Physical Society.

Contents

Foreword, vii

1 Introduction, 3
1.1 There Is Plenty of Room at the Bottom, 4
1.2 Classical Definition of Temperature and Heat, 9
1.3 Macroscopic Theory of Heat Transfer, 9
    1.3.1 Conduction, 9
    1.3.2 Convection, 11
    1.3.3 Radiation, 13
    1.3.4 Energy Balance, 16
    1.3.5 Local Equilibrium, 17
    1.3.6 Scaling Trends under Macroscopic Theories, 17
1.4 Microscopic Picture of Heat Carriers and Their Transport, 18
    1.4.1 Heat Carriers, 18
    1.4.2 Allowable Energy Levels of Heat Carriers, 22
    1.4.3 Statistical Distribution of Energy Carriers, 23
    1.4.4 Simple Kinetic Theory, 25
    1.4.5 Mean Free Path, 27
1.5 Micro- and Nanoscale Transport Phenomena, 28
    1.5.1 Classical Size Effects, 28
    1.5.2 Quantum Size Effects, 29
    1.5.3 Fast Transport Phenomena, 30
1.6 Philosophy of This Book, 32
1.7 Nomenclature for Chapter 1, 34
1.8 References, 35
1.9 Exercises, 37

2 Material Waves and Energy Quantization, 43
2.1 Basic Wave Characteristics, 44
2.2 Wave Nature of Matter, 46

- 2.2.1 Wave–Particle Duality of Light, 46
- 2.2.2 Material Waves, 48
- 2.2.3 The Schrödinger Equation, 49

2.3 Example Solutions of the Schrödinger Equation, 52
- 2.3.1 Free Particles, 52
- 2.3.2 Particle in a One-Dimensional Potential Well, 53
- 2.3.3 Electron Spin and the Pauli Exclusion Principle, 58
- 2.3.4 Harmonic Oscillator, 59
- 2.3.5 The Rigid Rotor, 63
- 2.3.6 Electronic Energy Levels of the Hydrogen Atom, 64

2.4 Summary of Chapter 2, 70
2.5 Nomenclature for Chapter 2, 72
2.6 References, 72
2.7 Exercises, 73

3 Energy States in Solids 77

3.1 Crystal Structure, 78
- 3.1.1 Description of Lattices in Real Space, 78
- 3.1.2 Real Crystals, 81
- 3.1.3 Crystal Bonding Potential, 84
- 3.1.4 Reciprocal Lattice, 87

3.2 Electron Energy States in Crystals, 91
- 3.2.1 One-Dimensional Periodic Potential (Kronig–Penney Model), 91
- 3.2.2 Electron Energy Bands in Real Crystals, 98

3.3 Lattice Vibration and Phonons, 100
- 3.3.1 One-Dimensional Monatomic Lattice Chains, 100
- 3.3.2 Energy Quantization and Phonons, 103
- 3.3.3 One-Dimensional Diatomic and Polyatomic Lattice Chains, 104
- 3.3.4 Phonons in Three-Dimensional Crystals, 105

3.4 Density of States, 105
- 3.4.1 Electron Density of States, 107
- 3.4.2 Phonon Density of States, 109
- 3.4.3 Photon Density of States, 110
- 3.4.4 Differential Density of States and Solid Angle, 111

3.5 Energy Levels in Artificial Structures, 111
- 3.5.1 Quantum Wells, Wires, Dots, and Carbon Nanotubes, 111
- 3.5.2 Artificial Periodic Structures, 114

3.6 Summary of Chapter 3, 117
3.7 Nomenclature for Chapter 3, 118
3.8 References, 119
3.9 Exercises, 121

4 Statistical Thermodynamics and Thermal Energy Storage 123

4.1 Ensembles and Statistical Distribution Functions, 124
    4.1.1 Microcanonical Ensemble and Entropy, 124
    4.1.2 Canonical and Grand Canonical Ensembles, 127
    4.1.3 Molecular Partition Functions, 130
    4.1.4 Fermi–Dirac, Bose–Einstein, and Boltzmann Distributions, 134
4.2 Internal Energy and Specific Heat, 137
    4.2.1 Gases, 138
    4.2.2 Electrons in Crystals, 141
    4.2.3 Phonons, 144
    4.2.4 Photons, 146
4.3 Size Effects on Internal Energy and Specific Heat, 148
4.4 Summary of Chapter 4, 150
4.5 Nomenclature for Chapter 4, 153
4.6 References, 154
4.7 Exercises, 155

5 Energy Transfer by Waves, 159

5.1 Plane Waves, 160
    5.1.1 Plane Electron Waves, 161
    5.1.2 Plane Electromagnetic Waves, 161
    5.1.3 Plane Acoustic Waves, 167
5.2 Interface Reflection and Refraction of a Plane Wave, 169
    5.2.1 Electron Waves, 169
    5.2.2 Electromagnetic Waves, 171
    5.2.3 Acoustic Waves, 178
    5.2.4 Thermal Boundary Resistance, 180
5.3 Wave Propagation in Thin Films, 185
    5.3.1 Propagation of EM Waves, 186
    5.3.2 Phonons and Acoustic Waves, 191
    5.3.3 Electron Waves, 193

xx   CONTENTS

   5.4   Evanescent Waves and Tunneling, 194
- 5.4.1 Evanescent Waves 194
- 5.4.2 Tunneling 195

   5.5   Energy Transfer in Nanostructures: Landauer Formalism, 198
   5.6   Transition to Particle Description, 204
- 5.6.1 Wave Packets and Group Velocity, 204
- 5.6.2 Coherence and Transition to Particle Description, 207

   5.7   Summary of Chapter 5, 216
   5.8   Nomenclature for Chapter 5, 218
   5.9   References, 220
   5.10 Exercises, 223

6 Particle Description of Transport Processes: Classical Laws, 227

   6.1   The Liouville Equation and the Boltzmann Equation, 228
- 6.1.1 The Phase Space and Liouville's Equation, 228
- 6.1.2 The Boltzmann Equation, 230
- 6.1.3 Intensity for Energy Flow, 233

   6.2   Carrier Scattering, 233
- 6.2.1 Scattering Integral and Relaxation Time Approximation, 234
- 6.2.2 Scattering of Phonons, 237
- 6.2.3 Scattering of Electrons, 240
- 6.2.4 Scattering of Photons, 240
- 6.2.5 Scattering of Molecules, 242

   6.3   Classical Constitutive Laws, 242
- 6.3.1 Fourier Law and Phonon Thermal Conductivity, 243
- 6.3.2 Newton's Shear Stress Law, 247
- 6.3.3 Ohm's Law and the Wiedemann–Franz Law, 249
- 6.3.4 Thermoelectric Effects and Onsager Relations, 254
- 6.3.5 Hyperpolic Heat Conduction Equation and Its Validity, 258
- 6.3.6 Meaning of Local Equilibrium and Validity of Diffusion Theories, 260

   6.4   Conservative Equations, 262
- 6.4.1 Navier–Stokes Equations, 263
- 6.4.2 Electrohydrodynamic Equation, 266
- 6.4.3 Phonon Hydrodynamic Equations, 268

   6.5   Summary of Chapter 6, 273
   6.7   Nomenclature for Chapter 6, 275
   6.8   References, 277
   6.9   Exercises, 279

7 Classical Size Effects, 282

7.1 Size Effects on Electron and Phonon Conduction Parallel to Boundaries, 283
    7.1.1 Electrical Conduction along Thin Films, 285
    7.1.2 Phonon Heat Conduction along Thin Films, 288

7.2 Transport Perpendicular to the Boundaries, 292
    7.2.1 Thermal Radiation between Two Parallel Plates, 292
    7.2.2 Heat Conduction across Thin Films and Superlattices, 299
    7.2.3 Rarefied Gas Heat Conduction between Two Parallel Plates, 302
    7.2.4 Current Flow across Heterojunctions, 307

7.3 Rarefied Poiseuille Flow and Knudsen Minimum, 308

7.4 Transport in Nonplanar Structures, 313
    7.4.1 Thermal Radiation between Concentric Cylinders and Spheres, 314
    7.4.2 Rarefied Gas Flow and Convection, 314
    7.4.3 Phonon Heat Conduction, 315
    7.4.4 Multidimensional Transport Problems, 316

7.5 Diffusion Approximation with Diffusion–Transmission Boundary Conditions, 317
    7.5.1 Thermal Radiation between Two Parallel Plates, 319
    7.5.2 Heat Conduction in Thin Films, 321
    7.5.3 Electron Transport across an Interface: Thermionic Emission, 322
    7.5.4 Velocity Slip for Rarefied Gas Flow, 327

7.6 Ballistic–Diffusive Treatments, 331
    7.6.1 Modified Differential Approximation for Thermal Radiation, 331
    7.6.2 Ballistic–Diffusive Equations for Phonon Transport, 333

7.7 Summary of Chapter 7, 336

7.8 Nomenclature for Chapter 7, 338

7.9 References, 340

7.10 Exercises, 344

8 Energy Conversion and Coupled Transport Processes, 348

8.1 Carrier Scattering, Generation, and Recombination, 349
    8.1.1 Nonequilibrium Electron–Phonon Interactions, 349
    8.1.2 Photon Absorption and Carrier Excitation, 358
    8.1.3 Relaxation and Recombination of Excited Carriers, 363
    8.1.4 Boltzmann Equation Revisited, 366

8.2 Coupled Nonequilibrium Electron–Phonon Transport without Recombination, 367
  8.2.1 Hot Electron Effects in Short Pulse Laser Heating of Metals, 369
  8.2.2 Hot Electron and Hot Phonon Effects in Semiconductor Devices, 370
  8.2.3 Cold and Hot Phonons in Energy Conversion Devices, 373
8.3 Energy Exchange in Semiconductor Devices with Recombination, 373
  8.3.1 Energy Source Formulation, 373
  8.3.2 Energy Conversion in a p–n Junction, 376
  8.3.3 Radiation Heating of Semiconductors, 384
8.4 Nanostructures for Energy Conversion, 386
  8.4.1 Thermoelectric Devices, 386
  8.4.2 Solar Cells and Thermophotovoltaic Power Conversion, 391
8.5 Summary of Chapter 8, 395
8.6 Nomenclature for Chapter 8, 396
8.7 References, 398
8.8 Exercises, 401

9 Liquids and Their Interfaces, 404

9.1 Bulk Liquids and Their Transport Properties, 405
  9.1.1 Radial Distribution Function and van der Waals Equation of State, 405
  9.1.2 Kinetic Theories of Liquids, 408
  9.1.3 Brownian Motion and the Langevin Equation, 411
9.2 Forces and Potentials between Particles and Surfaces, 416
  9.2.1 Intermolecular Potentials, 417
  9.2.2 Van der Waals Potential and Force between Surfaces, 419
  9.2.3 Electric Double Layer Potential and Force at Interfaces, 421
  9.2.4 Surface Forces and Potentials Due to Molecular Structures, 427
  9.2.5 Surface Tension, 428
9.3 Size Effects on Single-Phase Flow and Convection, 433
  9.3.1 Pressure-Driven Flow and Heat Transfer in Micro- and Nanochannels, 433
  9.3.2 Electrokinetic Flows, 436
9.4 Size Effects on Phase Transition, 438
  9.4.1 Curvature Effect on Vapor Pressure of Droplets, 439
  9.4.2 Curvature Effect on Equilibrium Phase Transition Temperature, 441
  9.4.3 Extension to Solid Particles, 441
  9.4.4 Curvature Effect on Surface Tension, 442
9.5 Summary of Chapter 9, 443
9.6 Nomenclature for Chaper 9, 445

9.7 References, 447
9.8 Exercises, 449

10 Molecular Dynamics Simulation 452

10.1 The Equations of Motion, 453
10.2 Interatomic Potential, 458
10.3 Statistical Foundation for Molecular Dynamic Simulations, 462
    10.3.1 Time Average versus Ensemble Average, 462
    10.3.2 Response Function and Kramers–Kronig Relations, 464
    10.3.3 Linear Response Theory, 466
    10.3.4 Linear Response to Internal Thermal Disturbance, 473
    10.3.5 Microscopic Expressions of Thermodynamic and Transport Properties, 476
    10.3.6 Thermostatted Ensembles, 479
10.4 Solving the Equations of Motion, 483
    10.4.1 Numerical Integration of the Equations of Motion, 483
    10.4.2 Initial Conditions, 485
    10.4.3 Periodic Boundary Condition, 485
10.5 Molecular Dynamics Simulation of Thermal Transport, 486
    10.5.1 Equilibrium Molecular Dynamics Simulation, 486
    10.5.2 Nonequilibrium Molecular Dynamics Simulations, 490
    10.5.3 Molecular Dynamics Simulation of Nanoscale Heat Transfer, 493
10.6 Summary of Chapter 10, 494
10.7 Nomenclature for Chapter 10, 496
10.8 References, 498
10.9 Exercises, 502

Appendix A: Homogeneous Semiconductors, 505

Appendix B: Seconductor p–n Junctions, 509

Index, 513

Units and Their Conversions, 530

Physical Constants, 531

Nanoscale Energy Transport
and Conversion

1

Introduction

The major sources of inspiration for this book are the recent rapid advancements in microtechnology and nanotechnology. Microtechnology deals with devices and materials with characteristic lengths in the range of submicron to micron scales (0.1–100 μm), while nanotechnology generally covers the length scale from 1 to 100 nm. For example, integrated circuits are now built on transistors with characteristic device length scales around 100 nm. The semiconductor industry roadmap predicts that, in 2010, the characteristic length in integrated circuits will further shrink to 25 nm (SEMATECH, 2002). In the late 1980s, microelectronics fabrication technology began to impact mechanical engineering, and the field of microelectromechanical systems, or MEMS, blossomed (Trimmer, 1997). Meanwhile, nanoscience and nanotechnology, explored by a few pioneers (Feynman, 1959, 1983), are currently generating much excitement across all disciplines of science and engineering. The fields of micro- and nanotechnologies are enormous in breadth and cannot be covered completely in any single book. In this book, we focus on microscopic mechanisms behind energy transport, particularly thermal energy transport. As the device or structure characteristic length scales (such as the gate length in field-effect transistors, used to build computers, and the film thickness in coatings) become comparable to the mean free path and the wavelength of energy and information carriers (mainly electrons, photons, phonons, and molecules), some of the classical laws are no longer applicable. By examining the microscopic pictures underlying transport processes, we will develop a consistent framework for treating thermal energy transport phenomena from the nanoscale to the macroscale.

In this chapter, we will first give a few examples of micro- and nanoscale transport phenomena from contemporary technologies to provide motivation for the rest of this book. We will then briefly summarize classical laws governing heat transfer processes and discuss the microscopic pictures behind heat transfer phenomena, followed by a simple derivation of the Fourier law, based on the kinetic theory, to demonstrate that many classical laws we have learned are actually not as fundamental as their names may suggest! The rest of the book will further expand on this chapter and answer in depth some of the questions we raise.

1.1 There Is Plenty of Room at the Bottom

Richard Feynman, who won the 1963 Nobel Prize in physics for his work in quantum electrodynamics, gave a visionary talk in 1959 entitled "There is plenty of room at the bottom" (Feynman, 1959, 1983). In this talk, Feynman described the possibilities of storing all the books in the world in a piece of dust, making micromachines that can go into the human body, shrinking computers, rearranging atoms, and so on. To put his ideas in a historic perspective, the possibility of integrated circuits was first demonstrated in 1958 and, at the time of Feynman's famous lecture, computers filled entire rooms and lasers were shown only to be theoretically possible. Feynman not only spoke of theoretical possibilities, but also provided potential approaches to realize his dreams. Although his talk was considered bold at the time, his insights on what was possible, or, better put, what was not impossible, were based on the laws of physics. Most of the audience at that 1959 talk did not take Feynman's visions seriously—rather, they thought he was trying to be humorous. Subsequent developments in micro- and nanotechnologies, however, have realized many of his dreams, and some have even followed the technical approaches that he envisioned.

Great visionaries like Richard Feynman point to directions and provide inspiration. Often, at the beginning of a revolutionary idea, only a small group appreciates the ideas and begins to work on practical ways to demonstrate the concepts. Some of them make breakthroughs along the way and prove that the idea works in principle; this attracts more attention from wider communities and eventually the general public. This small group is privileged because they have access to the ideas at an earlier stage, but, more importantly, they have the training and judgments to appreciate the importance of these ideas. Some of this small group have insights of their own, enabling them to realize these ideas and generate their own ideas along the way. With the rapid development of information technology, ideas propagate quickly nowadays. Academic training and scientific knowledge become even more important for one to filter through the flood of information for gold and to develop one's own ideas and visions. One objective of this book is to provide a knowledge base to its readers, assuming that most are not familiar with modern physics, with a foundation to understand energy transport and conversion processes, particularly thermal energy, from nanoscale up to macroscale, with an emphasis on nanoscale processes. We will give a few examples here to illustrate the importance of understanding nanoscale transport and energy conversion processes.

One major driver behind microtechnology and nanotechnology is information processing, which includes microelectronics, data storage, and data transmission. The information carriers are electrons in electrical circuits and photons in optical fibers. The transport of electrons and photons often generate unwanted heat. As more and more devices are compacted into a small area, heat generation density increases and thermal management becomes a major challenge for the microelectronics industry. A Pentium4

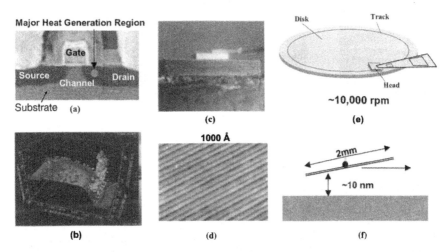

Figure 1.1 Nanoscale transport examples in information-oriented devices. (a) & (b): A MOSFET device (courtesy of IBM) and electron energy dissipation in a MOSFET (courtesy of Dr. S.E. Laux and Dr. M.V. Fischetti), indicating most heat is generated in nanometer region at the drain. Heat conduction from such small source cannot be described by Fourier law; (c) & (d): A InAs/AlSb based quantum cascade laser is made of many layers of thin films, each ranging from a few to hundreds Angstrom thick (courtesy of Dr. S. Pei). These films have thermal conductivity values significantly lower than those of their bulk materials; (e) & (f): A disk drive in magnetic data storage with the slider head hovering about 10 nm on top of the disk rotating at 10,000 rpm. Fluid flow through the gap between the slider and the disk is rarefied and cannot be described by the Newton shear stress law.

chip from Intel Corporation, for example, has an area of ~ 1 cm^2 and dissipates about 60 W of heat. The size of the fan used to maintain the chip temperature below its standard (typically 80–120°C) is much larger than the chip itself. As engineers develop various cooling solutions, it also becomes clearer that heat transfer characteristics must be considered at the device level (Cahill et al., 2002). The most important device in a computer chip is the metal-on-insulator field-effect transistor or MOSFET (Sze, 1981), as shown in figures 1.1(a) and (b). The source, drain, and channel are made of doped silicon (or other types of semiconductor). Electrons (or holes) flow from the source into the drain through the channel under an externally applied voltage between the source and the drain. The width of the channel is controlled by another voltage applied between the back of the substrate and the gate electrode, which is insulated from the channel by a very thin silicon dioxide layer. A MOSFET is thus like a variable resistor with its resistance controlled by the gate voltage. To make faster devices, the channel length (and thus the gate length) is shrinking by about 30% every 2 years, with the current gate length at around 90 nm. Electrons convert most of their energy into heat in a small region in the drain side [figure 1.1(b)]. Both modeling (Chen, 1996a) and experiments (Svedrup et al., 2001) suggest that the temperature rise due to heat generation in the small region is much higher than that predicted by the Fourier law, which can accelerate the failure of the device. As another example, semiconductor lasers used in telecommunication and data storage are often composed of multilayers of thin films, as shown

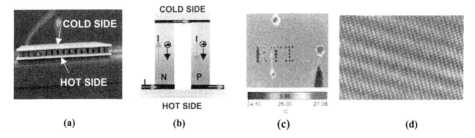

Figure 1.2 Illustration of thermoelectric devices, (a) a photograph of a thermoelectric cooler, (b) illustration of charge flow inside one pair of legs, (c) microcoolers fabricated based on superlattices (Venkatasubramanian et al., 2001; courtesy of Nature Publishing Groups), and (d) an example of a Si/Ge superlattice structure (courtesy of Dr. K.L. Wang).

in figures 1.1(c) and (d). Past studies have shown that the thermal conductivity of these structures is much lower than effective values calculated from their bulk materials on the basis of the Fourier heat conduction theory (Chen, 1996b). These lasers have severe heating problems that limit their performance. The reduced thermal conductivity calls for careful design of the lasers to minimize the number of interfaces. In a different example, figure 1.1(e) sketches the disk head in a magnetic disk drive. The separation between the slider and the magnetic disk [figure 1.1(f)] determines the data storage density. Currently, this separation distance is around 10–50 nm while the disk rotates at $\sim$ 10,000 rotations per minute (rpm). The relative motion between the slider and the disk is analogous to flying a Boeing 747 a few millimeters above the ground. The airflow between the slider and the disk is crucial in maintaining the slider–disk separation and is very different from the prediction of continuum fluid mechanics at this small spacing (Fukui and Kaneko, 1988). Some data storage processes are also based on heat transfer (Chen et al., 2004). Examples are rewritable CDs based on the phase change of the materials upon laser heating and thermomechanical data storage, where data bits are written on polymer substrates by heated atomic force microscope (AFM) tips. For such applications, it is desirable to limit the heating to a small domain. Nanoscale heat transfer effects including reduced thermal conductivity of thin films and nonlocal heat conduction surrounding nanoscale heating spots can be utilized to confine heat for writing smaller spots.

While heat in information technology is, in most cases, undesirable and needs to be "managed," it becomes a dominant factor in energy conversion technology. Nanoscale energy transport phenomena can be used for developing new strategies to improve the energy conversion efficiency. For example, the reduced thermal conductivity observed in semiconductor thin films can potentially be used to develop highly efficient thermoelectric materials for cooling and power generation (Tritt, 2001; Chen and Shakouri, 2002). Thermoelectric devices, as shown in figure 1.2(a–d), use electrons in solids as the media to carry energy from one location to another. Low thermal conductivity materials are required to reduce the thermal leakage between the hot and cold sides. At the same time, the materials must have good electron energy-carrying properties (Goldsmid, 1964). The use of nanostructures to control the thermoelectric transport properties for improving the electron energy carrying capability and reducing the thermal conductivity emerged over the last ten years as a very promising approach for realizing highly

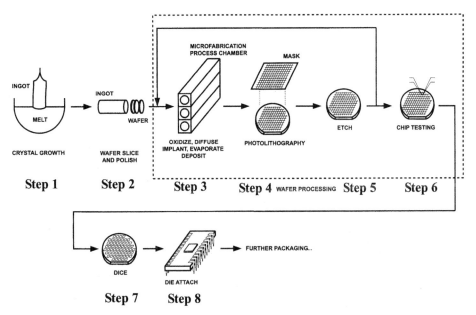

Figure 1.3 Illustration of major microelectronic device fabrication processes.

efficient thermoelectric devices (Dresselhaus et al., 1999; Tritt, 2001; Harman et al., 2002; Chen et al., 2003).

Radiation transport in micro- and nanostructures is also different from that in macrostructures because the wave properties of photons become dominant. For example, radiation exchange between two closely spaced vacuum gaps is much higher than that predicted on the basis of standard view-factor calculations because of tunneling and interference effects (Domoto et al., 1970), which potentially can be used to develop high power density thermophotovoltaic power generators (Whale and Cravalho, 2002). Photonic crystals, a concept developed in 1987 (Yablonovitch, 1987), can be used to design special thermal radiation surfaces with desirable properties (Fleming et al., 2002). Using microstructures, coherent thermal radiation was recently demonstrated (Greffet et al., 2002).

There are many outstanding nanoscale transport problems related to the fabrication of nanodevices and synthesis of nanomaterials. As an example, consider a typical fabrication process, shown in figure 1.3, for an integrated circuit. Important transport issues exist in almost every step and some of them are particularly relevant to the nanoscale transport discussed in this book. With regard to the process illustrated in figure 1.3, in step 1, heat transfer and fluid flow problems in crystal growth are in the continuum range and have been addressed extensively in literature. Many of the material deposition processes (step 3) occur at high temperatures and under low pressures. Atoms or gas molecules have long mean free paths at low pressures, and this must be considered in developing proper working conditions for filling trenches between devices. Lithography processes (step 4) should consider photon transport carefully. Optical interference and scattering effects can be either detrimental or useful for the lithography technology. Heat transfer issues arise in both the mask-making and lithography processes. For example, some candidates for next-generation lithography, such as extreme

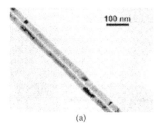 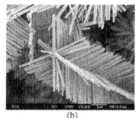 

Figure 1.4 Examples of nanowires and nanotubes synthesized by various methods: (a) a pair of bismuth nanowires obtained by pressure injection into a template (Dresselhaus et al., 2001; courtesy of Dr. M.S. Dresselhaus), (b) TiO_2 nanowires obtained by vapor condensation (courtesy of Dr. Z.F. Ren), (c) carbon nanotubes grown by plasma CVD (Ren et al., 1998; courtesy of AAAS).

ultra-violet lithography (EUV) and X-ray lithography, rely on multilayer structures for reflecting light (Hector and Mangat, 2001). The consequence of reduced thermal conductivity on mask reliability has yet to be investigated. The synthesis of nanoscale materials is a wide-open field and many nanomaterial and nanostructure synthesis methods being developed raise intriguing nanoscale transport questions. For example, nanowires and carbon nanotubes, shown in figure 1.4, have been synthesized with several different methods such as chemical or physical vapor deposition, filling of templates, plasma deposition, and laser ablation (Morales and Lieber, 1998; Ren et al., 1998; Dresselhaus et al., 2001). Understanding the transport processes during nanomaterial formation will allow better control of the final material quality.

The preceding examples emphasize the small length scales involved in nanodevices and nanomaterials. Short time scales are also becoming increasingly important. Similar questions can be raised for transport at short time scales as for the small length scales. Lasers can deliver a pulse as short as a few femtoseconds (1 fs = 10^{-15} s). Energy transduction mechanisms at such short time scales can differ significantly from those at macroscales (Qiu and Tien, 1993). Microelectronic devices are pushing to the tens of gigahertz clock frequency with much shorter transient times. The temperature rise of the device in such short time scales can be very different from that predicted by the Fourier law (Yang et al., 2002).

The examples given above illustrate a few of the motivations behind the rapid development in micro- and nanoscale heat transfer research over the last decade (see Tien and Chen, 1994; Tien et al., 1998). In the meantime, similar developments are occurring in various fields, as evidenced in the strong interest in nanoscience and nanotechnology from numerous fields of science and engineering, as well as from industry. Yet we are only at the entrance, and the room at the bottom is big. The convergence of interests from disparate fields into common subjects also creates confusion because different languages are used in each field for similar phenomena. For newcomers, these differences are often intimidating. One of the objectives of this book is to get the readers familiar with the terminologies. In fact, once they get involved, the readers will find that drastically different equations used in unrelated fields, such as the Fourier law for heat conduction and the drift-diffusion equation for electrical current flow, actually originate from the same principle. For this reason, the text will adopt a parallel treatment of different energy carriers whenever possible.

1.2 Classical Definition of Temperature and Heat

The classical definition of heat transfer from thermodynamics can be stated as follows: "Heat transfer is the energy flow across the boundaries of a system under a temperature difference." We can emphasize several points in this definition: heat transfer is a form of energy flow; heat transfer is associated with a temperature difference; and finally, heat transfer is a boundary phenomenon.

Since heat transfer is driven by temperature differences, it is necessary that we pay attention to the definition of temperature. In classical thermodynamics, temperature is defined on the basis of the concept of thermal equilibrium. If system A is in thermal equilibrium with system B, then system A and system B have the same temperature. In other words, temperature is a quantity that describes thermal equilibrium phenomena.

These definitions of temperature and heat transfer are independent of the material and serve well in establishing the universality of classical thermodynamics. Their strengths, however, are also their weaknesses. These definitions are devoid of the physical microscopic pictures underlying heat transfer processes and the meaning of temperature. This book aims to provide a more detailed picture of thermal energy transport processes. We will study how heat is transferred at the microscopic level and how temperature should be defined for transport processes that are intrinsically nonequilibrium.

1.3 Macroscopic Theory of Heat Transfer

There exist three basic modes of heat transfer: conduction, convection, and thermal radiation. We briefly review the classical laws that are used to describe these modes. Later in this book we will show how these laws can be derived and on what approximations they are built.

1.3.1 Conduction

Heat conduction represents the energy transfer processes through a medium, caused by a temperature difference due to the random motion of heat carriers in the substance. The key is that a medium is needed for heat conduction, and heat is the part of the energy that is carried around through random motion of heat carriers such as molecules. An example is heat transfer through a solid wall separating the inside and the outside of a room, which is due to the random vibrations of atoms within the wall materials. Heat conduction processes are usually modeled on the basis of the Fourier law (Joseph Fourier, 1768–1830) that relates the local heat flux to the local temperature gradient

$$\mathbf{q} = -k\nabla T \tag{1.1}$$

where k is the thermal conductivity, which is a temperature dependent material property and has units of [Wm^{-1}K^{-1}], $\mathbf{q}$ has units of [Wm^{-2}], and ∇ is the gradient operator such that

$$\nabla T \equiv \frac{\partial T}{\partial x}\hat{x} + \frac{\partial T}{\partial y}\hat{y} + \frac{\partial T}{\partial z}\hat{z} \tag{1.2}$$

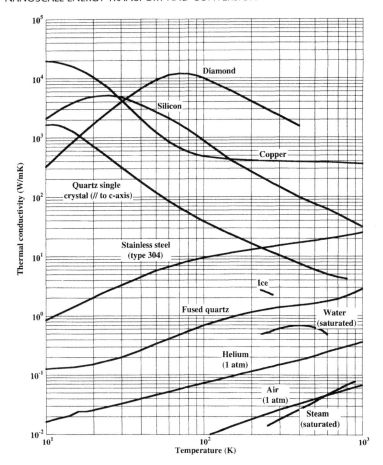

Figure 1.5 Thermal conductivity as a function of temperature for representative materials (data from Touloukian et al., 1970, and http://www.chrismanual.com/Default.htm).

where $\hat{x}$, $\hat{y}$, and $\hat{z}$ are the unit vectors along coordinate directions. Equation (1.1) is called the Fourier law in honor of the mathematician who first used it to solve heat conduction problems. It is called a law because, at the time of its creation, it was a postulate based on the observation of experimental results. This law applies to most engineering situations and is the foundation of classical heat transfer analysis. Thermal conductivity of materials is a very important material property. The higher the thermal conductivity, the better the material conducts heat. Figure 1.5 shows the thermal conductivity of some common materials. We notice that the temperature dependence of thermal conductivity of various materials is quite different. The value of the thermal conductivity spans several orders of magnitude, from 10^{-2} Wm^{-1}K^{-1} for gas to 10^5 Wm^{-1}K^{-1} for solids at low temperatures. Diamond is the best thermal conductor among naturally existing materials.

At this stage, we raise the following questions for interested readers. Why do thermal conductivities of various materials differ not only in magnitude, but also in their temperature dependence? Is the Fourier law applicable to nanostructures? Do

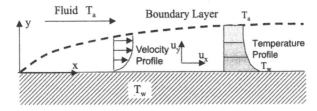

Figure 1.6 Forced convection over a solid surface. Fluids close to the boundary form a boundary layer in which temperature and velocity vary (thermal and momentum boundary layers may have different thicknesses) from their values at the wall to those outside the boundary layer.

nanostructures have the same thermal conductivity as their bulk counterparts? We will find answers to these questions throughout the text.

1.3.2 Convection

Convection heat transfer occurs when a bulk fluid motion overlaps a temperature gradient. When fluid molecules move from one place to another, they carry internal energy with them. In most situations, like the one shown in figure 1.6, we are interested in the heat transfer between a solid surface and the fluid. The convection heat transfer rate Q [W] between a solid surface at temperature T_w and a fluid at temperature T_a can be expressed by Newton's law of cooling (Isaac Newton, 1643–1727)

$$Q = hA(T_w - T_a) \quad (1.3)$$

where h [Wm^{-2}K^{-1}] is called the heat transfer coefficient and A is the surface area. Unlike the thermal conductivity, the heat transfer coefficient is not a material property. It is a flow property that depends on the flow field, fluid properties, and the geometry of the object over which the fluid flows. Convection is categorized into two types: natural convection in which the fluid motion is created by the buoyancy force due to the difference in the densities of hot and cold fluids, and forced convection in which the fluid is set into motion by some other means such as a pump or a fan. Heat transfer between a solid surface and a liquid undergoing a phase change, that is, boiling or condensation, is also characterized by a heat transfer coefficient.

Although Newton's law of cooling is simple in form, h is difficult to determine in general. The heat transfer coefficient is usually determined by experiment, although analysis and numerical simulation can be performed for certain simple geometries and flow conditions. Table 1.1 gives some empirical relations and ranges of heat transfer coefficients for simple geometries, mostly under laminar flow conditions. These empirical relations are expressed using nondimensional parameters such as the average Nusselt number ($\overline{Nu_L} = \overline{h}L/k$) (Wilhelm Nusselt, 1882–1957), the Reynolds number ($Re_L = uL/\nu$) (Osborne Reynolds, 1848–1912), and the Prandtl number ($Pr = \nu/\alpha$) (Ludwig Prandtl, 1875–1953), where L is a characteristic length, u is the fluid velocity, ν the kinematic viscosity [m^{2}s^{-1}], α the thermal diffusivity [m^{2}s^{-1}], and the bar indicates average properties.

In convection heat transfer analysis, it is usually assumed that the fluid molecules at the wall are stationary relative to the wall and have temperatures identical to that of the

Table 1.1 Convection heat transfer correlations for common configurations

Configuration	Correlation
1. Forced convection: fully developed laminar flow inside pipes with constant wall temperature	$\overline{Nu}_{D_h} = 3.66$ (circular pipe), 7.54 (parallel plates) $Re_{D_h} \lesssim 2500$, $\overline{Nu}_{D_h} = \overline{h}D_h/k$, $Re_{D_h} = \bar{u}D_h/\nu$, $D_h \equiv 4A_c/p$, A_c = cross-sectional area, p = perimeter; properties evaluated at the average of inlet and outlet temperatures.
2. Forced convection: laminar boundary layer on an isothermal flat plate	$\overline{Nu} = 0.664 \, Re_L^{1/2} Pr^{1/3}$ $10^3 < Re_L < 5 \times 10^5$, $Pr > 0.5$
3. Force convection: flow across a cylinder of external diameter D	$\overline{Nu}_D = \dfrac{1}{0.8237 - \ln(Re_D Pr)^{1/2}}$; $Re_D Pr < 0.2$ $\overline{Nu}_D = 0.3 + \dfrac{0.62 \, Re_D^{1/2} Pr^{1/3}}{[1+(0.4/Pr)^{2/3}]^{1/4}}$; $Re_D < 10^4$, $Pr > 0.5$
4. Forced convection: flow across a sphere of external diameter D	$\overline{Nu}_D = 2 + (0.4 \, Re_D^{1/2} + 0.06 \, Re_D^{2/3}) Pr^{0.4} (\mu/\mu_w)^{1/4}$; $3.5 < Re_D < 7.6 \times 10^4$; $0.7 < Pr < 380$ All properties except μ_w are evaluated at T_∞
5. Natural convection: boundary layer on an isothermal vertical wall of length L	$\overline{Nu}_L = 0.68 + \dfrac{0.670 \, Ra_L^{1/4}}{[1+(0.492/Pr)^{9/16}]^{4/9}}$; $Ra_L < 10^9$ where Rayleigh number $Ra_L \equiv g\beta(T_s - T_a)L^3/(\alpha \nu)$, g = gravitational acceleration; β = thermal expansion coefficient
6. Natural convection on a heated isothermal horizontal plate facing up, or a cooled plate facing down	$\overline{Nu}_L = 0.54 \, Ra_L^{1/4}$; $10^4 < Ra_L < 10^7$, Ra_L defined in (5) $L \equiv A_s/p$ = ratio of the plate surface area to perimeter
7. Natural convection on a heated isothermal horizontal plate facing down, or a cooled plate facing up	$\overline{Nu}_L = 0.27 \, Ra_L^{1/4}$; $10^5 < Ra_L < 10^{10}$ Ra_L and L defined in (5) and (6), respectively
8. Natural convection on an isothermal horizontal cylinder of external diameter D	$\overline{Nu}_D = 0.36 + \dfrac{0.518 \, Ra_D^{1/4}}{[1+(0.559/Pr)^{9/16}]^{4/9}}$; $10^{-4} < Ra_D < 10^9$
9. Natural convection on a sphere of external diameter D	$\overline{Nu}_D = 2 + \dfrac{0.589 \, Ra_D^{1/4}}{[1+(0.469/Pr)^{9/16}]^{4/9}}$; $Ra_D < 10^{11}$, $Pr \geq 0.7$

Note: All properties should be calculated at the average temperature of the wall and the fluid far away from the wall, unless otherwise mentioned. The Nusselt number is averaged over the area of the fluid–solid interface.

solid. This assumption is called the no-slip boundary condition. Referring to figure 1.6, the no-slip boundary condition is

$$u_x(x,y)|_{y=0} = u_y(x,y)|_{y=0} = 0 \tag{1.4}$$

$$T(x,y)|_{y=0} = T_w \tag{1.5}$$

where u_x and u_y are the fluid velocity components in Cartesian coordinates, T the fluid temperature distribution, and T_w the temperature of the solid surface. Because fluid particles are stationary on the surface, the heat transfer from the wall to the fluid in the vicinity of the surface is actually through heat conduction. We can calculate this heat transfer rate according to the Fourier law,

$$Q = -kA \frac{\partial T}{\partial y}\bigg|_{y=0} \tag{1.6}$$

Combining eqs. (1.3) and (1.6) leads to the expression for calculating the heat transfer coefficient

$$h = \frac{-k \frac{\partial T}{\partial y}\big|_{y=0}}{T_w - T_a} \tag{1.7}$$

The above equation furnishes a formula for determining the heat transfer coefficient, provided the fluid temperature gradient at the wall can be determined. This task usually requires solving the velocity and temperature distributions of the whole flow field on the basis of the Navier–Stokes equations. In typical heat transfer or fluid mechanics textbooks, the Navier–Stokes equations are derived on the basis of the conservation principles for mass, momentum, and energy, together with constitutive relations such as the Fourier law, which relates the heat flux to the temperature gradient, and the Newton shear stress law, which relates the local velocity gradient to the local shear stress. We will discuss the Navier–Stokes equations in greater detail further in chapter 6. In its simplest form, assuming that the velocity component u in figure 1.6 depends on y only, the Newton shear stress law can be written as

$$\tau_{xy} = \mu \frac{\partial u_x}{\partial y} \tag{1.8}$$

where the first subscript on τ denotes the direction of the shear stress and the second subscript denotes the plane of action of the shear stress (y = constant plane), and μ [N s m^{-2}] is the dynamic viscosity (or absolute viscosity). A popular unit for μ is P (poise), where 1 P = 0.1 N s m^{-2}. The ratio of the dynamic viscosity to the fluid density, μ/ρ, gives the kinematic viscosity, ν. Viscosity is generally regarded as an intrinsic property of the fluid.

Going back to the theme of this book, microfluidics, which deals with fluid flow at micro- and nanoscales, has attracted significant attention due to its applications in chemical and biological analysis (Ho and Tai, 1998). Many questions can be asked about fluid flow and heat transfer at such scales. Is the Newton shear stress law applicable to fluid flow at these length scales? Is the no-slip boundary condition always correct? In this book, we will answer these questions via the Boltzmann equation, surface force analysis, and molecular dynamics simulations.

1.3.3 Radiation

Thermal radiation, the third basic heat transfer mode, is different from conduction and convection. Heat transfer by thermal radiation does not require a medium and can propagate in vacuum, and the energy is carried by electromagnetic waves. A blackbody,

14 NANOSCALE ENERGY TRANSPORT AND CONVERSION

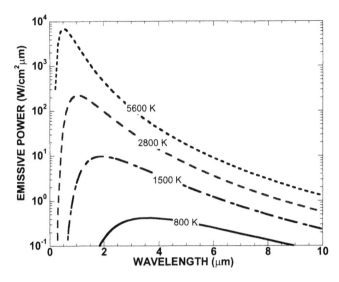

Figure 1.7 Blackbody emissive power as a function of wavelength at different temperatures.

which is an idealized object that emits the maximum amount of thermal radiation, radiates according to Planck's law

$$e_{b,\lambda} = \frac{C_1}{\lambda^5(e^{C_2/\lambda T} - 1)} \tag{1.9}$$

where $C_1 (= 37{,}413\,\text{W}\,\mu\text{m}^4\,\text{cm}^{-2})$ and $C_2 (= 14{,}388\,\mu\text{m}\,\text{K})$ are constants, λ is the wavelength of the radiation, and T is the absolute temperature. The spectral emissive power, $e_{b,\lambda}$, is defined as the radiated power per unit emitting area and per unit wavelength interval,

$$e_{b,\lambda} = \frac{\text{Power}}{\Delta A \Delta \lambda} \tag{1.10}$$

and has units of $\text{Wm}^{-2}\mu\text{m}^{-1}$. Examples of blackbody radiation spectra are shown in figure 1.7. The wavelength at which the maximum emissive power occurs is given by the Wien displacement law

$$\lambda_{\max} T = 2898\,\mu\text{m}\,\text{K} \tag{1.11}$$

Solar radiation has an equivalent blackbody temperature of 5600 K and peaks around 0.52 μm. Thus, the fact that the human visibility range is between 0.4 and 0.7 μm is not incidental.

Integrating Eq. (1.9) over all wavelengths, we obtain the total emissive power of a blackbody

$$e_b = \int_0^\infty e_{b,\lambda}\,d\lambda = \sigma T^4 \tag{1.12}$$

where $\sigma (= 5.67 \times 10^{-8}\,\text{Wm}^{-2}\,\text{K}^{-4})$ is the Stefan–Boltzmann constant. Equation (1.12) is called the Stefan–Boltzmann Law.

Real objects typically radiate less than a blackbody. The emissivity characterizes the thermal radiation characteristics of a surface. The spectral emissivity is defined as

$$\varepsilon_\lambda = e_\lambda / e_{b,\lambda} \tag{1.13}$$

where e_λ is the spectral emissive power of the surface.

As a form of electromagnetic wave, the propagation of thermal radiation can be described by Maxwell's equations. Calculating the radiative heat transfer, however, seldom requires the solution of these equations. Typically, we neglect the phase information carried by the electromagnetic waves and treat the thermal radiation as incoherent photon particles, or bundles of rays propagating in straight lines. These rays can be scattered, absorbed along the path, or enhanced by emission of the medium along the propagation direction. Upon reaching a surface, the thermal radiation can be reflected, absorbed, or transmitted. Calculating the radiation heat transfer between real surfaces requires information about the surface radiative properties, the geometrical arrangement of the surfaces, and the properties of the media between the surfaces. As an example, consider the simplest situation of two infinite, black, parallel walls separated by a vacuum. The radiation heat transfer per unit area, q, between the two surfaces is the difference of the energy carried by two groups of counter-propagating photons: one from the hot side toward the cold side $[\sigma T_1^4]$, and the other from the cold to the hot side $[\sigma T_2^4]$,

$$q = \sigma (T_1^4 - T_2^4) \tag{1.14}$$

In contrast to solving differential equations in heat conduction and convection to get the heat flux, in the case of thermal radiation we usually deal with the trajectory of photons through the use of view factor, or integral equations when scattering is involved. This approach is necessary because photons typically travel a long distance before they are scattered. In many radiation problems, photons collide more frequently with the walls rather than being scattered along their paths. In this sense, thermal radiation is always dealing with size or boundary effects. For heat conduction and convection in nanostructures, the heat carriers experience similar situations to photon transport in macrostructures because electrons, phonons, or molecules collide more often with the boundaries and interfaces than they collide with each other. Thus, many nanoscale transport processes can be understood through an analogy with photon transport on the macroscale. Such an analogy will be pursued throughout the text, whenever applicable.

The transport of photons in micro- and nanostructures usually differs from that in macrostructures because the wavelength becomes comparable to or even longer than the characteristic device dimension. Under such circumstances, the phase information can no longer be ignored and the wave properties of photons, such as interference, diffraction, and tunneling, become important. The treatment of the propagation of electromagnetic waves is well developed in the fields of optics and elecromagnetic waves (Born and Wolf, 1980). Many results in these fields can also be applied to thermal radiation in small spaces. In later chapters, we will see that the wave effects in micro- and nanostructures lead to significant deviations from the thermal radiation relations developed for macrostructures. For example, radiation heat transfer between nanoscale objects can be significantly higher than blackbody radiation (Domoto et al., 1970). An

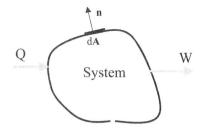

Figure 1.8 Energy conservation applied to a closed system.

understanding of the wave nature of photons provides a basis for comprehending the material waves and quantum size effects on other energy carriers such as electrons and phonons (Chen, 1996b).

1.3.4 Energy Balance

The equations that we have reviewed for different transport processes relate the heat flux to the temperature and the temperature gradient. Equations of this type are called constitutive equations. Each of them is a single equation with two unknowns: the heat flux and the temperature. In general, another equation is needed to solve for T and q. We use conservation principles to establish this other equation. For heat transfer, the most important conservation principle is the first law of thermodynamics (energy conservation), which states that heat transfer into a system minus the work output from a system equals the change of the internal energy of the system. Referring to a closed system, as shown in figure 1.8, the first law of thermodynamics is

$$Q - W = dU/dt \qquad (1.15)$$

where Q is the rate of heat transfer into the system, W is the power output, and U is the system energy, that is, the sum of the internal energy, the kinetic energy, and the potential energy. In many heat transfer situations, the changes in the kinetic and potential energy are usually negligible such that U represents the internal energy only.

One major difference between the constitutive equations and the conservation equations is that the former relate to specific materials and processes but may not be valid in all cases, while the latter are universal. For example, the Newton law of shear stress is valid for Newtonian fluids but not for non-Newtonian fluids. Similarly, the Fourier heat conduction equation may not be valid for all heat conduction processes, as we will soon show. On the other hand, no evidence exists that the first law of thermodynamics is not valid, although there are discussions in current literature on the validity of the second law at small length and short time scales (Wang et al., 2002). As we move to the micro- and nanoscale, what may change are the conservation equations. The conservation equations are expected to hold true.

As an example of using the conservation equations, we consider a region with heat conduction as the only mode of heat transfer with no work transfer and no internal heat generation. The heat conduction across the boundary into the solid can be obtained by integrating eq. (1.1) over the surface consisting of the boundary (referring to figure 1.8),

$$Q = -\oiint_S (-k\nabla T)\, d\mathbf{A} \qquad (1.16)$$

where $d\mathbf{A}$ is a differential area on the boundary with the norm pointing outward; thus a minus sign is added to eq. (1.16) to indicate heat conduction into the region. Substituting eq. (1.16) into eq. (1.15), we get

$$-\oiint_S (-k\nabla T) \bullet d\mathbf{A} = \frac{\partial}{\partial t} \int_V u \, dV \tag{1.17}$$

where u is the internal energy per unit volume and the integration on the right hand side is over the volume. The left-hand side of eq. (1.17) can be converted into volume integration using Gauss's divergence theorem and the right-hand side can be further related to temperature through the chain rule and the specific heat c,

$$\int_V \nabla \bullet (k\nabla T) dV = \int_V \rho c \frac{\partial T}{\partial t} dV \tag{1.18}$$

For this equation to be valid in any region, we must have

$$\nabla(k\nabla T) = \rho c \frac{\partial T}{\partial t} \tag{1.19}$$

which is the familiar heat diffusion equation that is used to solve heat conduction problems on the macroscale.

1.3.5 Local Equilibrium

In thermodynamics, we define equilibrium as a state of an isolated system in which no macroscopic change can be observed as a function of time. Quantities such as temperature and pressure are defined only under equilibrium conditions. Transport processes happen when the system is driven out of equilibrium. A system undergoing steady-state heat conduction is not in an equilibrium state. Although no change occurs in such a system, the steady state does not violate the definition of equilibrium, since the system is not isolated. However, the nonequilibrium state of the system does pose a problem because the temperature cannot be defined in accordance with thermodynamics. It would seem the constitutive relations we have introduced are meaningless. This dilemma can be resolved by employing the concept of local equilibrium. Although a system may be out of equilibrium globally, the deviation from equilibrium at each point is usually small. A small region surrounding each space point may be approximated as being in equilibrium, which allows us to define a local temperature, pressure, and chemical potential. We have not, however, established rigorous criteria based on when we can assume local equilibrium and we do not know yet how small this region should be. We can further ask what happens if the system is smaller than this minimal size.

1.3.6 Scaling Trends under Macroscopic Theories

The characteristic length scales at which the classical theories discussed in this section fail are typically on the order of submicrons, although the exact demarcation line depends on the type of energy carriers, the media, and the temperature. A wide range of microdevices exists with characteristic lengths larger than microns for which the classical transport theories are still applicable. It is, however, interesting to consider how the miniaturization will change the heat transfer characteristics. As the device size shrinks, the ratio of surface area to volume increases, leading to significant changes in thermal, electrical, and mechanical behavior. A well-known example in mechanics is the diminishing effect of inertia and the increasing importance of surface tension, as is evident in the fact that ants (with a characteristic dimension of millimeters) are not injured by falling off a table but can be easily trapped in a drop of water, which is contrary to what may happen to humans (with a characteristic length of meters) in similar situations. In table 1.2, we show some examples of the scaling trend for simple heat transfer configurations, assuming that macroscopic transport theories are still applicable.

1.4 Microscopic Picture of Heat Carriers and Their Transport

This section briefly addresses the following questions: (1) What carries heat? (2) How much energy do the carriers possess? (3) How fast do they travel? and (4) How far do they travel? The explanation will be brief here since these questions will be discussed in more detail in subsequent chapters.

1.4.1 Heat Carriers

People used to think that heat was carried by a special form of matter termed calories, which were supposedly massless and colorless. When two objects at different temperatures are in contact with each other, calories from the object at the high temperature would flow into the object at the lower temperature. This view prevailed in the early 19th century and contributed significantly to the development of classical thermodynamics. Of course, we now know that this is an incorrect picture. In fact, the caloric theory, despite its inception in the late 18th century, was completely abandoned by the middle of the 19th century. Yet most of the results on classical thermodynamics derived from such a wrong picture turned out to be correct and are still valid today because, as we mentioned before, classical thermodynamics does not consider detailed pictures of heat carriers.

Heat conduction actually results from the random motion of the material particles in the system carrying thermal energy from one location to another. These material particles are electrons, atoms, and molecules in gases, liquids, and solids. We use heat conduction in a gas, as shown schematically in figure 1.9, to illustrate the microscopic energy transport process. Gas molecules near the hot wall collide with the solid atoms of the wall often and gain a higher kinetic energy, that is, a higher random velocity. These molecules are in random motion and have the possibility of moving toward the lower temperature end. During this process, they collide with molecules having smaller random velocity (and thus cooler) and pass some of the excess energy to those molecules. Such a process cascades for all adjacent molecules until it reaches the molecules in the proximity of the cold wall. These molecules have a higher kinetic energy than the atoms in the solid wall and will impart their excess energy to the wall through collisions.

Table 1.2 Scaling trend under classical transport theories for representative microgeometries

Mode and Example	Quantity	Scaling Trend
Heat conduction (1) Along thin films (2) Perpendicular to thin films	Heat transfer rate: $Q = kA_c \Delta T/L$ Heat flux: $q = k\Delta T/L$ Thermal diffusion time: $\tau = L^2/a$ L: heat conduction path length L = film length (along film plane) L = film thickness (perpendicular) a: thermal diffusivity, m^2 s^{-1}	Case (1): Q small since A_c is small q larger than bulk, depending on L τ shorter than bulk, depending on L Case (2): Q depending on A_c/L q very large since L is small τ very short
Convection (1) Inside a microchannel (2) Across a microcylinder	Heat transfer rate: $Q = hA(T_w - T_a)$ Heat flux: $q = h(T_w - T_a)$ Nusselt No. for (1): $Nu = hd/k = $ const. h for case (1): $h \propto 1/d$ Nusselt No. for (2): $Nu - 0.3 \propto Re^{1/2}$ h for (2): $h \propto 1/d^{1/2}$ to $1/d$	For case (1): Q independent of d q increases as $1/d$ h increases as $1/d$ For case (2): Q decreases with d^0 to $d^{1/2}$ q increases with $1/d^{1/2}$ to $1/d$ h increases with $1/d^{1/2}$ to $1/d$
Radiation Between two black parallel plates	Heat transfer rate: $Q = \sigma A(T_1^4 - T_2^4)$ Heat flux: $q = \sigma(T_1^4 - T_2^4)$	Q decreases since A decreases q remains constant
Coupled conduction and convection (1) Heat conduction along fins (2) Lumped capacitance	Case (1) Temperature: $\frac{[T_b - T(x)]}{T_b - T_a} = e^{-\gamma x}$ Heat transfer rate: $Q = kA_c\gamma(T_b - T_a)$ fin parameter: $\gamma = [hp/(kA_c)]^{1/2}$ Case (2) Temperature: $\frac{[T(t) - T_a]}{T_i - T_a} = e^{-t/\tau}$ Time Constant: $\tau = \rho c V/(hA)$ p = fin perimeter; T_i = initial temperature T_b = fin base temperature	For case (1): γ increases, temperature decreases rapidly along fins Q decreases due to decreasing A_c For case (2): τ decreases due to decreasing V/A and increasing h

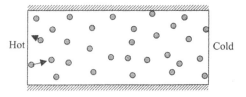

Figure 1.9 Microscopic heat conduction process through a gas.

The resultant effect is that a net energy flows from the hot wall to the cold wall due to the temperature difference between the two walls. It should become clear that the net energy flow is due to the random motion of the molecules and that the molecules do not necessarily move from the hot to the cold side.

In dielectric solid materials (electrical insulators), heat is conducted through the vibration of atoms. The atoms are bonded to each other in a dielectric material through interatomic force interactions. Figure 1.10 shows a schematic of the interatomic potential, ϕ, between two atoms as a function of their separation, x. The force interaction between two atoms is the derivative of such an interatomic potential

$$F = \frac{-d\phi}{dx} \qquad (1.20)$$

If the two atoms are far apart, an attractive force exists between the atoms because the electrons of one atom attract the nucleus of the other. When the atoms are close, the interaction force becomes repulsive because the electron orbits, or the nuclei, of different atoms begin to overlap. The minimum potential defines the equilibrium positions of the atoms. Each atom in a solid vibrates around its equilibrium position. The motion of each atom is constrained by its neighboring atoms through the interatomic potential. A simplified picture of the interatomic interactions in crystals can be represented by a mass-spring system, as shown in figure 1.11(a). In such a system, the vibration of any one atom can cause the vibration of the whole system by creating lattice waves in the system. The propagation of sound in a solid is due to long-wavelength lattice waves. If one side of the solid is hotter, the atoms near the hot side will have larger vibrational amplitudes, which will be felt by the atoms on the other side of the system through the propagation and interaction of lattice waves. Quantum mechanical principles dictate that the energy of each lattice wave is discrete and must be a multiple of $h\nu$ (except for a small modification called zero point energy that equals $h\nu/2$), where ν is the frequency of the lattice wave

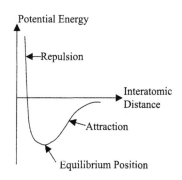

Figure 1.10 A typical interatomic potential profile.

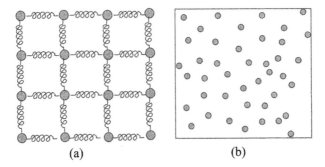

Figure 1.11 (a) A mass-spring system representing interconnected atoms in a crystal, and (b) phonon gas model replaces the solid atoms in a crystal.

and h the Planck constant (6.6×10^{-34} J s). This minimum energy $h\nu$ of a quantized lattice wave is called a phonon. A phonon at a specific frequency and wavelength is a wave that extends through the entire crystal. The superposition of phonons of multiple frequencies existing in the solid forms wave packets that have a narrow spatial extent. These wave packets can be considered as particles as long as they are much smaller than the crystal size. Using this phonon particle picture, the spring system in figure 1.11(a) can be replaced by a box of phonon particles as in figure 1.11(b). Heat in a dielectric crystal is conducted by such a phonon gas similar to that in a box of gas molecules as shown in figure 1.9. The collision of phonon particles is due to the interaction of phonon waves, which can be further attributed to the nonparabolicity (or anharmonic) potential profile as shown in figure 1.10. A parabolic potential, according to eq. (1.20), would lead to an ideal spring with force linearly proportional to displacement from the minimum potential position. Phonons in such an ideal potential do not collide with each other. In chapters 3 and 5 we will explain in more detail the phonon concept and when we can treat phonons as particles rather than waves.

In metals, heat is conducted by free electrons as well as by phonons. When atoms are bonded together to form a metal, some of the electrons in the outer orbits of the nuclei become free from the bonding of the nuclei. These free electrons can travel a distance much longer than the interatomic distance until they are scattered by either atoms, electrons, or impurities. Under this picture, the free electrons inside a metal can also be thought of as a gas—an electron gas. The heat conduction process for the electron gas is again similar to that shown in figure 1.9 for molecules. Electrons in a metal travel at a velocity typically three orders of magnitude larger than phonons do. Thus, compared to the energy carried by lattice waves, the energy flux carried by electrons is in general much larger. Therefore, in metals, electrons are usually the dominant heat carriers.

One may guess that, in semiconductors, heat is probably conducted partially by phonons and partially by electrons. That assumption is not completely correct. In fact, heat is carried dominantly by phonons in most semiconductors, because the free electron density in a normal semiconductor is much smaller than that in a metal. As an example, the electron carrier density in a metal is $\sim 10^{23}$ cm^{-3} while in a semiconductor it is typically less than 10^{18} cm^{-3}. In lightly and moderately doped semiconductors, phonons are the dominant heat carriers. However, the electron contribution in a heavily doped semiconductor can be appreciable.

Given the above pictures of heat carriers for heat conduction processes, the convection of heat can be understood rather easily since the only difference of convection from heat conduction is that now heat carriers have a nonvanishing average velocity superimposed on their random velocity. When a liquid or gas molecule moves from one place to another due to its nonzero average velocity, it also carries its internal energy directly from one place to another. This direct motion of internal energy in convection is very different from heat conduction process. In the latter, heat is transferred due to the energy exchange of heat carriers in the collision process. Heat conduction process exists even in convection process because molecules are still doing random motion. In fact, the random motion velocity of the molecules is usually much higher than the nonvanishing average velocity in convection. However, convection is more effective than heat conduction because energy moves directly from one place to another.

Thermal radiation involves another heat carrier, that is, electromagnetic waves. The propagation of electromagnetic waves emitted from a thermal source does not differ fundamentally from those carrying TV and radio signals because all are governed by the Maxwell equations. The major difference lies in how these waves are generated. Thermal radiation typically refers to the electromagnetic waves that are generated by a heat source, while TV and radio signals are generated by an artificial source such as the oscillation of current in a circuit. At the microscopic level, thermal radiation is due to the oscillation of charges in the atoms and crystals. Similarly to lattice waves, electromagnetic waves are also quantized as a result of quantum mechanics. An electromagnetic wave at frequency ν can have energy that is only a multiple of $h\nu$. This smallest energy quantum of an electromagnetic field, $h\nu$, is called a photon. In fact, most people are probably more familiar with the terminology of photon than phonon. Both of them are the basic quantum of a wave; one for electromagnetic waves and the other for lattice vibrational waves.

1.4.2 Allowable Energy Levels of Heat Carriers

The transport of heat is due to the motion of the energy carriers discussed above and the associated energy they carry. To describe heat transfer quantitatively, we need to know the energy associated with these heat carriers. The possible energy states of heat carriers are determined by quantum mechanical principles, which we will cover in chapters 2 and 3. Here, we will only give the reader some brief idea. For individual atoms and molecules, the energy levels are typically discrete. For example, the allowed energy levels of a harmonic oscillator, which is a good model for the vibrations of a diatomic molecule such as H_2, are given by

$$E_n = h\nu(n + 1/2) \qquad (n = 0, 1, 2, 3, \ldots) \qquad (1.21)$$

where ν is the fundamental vibration frequency. The amplitude of the vibration must be such that the total energy of the molecule fits into one of the above discrete energy levels. In crystalline solids, the allowable energy of electrons and phonons is a function of the wavevector. Such functional relations are also called dispersion relations. A wavevector **k** points in the direction of wave propagation (electron and phonon waves) and its magnitude equals 2π divided by the wavelength. Figures 1.12(a) and (b) show examples

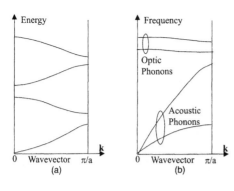

Figure 1.12 Illustration of typical dispersion relation for (a) electrons and (b) phonons in crystals.

of the electron and phonon energy levels, represented by either energy or frequency, in solids along a specific crystallographic direction, where a is the periodicity of the atoms in the direction of the wave propagation and **k** is the wavevector. The allowable energy levels form bands. Within each band, the energy is quasi-continuous because the wavevector is discrete, as we will discuss in more detail in chapter 3. However, there can be substantial gaps between different bands. The filling of the electronic bands by electrons and the magnitude of the gaps determine whether a solid is a metal, a semiconductor, or an insulator. The phonon band differs significantly from that for electrons. Electron bands are typically approximated by a parabolic relationship between the energy and the wavevector, whereas the dispersion relation between the phonon wavelength and the wavevector is often approximated by a linear relationship.

1.4.3 Statistical Distribution of Energy Carriers

Not all the possible energy levels of energy carriers (molecules, electrons, and phonons) will be actually occupied by the carriers. Classical thermodynamics tells us that a non-isolated system at equilibrium tends to minimize its energy. In a system with different allowable energy levels, heat carriers will fill the lowest energy levels at zero temperature. At higher temperatures, some of the carriers will have higher energy levels. The most probable energy distributions of the carriers are governed by statistical principles. Classical statistical theory gives the probability density $f(E)$, defined as the probability of finding the carrier at energy E per energy interval surrounding E, for a particle in an equilibrium system at a temperature T,

$$f(E) = Be^{-E/(\kappa_B T)} \tag{1.22}$$

where κ_B ($= 1.38 \times 10^{-23}$ J K^{-1}) is the Boltzmann constant, T is the absolute temperature, and B is a normalization factor. Equation (1.22) is the famous Boltzmann distribution (Ludwig Boltzmann, 1844–1906). For a monoatomic ideal gas system, the only energy of each atom is its translational kinetic energy

$$E = \frac{m}{2}(v_x^2 + v_y^2 + v_z^2) \tag{1.23}$$

where m is the mass of the atom and v_x, v_y, and v_z are the components of its random velocity. Since the probability of finding this particle to have energy between zero and infinite speed must be one, we have

$$\int_{-\infty}^{\infty} dv_x \int_{-\infty}^{\infty} dv_y \int_{-\infty}^{\infty} B \exp\left[-\frac{m(v_x^2 + v_y^2 + v_z^2)}{2\kappa_B T}\right] dv_z = 1 \quad (1.24)$$

Carrying out the above integration leads to

$$B = \left(\frac{m}{2\pi\kappa_B T}\right)^{3/2} \quad (1.25)$$

The probability density of a monatomic gas is thus

$$f(\mathbf{v}) = \left(\frac{m}{2\pi\kappa_B T}\right)^{3/2} \exp\left[-\frac{m(v_x^2 + v_y^2 + v_z^2)}{2\kappa_B T}\right] \quad (1.26)$$

which is called the Maxwell distribution. With this probability density, we can calculate other expectation values (or average values). For example, the average energy (internal energy) of a monatomic gas molecule is

$$\langle E \rangle = \int_{-\infty}^{\infty} dv_x \int_{-\infty}^{\infty} dv_y \int_{-\infty}^{\infty} \frac{m}{2}(v_x^2 + v_y^2 + v_z^2) \left(\frac{m}{2\pi\kappa_B T}\right)^{3/2}$$

$$\exp\left[-\frac{m(v_x^2 + v_y^2 + v_z^2)}{2\kappa_B T}\right] dv_z \quad (1.27)$$

Transforming the Cartesian coordinates v_x, v_y, v_z into spherical coordinates makes the above integration much easier. The final result is

$$\langle E \rangle = \frac{3}{2}\kappa_B T \quad (1.28)$$

The above expression means that temperature is a measure of the average kinetic energy for a monatomic gas. Equation (1.28) is a very useful result to remember and is an example of the equipartition theorem in statistical thermodynamics. The equipartition theorem for classical systems states that at sufficiently high temperatures (such that the Boltzmann distribution is valid) each quadratic term of the molecular energy contributes to the molecule an average energy $\kappa_B T/2$. For a monatomic gas, each molecule has three quadratic energy components, from its translational motion, as indicated by eq. (1.23), so that the total average kinetic energy is $3\kappa_B T/2$.

Example 1.1 *Speed and specific heat of gas molecules*

Estimate the average speed of helium atoms as an ideal gas at 300 K, and also estimate its specific heat at constant volume.

Solution: The average speed of molecules obeying the Maxwell distribution is

$$\langle v \rangle = \int_{-\infty}^{\infty} dv_x \int_{-\infty}^{\infty} dv_y \int_{-\infty}^{\infty} v \left(\frac{m}{2\pi\kappa_B T}\right)^{3/2} \exp\left[-\frac{m(v_x^2 + v_y^2 + v_z^2)}{2\kappa_B T}\right] dv_z$$

$$= \int_0^{\infty} v \left(\frac{m}{2\pi\kappa_B T}\right)^{3/2} \exp\left(-\frac{mv^2}{2\kappa_B T}\right) 4\pi v^2 dv = \sqrt{\frac{8\kappa_B T}{\pi m}}$$

The mass of a helium atom is $4 \times 1.67 \times 10^{-27}$ kg, where 4 comes from the fact that a helium atom has two protons and two neutrons, and 1.67×10^{-27} kg is the rest mass of a proton (a neutron has approximately the same weight as a proton). We thus obtain the average velocity of helium atoms as 1257 m s^{-1}. To obtain the specific heat, we know that a mole contains $N_A = 6.02 \times 10^{23}$ (Avogadro's constant) molecules. The total energy of a mole of helium atoms is thus $u = 3N_A\kappa_B T/2$. The specific heat at constant volume is $c_v = \partial u/\partial T = 3\kappa_B N_A/2 = 3R_u/2 = 12.5$ J K^{-1} mol^{-1}, where $R_u = \kappa N_A = 8.314$ J K^{-1} mol^{-1} is the universal gas constant. The actual specific heat of helium is in agreement with this number.

Comment: A quick order-of-magnitude estimation of the average speed can be obtained by setting the average kinetic energy of each atom, $3\kappa_B T/2$, equal to $mv^2/2$, which leads to $v = (3\kappa_B T/m)^{1/2} \approx 1364$ m s^{-1}, which is slightly higher than $\langle v \rangle$.

When energy levels are not continuous, the normalization factor B in eq. (1.22) can no longer be determined as is done in deriving the Maxwell distribution. We will give the distribution functions for electrons and phonons here, leaving the details to chapter 4. The electron probability density is governed by the Fermi–Dirac distribution and that of phonons and photons by the Bose–Einstein distribution,

$$\text{Fermi–Dirac distribution: } f(E) = \frac{1}{\exp\left(\frac{E-\mu}{\kappa_B T}\right) + 1} \quad (1.29)$$

$$\text{Bose–Einstein distribution: } f(\nu) = \frac{1}{\exp\left(\frac{h\nu}{\kappa_B T}\right) - 1} \quad (1.30)$$

where μ is the chemical potential. The Bose–Einstein distribution is expressed in terms of the phonon energy $E = h\nu$ rather than E directly.

From the above discussion, we see that temperature is only meaningful when we deal with a large number of molecules. At equilibrium, the temperature alone determines the statistical distribution of all the heat carriers in the system. It is meaningless to speak of the temperature of one single particle. But it is meaningful to talk about the energy of one particle and the average energy of a cluster of particles, even if the particles are out of equilibrium such that they do not obey the Boltzmann distribution (or other kinds of expected statistical distributions). Later, we will use the concept of temperature for highly nonequilibrium systems. In these situations, temperature should be understood as a measure of the local average energy density, but one cannot determine the statistical distributions of the particles on the basis of temperature alone.

1.4.4 Simple Kinetic Theory

By definition, heat transfer involves the motion of heat carriers generated by temperature differences. Statistically, heat carriers generated by thermal sources are randomly distributed in all directions. Given the position and velocity of all heat carriers, their subsequent motion determines the energy transport. Although, in principle, the trajectory of these energy carriers can be traced on an individual basis, such an analytical approach is usually impractical due to the large number of carriers existing in the medium. With the rapid advancement of computational power, however, some problems are within

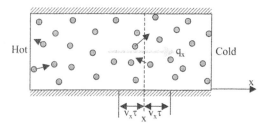

Figure 1.13 Simplified derivation of the Fourier law based on kinetic theory.

the reach of this direct approach. We will devote a chapter (chapter 10) to molecular dynamics, which is based on tracing the trajectory of individual molecules. In most cases, some kinds of averaging are necessary for practical mathematical descriptions of the heat carrier motion. The Fourier law for heat transfer, the Fick law for mass diffusion, and the Ohm law for electrical conduction are the results of averaging the microscopic motion in a sufficiently large region and over a sufficiently long time. The laws are the correct representations of the average behavior of the energy, mass, and current flow in macroscopic systems going through relatively slow processes. Such averaging may no longer be valid in microscale and nanoscale domains and for high-speed processes because the conditions for the average behavior are no longer satisfied. In subsequent chapters, we will take a closer look at the averaged motion of heat carriers in micro- and nanoscale systems. Here, we first introduce a simple derivation of the Fourier law from kinetic theory.

We consider a one-dimensional model as shown in figure 1.13. If we take an imaginary surface perpendicular to the heat flow direction, the net heat flux across this surface is the difference between the energy fluxes associated with all the carriers flowing in the positive and negative directions. Considering the positive direction, the carriers within a distance $v_x \tau$ can go across the interface before being scattered. Here v_x is the x component of the random velocity of the heat carriers and τ is the relaxation time—the average time a heat carrier travels before it is scattered and changes its direction. Thus, the net heat flux carried by heat carriers across the surface is

$$q_x = \frac{1}{2}(nEv_x)\Big|_{x-v_x\tau} - \frac{1}{2}(nEv_x)\Big|_{x+v_x\tau} \tag{1.31}$$

where n is the number of carriers per unit volume and E is the energy of each carrier. The factor 1/2 implies that only half of the carriers move in the positive x direction while the other half move in the negative x direction. Using a Taylor expansion, we can write the above relation as

$$q_x = -v_x \tau \frac{d(Env_x)}{dx} \tag{1.32}$$

Now we assume that v_x is independent of x, and $v_x^2 = (1/3)v^2$, where v is the average random velocity of the heat carriers. The above equation becomes

$$q_x = -\frac{v^2 \tau}{3} \frac{dU}{dT} \frac{dT}{dx} \tag{1.33}$$

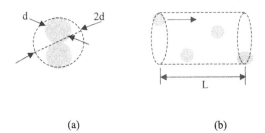

Figure 1.14 Mean free path estimation for a box of gas molecules with effective diameter d for each molecule: (a) effective diameter of two molecules to scatter is $2d$, and (b) mean free path is the average distance between two consecutive scattering.

where $U = nE$ is the local energy density per unit volume and dU/dT is the volumetric specific heat C of the heat carriers at constant volume, which equals the mass specific heat c times the density ρ, i.e., $C = \rho c$. This formulation leads to the Fourier law

$$q_x = -(Cv^2\tau/3)dT/dx = -k dT/dx \tag{1.34}$$

Although the above model is fairly crude, the expression for the thermal conductivity is actually a surprisingly good approximation,

$$k = Cv^2\tau/3 = Cv\Lambda/3 = \rho c v\Lambda/3 \tag{1.35}$$

where $\Lambda = v\tau$ is the mean free path—the average distance a heat carrier travels before it loses its excess energy due to scattering. Very often, the above equation is used to estimate the mean free path on the basis of experimental results for the other parameters in the equation.

1.4.5 Mean Free Path

We now give a very crude derivation of the mean free path for gas molecules. By definition, the mean free path is the average distance that a gas molecule travels between successive collisions. It is not the distance that separates individual molecules. Suppose that the effective diameter of an atom (or molecule) is d, as shown in figure 1.14.* The effective diameter for two atoms to collide is $2d$ and the collision cross-section is $\pi(2d)^2/4$. If the molecule travels a distance L, it sweeps out a volume $\pi d^2 L$. If the molecular number concentration is n, then the number of molecules that this particle will collide with is $n\pi d^2 L$. The average distance between each collision is (Kittel and Kroemer, 1980)

$$\Lambda = \frac{L}{n\pi d^2 L} = \frac{m}{\pi \rho d^2} \tag{1.36}$$

where we have invoked the relation that the number of molecules per unit volume is the density divided by the molecular weight m. This expression assumes that the molecule being considered moves but the other molecules in the volume are stationary. Dropping this assumption leads to a more accurate expression (Tien and Lienhard, 1979),

$$\Lambda = \frac{m}{\pi \sqrt{2}\, \rho d^2} \tag{1.37}$$

---

*This is not the diameter of the ion or the electron orbit. It represents the force range that a molecule exerts on its surroundings.

For an ideal gas, $P = \rho \kappa_B T/m$, so

$$\Lambda = \frac{\kappa_B T}{\pi \sqrt{2}\, P d^2} \tag{1.38}$$

Estimating the effective diameter d of an atom or molecule requires detailed information regarding the molecular and electronic structures of the molecule and the atom. We can readily reason that d is proportional to the number of atoms in the molecule and the number of electrons in the atom. Let us take an order-of-magnitude value for d of 2.5 Å. At $P = 101$ kPa, $T = 300$ K, Λ is approximately 0.14 µm. At a low pressure, for example, at $P = 10^{-8}$ torr $= 1.32 \times 10^{-6}$ Pa, $\Lambda \approx 14{,}000$ m, which is a very long traveling distance and means that no collision occurs in practical systems under such a vacuum condition. In thin-film deposition processes, vacuum conditions are often used to avoid contamination and the long mean free path ensures that atoms will travel uninterrupted (ballistically) from the source to the depositing surface. Equation (1.38) shows that the mean free path of gas molecules increases with increasing temperature. In addition, the gas random velocity increases with the square root of the temperature. At room temperature and higher, the molar specific heat remains constant and the density is proportional to P/T. This effect leads to the conclusion that the thermal conductivity increases with $T^{1/2}$ for gases and is independent of pressure. Check figure 1.5 for this trend of thermal conductivity for various gases.

Example 1.2 *Thermal conductivity of gas*

Estimate the thermal conductivity of air.

Solution: We will use eq. (1.35) to estimate the thermal conductivity of air. We can estimate the random velocity of air molecules from $mv^2/2 = 3\kappa_B T/2$, where m is the average mass of an air molecule (average molar weight of air is 29). This gives $v = 524$ m s^{-1}. At room temperature, $\rho = 1.16$ kg m^{-3} and $c = 1007$ J kg^{-1}K^{-1}. Substituting these values and a mean free path of 0.14 µm into eq. (1.35), we obtain $k = 0.028$ W m^{-1}K^{-1}. This is not far from the actual thermal conductivity of air at 300 K, which is 0.026 W m^{-1}K^{-1}.

1.5 Micro- and Nanoscale Transport Phenomena

At small scales, many macroscopic descriptions of heat transfer become invalid. In this section, we will provide examples to illustrate microscale transport phenomena.

1.5.1 Classical Size Effects

One example of the classical size effect is rarefied gas heat conduction. This size effect occurs when the mean free path of gas molecules becomes comparable to or larger than the size of the system. We have seen that the mean free path increases with decreasing gas pressure. The air pressure in the outer atmosphere is very low and thus the mean free path of the molecules is long. A spacecraft going through the outer atmosphere cannot be modeled on the basis of continuum theory; this sparked a substantial amount of work

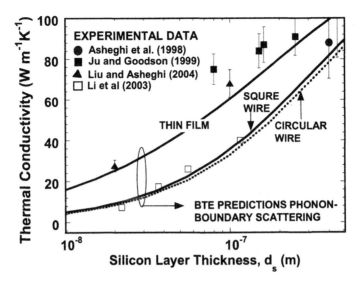

Figure 1.15 Thermal conductivity of silicon films as a function of the film thickness or wire diameter. (Courtesy of M. Ashegli).

in the past on rarefied gas flow and heat transfer. Certain semiconductor manufacturing processes are often performed in vacuum environments, for which heat transfer and fluid flow may fall into the rarefied gas regime.

Past research interests in rarefied gas dynamics and heat transfer were stimulated by the increasing mean free path encountered in low-pressure environments. Microfabrication and nanotechnology led to micrometer and nanometer structures with the characteristic length comparable to the mean free path of gas molecules, even at normal atmospheric pressures. The rarefaction effects must be considered for gas flow and heat transfer in such structures. The small size also brings in additional factors. For example, despite the fact that liquids molecules have a mean free path only on the order of angstroms, the surface charges that build up may significantly affect the heat transfer and fluid flow in submicron channels.

Size effects, which are well studied for gases, can also be expected for electrons and phonons, since both electrons and phonons can be considered as gases existing within solids (electron gas and phonon gas). When the mean free paths of electrons and phonons become comparable to or larger than an object's characteristic length, heat conduction in solids can deviate significantly from the predictions of the Fourier law. The thermal conductivity of thin films or nanowires is no longer solely a material property, but also depends on the film thickness or the wire diameter. Figure 1.15 shows the thermal conductivity of Si single crystal thin films and nanowires as a function of the film thickness and wire diameter.

1.5.2 Quantum Size Effects

According to quantum mechanics heat carriers such as electrons and phonons are also material waves. The finite size of the system can influence the energy transport by altering

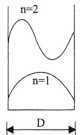

Figure 1.16 Standing waves in a quantum well, the two lowest energy levels.

the wave characteristics, such as forming standing waves and creating new modes that do not exist in bulk materials. For example, electrons in a thin film can be approximated as standing waves inside a potential well of infinite height as shown in figure 1.16. The condition for the formation of such standing waves is that the wavelength λ satisfies the following relation

$$n\lambda/2 = D \ (n = 1, 2, \ldots) \tag{1.39}$$

where D is the width of the potential well. Given the above electron wavelength, we can calculate its momentum according to the de Broglie relation between wavelength λ and momentum p,

$$p = h/\lambda \tag{1.40}$$

where h is the Planck constant ($h = 6.6 \times 10^{-34}$ J s). The energy of the electron is thus $E = p^2/2m$,

$$E_n = \frac{h^2}{8m} \left[\frac{n}{D}\right]^2 (n = 1, 2, \ldots) \tag{1.41}$$

For a free electron, $m = 9.1 \times 10^{-31}$ kg, and $D = 1$ μm, $E_n = 5.9 \times 10^{-25} n^2$ J, so that the energy separation between the $n = 1$ and $n = 2$ levels is 1.8×10^{-24} J. At room temperature, this energy separation is too small compared to the thermal fluctuation energy $\kappa_B T = 41.6 \times 10^{-22}$ J to be distinguishable from thermal fluctuation. In addition, the electron mean free path at room temperature is usually much smaller than 1 μm. Scattering of electrons destroys the condition for forming standing waves, making eq. (1.39) inapplicable. However, as the film size is further reduced, say to $D \approx 100$ Å, scattering is negligible and energy quantization becomes observable in comparison with thermal fluctuation. The electron energy quantization will affect the electrical, optical, and thermal properties of nanostructures and nanomaterials.

1.5.3 Fast Transport Phenomena

Size effects create deviations from commonly used classical laws by imposing new conditions at the boundaries. Transport at short time scales may also differ significantly from that at the longer time scales that we are used to. This difference is because the classical laws are commonly derived for time scales much longer than the time scale of

Table 1.3 Basic characteristics of energy carriers

	Free Electrons	Phonons	Photons	Molecules
Source	Freed from nucleic bonding	Lattice vibration	Electron and atom motion	Atoms
Propagation media	In vacuum or media	In media	In vacuum or media	In vacuum or media
Statistics	Fermi–Dirac	Bose–Einstein	Bose–Einstein	Boltzmann
Frequency or energy range	0–infinite	Debye cutoff	0–infinite	0–infinite
Velocity (m s^{-1})	$\sim 10^6$	$\sim 10^3$	$\sim 10^8$	$\sim 10^2$

microscopic processes. We can quickly estimate, for example, the average time interval τ between successive collisions of phonons as

$$\tau = \Lambda/v \qquad (1.42)$$

where Λ is the phonon mean free path and v the average phonon velocity. For many materials, this relaxation time is of the order 10^{-12} to 10^{-10} s. Although this appears to be an amazingly short time, a laser pulse can be as short as a few femtoseconds (1 fs = 10^{-15} s). Clearly, if we deal with processes shorter than the relaxation time, the classical Fourier diffusion law will no longer hold true, since the diffusion process is established by considering the multiple collisions of the heat carriers such that their motion is almost random. In addition to the relaxation time, there are also other time scales that need to be considered, such as the time characterizing the energy exchange between electrons and phonons. An example of the latter is the femtosecond laser heating of metals (Qiu and Tien, 1993).

Table 1.3 summarizes the basic characteristics of energy carriers, including:

- Free electrons in solids, which are released from the bonding of the nuclei and can propagate in solids as well in a vacuum;
- Phonons, which are due to atom vibration in crystals and cannot propagate in a vacuum;
- Photons, which are generated from the electron and atom motion and can propagate in a vacuum;
- Molecules, which form due to bonding of atoms.

Electrons obey the Fermi–Dirac statistics and are called fermions, while photons and phonons obey the Bose–Einstein statistics and are called bosons. Molecules obey the Boltzmann statistics (classical) under most temperatures except when approaching absolute zero, when Bose–Einstein statistics must be considered. The quantum mechanically allowable energy (or frequency) of one carrier spans a wide range, from zero to infinite, except for phonons for which the maximum is capped. The allowable energy only tells what is possible, but not the average, which is determined by the temperature and the statistics. The random average velocity of the heat carriers increases from molecules, to phonons, to electrons, and to photons, approximately.

Table 1.4 illustrates the transport regimes of energy carriers. This regime table can be best understood after reading through chapters 5–7. It divides the transport into

Table 1.4 Transport regimes of energy carriers; O represents order of magnitude

Important Length Scales		Regimes	Photon	Electron	Phonon	Fluids
Coherence length, ℓ_c	Wave regime	$D < O(\ell_p)$ $D < O(\ell_c)$	Maxwell EM theory	Quantum mechanics	Quantum mechanics	Super fluidity
Phase-breaking length, ℓ_p ℓ_c: for photon: μm–km for phonon: 10 Å for electron: 100 Å $\ell_p \gtrsim$ Mean free path	Transition regime	$D \sim O(\ell_p)$ $D \sim O(\ell_c)$	Coherence theory	Quantum Boltzmann equation		
	Particle regime $D > O(\ell_c), D > (\ell_p)$	$D < O(\Lambda)$ ballistic	Ray tracing	Ballistic transport	Ray tracing	Free molecular flow
Mean free path, Λ Photon: 100 Å–1 km Electron: 100–1000 Å Photon: 100–1000 Å		$D \sim O(\Lambda)$ quasi-diffusive	Radiative transfer equation	Boltzmann transport equation	Boltzmann transport equation	Boltzmann transport equation
		$D > O(\Lambda)$ diffusive	Diffusion approximation	Ohm's law	Fourier's law	Newton's shear stress

several regimes. The wave regime is where the phase information of the energy carriers must be considered and the transport is coherent, and the particle regime is where the phase information can be neglected and the transport is incoherent. In between is the partially coherent transition regime from wave description to the particle description. The phase-breaking length is the distance needed to completely destroy the phase of the heat carriers through various collision processes such as phonon–phonon collision and phonon–electron collision, and it is usually comparable to or slightly longer than the mean free path. The coherence length measures the distance beyond which waves from the same source can be superimposed without considering the phase information. The overlapping length scales in table 1.4 hint at the complexity in judging when to treat them as waves and when to treat them as particles, but this should become clearer after we treat wave and particle size effects in chapters 5 and 7.

Most engineering courses teach only the classical transport theories; that is, the bottom row of table 1.4. Some engineering disciplines may be more familiar with electromagnetic waves and photon radiative transfer: in other words, the column for photons. A wide range of transport problems fall into territories that are not familiar to classical engineering disciplines but are becoming increasingly important in contemporary technology. This book covers these unfamiliar domains as well as the more familiar regimes to help the reader solve problems on all scales.

1.6 Philosophy of This Book

The introductory discussion thus far suggests that, to deal with micro- and nanoscale thermal energy transport processes, one needs to have a clear picture of the motion of energy carriers and the thermal energy associated with their motion. This book aims

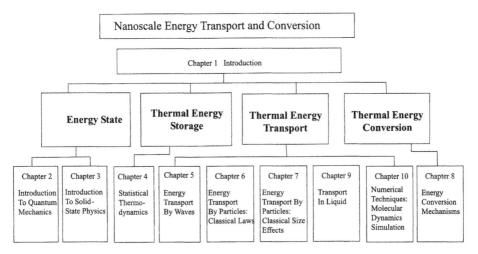

Figure 1.17 Structure of this book.

to develop a unified microscopic picture on energy transport processes for elementary heat carriers including electrons, phonons, photons, and molecules. Figure 1.17 shows the structure of this book. We will began with the energy states of energy carriers, which are established on quantum mechanical principles, solid-state physics, and electrodynamics (chapters 2 and 3). The association of these energy states with heat transfer comes about through the statistical distribution of heat carriers at equilibrium, where temperature enters the picture. Statistical thermodynamics is discussed in chapter 4. Heat carriers have both wave and particle characteristics, as is implied by quantum-mechanical wave-particle duality. The transport of thermal energy can be analyzed from either the wave or the particle picture. It is important to understand when to treat the carriers as waves and when as particles. A key question is whether the phase information of the carriers is maintained or not in the transport process. Energy transport as waves is treated in chapter 5, together with a discussion on when the wave characteristics can be ignored so that the particle description alone is sufficient. Energy transfer by particles is treated in chapter 6 on the basis of the Boltzmann equation, from which classical laws valid at macroscales are derived, accompanied by discussions on the approximations used behind the derivations. Classical size effects are treated in chapter 7. Energy conversion between different carriers is discussed in chapter 8. Throughout the entire text, attempts are made to treat all the energy carriers—electrons, phonons, photons, and molecules—in parallel so that readers can draw analogies from their previous engineering and scientific backgrounds. Liquid molecules, however, defy such a unified treatment and thus chapter 9 is devoted to a discussion on liquids. Direct simulation tools based on molecular dynamics are becoming increasingly useful. Thus, chapter 10 introduces molecular dynamics techniques, with an emphasis on the statistical foundation for analyzing the trajectory of the molecules obtained in a typical molecular dynamics simulation. It should be noted that the linear response theory introduced in this chapter has much broader applications than just to molecular dynamics.

The text strives to develop connections and analogies among the energy carriers. The transport of these carriers is often discussed in various disciplines and is often considered to be governed by different laws. The underlying principles, however, are either very similar or identical, as is indicated by table 1.4, which justifies the attempt to develop a parallel treatment. In this table, we divide transport into generally the wave regime and the particle regime because all matter has both wave and particle characteristics. In the wave regime, the phase information of the energy carriers must be considered, whereas in the particle regime, only the trajectory is important. The propagation of material waves or electromagnetic waves shares many similarities despite the differences in the governing equations. In the particle regime, the phase information of the energy carriers can be ignored. Particle transport, under most conditions, can be described by the Boltzmann equation. The bottom line of table 1.4 lists the diffusion theories that most readers with an engineering background are familiar with. Some readers may also be familiar with propagation of electromagnetic waves or photon transport (or an acoustic wave equivalent). Other regimes in the table are most likely less familiar for readers in engineering but are often encountered in dealing with nanoscale transport. This book aims to cover the transport in all of the aforementioned regimes, while developing a parallel treatment for all carriers so that readers with different backgrounds can draw on their prior knowledge in exploring the nano-territory. Because of the wide range of topics covered in this book, however, readers should be prepared to embrace new terminologies. I believe that familiarity with different terminologies is a necessity for interdisciplinary work, which is becoming increasingly important in our exploration of the "ample room at the bottom."

1.7 Nomenclature for Chapter 1

a	thermal diffusivity, $m^2 s^{-1}$	E	allowed energy level, J
A	cross-sectional area for conduction, and surface area for convection, m^2	f	probability distribution function
		F	interatomic force, N
		g	gravitational acceleration, $m\,s^{-2}$
		h	convection heat transfer coefficient, $W\,m^{-2}\,K^{-1}$; Planck constant, J s
B	normalization factor		
c	specific heat, $J\,kg^{-1}K^{-1}$	k	thermal conductivity, $W\,m^{-1}\,K^{-1}$
C	volumetric specific heat, $J\,m^{-3}K^{-1}$	$\mathbf{k}$	wavevector, m^{-1}
d	effective diameter of molecules, m	L	characteristic length, m
		m	mass, kg
D	diameter of cylinders or spheres; separation of plates, m	n	integer; Particular number density, m^{-3}
		N_A	Avogadro constant, mol^{-1}
D_h	hydraulic diameter, m	Nu	Nusselt number
e	emissive power per unit area, $W\,m^{-2}$	p	perimeter, m; momentum, $kg\,m\,s^{-1}$
e_λ	spectral emissive power per wavelength interval, $W\,m^{-2}\mu m^{-1}$	Pr	Prandtl number
		$\mathbf{q}$	heat flux vector, $W\,m^{-2}$

Q	heat transfer rate, W	ν	kinematic viscosity, $m^2 s^{-1}$; frequency of phonons and photons, s^{-1}
Ra	Rayleigh number		
Re	Reynolds number	ρ	density, $kg\, m^{-3}$
t	time, s	σ	Stefan–Boltzmann constant, $W\, m^{-2} K^{-4}$
T	temperature, K	τ	relaxation time, s; time constant, s
u	internal energy per unit volume, $J\, m^{-3}$	τ_{xy}	shear stress, $N\, m^{-2}$
		ϕ	interatomic potential, J
u	fluid velocity, $m\, s^{-1}$		
U	system or internal energy, J		

Subscripts

v	molecular instantaneous random velocity, $m\, s^{-1}$	a	ambient
V	volume, m^3	b	blackbody
W	power output, W	c	cross-section
β	thermal expansion coefficient, K^{-1}	λ	per unit wavelength
γ	fin parameter, m^{-1}	w	wall
κ_B	Boltzmann constant, $J\, K^{-1}$	x, y, z	Cartesian components
λ	wavelength, m		
Λ	mean free path, m		Superscripts
μ	dynamic viscosity, $N\, s\, m^{-2}$; chemical potential, J	$-$	average

1.8 References

Asheghi, A., Touzelbaev, M.N., Goodson, K.E., Leung, Y.K., and Wong, S.S., 1998, "Temperature-Dependent Thermal Conductivity of Single-Crystal Silicon Layers in SOI Substrates," *Journal of Heat Transfer*, vol. 120, pp. 30–36.

Born, M. and Wolf, E., 1980, *Principles of Optics*, 6th ed., Pergamon Press, Oxford.

Cahill, D.G., Goodson, K.E., and Majumdar, A., 2002, "Thermometry and Thermal Transport in Micro/Nanoscale Solid-State Devices and Structures," *Journal of Heat Transfer*, vol. 124, pp. 223–241.

Chen, G., 1996a, "Nonlocal and Nonequilibrium Heat Conduction in the Vicinity of Nanoparticles," *Journal of Heat Transfer*, vol. 118, pp. 539–545.

Chen, G., 1996b, "Heat Transfer in Micro- and Nanoscale Photonic Devices," *Annual Review of Heat Transfer*, vol. 7, pp. 1–57.

Chen, G., and Shakouri, A., 2002, "Nanoengineered Structures for Solid-State Energy Conversion," *Journal of Heat Transfer*, vol. 124, pp. 242–252.

Chen, G., Borca-Tasciuc, D., and Yang, R.G., 2004, "Nanoscale Heat Transfer," in *Encyclopedia of Nanoscience and Nanotechnology*, ed. Nalwa, H.S., IAP Press, La Jolla, CA.

Chen, G., Dresselhaus, M.S., Fleurial, J.-P., and Caillat, T., 2003, "Recent Developments in Thermoelectric Materials," *International Materials Review*, vol. 48, pp. 45–66.

Domoto, G.A., Boehm, R.F., and Tien, C.L., 1970, "Experimental Investigation of Radiative Transfer between Metallic Surfaces at Cryogenic Temperatures," *Journal of Heat Transfer*, vol. 92, pp. 412–417.

Dresselhaus, M.S., Dresselhaus, G., Sun, X., Zhang, Z., Cronin, S.B., Koga, T., Ying, J.Y., and Chen, G., 1999, "The Promise of Low-Dimensional Thermoelectric Materials," *Microscale Thermophysical Engineering*, vol. 3, pp. 89–100.

Dresselhaus, M.S., Lin, Y.M., Cronin, S.B., Rabin, O., Black, M.R., Dresselhaus, G., and Koga, T., 2001, "Quantum Wells and Quantum Wires for Potential Thermoelectric Applications," *Semiconductors and Semimetals*, vol. 71, pp. 1–121.

Feynman, R.P., 1959, "There's Plenty of Room at the Bottom," Talk at Annual Meeting of American Physical Society, December 26; see *Journal of Microelectromechanical Systems*, vol. 1, pp. 60–66 (1992) or *Science*, vol. 254, pp. 1300–1301 (1991).

Feynman, R.P., 1983, "Infinitesimal Machinery," Talk at Jet Propulsion Laboratory, February 23; see also *Journal of Microelectromechanical Systems*, vol. 2, pp. 4–14 (1993).

Fleming, J.G., Lin, S.Y., El-Kady, I., Biswas, R., and Ho, K.M., 2002, "All-Metallic Three-Dimensional Photonic Crystals with a Large Infrared Bandgap," *Nature*, vol. 417, pp. 52–55.

Fukui, S., and Kaneko, R., 1988, "Analysis of Ultra-Thin Gas Film Lubication Based on Linearized Boltzmann Equation: First Report—Derivation of a Generalized Lubrication Equation Including Thermal Creep Flow," *ASME Journal of Tribology*, vol. 110, pp. 253–263.

Goldsmid, H.J., 1964, *Thermoelectric Refrigeration*, Plenum Press, New York.

Greffet, J.-J., Carminati, R., Joulain, K., Mulet, J.-P., Malnguy, S., and Chen, Y., 2002, "Coherent Emission of Light by Thermal Sources," *Nature*, vol. 416, pp. 61–64.

Harman, T.C., Taylor, P.J., Walsh, M.P., and LaForge, B.E., 2002, "Quantum Dot Superlattice Thermoelectric Materials and Devices," *Science*, vol. 297, pp. 2229–2232.

Hatta, I., 1990, "Thermal Diffusivity Measurement of Thin Films and Multilayered Composites," *International Journal of Thermophysics*, vol. 11, pp. 293–302.

Hector, S., and Mangat, P., 2001, "Review of Progress in Extreme Ultraviolet Lithography Masks," *Journal of Vacuum Science and Technology B*, vol. 19, pp. 2612–2616.

Ho, C.M., and Tai, Y.-C., 1998, "Micro-Electro-Mechanical Systems (MEMS) and Fluid Flow," *Annual Review of Fluid Mechanics*, vol. 30, pp. 579–612.

Ju, Y.S., and Goodson, K.E., 1999, "Phonon Scattering in Silicon Films with Thickness of Order of 100 nm," *Applied Physics Letters*, vol. 74, pp. 3005–3007.

Kittel, C., and Kroemer, H., 1980, *Thermal Physics*, Wiley, New York.

Lee, S.M., and Cahill, D.G., 1997, "Heat Transport in Thin Dielectric Films," *Journal of Applied Physics*, vol. 81, pp. 2590–2595.

Li, D., Wu, Y., Shi, L., Yang, P., and Majumdar, A., 2003 "Thermal Conductivity of Individual Silicon Nanowires," *Applied Physics Letters*, vol. 83, pp. 2934–2936.

Liu, W. and Asheghi, A., 2004 "Phonon-Boundary Scattering in Ultrathin Single-Crystal Silicon Layers," *Applied Physics Letters*, vol. 84, pp. 3819–3821.

Morales, A.M., and Lieber, C.M., 1998, "A Laser Ablation Method for the Synthesis of Crystalline Semiconductor Nanowires," *Science*, vol. 279, pp. 208–211.

Qiu, T.Q., and Tien, C.L., 1993, "Heat Transfer Mechanisms during Short-Pulse Laser Heating of Metals," *Journal of Heat Transfer*, vol. 115, pp. 835–841.

Ren, Z.F., Huang, Z.P., Xu, J.W., Wang, J.H., Bush, P., Siegal, M.P., and Provencio, P.N., 1998, "Synthesis of Large Arrays of Well-Aligned Carbon Nanotubes on Glass," *Science*, vol. 282, pp. 1105–1107.

SEMATECH, 2001, International Technology Roadmap for Semiconductors, http://www.sematech.org

Sverdrup, P.G., Sinha, S., Asheghi, M., Uma, S., and Goodson, K.E., 2001, "Measurement of Ballistic Phonon Heat Conduction near Hot Spots in Silicon," *Applied Physics Letters*, vol. 78, pp. 3331–3333.

Sze, S.M., 1981, *Physics of Semiconductor Devices*, John Wiley, New York.

Tien, C.L., and Chen, G., 1994, "Challenges in Microscale Conductive and Radiative Heat Transfer," *Journal of Heat Transfer*, vol. 116, pp. 799–807.

Tien, C.L., and Lienhard, J.H., 1979, *Statistical Thermodynamics*, pp. 344–346, Hemisphere, Washington.

Tien, C.-L., Majumdar, A., and Gerner, F.M., 1998, *Microscale Energy Transport*, Taylor & Francis, Washington.

Touloukian, Y.S., Powell, R.W., Ho, C.Y., and Klemens, P.G., 1970, "Thermal Physics Properties of Matter," *TPRC Data Series*, Ifi/Plenum Press, New York.

Trimmer, W., 1997, *Micromechanics and MEMS: Classical and Seminal Papers to 1990*, IEEE Press, New York.

Tritt, T.M., ed., 2001, Recent Trends in Thermoelectric Materials Research, *Semiconductors and Semimetals*, vols. 69–71, Academic Press, San Diego.

Venkatasubramanian, R., Silvona, E., Colpitts, T., and O'Quinn, B., 2001, "Thin-Film Thermoelectric Devices with High Room-Temperature Figures of Merit," *Nature*, vol. 413, pp. 597–602.

Wang, G.M., Sevick, E.M., Mittag, E., Searles, D.J., and Evans, D.J., 2002, "Experimental Demonstration of Violations of the Second Law of Thermodynamics for Small Systems and Short Time Scales," *Physical Review Letters*, vol. 89, pp. 050601/1–4.

Whale, M.D., and Cravalho, E.G., 2002, "Modeling and Performance of Microscale Thermophotovoltaic Energy Conversion Devices," *IEEE Transactions on Energy Conversion*, vol. 17, pp. 130–142.

Yablonovitch, E., 1987, "Inhibited Spontaneous Emission in Solid-State Physics and Electronics," *Physical Review Letters,* vol. 58, pp. 2059–2062.

Yang, R.G., Chen, G., and Taur, Y., 2002, "Ballistic-Diffusive Equations for Multidimensional Nanoscale Heat Conduction," *Proceedings of the 2002 International Heat Transfer Conference (IHTC 2002)*, vol. 1, pp. 579–584, Grenoble, France.

1.9 Exercises

1.1 *Membrane method for thin-film thermal conductivity measurement.* One technique for measuring the thermal conductivity of thin films is to create a freestanding film by removing part of the substrate, as shown in figure P1.1. A thin-film heater is deposited at the center of the film. A thin layer of electrical insulator (such as SiO_2 or Si_3N_4) is used between the film and the heater if the film itself is electrically conducting. The temperature rise of the heater is determined by measuring the change in its electrical resistance. The substrate temperature is assumed to be uniform (alternatively, another temperature sensor can be deposited at the edge of the film and the substrate). Thermal conductivity along the film can be measured but one must be very careful to address various factors that may affect the final results. Answer the following questions.

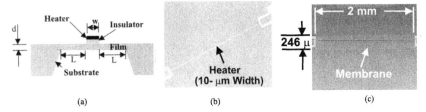

Figure P1.1 A thin-film conductivity measurement method (a) schematic of the cross-section, (b) photograph of a fabricated heater, and (c) photograph of a the free-standing silicon membrane.

(a) Derive an expression for determining the thermal conductivity of the film, given the power input to the heater, the temperature rise of the heater, the temperature of the substrates and the geometries, under the following assumptions: (1) heat conduction is one-dimensional, (2) heat losses along the film are negligible, and (3) the thermal resistance of the insulating film is negligible.

(b) For a 3 μm thick silicon membrane, the thermal conductivity at room temperature is 145 W m^{-1}K^{-1}. The measured temperature rise of the heater is 2°C. Given the geometries in figure P1.1(b) and (c), estimate how much power input is needed.

(c) For the silicon thermal conductivity measurement, an insulating layer must be placed between the heater and the silicon film for electrical isolation. Assuming a 200 nm thick SiO$_2$ film with a thermal conductivity of 1.2 W m^{-1}K^{-1} is used, estimate what thermal conductivity you will get if the thermal resistance of the SiO$_2$ layer is not taken into account in analyzing the experimental data, based on the power input condition given in (b).

(d) Now consider heat losses along the film. The combined heat transfer coefficient due to convection and radiation is 10 W m^{-2}K^{-1}. For the silicon membrane example given in (b), determine how much additional power input is needed as a result of this heat loss.

(e) One concern in measuring low thermal conductivity membranes is the heat loss along the heater, which has a relatively high thermal conductivity. Assuming that the heater thickness is $t = 200$ nm and the material is gold with a thermal conductivity of 315 W m^{-1}K^{-1}, develop a model to estimate the heat loss along the heater to the substrate.

1.2 *AC calorimetry method for measuring thin-film thermal diffusivity* (Hatta, 1990). One thin-film thermal diffusivity measurement method is to use a modulated light source to heat up a membrane, as shown in figure P1.2. A small temperature sensor, either a tiny thermocouple or a microfabricated sensor, is placed onto the film. The distance between the film and the temperature sensor is controlled by a mask that blocks part of the light source. Amplitude and phase of the temperature response are measured as a function of the distance L between the temperature sensor and the edge of the light source. This is equivalent to measuring the distribution of the amplitude and phase as a function of x. Derive an expression for determining the thermal diffusivity on the basis of (a) phase signal and (b) amplitude signal. List all the assumptions made in establishing the model.

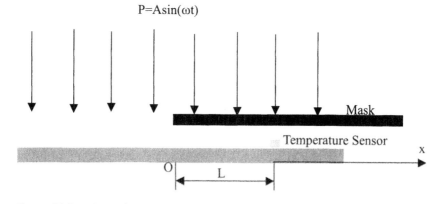

Figure P1.2 AC calorimetry method for determining thermal diffusivity of thin films.

1.3 *Cross-plane thermal conductivity measurement of thin films: steady-state method.* The measurement of the thermal conductivity perpendicular to a thin-film plane (cross-plane) is difficult because the temperature drop across a thin film is small unless a high heat flux is applied. Since thin films are usually deposited on a substrate, one-dimensional heating is unfavorable because a large temperature drop would occur across the substrate rather than the film. To avoid this situation, one solution is to use a narrow heater patterned directly on the film, as shown in figure P1.3. In this case, the heat flux through the film is high but the heat spreading inside the substrate lowers the heat flux, leading to a relatively large temperature drop across the film compared to that in the substrate. For the given configuration, we can assume that the heat flux is uniform from the heater into the substrate. If the substrate thermal conductivity is known, the thin-film thermal conductivity can be determined from the measured heater temperature rise. In this problem, the substrate is silicon with a thermal conductivity of 145 W m^{-1}K^{-1} and the film is SiO$_2$ with a thermal conductivity of 1.2 W m^{-1}K^{-1}.

(a) Assuming heat conduction is two-dimensional, derive an expression for the average temperature rise of the heater.

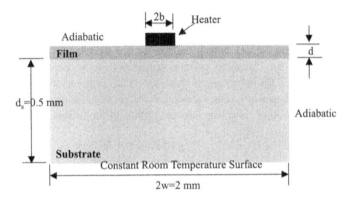

Figure P1.3 Figure for problem 1.3

(b) The ideal case is that heat conduction through the film is one-dimensional. For a 400 nm SiO$_2$ film on a silicon substrate and a heater 10 μm wide and 2 mm long with a power input of 40 mW, compare the exact solution for the average heater temperature with the approximation that heat conduction inside the SiO$_2$ is one-dimensional.

(c) Determine the temperature drop across the film and inside the substrate.

(d) Estimate the heat loss through radiation and convection and compare its magnitude with that of heat conduction.

1.4 *Cross-plane thermal conductivity measurement of thin films: 3ω method* (Lee and Cahill, 1997). One disadvantage of the steady-state method in problem 1.3 is that the backside temperature of the substrate must be known. In reality, the substrate is placed on a heat sink and the backside may not be at uniform temperature. In addition, the thermal resistance between the substrate and the heat sink also changes the temperature of the substrate. To avoid this situation, heat input into the heater can be modulated by a sine input current at angular frequency ω. In this case, power will be modulated at 2ω, leading to a 2ω temperature oscillation and corresponding electrical resistance oscillation of the heater, due to the temperature dependence of the resistance. This resistance oscillation at 2ω leads to a third harmonic component in the heater voltage. By measuring the phase and amplitude of this third harmonic, the temperature rise of the heater due to the modulating power input can be determined and, from this, the thermal conductivity of the film can be determined. This is called the 3ω method.

(a) Assuming a power input of the form of A sin $(2\omega t)$, derive an expression for the in-phase (sine function) and out-of-phase (cosine function) components of the heater temperature rise. Assume that all the thermal properties (thermal diffusivity, thermal conductivity, and specific heat) of the film and the substrate are known.

(b) One additional advantage of the 3ω method is that the substrate thermal conductivity can be determined from the frequency dependency of the temperature response under appropriate conditions. Try to identify these conditions.

(c) Another advantage of the 3ω method is that the radiation loss can be minimized, which is particularly important for low thermal conductivity materials and measurements at extreme temperatures (low and high). Explain why.

1.5 *Thermal diffusivity determination of thin films: laser pulse method.* One method for determining the thermal diffusivity of a thin film is to use a short laser pulse to heat up the front side of the film and to measure the decay of the front side temperature by monitoring the change in reflectance of a probe laser beam (see figure P1.5). The short pulse concentrates temperature drop in the film rather than across the substrate. In this case, it is not the absolute surface temperature rise that is measured but the normalized profile of the surface temperature decay as a function of time. For a heating pulse of the following profile,

$$q = \begin{cases} 0 & t < 0 \\ q_0 & 0 < t < t_p \\ 0 & t > t_p \end{cases}$$

(a) Derive an expression for the temporal response of the front surface temperature, assuming all thermal properties (thermal conductivity k, thermal diffusivity a, and specific heat c) are known for both the film and the substrate.

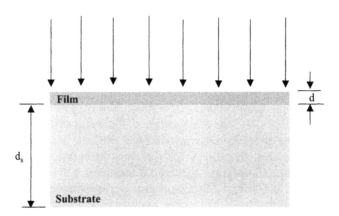

Figure P1.5 Figure for problem 1.5.

(b) What are the requirements on the pulse width that will maximize the sensitivity for measuring the thermal diffusivity of the film?

1.6 *Lumped heat capacitance and time constant.* Develop a lumped capacitance model for a solid sphere at uniform temperature T_i that is suddenly immersed inside a liquid at temperature T_0. In such a model, the temperature of the solid is assumed to be uniform, and the heat transfer coefficient between the solid object and the fluid is taken to be h. Other known parameters are the surface area A, the volume V, the density ρ, and specific heat c of the solid.
   (a) Derive the differential equation governing the temperature history of the solid.
   (b) Solve the equation and find the time constant of the process.
   (c) Investigate how the time constant varies with the diameter of the solid sphere.

1.7 $\kappa_B T$ *energy.* One unit for energy is the electron-volt (eV). It is the energy difference of one electron under a potential difference of 1 V. Convert 1 $\kappa_B T$ at 300 K into milli-eV (meV).

1.8 *Thermal conductivity of gases.* Estimate the thermal conductivity of air and argon as a function of temperature between 300 K and 1000 K at 1 atm.

1.9 *Mean free path in air.* Estimate the mean free path of air molecules as a function of temperature at atmospheric pressure on the basis of (a) kinetic theory and (b) experimental data on the thermal conductivity and specific heat of air.

1.10 *Speed of electrons.* Estimate the average random speed of an electron gas in a semiconductor at 300 K.

1.11 *Thermal conductivity of liquid.* Although the application of kinetic theory to a dense liquid is questionable, estimate the thermal conductivity of water at room temperature on the basis of a simple derivation for the mean free path and the results from the kinetic theory. This estimation is typically smaller than experimental values because, for liquid, potential energy exchange contributes to heat conduction.

1.12 *Phonon mean free path and relaxation time.* Given the thermal conductivity of Si at room temperature as 145 W m^{-1}K^{-1}, the speed of sound as 6400 m s^{-1}, the volumetric specific heat as 1.66×10^6 J m^{-3}K^{-1},
   (a) Estimate the phonon mean free path in Si at room temperature from the kinetic theory. In reality, this estimation usually leads to a much shorter mean free path (about a factor of 10 shorter) than with more sophisticated modeling.
   (b) Estimate the relaxation time of phonons in silicon.

1.13 *Fick's law of diffusion.* Using a simple kinetic argument that is similar to the derivation of the Fourier law, derive the Fick law of diffusion, which gives the mass flux for species i under a concentration gradient as

$$J_i = -\rho D \frac{dm_i}{dx},$$

where D is the mass diffusivity, ρ is the density of the mixture, and m_i the local mass fraction of species i.

1.14 *Newton's shear stress law.* Using a simple kinetic argument that is similar to the derivation of the Fourier law, derive the Newton law of shear stress (in one-dimensional form). Hint: consider the momentum exchange across a plane parallel to the flow.

1.15 *Energy quantization.*

(a) Assuming a person weighing 100 kg trapped deep inside a two-dimensional ditch 1 m in width, estimate the energy difference between the first and second quantized energy levels. Compare this energy difference with the thermal fluctuation energy $\kappa_B T$ for $T = 300$ K.

(b) Assuming an electron of mass 9.1×10^{-31} kg is trapped inside a two-dimensional infinitely high potential well, plot the first and second energy levels of the electron as a function of well width between 10 and 100 Å. Also mark the thermal energy $\kappa_B T$ on the graph for $T = 300$ K.

2

Material Waves and Energy Quantization

For macroscopic systems, we take the continuity of many variables for granted, including the continuity in energy. For example, the heat flux along a rod through conduction, according to the Fourier law, can be continuously varied to any desired value by controlling the temperature difference and the material properties. The microscopic picture of energy, however, is entirely different. According to quantum mechanical principles, the permissible energy levels of matter (electrons, crystals, molecules, and so on) are often discontinuous. Differences in allowable energy levels among materials are major factors that distinguish them from each other. For example, why is glass transparent in the visible light range but not silicon, and why are some materials electrical insulators but others are conductors?

In this chapter, we introduce the basic quantum mechanical concepts necessary to appreciate various energy states found in different materials. It should be remembered that these energy states represent the range of possibilities for the matter but do not tell which state the matter will be in. The latter depends on the temperature, a topic we will discuss in chapter 4. Important concepts that should be mastered through this chapter include the wave–particle duality, the Schrödinger equation and the meaning of the wavefunction, the Pauli exclusion principle, quantum states, and degeneracy. Solutions of the Schrödinger equation for various simple yet very common potentials will be given. Key concepts and results of this chapter are summarized in the last section of the chapter.

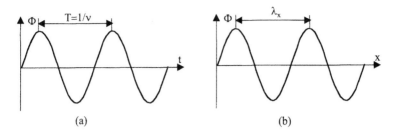

Figure 2.1 Traveling wave: (a) temporal variation at a fixed point; and (b) spatial variation at a fixed time.

2.1 Basic Wave Characteristics

Let's start by reviewing some basic characteristics of waves. We consider a harmonic wave (such as an electric or a magnetic field) represented by a sine function traveling along the positive x-direction,*

$$\mathbf{\Phi} = A\sin\left(2\pi\nu t - \frac{2\pi x}{\lambda_x}\right)\hat{\mathbf{y}} = A\sin(\omega t - k_x x)\hat{\mathbf{y}} \qquad (2.1)$$

where A is the amplitude and $\hat{\mathbf{y}}$ is a unit vector in the y-coordinate direction. Such a wave has two kinds of periodicity: one in space and one in time. The periodicity in time is characterized by the frequency ν, which equals the inverse of the period in time. The angular frequency $\omega = 2\pi\nu$ is often used instead of frequency to avoid writing the 2π factor. At any fixed point, the temporal variation of the field is a sine function, as shown in figure 2.1(a). The periodicity along the x-direction is characterized by the wavelength λ_x. Taking a snapshot of the field in space at any fixed time, the field is a sine function as shown in figure 2.1(b). The inverse of the wavelength $1/\lambda_x$ is called the wavenumber. The wave represented by eq. (2.1) is propagating along the x-direction, but the field is vibrating along the y-direction. When the field vibration direction (the direction of the electric field oscillation or the atomic displacement) is perpendicular to the wave propagation direction, the wave is said to be a transverse wave. When the wave propagation and the field vibration are along the same direction, the wave is called a longitudinal wave. The wavevector, $\mathbf{k}$, represents the wave propagation direction and has a magnitude of $k_x = 2\pi/\lambda_x$ so that for a wave propagating along the x-direction as shown in figure 2.1(b),

$$\mathbf{k} = \frac{2\pi}{\lambda_x}\hat{\mathbf{x}} = k_x\hat{\mathbf{x}} \qquad (2.2)$$

For the wave represented by Eq. (2.1), the constant phase plane in the x–t space is

$$\omega t - k_x x = \text{const.} \qquad (2.3)$$

*We will discuss waves in more detail in chapter 5.

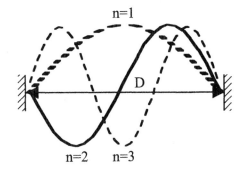

Figure 2.2 Standing waves, with vanishing amplitude at the boundaries.

This constant phase plane propagates along the positive x direction at the speed

$$v_{p,x} = \frac{dx}{dt} = \frac{\omega}{k_x} = \nu \lambda_x \qquad (2.4)$$

which is called the phase velocity. Eq. (2.1) therefore represents a transverse *traveling wave* along the positive x-direction. For such a simple wave, the constant phase plane is also the constant amplitude plane, as is shown by substituting eq. (2.3) into (2.1). The frequency and wavevector of a wave are not independent of each other. The relationship between ω and $\mathbf{k}$, or $\omega(\mathbf{k})$, is called the dispersion relation and may be different along different wavevector directions. For electromagnetic waves, we know that v_p is the speed of light, c, and eq. (2.4) gives $\omega = ck$. The dispersion relations for electrons and phonons are not this simple, as we indicated in table 1.3 and will discuss in more detail in chapter 3.

Sometimes, it is convenient to use the complex representation of the sine and cosine functions

$$\mathbf{\Phi}_c = A e^{-i(\omega t - k_x x)} \hat{\mathbf{y}} = A[\cos(\omega t - k_x x) - i \sin(\omega t - k_x x)]\hat{\mathbf{y}} \qquad (2.5)$$

where $i = \sqrt{-1}$ is the unit imaginary number. This is because mathematical operations with the exponential function are much easier to manipulate than those with sine and cosine functions. In a typical mathematical operation using the complex representation, it is implicitly assumed that either the real or the imaginary part of the final solution is the true solution to the problem of interest. Which one of the two parts is the desired solution depends on whether the input (such as the initial or the boundary conditions) is in terms of a sine (imaginary part) or a cosine (real part).

A *standing wave* has fixed boundary points, as shown in figure 2.2. We can create such a standing wave by superimposing two traveling waves along the positive and negative x-directions (assuming that the problem is linear such that the superposition principle applies),

$$\mathbf{\Phi} = A[\sin(\omega t - k_x x) + \sin(\omega t + k_x x)]\hat{\mathbf{y}} = -2A \cos(\omega t) \sin(k_x x)\hat{\mathbf{y}} \qquad (2.6)$$

Unlike a traveling wave, eq. (2.6) has fixed nodes in space such that $\mathbf{\Phi} = 0$ at all times. Also, we see that the magnitude of $\mathbf{\Phi}$ at different locations is a cosine function in time. Equation (2.6) is a simple form of a standing wave. It is a good representation of a wave inside a cavity of length D that requires the amplitude of the wave to vanish at the cavity boundaries, that is, $\Phi(x=0) = \Phi(x=D) = 0$, which leads to

$$\sin\left(\frac{2\pi}{\lambda_x}D\right) = 0 \tag{2.7}$$

or

$$D = \frac{n\lambda_x}{2} \quad (n = 1, 2, 3, \ldots) \tag{2.8}$$

Thus, for a stable wave to form inside a cavity that vanishes completely outside the cavity, the cavity length D must be multiples of the half wavelength.

The energy contained in a wave is usually proportional to the square of the field,

$$U \propto |\Phi|^2 \tag{2.9}$$

One can understand this point intuitively by imagining that eq. (2.1) represents the instantaneous displacement of a particle. Its velocity is the derivative of this displacement with respect to time and the kinetic energy is proportional to the square of this velocity. Classically, the allowable energy of the wave can change continuously since there is no limit on the amplitude of vibration. This picture, however, is no longer true under quantum mechanical principles.

2.2 Wave Nature of Matter

From the previous section, we see that a wave is characterized by its frequency and wavelength, and its energy is determined by the magnitude of the wave. We also know that a particle is characterized by its energy and momentum. Waves and particles are two completely different and unrelated phenomena in classical mechanics and electrodynamics. In quantum mechanics, however, they are interrelated and are two aspects of matter.

2.2.1 Wave–Particle Duality of Light

Quantum mechanics started with the explanation of blackbody radiation and the absorption spectra of gases. By the end of the 19th century, classical Newtonian mechanics and electromagnetism were well established as two separate entities: Newtonian mechanics is based on the particle picture of materials, and electromagnetism is based on the wave picture. Interestingly, Sir Isaac Newton believed that radiation was particle-like in nature rather than wave-like, as we are more familiar with today. It was the discovery and explanation of interference and diffraction phenomena, from the work of Christian Huygens (1629–1695), Thomas Young (1773–1829), Augustin Jean Fresnel (1788–1827), and others, followed by Maxwell (1831–1879) and his celebrated equations, that solidified the foundation of the wave nature of the electromagnetic field.

The Maxwell equations, however, fail to explain the emission and absorption processes, such as the experimentally observed fine spectra of absorption in various gases, and the blackbody radiation (figure 2.3). According to classical theory, the blackbody

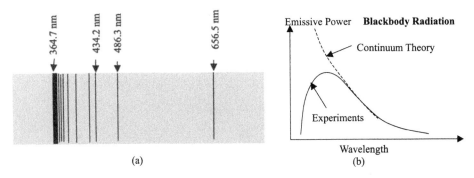

Figure 2.3 (a) An example of the hydrogen emission spectrum for the final state of $n = 2$. Classical mechanics fails to explain discrete lines in the emission spectrum. (b) Experimental measurement on blackbody radiation contradicts the predictions of continuum theory.

emissive power should be proportional to $(\lambda T)^{-5}$, which approaches infinity at short wavelengths, while the experimental blackbody spectrum is reduced to zero as the wavelength decreases, as shown in figure 1.7. The discrete absorption lines in the hydrogen spectrum also cannot be explained by continuum mechanics. To explain blackbody radiation, Max Planck (1858–1948) introduced a radical hypothesis that the allowable energy of the electromagnetic field at a frequency ν_p is not continuous, but is a multiple of the following basic energy unit*

$$E_p = h\nu_p \tag{2.10}$$

where h is called the Planck constant and has a value $h = 6.6 \times 10^{-34}$ J s. We will show later how the idea of photon energy quantization leads to the Planck law. His success led Albert Einstein (1879–1955) to consider that the electromagnetic field also has particle (granular or corpuscular) characteristics such as momentum (Einstein, 1905, 1906).** The basic energy unit as given by eq. (2.10) was later called a photon (Lewis, 1926). Einstein used the corpuscular characteristics of electromagnetic radiation to explain some puzzling results from the basic photoelectricity experiment shown in figure 2.4. It was found that when light is incident on one of two metal electrodes separated by a vacuum, a current can be generated in the loop. The current generation, however, occurs only when the wavelength is shorter than a certain value. No current can be generated for wavelengths longer than this value, even at high light intensities. This experimental observation could not be explained from the classical wave point of view, according to which the energy of an electromagnetic wave is proportional to its intensity, as implied by eq. (2.8). On the basis of the photon particle concept, Einstein reasoned that one photon can excite an electron out of the metal surface only when the photon energy is higher than the electrode workfunction $A(= E_v - E_f)$, which is the energy difference between electrons at the vacuum level, E_v, and inside the metal, E_f,

$$h\nu_p \geq E_v - E_f \tag{2.11}$$

---

*We will neglect the zero point energy in the discussion here.

**Einstein developed theories on special relativity, particle characteristics of photons, and Brownian motion before age 26, while he worked at a patent office.

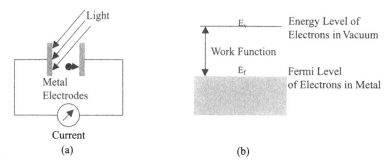

Figure 2.4 (a) Electron emission due to light excitation is called the photoelectric effect. The effect was explained by Einstein through the introduction of corpuscular properties of light. (b) Electrons in a metal have energy close to the Fermi level and their emission out of the metal surface into vacuum is possible only when the photon energy is larger than the work function.

Einstein reasoned that although a photon does not have a rest mass it has a moving mass determined through the relation $E = mc^2$. Its corresponding momentum is $p = mc = E_p/c = h\nu/c$, such that

$$p = \hbar k = \frac{h}{\lambda} \qquad (2.12)$$

where $\hbar = h/(2\pi)$. This $\hbar$ is used more often than the Planck constant h because angular frequency and wavevector include the 2π factor. Equations (2.10) and (2.12) are called the Planck–Einstein relations. These two relations thus relate the energy and momentum, which we normally associate with particles, to the frequency and wavevector, which we normally associate with waves. Electromagnetic radiation, and thus photons, has both wave and particle characteristics. This wave–particle duality of light led to the development of quantum mechanics. Einstein also discovered many new properties of photons, such as stimulated emission which forms the basis of all lasers.

2.2.2 Material Waves

The wave–particle duality of light triggered de Broglie, who was a graduate student then, to postulate that a material particle also has wave properties (Broglie, 1925).* On the basis of an analogy with the Planck–Einstein relations, he proposed that the wavelength of any particle is

$$\lambda = h/p \qquad (2.13)$$

---

*Quantum mechanics was developed by a group of young researchers. Louis de Broglie (1892–1987) developed the material wave concept in 1923 when he was doing his Ph.D. research. He received the Nobel prize in 1929 at age 38. Werner Heisenberg (1901–1976) developed the matrix formulation of quantum mechanics in 1925, immediately after he finished his Ph.D. thesis on turbulence in 1923, and won the Nobel prize for his work in quantum mechanics in 1932 at age 31. Paul Dirac (1902–1984) developed relativistic quantum mechanics and won the Nobel prize in 1933 at age 31. Erwin Schrödinger (1887–1961) developed his famous equation in 1926 and received the Nobel prize in 1933 at age 46.

where p is the magnitude of the particle momentum. To see how large this wavelength is for a macroscopic object, let's assume $p = mv \approx 1$ kg m s^{-1}, leading to $\lambda \approx 6.6 \times 10^{-34}$ m which is impossible to detect even with current technology. On the other hand, an electron with a velocity of 1 m s^{-1} and a mass of 9.1×10^{-31} kg yields $\lambda \approx 0.7 \times 10^{-3}$ m; a quite long wavelength. The first proof of the wave properties of particles came from the electron diffraction experiment performed by Davisson and Germer (1927).

Now let's consider a simple example to illustrate the consequence of material waves. Consider an electron as a wave that is situated inside a one-dimensional cavity of length D surrounded by an infinite potential. Outside this cavity, the wave amplitude must be zero since an infinitely high potential means that no electrons can have an energy larger than this potential height. This means that the electron wave inside the cavity must be a standing wave and its wavelength must satisfy eq. (2.8),

$$\lambda = \frac{2D}{n} \qquad (n = 1, 2, 3, \ldots) \tag{2.14}$$

The momentum and energy of the electron are

$$p_n = \frac{nh}{2D} \text{ and } E_n = \frac{p^2}{2m} = \frac{1}{2m}\left(\frac{nh}{2D}\right)^2 \qquad (n = 1, 2, 3, \ldots) \tag{2.15}$$

which are discontinuous, or quantized. Later, we will derive the same result from solving the Schrödinger equation.

2.2.3 The Schrödinger Equation

Two basic methods have been developed to describe the material waves. The first was the matrix method developed by Heisenberg (1925). Shortly after, Schrödinger developed the famous equation that bears his name. These two descriptions are equivalent among themselves, so we will focus on the Schrödinger equation (Schrödinger, 1926), which states that the wavefunction of any matter obeys the following:

$$-\frac{\hbar^2}{2m}\nabla^2 \Psi_t + U\Psi_t = i\hbar \frac{\partial \Psi_t}{\partial t} \tag{2.16}$$

where m is the mass, t is the time, U is the potential energy constraint that the matter is subject to and $\Psi_t(t, \mathbf{r})$ is called the wavefunction of the matter and is a function of time and coordinate $\mathbf{r}$. If $U = 0$, that is, a matter with no potential constraint, the Schrödinger equation becomes

$$-\frac{\hbar^2}{2m}\nabla^2 \Psi_t = i\hbar \frac{\partial \Psi_t}{\partial t} \tag{2.17}$$

One may think that the equation is a parabolic type of equation similar to the transient heat conduction equation [eq. (1.19)], but the "magic" imaginary unit i really gives rise to wave behavior. Schrödinger himself did not come up with an explanation for the meaning of wavefunction. The right explanation was given by Born, who suggested that Ψ_t itself is not an observable quantity, but that $\Psi_t \Psi_t^*$ is the probability density function to find the matter at location **r**, where "*" means complex conjugate. The normalization requirement for the probability function is then

$$\int_{-\infty}^{\infty} \Psi_t \Psi_t^* dx = 1 \quad (2.18)$$

for the one-dimensional case. For three-dimensional problems, the integration should be over the volume. The probabilistic interpretation is difficult to appreciate since we are most used to deterministic events in mechanics. Einstein, and even Schrödinger himself, rejected this interpretation. However, this interpretation has endured the test of experiments and time. Since $\Psi_t \Psi_t^*$ is a probability, the quantum world is full of uncertainties. Any quantities, such as energy, momentum, and location, are no longer a deterministic quantity but have an average, or expectation, value and uncertainties. The expectation value (or most probable value) of any quantity can be calculated from

$$\langle \Omega \rangle = \int_{-\infty}^{\infty} \Psi_t^* \Omega \Psi_t dx \quad (2.19)$$

where $\langle \Omega \rangle$ is the expectation value and Ω is the operator for this quantity. The operators for position, momentum, and energy of matter are

position operator:

$$\Omega = \mathbf{r} \quad (2.20)$$

momentum operator:

$$\Omega = \mathbf{p} = -i\hbar \nabla$$
$$= -i\hbar \left(\frac{\partial}{\partial x} \hat{\mathbf{x}} + \frac{\partial}{\partial y} \hat{\mathbf{y}} + \frac{\partial}{\partial z} \hat{\mathbf{z}} \right) \quad (2.21)$$
$$= p_x \hat{\mathbf{x}} + p_y \hat{\mathbf{y}} + p_z \hat{\mathbf{z}}$$

and the energy operator:

$$\Omega = H = \frac{\mathbf{p} \cdot \mathbf{p}}{2m} + U = \frac{\mathbf{p}^2}{2m} + U$$
$$= -\frac{\hbar^2}{2m} \nabla^2 + U = -\frac{\hbar^2}{2m} \left(\frac{\partial^2}{\partial x^2} + \frac{\partial^2}{\partial y^2} + \frac{\partial^2}{\partial z^2} \right) + U \quad (2.22)$$

The first term in eq. (2.22) corresponds to the kinetic energy operator and the second term to the potential energy. In classical mechanics, the kinetic energy plus the potential energy of an energy-conserve system is called the Hamiltonian of the system. In quantum mechanics, the Hamiltonian becomes an operator, according to eq. (2.22). In a Cartesian coordinate system, the gradient operator ∇ and the Laplace operator ∇^2 are given by

$$\nabla = \hat{\mathbf{x}} \frac{\partial}{\partial x} + \hat{\mathbf{y}} \frac{\partial}{\partial y} + \hat{\mathbf{z}} \frac{\partial}{\partial z}, \quad \nabla^2 = \frac{\partial^2}{\partial x^2} + \frac{\partial^2}{\partial y^2} + \frac{\partial^2}{\partial z^2} \quad (2.23)$$

The differentials in the operators are applied to the function immediately following the operator. Thus the order cannot be exchanged, which is similar to the matrix operation. Heisenberg's matrix formulation of quantum mechanics naturally possesses such characteristics. In eq. (2.19), $\Psi_t^* \Omega \Psi_t$ inside the integral means that the operator Ω is first applied to Ψ_t and the obtained function is multiplied by Ψ_t^*. As another example,

$$p_x p_x \Psi_t = -i\hbar \frac{\partial}{\partial x} \left(-i\hbar \frac{\partial \Psi_t}{\partial x} \right) = -\hbar^2 \frac{\partial^2 \Psi_t}{\partial x^2} \quad (2.24)$$

which explains the way of expressing H in terms of **p** in eq. (2.22).

The Schrödinger equation is time dependent. When the potential energy is independent of time, we can derive the steady-state Schrödinger equation using the separation-of-variables method. Assuming $\Psi_t(\mathbf{r}, t) = \Psi(\mathbf{r}) Y(t)$ and substituting into the Schrödinger equation, we get

$$\frac{1}{\Psi} \left[-\frac{\hbar^2}{2m} \nabla^2 \Psi + U\Psi \right] = i\hbar \frac{1}{Y} \frac{dY}{dt} = E \quad (2.25)$$

where E is a constant (eigenvalue) since Ψ depends on **r** only and Y depends on t only, and its meaning will be explained later. Solving for Y leads to

$$Y = C_1 \exp\left[-i\frac{E}{\hbar} t \right] \quad (2.26)$$

The governing equation for $\Psi(\mathbf{r})$ is called the steady-state Schrödinger equation

$$-\frac{\hbar^2}{2m} \nabla^2 \Psi + (U - E)\Psi = 0 \quad (2.27)$$

This is an eigenvalue equation with the eigenvalue E and eigenfunction Ψ determined by the potential energy profile U and the boundary conditions. On the basis of eqs. (2.19) and (2.22), we can prove that the expected energy of a system is

$$\langle H \rangle = \int_{-\infty}^{\infty} \Psi_t^* H \Psi_t \, dx = E \quad (2.28)$$

So the separation-of-variable constant E, or the eigenvalue, actually represents the energy states of the system. Correspondingly, we could write the time-dependent part as $Y = e^{-i\omega t}$, with $E = \hbar\omega$. So the material waves obey the Planck–Einstein relation, eq. (2.10).

Because $\Psi_t \Psi_t^*$ is a probability and the physical observable quantities are only the expectation values, there are also standard deviations for these expectations, such as the standard deviations in location Δx, momentum Δp, energy ΔE, and time Δt. It can be proven that, for any solution of the Schrödinger equation, the following relationship holds

$$\Delta p_x \bullet \Delta x \geq \hbar/2 \quad \text{and} \quad \Delta E \bullet \Delta t \geq \hbar/2 \quad (2.29)$$

This is the famous Heisenberg uncertainty principle, which means that position and momentum, or energy and time, cannot be accurately determined simultaneously in the quantum world. Because $\hbar$ is a very small number, the uncertainty represented by eq. (2.29) for a macroscopic object is very small. For example, if we decide that an object with a momentum of 1 kg m s^{-1} has an uncertainty of 10^{-10} kg m s^{-1}, the corresponding uncertainty in determining its position is $\sim 10^{-24}$ m, a negligible quantity. This uncertainty, however, becomes quite appreciable for small particles such as electrons.

For our further use, we need also to have an expression for the flux of the matter being considered. This can be obtained by (1) first multiplying the Schrödinger equation, (2.16), by Ψ_t^*, (2) taking the complex conjugate of the Schrödinger equation and multiplying the obtained equation by Ψ_t, and (3) subtracting the two resulting equations, which leads to

$$\frac{\partial |\Psi_t|^2}{\partial t} + \nabla \bullet \mathbf{J} = 0 \tag{2.30}$$

where $\mathbf{J}$ is

$$\mathbf{J} = \frac{i\hbar}{2m}(\Psi_t \nabla \Psi_t^* - \Psi_t^* \nabla \Psi_t) \tag{2.31}$$

Since the first term in eq. (2.30) is the rate of the change of the probability of finding the matter at each location, the second term in eq. (2.30) must be the net rate of matter flowing out of the point. Equation (2.30) is the particle conservation equation and $\mathbf{J}$ [m^{-2}s^{-1}] is understood as the current density (or flux) of the material wave.

The wavefunction is a difficult concept to grasp at first sight and this is not strange, since even Schrödinger himself was not able to explain the meaning of the wavefunction. However, Schrödinger was successful in using the equation to show that the energy states of electrons are quantized, as we will see later. Born's explanation of the wavefunction products $\Psi_t \Psi_t^*$ as a probability density of matter implies that material particles have spatial extent with some ambiguity, as we will see from the example solutions of the Schrödinger equation.

2.3 Example Solutions of the Schrödinger Equation

In this section, we will give solutions to the Schrödinger equation for several important cases that we will use later.

2.3.1 Free Particles

A free particle is one that it is not subject to any potential constraints; that is, $U = 0$. We can think of this free particle as a free electron. For the particle traveling along the x-direction, eq. (2.27) becomes

$$-\frac{\hbar^2}{2m}\frac{d^2\Psi}{dx^2} - E\Psi = 0 \tag{2.32}$$

The solution of the above equation is

$$\Psi(x) = A\exp(-ikx) + B\exp(ikx) \tag{2.33}$$

MATERIAL WAVES AND ENERGY QUANTIZATION   53

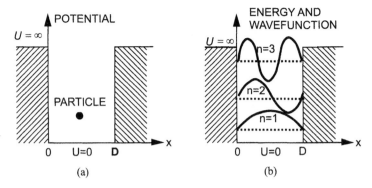

Figure 2.5 (a) One-dimensional potential well with infinite potential heights on both sides and zero potential inside the box. (b) Particle energy quantization in the box and wavefunction for the first three levels.

where $k = \sqrt{2Em/\hbar^2}$. When combined with the time-dependent factor, eq. (2.26), the time-dependent wavefunction is found to be

$$\Psi_t(x,t) = Ae^{-i(\omega t + kx)} + Be^{-i(\omega t - kx)} \tag{2.34}$$

where C_1 in eq. (2.26) is absorbed into A and B. The first term represents a free particle traveling in the negative x-direction and the second term along the positive x-direction.

Interested readers may ask what the Heisenberg uncertainty principle means for a free electron with a given momentum and energy. Equation (2.34) shows that the wavefunction for a right traveling wave extends from $x = -\infty$ to $x = \infty$, with equal probability everywhere, which means that its position is not determined at all. Similarly, the wave has fixed energy but spans time from negative infinity to positive infinity, that is, the whole time history. Thus, the Heisenberg uncertainty principle holds true for this simple case.

2.3.2 Particle in a One-Dimensional Potential Well

On the basis of the requirement for standing waves, we derived eq. (2.15) which shows the quantization of the allowable energy levels for a material wave inside a one-dimensional cavity of length D. Now let's start from the Schrödinger equation and demonstrate that eq. (2.15) is a natural solution of the equation. We consider the case of a particle in a one-dimensional potential well, which can be, for example, an electron subject to an electric potential field as shown in figure 2.5(a). The steady-state Schrödinger equation for the particle in such a potential profile is

$$-\frac{\hbar^2}{2m}\frac{d^2\Psi}{dy^2} + (U - E)\Psi = 0 \tag{2.35}$$

The solution of the above equation is

$$\Psi = A\exp\left[-ix\sqrt{\frac{2mE}{\hbar^2}}\right] + B\exp\left[ix\sqrt{\frac{2mE}{\hbar^2}}\right] \quad \text{(for } 0 < x < D\text{)} \tag{2.36}$$

$$\Psi = 0 \text{ (for other } x \text{ where } U \to \infty) \tag{2.37}$$

The general boundary conditions are the continuity of the wavefunctions and their first-order derivatives at the boundaries. The continuity of the wavefunction states that the probability density of finding the matter cannot be double valued at the same location. The continuity of the first-order derivative of the wavefunction can be derived by integrating eq. (2.16) over an infinitely thin control volume encompassing the boundary, and this condition implies the continuity of the particle flux. For the current problem, the continuity of the first derivatives is not required because the wavefunction at the boundaries is already known to be zero. Applying the continuity of the wave function for $x = 0$ and $x = D$, we have at $x = 0$

$$A + B = 0 \tag{2.38}$$

at $x = D$

$$A \exp\left[-iD\sqrt{\frac{2mE}{\hbar^2}}\right] + B \exp\left[iD\sqrt{\frac{2mE}{\hbar^2}}\right] = 0 \tag{2.39}$$

Simultaneous solution of eqs. (2.38) and (2.39) yields

$$\sin\left(D\sqrt{\frac{2mE}{\hbar^2}}\right) = 0 \tag{2.40}$$

so that multiple solutions for E exist at

$$D\sqrt{\frac{2mE_n}{\hbar^2}} = n\pi \quad (n = 1, 2, 3...) \tag{2.41}$$

Here we take only positive values of n since the negative values give the same electron probability distribution functions and are thus identical to the positive solutions. The value $n = 0$ is excluded since taking this value will lead to $\Psi = 0$, which means no particle exists inside the region. The integer n is called the quantum number. Each n corresponds to a wavefunction and an energy level. For multidimensional problems, which we will encounter later, there will be more quantum numbers, including the spin quantum number for electrons. From eq. (2.41), the allowable energy levels are

$$E_n = \frac{1}{2m}\left(\frac{\pi\hbar n}{D}\right)^2 = \frac{1}{2m}\left(\frac{hn}{2D}\right)^2 \quad (n = 1, 2, 3, \ldots) \tag{2.42}$$

which is the same result as eq. (2.15). The material wave function inside the potential well is

$$\Psi_n = -2iA \sin\left(\frac{n\pi x}{D}\right) \tag{2.43}$$

To find the coefficient A, we use the normalization condition

$$\int_{-\infty}^{\infty} \Psi_n \Psi_n^* dx = 1 \tag{2.44}$$

which gives $A = i/\sqrt{2D}$, and thus

$$\Psi_n = \sqrt{\frac{2}{D}} \sin\left(\frac{n\pi x}{D}\right) \tag{2.45}$$

These wavefunctions are standing waves, as shown in figure 2.5(b). The separation between successive energy levels depends on the width D of the potential well, the mass, and the order of the energy level. When D is large, the energy separation is very small. The observation of such energy separation requires sensitive tools that can discern small energy separation. When the energy separation is larger than the thermal energy fluctuation, the quantization effect can be easily observed. For electrons with a small mass, this requires in general that D is smaller than 100 Å.

Let us now show that the Heisenberg principle is satisfied for $n = 1$. The most probable position and the standard deviation in its position can be calculated following eq. (2.19):

$$\langle x \rangle = \int_0^D \Psi_1^* x \Psi_1 dx = 4A^2 \int_0^D x \sin^2\left(\frac{\pi x}{D}\right) dx = \frac{D}{2} \tag{2.46}$$

$$\Delta x = \langle (x - \langle x \rangle)^2 \rangle^{1/2} = \left[\int_0^D \Psi_1^*(x - \langle x \rangle)^2 \Psi_1 dx \right]^{1/2}$$

$$= D\sqrt{\frac{1}{12} - \frac{1}{2\pi^2}} \tag{2.47}$$

Similarly, the most probable momentum and the uncertainty standard deviation in its momentum are

$$\langle p \rangle = -\int_0^D \Psi_1^* i\hbar \frac{d\Psi_1}{dx} dx = 0 \tag{2.48}$$

$$\Delta p = \langle (p - \langle p \rangle)^2 \rangle^{1/2} = \left[\int_0^D \Psi_1^*(-\hbar^2) \frac{d^2\Psi_1}{dx^2} dx \right]^{1/2} = \frac{\pi \hbar}{D} \tag{2.49}$$

From eqs. (2.47) and (2.49), we obtain $\Delta x \Delta p = 0.57\hbar$, thus satisfying the Heisenberg uncertainty principle. This example shows although the electron at $n = 1$ energy level is most probably positioned in the middle of the potential well ($x = D/2$) and has zero average momentum, it can also be at other locations, as the wavefunction suggests.

Although the above solution for an electron in a potential well is one of the simplest solutions of the Schrödinger equation, the experimental realization of such a system came only in the 1970s, after the concept of superlattices was proposed (Esaki and Tsu, 1970) and the molecular-beam-epitaxy (MBE) thin-film growth technique was invented. The MBE technique allows the controlled growth of thin films to an accuracy of one atomic layer or less. Since then, studies of man-made quantum structures have become one of

Figure 2.6 (a) Quantum well laser (Yariv, 1989; courtesy of Wiley). (b) A carbon nanotube (courtesy of Dr. Z.F. Ren). (c) Ge quantum dots in silicon (Liu et al., 2001).

the most active areas of research. Figure 2.6(a) shows a practical quantum well made of a very thin solid layer of gallium arsenide (GaAs) semiconductor. It is sandwiched between other materials in which the potential energy for the electrons is higher than that in the GaAs layer. The allowed electron energy states inside the GaAs layer become discrete. This quantum size effect has been used to make better semiconductor lasers and detectors. In addition to thin films, in which the quantum effect occurs in the direction of film thickness, other quantum structures, such as quantum wires, carbon nanotubes, and dots (quantum dots), are also being actively studied, as shown in figures 2.6(b) and (c), respectively. The potentials in the surrounding for these cases may not be infinite as in the preceding simple example. Thus, the energy levels may be more complicated than given by eq. (2.42).

Example 2.1 *Electron energy levels inside a square nanowire*

Determine the allowable energy levels of an electron in a two-dimensional square quantum wire, assuming the potential inside the quantum wire is $U = 0$ and outside is $U = \infty$.

Solution: We establish a coordinate system as shown in figure E2.1. Clearly, outside the potential well, we have $\Psi = 0$ because $U = \infty$. We thus focus on the solution inside the potential well. The Schrödinger equation inside the well ($U = 0$) is

$$-\frac{\hbar^2}{2m}\left(\frac{\partial^2 \Psi}{\partial x^2} + \frac{\partial^2 \Psi}{\partial y^2}\right) - E\Psi = 0 \qquad (E2.1.1)$$

with the following boundary conditions

$$\begin{array}{ll} x = 0 \text{ or } D, & \Psi = 0 \\ y = 0 \text{ or } D, & \Psi = 0 \end{array} \qquad (E2.1.2)$$

We use the separation-of-variables technique. Assuming $\Psi(x, y) = X(x)Y(y)$ and substituting into the above equation lead to

$$\frac{1}{X}\frac{d^2 X}{dx^2} + \frac{1}{Y}\frac{d^2 Y}{dy^2} + \frac{2mE}{\hbar^2} = 0 \qquad (E2.1.3)$$

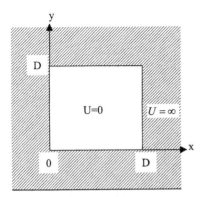

Figure E2.1 A square potential well.

In this equation, the first term depends on x and the second term on y. The third term is a constant. This leads to the requirement that both the first and the second term must be a constant. Since E is positive, we can show that neither of the first two terms can be positive (see what happens if you assume one of them is positive). Thus, we write

$$\frac{1}{X}\frac{d^2 X}{dx^2} = -\alpha^2 \tag{E2.1.4}$$

$$\frac{1}{Y}\frac{d^2 Y}{dy^2} = -\beta^2$$

The solution for X is

$$X(x) = A\sin(\alpha x) + B\cos(\alpha x) \tag{E2.1.5}$$

To satisfy the boundary condition that $\Psi = 0$ at $x = 0$ and $x = D$, we must have $X = 0$ at $x = 0$ and $x = D$. Applying these boundary conditions, we see that

$$\alpha = \frac{n\pi}{D} \quad (n = 1, 2, \ldots) \tag{E2.1.6}$$

$$X_n(x) = A_n \sin\left(\frac{n\pi}{D}x\right) \tag{E2.1.7}$$

Similarly, for Y, we have

$$\beta = \frac{\ell\pi}{D} \quad (\ell = 1, 2, \ldots) \tag{E2.1.8}$$

$$Y_\ell(y) = B_\ell \sin\left(\frac{\ell\pi}{D}y\right) \tag{E2.1.9}$$

On the basis of the allowable values of α and β, we can determine the allowable energy levels as

$$E_{\ell n} = \frac{(\ell^2 + n^2)\pi^2 \hbar^2}{2mD^2} \quad (\ell, n = 1, 2, \ldots) \quad \text{(E2.1.10)}$$

Corresponding to each set of ℓ and n, there is a distinct wavefunction

$$\Psi_{\ell n} = C_{\ell n} \sin\left(\frac{n\pi x}{D}\right) \sin\left(\frac{\ell \pi y}{D}\right) \quad \text{(E2.1.11)}$$

We can further determine that the constant $C_{\ell n} = 2/D$ using normalization condition (2.44)

Comment: For the above two-dimensional problem, we obtain two quantum numbers ℓ and n. Typically, the number of the quantum numbers is identical to the dimensionality of the problem. For a three-dimensional problem, there are three quantum numbers, as we will see later. Each set of quantum numbers determines a unique wavefunction. The energy levels for different sets of ℓ and n, however, can be identical. For example, the wavefunction corresponding to $\ell = 1$ and $n = 2$ has the same energy as that corresponding to $\ell = 2$ and $n = 1$. These states that have different wavefunctions but the same energy are said to be degenerate.

2.3.3 Electron Spin and the Pauli Exclusion Principle

Each wavefunction obtained from the Schrödinger equation, as exemplified in the previous sections represents a possible quantum mechanical state at which a particle can exist under the given potential. The solutions of the Schrödinger equation, however, do not tell the entire story regarding the quantum state of a particle. For example, the equation cannot predict the spin of a particle. The spin is a property that preserves the particle's rotational symmetry and can only be derived from relativistic quantum mechanics, developed originally by Dirac. It is an intrinsic property of the particle and should not be understood, simply for example, as the rotation of an electron around a nucleus. For electrons, corresponding to each wavefunction obtained from the Schrödinger equation, there are two quantum states (or two relativistic wavefunctions), which are usually denoted by an additional quantum number s that can have the following values

$$s = \frac{1}{2} \text{ or } -\frac{1}{2} \quad (2.50)$$

where $s = 1/2$ is called spin up and $s = -1/2$ is called spin down. The spin quantum numbers for other types of particle are different. Interested readers should consult quantum mechanics textbooks (Feynman, 1965; Cohen-Tannoudji et al., 1977; Landau and Liftshitz, 1977).

We can combine this spin quantum number with the wavefunctions obtained from the Schrödinger equation to denote the complete set of wavefunctions that a particle can have. For an electron in a one-dimensional box, the wavefunction can be denoted as $\Psi_{n,s}$. Each set of quantum numbers n and s represents a quantum state. For each n, there are two quantum states ($s = 1/2$ or $s = -1/2$) with identical energy. The number of wavefunctions, that is, the number of quantum states, at an identical energy level is called the level degeneracy. Thus, the electron in a one-dimensional box at any energy level n has a degeneracy of two. The *Pauli exclusion principle* says that each quantum state can be occupied by at most one electron. This principle determines how the allowable energy levels, such as those given by eqs. (2.42) and (E2.1.10), will actually be occupied by electrons. Although eq. (2.42) [or (E2.1.10)] is derived under the assumption of one electron in an infinite potential, the solution should be valid if there is more than one electron inside the well, under the approximation that the interactions of these electrons do not change the potential profile. From the principles of thermodynamics, we know that electrons tend to occupy the lowest energy states. If there are five electrons in the potential well and the temperature is at absolute zero (to avoid thermal fluctuations), two electrons will take the lowest E_1 energy level ($\Psi_{1,1/2}$ and $\Psi_{1,-1/2}$ quantum states), and another two will occupy the next lowest level E_2 ($\Psi_{2,1/2}, \Psi_{2,-1/2}$ quantum states). The fifth electron will occupy either the $\Psi_{3,1/2}$ or the $\Psi_{3,-1/2}$ state. The occupation of a quantum state at non-zero temperatures will be discussed in chapter 4.

2.3.4 Harmonic Oscillator

Two main factors govern the Schrödinger equation and the consequent energy eigenvalues and wavefunctions: the potential distribution and the boundary conditions. The rectangular potential profile for the particle-in-a-box (problem discussed in section 2.3.2) gives energy eigenvalues that depend on the square of the quantum numbers, as represented by eqs. (2.42) and (E2.1.10). In this section, we will study the energy levels and eigenfunctions of a harmonic oscillator, which is a very useful model for describing the atomic vibrations in simple molecules, such as H_2 and CO, or the atomic vibrations in solids. We can appreciate this by considering the general shape of the interatomic potential, shown in figure 2.7(a). If the vibrational amplitude of the atom is not large, we can expand the potential around the equilibrium position, x_0, such that

$$U(x') = U(x_0) + \frac{1}{2}\left[\frac{d^2U}{dx'^2}\right]_{x'=x_0}(x'-x_0)^2 + O[(x'-x_0)^3] \qquad (2.51)$$

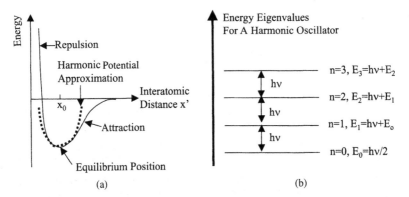

Figure 2.7 (a) The harmonic oscillator model approximates the potential at equilibrium by a parabola. (b) The energy levels of a harmonic oscillator.

where we have used the fact that $dU/dx' = 0$ at the equilibrium point $x' = x_0$ (the minimum of a curve). Neglecting the higher order term $O[(x' - x_0)^3]$, the force acting on the atom thus becomes

$$F = -dU/dx' = -K(x' - x_0) = -Kx \quad (2.52)$$

where $x(= x' - x_0')$ represents the displacement from the equilibrium position rather than the separation between atoms, and $K = (d^2U/dx'^2)_{x=x_0}$ is the spring constant such that $U = Kx^2/2$. This potential represents a classical mass–spring system, the Schrödinger equation for the system is

$$-\frac{\hbar^2}{2m}\frac{d^2\Psi}{dx^2} + \left(\frac{Kx^2}{2} - E\right)\Psi = 0 \quad (2.53)$$

Boundary conditions for the above equation are

$$\Psi(x \to \infty) = \Psi(x \to -\infty) = 0 \quad (2.54)$$

because as $x \to \pm\infty$, the potential $U \to \infty$, which requires vanishing wavefunction values in these limits. Solving eq. (2.53) involves a coordinate transformation and a series expansion, and will not be pursued here (see Landau and Lifshitz, 1977). Final results for the energy levels and wavefunctions are

$$E_n = \left(n + \frac{1}{2}\right)h\nu = \left(n + \frac{1}{2}\right)\hbar\omega \quad (n = 0, 1, 2, \ldots) \quad (2.55)$$

$$\psi_n(x) = \left[\sqrt{\frac{m\omega}{\pi\hbar}}\frac{1}{2^n n!}\right]^{1/2} H_n\left[\left(\frac{m\omega x^2}{\hbar}\right)^{\frac{1}{2}}\right]\exp\left(-\frac{m\omega x^2}{2\hbar}\right) \quad (2.56)$$

where

$$\nu = \frac{1}{2\pi}\sqrt{\frac{K}{m}} \quad (2.57)$$

$\omega = 2\pi\nu$ is the angular frequency, and H_n is a standard function called the Hermite polynomial, given by

$$H_n(\xi) = (-1)^n \exp(\xi^2)\frac{d^n}{d\xi^n}[\exp(\xi^{-2})] \quad (2.58)$$

We see that ν is the fundamental vibration frequency that we get from classical mechanics for a mass–spring system. In classical mechanics, however, the mass–spring system energy, which equals the sum of the kinetic and the potential energy, can be a continuous function of the amplitude of the oscillator. Equation (2.55) says, however, that the energy of the harmonic oscillator is quantized and can only be a multiple number of $h\nu$ plus $\frac{1}{2}h\nu$, which means that the separation between adjacent energy levels is constant and equal to $h\nu$, as shown in figure 2.7(b). The $\frac{1}{2}h\nu$ term in eq. (2.55a) is called the zero point energy, which is unimportant for most heat transfer problems. It is a manifestation of the Heisenberg uncertainty principle. In figure 2.8, we show the wavefunctions (Ψ)

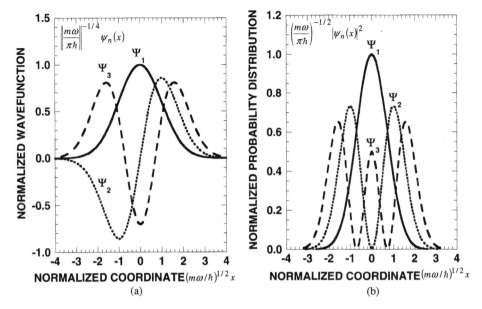

Figure 2.8 (a) Normalized wavefunction and (b) normalized probability distribution for a harmonic oscillator.

and the probability density distributions ($|\Psi|^2$) of the harmonic oscillator states. The latter shows the probability of finding the vibrating atom at position x. As the quantum number n increases, $|\Psi|^2$ spreads wider which means that the potential energy grows with increasing n. The kinetic energy also increases with increasing n.

The harmonic oscillator model is important for understanding the absorption characteristics of gases in the infrared spectrum. When a photon interacts with the gas, absorption occurs only when the photon energy equals the energy difference between the final state and the initial state of the molecule,

$$E_p = h\nu_p = E_f - E_i \tag{2.59}$$

where the subscript p represents photons. For the vibrational modes of a molecule, eqs (2.55) and (2.59) lead to $\nu_p = \nu \Delta n$. Further quantum mechanical consideration limits $\Delta n = \pm 1$ (the minus sign corresponds to emission), which is called the selection rule. The absorption or emission of a photon occurs when the photon frequency equals the molecular vibration frequency (also called an absorption line),

$$\nu_p = \frac{1}{2\pi}\sqrt{\frac{K}{m}} \tag{2.60}$$

The relative vibration of atoms in a polyatomic molecule can be modeled as a harmonic oscillator. When applying the above expression to a diatomic molecule with two atoms of mass m_1 and m_2, the reduced mass should be used,

$$m = \frac{m_1 m_2}{m_1 + m_2} \tag{2.61}$$

62 NANOSCALE ENERGY TRANSPORT AND CONVERSION

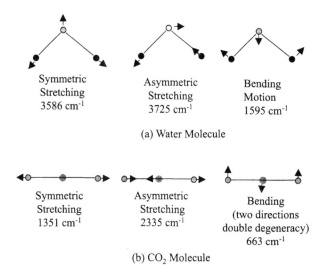

Figure 2.9 Vibrational normal modes of (a) H_2O and (b) CO_2 molecules. Arrows represent atom vibration direction at one instant.

Consider, for example, an H_2 molecule having $K = 1113$ Nm^{-1} and a reduced mass of 1.67×10^{-27} kg. The corresponding vibrational frequency is 1.319×10^{14} Hz, or, in terms of wavelengh, $\lambda = 2.3$ µm, and wavenumber, $\eta = 1/\lambda = 4401$ cm^{-1} (wavenumber equals the inverse of wavelength). Experimentally, we can determine the absorption frequency using spectroscopy techniques and thus use eq. (2.60) to estimate the effective spring constant of the interatomic bonds in such molecules.

For polyatomic molecules (larger than two atoms), there exists more than one vibrational frequency. In general, the complex vibrational patterns can be decomposed into normal modes. Each normal mode can be thought of as one harmonic oscillator with a corresponding fundamental frequency. Examples of the normal modes for water and carbon dioxide (CO_2) are shown in figures 2.9(a) and (b). These fundamental normal modes can be superimposed to form new absorption lines. For example, the difference between the asymmetric and symmetric stretching gives the familiar absorption line of CO_2 at ~ 10 µm, which is a major factor in global warming because this absorption line is near the peak of terrestrial thermal radiation.

The harmonic oscillator model also represents quantized electromagnetic fields and atomic vibrations in solids. For an electromagnetic field at frequency ν, the allowable energy levels are

$$E_p = h\nu_p \left(n + \frac{1}{2}\right) \quad (n = 0, 1, 2, \ldots) \quad (2.62)$$

where n is the number of photons in the field having frequency ν_p. This expression is consistent with our previous discussion on the quantized electromagnetic field eq. (2.10). In the next chapter, we will show that atomic vibrations in a crystalline solid can also be decomposed into normal modes and that the energy of each mode obeys eq. (2.62) as well.

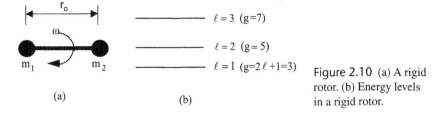

Figure 2.10 (a) A rigid rotor. (b) Energy levels in a rigid rotor.

2.3.5 The Rigid Rotor

In diatomic and polyatomic molecules, the atoms can vibrate relative to each other or rotate as a whole. The relative vibration can be treated by the harmonic oscillator approximation. Let us consider the rotation only and assume that the distance between the two masses of a diatomic molecule is constant (rigid rotation), as shown in figure 2.10(a).

In classical mechanics, a quantity often used to describe the rotation is the moment of inertia. For the two-mass system shown in figure 2.10 rotating relative to its mass center, the moment of inertia is

$$I = \frac{m_1 m_2 r_0^2}{m_1 + m_2} \tag{2.63}$$

where r_0 is the effective separation between the two atoms. In classical mechanics, we often use the angular momentum equations to solve rotational problems. In quantum mechanics, there are also corresponding angular momentum equations that govern the wavefunctions and energy eigenvalues for rotation. Here, we will skip the details but give the solutions for the wavefunction (Landau and Lifshitz, 1977)

$$Y_\ell^m(\theta, \varphi) = \varepsilon i^\ell \left[\frac{(2\ell + 1)(\ell - |m|)!}{4\pi(\ell + |m|)!} \right]^{1/2} P_\ell^{|m|}(\cos\theta) e^{im\varphi} \tag{2.64}$$

θ and φ are polar and azimuthal angles, respectively, in a spherical coordinate system, where $\varepsilon = (-1)^m$ for $m \geq 0$ and $\varepsilon = 1$ for $m \leq 0$, and P is the associated Legendre polynomial

$$P_\ell^m(\xi) = \frac{(1 - \xi^2)^{m/2}}{2^\ell \ell!} \frac{d^{\ell+m}}{d\xi^{\ell+m}} (\xi^2 - 1)^\ell$$

The functions $Y_\ell^m(\theta, \phi)$ are called the spherical harmonics. The energy eigenvalues are

$$E_\ell = \frac{\hbar^2}{2I} \ell(\ell + 1) = hB\ell(\ell + 1) \quad \text{(for } |m| \leq \ell, \ell = 0, 1, 2, \ldots) \tag{2.65}$$

where $B = h/(8\pi^2 I)$ [Hz] is called the rotational constant. For H_2, $B = 1.8 \times 10^{12}$ Hz. The corresponding wavelength for this rotational state alone is very small (~ 100 μm). Because there are two degrees of freedom for the rigid rotational motion (polar and azimuthal directions in a spherical coordinate system), we see that two quantum numbers emerge, that is, ℓ and m. Each set of ℓ and m gives a unique wavefunction and thus a unique quantum state (spin quantum number not included). The energy levels given by eq. (2.65), however, do not depend on the magnitude of m. For each ℓ, there are $(2\ell+1)$ values of m, that is, $(2\ell+1)$ wavefunctions having the same energy level. The energy level E_ℓ is thus $(2\ell+1)$-fold degenerate, similar to the degeneracy of a particle in a two-dimensional box discussed in example 2.1, because there are $(2\ell+1)$ rotational orbits. We use g to denote the *degeneracy* so that, for the rotational level at energy E_ℓ we have

$$g(\ell) = 2\ell + 1 \quad (2.66)$$

Similar to the case of harmonic oscillators, the photon absorption by rigid rotors occurs only when the separation of the energy levels of the oscillators matches the photon energy. The selection rule in quantum mechanics further limits the photon to two adjacent energy levels ($\Delta\ell = 1$),

$$\nu_p = (E_{\ell+1} - E_\ell)/h = 2B(\ell+1) \quad (2.67)$$

where B in the unit of wavelength is of the order of 100 μm for hydrogen molecules; thus pure rotational modes have long wavelengths and are typically unimportant in thermal energy transfer, but they can be important in the microwave range. Since a diatomic molecule can have both vibrational and rotational modes, we can approximate their allowable energy states as the superposition of the rotational and vibrational energy levels, forming vibrational–rotational states. Assuming, for simplicity, that the rotational and vibrational motions are independent, we can then write

$$E(\text{vib} + \text{rot}) = E_n(\text{vib}) + E_\ell(\text{rot}) \quad (2.68)$$

and, correspondingly, the absorption lines of the combined vibrational–rotational states can be written as

$$\nu_p = \nu \pm 2B(\ell+1) \quad (2.69)$$

where the positive sign means that accompanying the increase of the vibrational energy level due to the photon absorption, the rotational energy also increases by one level, while the negative sign means that the rotational energy decreases by one level. Thus, surrounding each fundamental vibrational frequency ν of polyatomic molecules, spectral lines with fine structures due to molecular rotations are formed. These lines often overlap due to various broadening and interaction mechanisms (such as thermal vibrations) so that the absorption of gases effectively occurs over certain bandwidths (called bands), rather than only at sharp discrete lines. Figure 2.11 illustrates the absorption bands of CO_2 molecules at 0.5 atm and 300 K.

2.3.6 Electronic Energy Levels of the Hydrogen Atom

The vibrational and rotational energy levels that we previously obtained are for the atomic motion of polyatomic molecules. Now let's consider the electronic energy levels of an atom. We use hydrogen atom as an example since there is only one electron surrounding the nucleus. The solution we will find, however, can be used as a basis for understanding the energy levels of other atoms having multiple electrons. The nucleus of a hydrogen

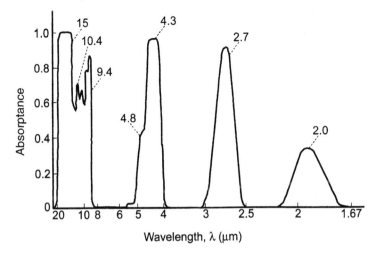

Figure 2.11 Vibrational-rotational absorption bands of CO_2 molecules (Siegel and Howell, 1992; courtesy of J.R. Howell).

atom can be treated as stationary because its mass is relatively heavy compared to that of the electron. The interaction between the nucleus and the orbiting electron is governed by the Coulomb electrostatic force

$$F = \frac{1}{4\pi\varepsilon_0}\frac{e^2}{r^2} = \frac{c_1}{r^2} \quad (2.70)$$

where $c_1 = e^2/4\pi\varepsilon_0$ and the electron charge is $e = 1.6 \times 10^{-19}$[C], while $\varepsilon_0 = 1.124 \times 10^{-10}/4\pi$ [C^2 m^{-2} N^{-1}] is the electrical permittivity of the vacuum. The potential is

$$U = -\int_r^\infty F\,dr = -\frac{c_1}{r} \quad (2.71)$$

Since the nucleus is stationary, we can take it as the origin of the coordinate system. The Schrödinger equation then becomes

$$-\frac{\hbar^2}{2m}\nabla^2\Psi + \left(-\frac{c_1}{r} - E\right)\Psi = 0 \quad (2.72)$$

In spherical coordinates, the Laplace operator is

$$\nabla^2 = \frac{1}{r^2}\frac{\partial}{\partial r}\left(r^2\frac{\partial}{\partial r}\right) + \frac{1}{r^2\sin\theta}\frac{\partial}{\partial \theta}\left(\sin\theta\frac{\partial}{\partial \theta}\right) + \frac{1}{r^2\sin^2\theta}\frac{\partial^2}{\partial \varphi^2} \quad (2.73)$$

The solution of eq. (2.72) can then be separated into a radial wavefunction $R_{n\ell}(r)$ and into spherical harmonics Y_ℓ^m (Griffiths, 1994)

$$\Psi_{n\ell m} = R_{n\ell}(r)Y_\ell^m(\theta, \varphi) \quad (2.74)$$

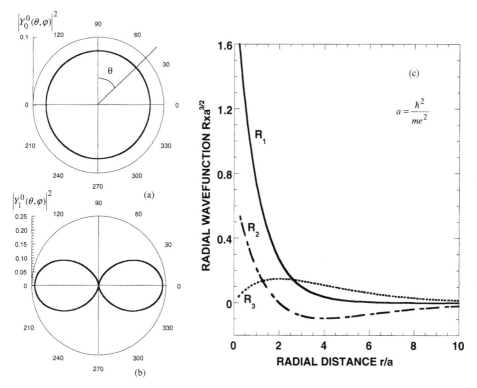

Figure 2.12 (a) and (b) Amplitude square of spherical harmonic functions in polar coordinate θ. The φ direction is axisymmetric around the $\theta = 0$ axis and (c) normalized radial wavefunction of the hydrogen atom.

where Y_ℓ^m is given by eq. (2.64). Because of the three degrees of freedom in space, there are three quantum numbers n, ℓ, and m in the wavefunction. We will skip the analytical expressions for the radial wavefunction, but give a few graphical examples of $R_{n\ell}(r)$ and $\left|Y_\ell^m\right|^2$ in figure 2.12. The allowable energy levels of the electron–nucleus system are

$$E_n^{\text{el}} = -\frac{mc_1^2}{2\hbar^2 n^2} = -\frac{13.6 \text{ eV}}{n^2} (n \geq 1, n \geq \ell + 1, \text{ and } |m| \leq \ell, \ell = 0, 1, 2, \ldots) \tag{2.75}$$

where n is an integer called principal quantum number. We note that the energy levels depend on n but not on ℓ or m. The allowable values of ℓ and m are, however, restricted by the value of n. Let us recall the concept of degeneracy; this means that the energy levels are degenerate. The three quantum numbers are as follows: n is the principal quantum number, ℓ is the quantum number of total angular momentum, and m is the magnetic quantum number. Corresponding to each set of n, l, m quantum numbers, there are two wave functions $\Psi_{n,l,m,s}$, where $s = \pm 1/2$ is for the electron spin, which determine two quantum states. For $n = 1$, the allowable values for ℓ and m are 0. The wave functions for this set of n, l, m values are $\Psi_{1,0,0,s}$ and are called the 1s-orbital (s-shell),

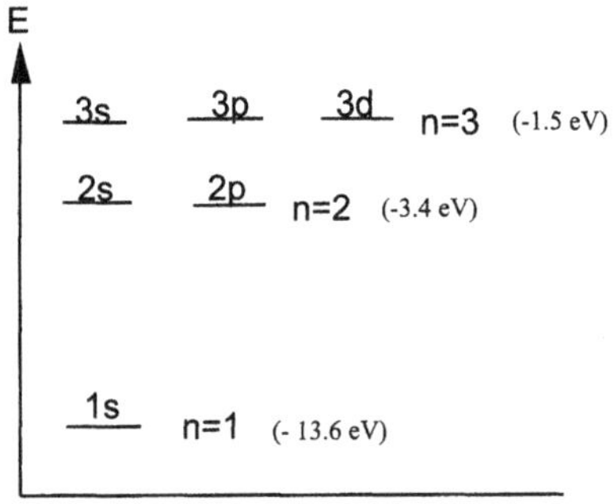

MATERIAL WAVES AND ENERGY QUANTIZATION 67

```
E
▲
   3s    3p    3d   n=3  (-1.5 eV)

   2s    2p         n=2  (-3.4 eV)



   1s          n=1  (-13.6 eV)
```

Figure 2.13 Illustration of first three energy levels of the hydrogen atom.

where "1" in front of s represents $n = 1$ and "s" following 1 represents $\ell = 0$. Since a maximum of one electron is allowable for each quantum state (Pauli exclusion principle), the 1s-orbital can have a maximum of two electrons as a result of the two possible values of the spin ($s = 1/2, -1/2$). For $n = 2$, the following values of ℓ and m are possible

$$n = 2 \rightarrow \begin{cases} l = 0 & m = 0 & 2s \text{ orbital} \\ l = 1 & m = -1, 0, 1 & 2p \text{ orbital} \end{cases}$$

Two electrons are allowed for the 2s-orbital and six electrons in total are allowable for the three quantum states in the 2p-orbital (p-shell). A total of eight electrons is thus allowed for the $n = 2$ states. In general, the degeneracy for a hydrogen atom of any arbitrary n is

$$g = 2n^2 \tag{2.76}$$

Figure 2.13 shows the energy levels of a hydrogen atom. An alphabetical symbol is assigned for each value of ℓ, which is inherited from the historical studies of absorption spectra of hydrogen before quantum mechanics was developed:

$$\ell = 0 \leftrightarrow s; \ \ell = 1 \leftrightarrow p; \ \ell = 2 \leftrightarrow d; \ \ell = 3 \leftrightarrow f; \ \ell = 4 \leftrightarrow g$$

The allowable hydrogen energy levels provide a framework for understanding the periodic table. Since the electron–ion interaction dominates the potential, the existence of more electrons in other atoms does not alter very much the major picture of the allowable energy levels and we can assume that the energy levels of a hydrogen atom also apply to other atoms for qualitative understanding. As a start, let's fill the electronic energy states, as shown in figure 2.13, according to the thermodynamics principle that the lowest level will be occupied first and each quantum state can have at most one electron. The one electron in hydrogen will occupy one of the two 1s orbitals. The two electrons in a helium atom will fully occupy the two 1s quantum states. Lithium has three electrons and the first two will fill the 1s-orbital and its third electron will fill one 2s-orbital, and so on.

Table 2.1 lists the electron orbital occupancy of elemental atoms. Up to the argon atom, it is always the case that quantum states with smaller orbital quantum number

Table 2.1 Electron configuration of first 30 elements in the periodic table

Atomic Number	Element		Electron Configuration									
			1s	2s	2p	3s	3p	3d	4s	4p	4d	4f
1	H	hydrogen	1									
2	He	helium	2									
3	Li	lithium		1								
4	Be	beryllium		2								
5	B	boron		2	1							
6	C	carbon	Filled	2	2							
7	N	nitrogen	(2)	2	3							
8	O	oxygen		2	4							
9	F	fluorine		2	5							
10	Ne	neon		2	6							
11	Na	sodium				1						
12	Mg	magnesium				2						
13	Al	aluminum				2	1					
14	Si	silicon				2	2					
15	P	phosphor				2	3					
16	S	sulfur				2	4					
17	Cl	chlorine				2	5					
18	A	argon	Filled	Filled		2	6					
19	K	potassium	(2)	(8)		2	6		1			
20	Ca	calcium				2	6		2			
21	Sc	scandium				2	6	1	2			
22	Ti	titanium				2	6	2	2			
23	V	vanadium				2	6	3	2			
24	Cr	chromium				2	6	5	1			
25	Mn	manganese				2	6	5	2			
26	Fe	iron				2	6	6	2			
27	Co	cobalt				2	6	7	2			
28	Ni	nickel				2	6	8	2			
29	Cu	copper				2	6	10	1			
30	Zn	zinc				2	6	10	2			

are filled first. Starting with potassium, however, we begin to see deviations from this trend. Rather than filling the 3d-orbital, one electron actually fills the 4s-orbital first. This effect is due to electron–electron interaction such that the energy levels for the same n (for example, 3s, 3p, and 3d have the same $n = 3$) are no longer the same. This change is called the lifting of the degeneracy, and is because the 3d levels have a slightly higher energy than the 4s level. Therefore, the extra electron in potassium will fill the 4s-orbital rather than occupy one of the 3d-orbitals.

The filling of the electronic states determines the chemical activity of each atom. If all orbitals of the same principal quantum number n are filled, the atom is inert because the energy difference to the next level of n is much larger than $\kappa_B T$ (26 meV at room temperature). Otherwise, the vacant quantum states within the same principal quantum number can accept other electrons or lose electrons to form a more stable state. For example, the H_2 molecule has two electrons sharing the two 1s quantum states in the hydrogen atom, such that each atom "feels" that it has two electrons. In various acids, the hydrogen atom is also happy to give up its electrons. We call the electrons in the outermost principal orbitals the valence electrons.

Using eq. (2.75), we can calculate the absorption lines of hydrogen atoms as

$$h\nu_p (n_1 \rightarrow n_2) = 13.6 \left(\frac{1}{n_2^2} - \frac{1}{n_1^2} \right) \text{eV} \qquad (2.77)$$

Example 2.2

Determine the photon frequency and wavelength for series of allowable emission from all other states to the $n = 1$ states from the hydrogen atom. This series is called the Lyman series.

Solution: The emission occurs when the energy of the hydrogen atom drops from a high energy state to a low one. From eq. (2.77), the emission spectrum is

$$\begin{aligned} \nu_p &= \frac{13.6 \text{ eV}}{h} \left(1 - \frac{1}{n^2} \right) \\ &= 3.288 \times 10^{15} \left(1 - \frac{1}{n^2} \right) \text{Hz}(n = 2, 3, 4, \ldots) \end{aligned} \qquad (E2.2.1)$$

The emitted photon frequency and wavelength are listed in the following table. These numbers are in excellent agreement with experiments.

n	$\nu_p(n \rightarrow 1) \times 10^{15}$ Hz	$\lambda_p(n \rightarrow 1)$(nm)
2	2.466	121.57
3	2.9227	102.57
4	3.8025	97.255
$\vdots$	$\vdots$	$\vdots$
∞	3.288	91.177

Now we are in a position to discuss the total energy of an atom or molecule. The total energy can be approximated as the summation of translational, vibrational, rotational, and electronic energies:

$$E^{\text{tot}} = E^{\text{trans}} + E^{\text{el}} + E^{\text{vib}} + E^{\text{rot}} \qquad (2.78)$$

For a monatomic gas, there are no vibrational or rotational energy levels. Although we did not discuss the translational energy levels much, the particle in a potential well model describes the allowable translational energy levels of an atom or molecule. Parallel to example 2.1, the translational energy level of an atom (or a molecule) in a three-dimensional potential well can be expressed as

$$E^{\text{trans}} = \frac{h^2}{8m}\left(\frac{n_x^2}{D_x^2} + \frac{n_y^2}{D_y^2} + \frac{n_z^2}{D_z^2}\right) \quad (n_x, n_y, n_z = 1, 2, 3, \ldots) \tag{2.79}$$

where D_x, D_y, D_z are the lengths of the potential well in the x, y, and z directions, respectively. Since the mass of an atom is typically much larger than that of an electron, the separations between translational energy levels are very small. Thus the translational energy can be considered as a continuous variable and is often simply expressed as the kinetic energy;

$$E^{\text{trans}} = \frac{m}{2}(v_x^2 + v_y^2 + v_z^2) \tag{2.80}$$

Comparing eqs. (2.80) and (2.79), we see that

$$mv_x = \frac{hn_x}{2D_x} \tag{2.81}$$

and this is similar for y and z directions. The left-hand side of eq. (2.81) is momentum p_x. The right-hand side, according eqs. (2.8) or (1.38), is h/λ_x. Thus, eq. (2.81) is a consequence of the Planck–Einstein relation, eq. (2.12), between momentum and wavelength.

2.4 Summary of Chapter 2

In this chapter, we have introduced the wave–particle duality of electromagnetic radiation. It was through the work of Planck and Einstein that the particle characteristics of electromagnetic radiation were revealed. Planck suggested that the energy of an electromagnetic wave at frequency ν must be an integral multiple of $E = h\nu$. Einstein further showed that this basic energy unit has particle characteristics, and this basic quantum of energy was eventually named photon. The momentum and energy relations between waves and particles are called the Planck–Einstein relations

$$E = h\nu, \quad p = \frac{h}{\lambda}$$

On the basis of the wave–particle duality of light, de Broglie further suggested that matter has wave characteristics that follow the same Planck–Einstein relations. This suggestion led to the development of quantum mechanics. The Schrödinger equation describes material waves,

$$-\frac{\hbar^2}{2m}\nabla^2 \Psi_t + U\Psi_t = i\hbar \frac{\partial \Psi_t}{\partial t}$$

where Ψ_t is the wavefunction. The meaning of the wavefunction is that $\Psi_t \Psi_t^*$ gives the probability that matter will be found at location $\mathbf{r}$ and time t. In the quantum mechanics world, things are uncertain and the most probable value of any quantity is calculated from the operator for that quantity, including its location, momentum, energy, and so on. The uncertainties of the location and momentum, and of time and energy, obey the Heisenberg uncertainty principle

$$\Delta p \Delta x \geq \hbar/2 \text{ and } \Delta E \Delta t \geq \hbar/2$$

Solution of the Schrödinger equation leads to eigenvalues that are identified as the (most probable) energies of the system. We have given solutions for the following potential fields:

1. Free electrons. The free electron energy level is a continuous variable.
2. Particle in an infinite one-dimensional potential well. The particle energies are quantized and their wavefunctions form standing waves. Research on artificial quantum structures has become a mainstream research field and has led to many exciting applications.
3. Harmonic oscillator. The energy levels of a harmonic oscillator are quantized according to

$$E^{\text{vib}} = h\nu \left(n + \frac{1}{2}\right) \text{ where } \nu = \frac{1}{2\pi}\sqrt{\frac{K}{m}} \quad (n = 0, 1, 2, \ldots)$$

The harmonic oscillator represents a wide range of phenomena such as the vibrational energy levels in polyatomic molecules, an electromagnetic field, and atom vibrations in solids.

4. Rigid rotor. The energy levels of a rigid rotor are given by

$$E_\ell = \frac{\hbar^2}{2I}\ell(\ell+1) \quad (\text{for } |m| \leq \ell, \ell = 0, 1, 2, \ldots)$$

where ℓ and m are integers. Because there are multiple ℓ and m values that give the same energy, and each set of ℓ and m represents one quantum state, the rotational energy levels are degenerate. The degeneracy is $g(\ell) = 2\ell + 1$. The energy separation between rotational energy levels is very small. They are typically observed together with vibrational energy levels.

5. Hydrogen atom. The electron energy levels of a hydrogen atom are

$$E_n^{\text{el}} = -\frac{mc_1^2}{2\hbar^2 n^2} = -\frac{13.6 \text{ ev}}{h^2} \quad (n \geq 1, n \geq \ell + 1, \text{ and } |m| \leq \ell, \ell = 0, 1, 2, \ldots)$$

where, again, each set of (n, m, ℓ), plus the spin quantum number s, determines a quantum state. Because the energy level depends on n only, the energy levels are degenerate, the degeneracy being $g(n) = 2n^2$. The electron energy levels in the hydrogen atom provide a basis for understanding the periodic table and the chemical activity of atoms.

6. For an atom or molecule, the total energy is the sum of translational, electronic, rotational, and vibrational energy levels (the latter two are for polyatomic molecules only).

7. A photon interacts with matter (absorption or emission) only when the photon energy and allowable energy levels of the matter satisfy the following relation

$$E_{\text{photon}} = h\nu_{\text{photon}} = E_f - E_i$$

In addition, we should understand that each wavefunction determines one quantum state. Electrons also have spin, which cannot be obtained from solving the Schrödinger equation. The two spin quantum numbers for an electron are $s = 1/2$ and $-1/2$. The Pauli exclusion principle dictates that each quantum state can have a maximum of one electron.

2.5 Nomenclature for Chapter 2

A	wave amplitude	α	separation of variable constant
B	rotational constant, Hz; or undetermined coefficient in the wavefunction	β	separation of variable constant
		Δx	standard deviation in x from the expected value
c	constant in eq. (2.70), N m^2	∇	gradient operator
D	width of potential well, m	∇^2	Laplace operator
e	charge per electron, C	ε_0	electrical permittivity of vacuum, C^2 N^{-1} m^{-2}
F	force, N		
g	degeneracy	κ_B	Boltzmann constant, J K^{-1}
h	Planck constant, J s	λ	wavelength, m
$\hbar$	Planck constant divided by 2π, J s	ν	frequency, s^{-1}
H	system Hamiltonian, J	$\mathbf{\Phi}$	vector waveform
I	moment of inertia, kg m^2	$\mathbf{\Phi}_c$	complex waveform
J	particle flux, m^{-2} s^{-1}	Ψ	wavefunction
k	magnitude of wavevector, m^{-1}	ω	angular frequency, rad s^{-1}
$\mathbf{k}$	wavevector, m^{-1}	Ω	operator
K	spring constant, N m^{-1}	$\langle\rangle$	expectation value
ℓ	quantum number		
m	mass, kg; quantum number	**Subscripts**	
n	quantum number		
p	magnitude of the momentum, kg m s^{-1}	0	equilibrium position
		f	Fermi level
$\mathbf{p}$	momentum operator, kg m s^{-1}	ℓ	quantum number
r_0	effective distance between two atoms	m	quantum number
		n	quantum number
r	radial distance from origin, m	p	photon
$\mathbf{r}$	position vector	s	spin quantum number
s	spin quantum number	t	total
t	time, s	v	vacuum level
U	system energy or potential energy, J	x, y, z	Cartesian components
x, y, z	Cartesian coordinates		
X	separation of variable component	**Superscripts**	
$\hat{\mathbf{y}}$	unit vector in y direction		
Y	separation of variable component; spherical harmonics	$\hat{}$	unit vector
		$*$	complex conjugate

2.6 References

Broglie, L. de, 1925, "Recherches sur la Théorie des Quanta," *Annales de Physique*, vol. 3, p. 22.

Cohen-Tannoudji, C., Diu, B., and Laloë, F., 1977, *Quantum Mechanics*, vol. 1, p. 502, Wiley, New York.

Davisson, C., and Germer, L.H., 1927, "Diffraction of Electrons by a Crystal of Nickel," *Physical Review*, vol. 30, pp. 705–740.

Einstein, A., 1905, "On a Heuristic Point of View Concerning the Production and Transformation of Light," *Annalen der Physik*, vol. 17, pp. 132–148; see also *Collected Papers of Albert Einstein*, vol. 2, Princeton University Press, Princeton, NJ, 1989.

Einstein, A., 1906, "On the theory of Light Production and Light Absorption," *Annalen der Physik*, vol. 20, pp. 199–206; see also *Collected Papers of Albert Einstein*, vol. 2, Princeton University Press, Princeton, NJ, 1989.

Esaki, L., and Tsu, R., 1970, "Superlattice and Negative Differential Conductivity in Semiconductors," *IBM Journal of Research and Development*, vol. 14, pp. 61–65.

Feynman, R., 1965, *Lectures on Physics*, vol. 3, chapters 1–3, Addison-Wesley, Reading, MA.

Griffiths, D.J., 1994, *Introduction to Quantum Mechanics*, Prentice Hall, Englewood Cliffs, NJ.

Heisenberg, W., 1925, "Über Quantentheoretische Umdeutung Kinematischer und Mechanischer Beziehungen," *Zeitschrift für Physik*, vol. 33, pp. 879–893.

Landau, L.D., and Lifshitz, E.M., 1977, *Quantum Mechanics*, Pergamon, New York.

Lewis, G.N., 1926, Letters to Editor, *Nature*, vol. 118, pt. 2, pp. 874–875.

Liu, W.L., Borca-Tasciuc, T., Chen, G., Liu, J.L., and Wang, K.L., 2001, "Anisotropic Thermal Conductivity of Ge-Quantum Dot and Symmetrically Strained Si/Ge Superlattice," *Journal of Nanoscience and Nanotechnology*, vol. 1, no. 1, pp. 39–42.

Schrödinger, E., 1926, "Quantisation as a Problem of Characteristic Values," *Annalen der Physik*, vol. 79, pp. 361–376, 489–527.

Siegel, R., and Howell, J.R., 1992, *Thermal Radiation Heat Transfer*, Hemisphere, Washington, DC.

Yariv, A., 1989, *Quantum Electronics*, Wiley, New York.

2.7 Exercises

2.1 *Planck–Einstein relations.* (a) An argon laser emits light at 514 nm and at a power of 1 W. Calculate (1) the frequency of the photons in Hz, (2) their wavelength, expressed as a wavenumber, (3) the energy of each photon, (4) the momentum of each photon, and (5) the number of photons generated per second.
(b) If the photons are completely absorbed by a 1 mm^2 surface, calculate (1) the pressure exerted on the surface by the photons, and (2) the heat flux generated by the photon absorption.

2.2 *Transmission electron microscope.* Electron beams are used to study the atomic structure of crystals, as in the transmission electron microscope (TEM). The resolution of the microscope depends on the energy of the electrons, which determines the corresponding wavelength of the electrons. The minimum focal point of the electron beam depends on its wavelength. Determine the electron wavelength if they have an energy of (a) 100 keV and (b) 1 MeV.

2.3 *Spring constant and interatomic distance between H atoms in H_2.* The fundamental vibrational frequency of the H_2 molecule is 4401 cm^{-1} and the rotational constant is 59.32 cm^{-1}. Estimate the effective spring constant and the interatomic distance between the two hydrogen atoms. What are the photon wavelength and frequency corresponding to the vibration transition?

2.4 *Expectation value of Hamiltonian.* Prove that E in the separation of variables of the time-dependent Schrödinger equation represents the system energy; in other words, prove eq. (2.28).

2.5 *Particle Flux.* Derive the material wave continuity equation (2.30) and the flux expression (2.31).

2.6 *Photon emission wavelength.* Calculate the emitted photon wavelength if an electron falls from the $n = 2$ state into $n = 1$ state inside an infinite potential quantum well of width $D = 20$ Å.

2.7 *Heisenberg uncertainty principle.* For the $n = 2$ state of an electron inside an infinite potential well, prove that the Heisenberg uncertainty relation $\Delta p \Delta x \geq \hbar/2$ is satisfied.

2.8 *Spring constant of C—O bonds.* The absorption by a CO molecule at 5.61 μm is its fundamental vibrational mode. Determine the effective spring constant of C—O bonds.

2.9 *Vibrational-rotational energy levels.* The fundamental vibrational frequency of the H_2 molecule is 4401 cm^{-1} and its rotational constant is 59.32 cm^{-1}. Determine the photon emission wavelengths due to combined vibrational–rotational modes in H_2 near the fundamental vibrational mode.

2.10 *Electron reflection.* As shown in figure P2.10, an electron of energy E moving from left to right encounters a potential barrier of height δ. The electron wave can be reflected or transmitted.

(a) Show that the proper forms of the incoming, reflected, and transmitted wave functions are

$$\Psi_i = Ae^{-i(\omega t - k_1 x)}, \quad \Psi_r = Be^{-i(\omega t + k_1 x)}, \quad \Psi_t = Ce^{-i(\omega t - k_2 x)},$$

respectively, where

$$k_1 = \sqrt{\frac{2mE}{\hbar^2}} \text{ and } k_2 = \sqrt{\frac{2m(E-\delta)}{\hbar^2}}$$

and A, B, C are constants to be determined from the interface conditions at $x = 0$.

(b) At the interface $x = 0$, the wavefunction and its first derivative must be continuous. Using these conditions, derive expressions for B/A and C/A.

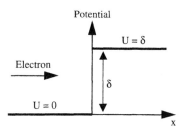

Figure P2.10 Figure for problem 2.10.

(c) The reflectivity R is defined as the ratio of the reflected particle flux divided by the incoming particle flux, and similarly for transmissivity T,

$$R = \frac{J_r}{J_i} \text{ and } T = \frac{J_t}{J_i}$$

Derive expressions for R and T.

(d) For $E = 1.2$ eV and $\delta = 1$ eV, calculate the electron reflectivity and transmissivity.

(e) For $E = 1$ eV and $\delta = 1.2$ eV, show that the transmissivity T is zero, and also show that Ψ_t is not zero. This nonzero wavefunction that does not carry a material flux is called an evanescent wave.

2.11 *Electron tunneling through a potential barrier.* An electron moving from left in region I encounters a potential barrier of finite width D and barrier height

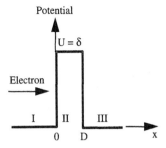

Figure P2.11 Figure for problem 2.11.

δ (region II), as shown in figure P2.11. The electron has a certain probability of traversing the potential barrier and entering region III. The objective of this exercise is to derive expressions for the electron reflectivity and transmissvity. To complete the derivation, go through the following steps:

(a) Solve the Schrödinger equation for each region. Identify which parts of the solution represent the incoming, reflected, and transmitted waves.

(b) Use the continuity of the wavefuction and its first-order derivative at the two interfaces to relate the solutions in the three regions.

(c) Use the definitions of reflectivity and transmissivity that are discussed in problem 2.10 to derive expressions for the electron reflectivity and transmissivity through the barrier region.

(d) Examine the solutions and show that even if the incoming electron energy E is lower than the Barrier height δ, there is still a nonzero probability that the transmissivity is not zero. The phenomenon that an electron with energy lower than the barrier height can transverse the barrier is called tunneling.

2.12 *Electron energy quantization in a potential well of finite barrier height.*

(a) Derive expressions determining the electron energy levels in a potential well surrounded by a barrier of finite height, as shown in figure P2.12.

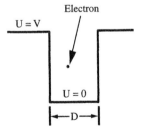

Figure P2.12 Figure for problem 2.12.

(b) For $D = 50$ Å, determine the first three energy levels for $V = 0.5$ eV and 1 eV.

2.13 *Electron energy states in a quantum dot.* Determine (a) the energy levels of an electron in a cubic quantum dot of length D, assuming an infinitely high potential barrier around the cube, (b) the allowable energy levels for $D = 100$ Å, and (c) the degeneracy of the first four energy levels.

2.14 *Electron energy states in a potential wire.* Determine (a) the energy levels of an electron in a two-dimensional square box of length D, assuming an infinitely high potential barrier around the box, (b) the allowable energy levels for $D = 50$ Å, and (c) the degeneracy of the first four energy levels.

2.15 *Wave function in a one-dimensional potential well.* Plot the most probable electron distribution in a one-dimensional infinite potential well for $n = 1, 2, 3$.

2.16 *Degeneracy of electron energy levels in a hydrogen atom.* Prove that the electron degeneracy in a hydrogen atom is $2n^2$.

2.17 *Translational energy level.* A 10 cm^3 box contains a H_2 molecule at 300 K.
    (a) Estimate the average translational energy of the H_2 molecule.
    (b) How do the first few translational energy levels compare with $\kappa_B T$?
    (c) Can you think about a way to count the degeneracy of the translational energy levels corresponding to this average energy?

3

Energy States in Solids

The previous chapter introduced the energy levels in simple potential fields, such as quantum wells, harmonic oscillators, atoms, and molecules. In this chapter, we will discuss energy levels in solids. We focus our discussion on single crystals, which are the simplest form of solids because the atoms are regularly arranged. As we will see, crystal periodicity plays a central role in determining the energy levels. So we will start by discussing crystal structures, including lattices and the potentials binding the atoms into a crystal. Since atoms in solids are packed closely, the electron wavefunctions of individual atoms overlap and form new wavefunctions and, correspondingly, new electron energy levels. The interatomic forces bond nuclei together so that the vibration of the atoms inside the crystal is strongly coupled. The collective atomic vibration can be decomposed into normal modes extending over the crystal and the basic energy quantum of each normal mode is called a phonon, in the same way of that the basic energy quantum of an electromagnetic mode is called a photon. Each electron and phonon wavefunction is characterized by a frequency (or energy) and a wavevector. The relationship between the energy and the wavevector is called the dispersion relation, which plays a central role in determining the properties of the crystal. We limit mathematical derivations to the dispersion relations of electrons and phonons in one-dimensional periodic structures and explain the energy levels in real crystals without a more detailed mathematical derivation, because the dispersion relation in a real crystal can be appreciated on the basis of a sound understanding of the behavior of a one-dimensional periodic system. The energy levels in crystals are highly degenerate. A very useful tool that takes into account the degeneracy of the energy states is the density of states, which will be used repeatedly throughout the book and should be mastered.

3.1 Crystal Structure

A perfect bulk crystal is a three-dimensional periodic arrangement of atoms. To describe a crystal, we use an atomless lattice—a periodic array of mathematical points that replicate the inherent periodicity of the actual crystal. Every point in the lattice is identical to other points. To form an actual crystal, a basis consisting of one or several atoms (or a molecule) is attached to each lattice point, i.e.

$$\text{crystal} = \text{lattice} + \text{basis} \qquad (3.1)$$

The exact position of the basis relative to the lattice point is not important, as long as the relative position between the basis and the lattice point is the same for all the lattice points. Many crystals have the same lattice structure; in fact, there is only a limited number of possible lattice types. Thus, we will first discuss the description of lattices in real space, followed by an introduction of the concept of reciprocal lattice, which is the Fourier transform of the real space lattice. The binding between atoms in real crystals will then be discussed.

3.1.1 Description of Lattices in Real Space

Consider a two-dimensional lattice as shown in figure 3.1. From a mathematical point of view, the location of each a point can be described by a vector. Because of the periodic arrangement of lattice points, we can choose a basic set of vectors, called the *primitive lattice vectors*, to construct all other vectors in the lattice. In a three-dimensional lattice, a_1, a_2, a_3 are primitive lattice vectors if, from any point, we could

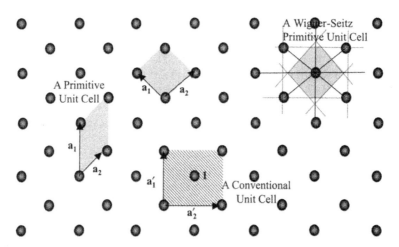

Figure 3.1 A two-dimensional lattice. Different choices of primitive lattice vectors a_1 and a_2 and primitive unit cells (gray areas) are possible. The Wigner-Seitz primitive unit cell is one way to uniquely construct a primitive unit cell. Vectors a'_1 and a'_2 are not a set of primitive lattice vectors and the shaded area is not a primitive unit cell. This area is, however, often used due to its regular shape and is called a conventional unit cell.

reach all other lattice points by a proper choice of integers through the following construction

$$\mathbf{R} = n_1\mathbf{a}_1 + n_2\mathbf{a}_2 + n_3\mathbf{a}_3 \qquad (n_1, n_2, n_3 \text{ cover all integers}) \qquad (3.2)$$

The magnitudes of $\mathbf{a}_1$, $\mathbf{a}_2$, and $\mathbf{a}_3$ are called the lattice constants. A lattice constructed according to eq. (3.2) is often called a Bravais lattice. Primitive lattice vectors are not unique. We have drawn two sets of primitive lattice vectors in figure 3.1 with primitive unit vectors denoted by $\mathbf{a}_1$ and $\mathbf{a}_2$. The other set of vectors, $\mathbf{a}'_1$ and $\mathbf{a}'_2$, are not primitive lattice vectors because we cannot use them to construct all other lattice points by a two-dimensional version of eq. (3.2). For example, we cannot reach point 1 through any linear integer combination of $\mathbf{a}'_1$ and $\mathbf{a}'_2$.

A *primitive unit cell* is the parallelepiped defined by the primitive lattice vectors. There is only one lattice point (equivalently speaking) per primitive unit cell. For example, each of the four lattice points in the two parallelograms formed by the two sets of primitive lattice vectors in figure 3.1 is shared by four unit cells and thus the number of equivalent lattice points in each parallelogram is one. These are thus primitive unit cells. On the other hand, the shaded rectangle formed by $\mathbf{a}'_1$ and $\mathbf{a}'_2$ is not a primitive unit cell because there are two lattice points in such a rectangle: the center point plus the four corners, each of the latter being shared by four cells. Because the choice of primitive lattice vectors is not unique, there can be different ways to draw a primitive unit cell, as shown by the two examples in figure 3.1. One method to construct a unit cell uniquely is the Wigner–Seitz cell (see figure 3.1), which is constructed by connecting all the neighboring points surrounding an arbitrary lattice point (as shown by the solid lines in figure 3.1) and drawing the bisecting plane (shown by dashed lines in the figure) perpendicular to each connection line. The smallest space formed by all the bisecting planes is a Wigner–Seitz cell, as indicated in the figure.

Sometimes, it is more convenient to describe a lattice by the *conventional unit cell*. For example, in figure 3.1, the rectangle formed by $\mathbf{a}'_1$ and $\mathbf{a}'_2$ is more convenient than the parallelogram formed by the primitive lattice vectors. This unit cell has two lattice points and is called a conventional unit cell. The crystal can also be constructed by repeating such a cell.

A general unit cell in the three-dimensional space is designated by three lattice vectors and the three angles formed between them. In the most general case, these three lattice vectors are of different lengths and the three angles are all oblique, as shown in figure 3.2(a). This lattice is called a triclinic lattice and does not have much symmetry. The symmetry of a lattice is often characterized by the symmetry operations, which include rotation of the unit cell around a fixed lattice point, reflection of the unit cell along a specific plane, and inversion with respect to a lattice point. A fundamental requirement on the lattice is that one can fill the entire space by placing a primitive unit cell at every lattice point. This requirement puts a limitation on the symmetric operations of a lattice. For example, the allowable rotational symmetry operations are 2π, π, $2\pi/3$, $2\pi/4$, and $2\pi/6$. No lattice, however, can have $2\pi/7$ or $2\pi/5$ rotational symmetry.* Given these conditions, it turns out that there are 13 other types of lattice that have special symmetry operations on top of the π and 2π rotational symmetry of

---

*Some quasicrystals can have five-fold symmetry patterns but they do not satisfy the definition of a crystal discussed in this section (Kittel, 1996).

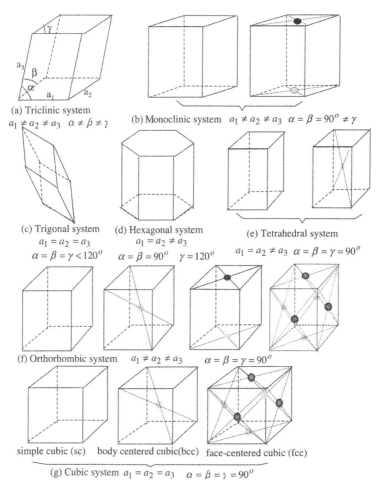

Figure 3.2 Fourteen Bravais lattices and the associated seven systems.

a triclinic lattice, so that there are in total 14 types of Bravais lattice. These 14 lattices can be further grouped into seven types of point symmetry operations (seven crystal systems) or operations around a fixed lattice point (such as rotation and inversion), as shown in figure 3.2. For example, in the cubic system, there are three types of Bravais lattices, the simple cubic (sc), the body-centered cubic (bcc), or the face-centered cubic (fcc). The cube is a primitive unit cell only for the sc lattice and is a conventional unit cell for the bcc and fcc lattices.

Certain crystal planes inside a lattice are identical. These planes are parallel to each other and equally spaced. A common convention for indexing the crystal planes and the high symmetry directions is in terms of the Miller indices. The *Miller indices of crystal planes*, represented by a set of integers in parentheses (hkl), are obtained in accordance with the following steps:

1. Find the intercepts of the crystal plane with the axes formed by the lattice vectors $\mathbf{a}_1$, $\mathbf{a}_2$, $\mathbf{a}_3$ in terms of the lattice constants. The origin of the lattice vectors can be

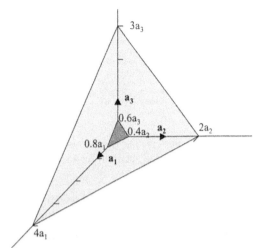

Figure 3.3 Two identical planes with Miller indices (364) in a crystal.

at any lattice point. One can choose any crystal plane that is convenient to use. For example, in figure 3.3 we have two crystallographically identical planes; one intercepts the axis at $0.8a_1, 0.4a_2, 0.6a_3$, and the other at $4a_1, 2a_2, 3a_3$.

2. Take the reciprocal of the intercepts and reduce these reciprocals to the three smallest integers that have the same ratio as the original set. The result is enclosed in parentheses (hkl) and this set of numbers is called the Miller indices of the plane. The example in figure 3.3 yields

$(1/0.8, 1/0.4, 5/3)$ (for inner plane) or

$(1/4, 1/2, 1/3)$ (for outer plane) $\rightarrow$ (364)

If the plane intersects at the negative side of the chosen primitive lattice vector, a line is placed above the number. For example, the six square faces of a cubic unit cell [figure 3.2(g)] are (100), (010), (001), ($\bar{1}$00), (0$\bar{1}$0), (00$\bar{1}$). We can use the sign {100} to denote all the six equivalent planes. The direction index in a crystal is denoted by a set of smallest integers $[uvw]$ proportional to the unit vector in the desired direction. All equivalent directions can be denoted by $\langle uvw \rangle$. Based on these definitions, one can prove that the (hkl) plane is perpendicular to the $[hkl]$ direction.

Crystal planes and directions are often determined by using X-rays or through transmission electron microscopy. Semiconductor wafers are sold with the major crystallographic directions and dopant types marked by the wafer flats. A wafer typically has a primary flat representing a crystal plane and a secondary flat that is positioned to denote the dopant type and the surface crystallographic direction, as shown in figure 3.4 for a 4 in. silicon wafer.

3.1.2 Real Crystals

By attaching a basis, which can be one or several atoms or a molecule, to each lattice point, real crystals are formed. For example, silicon has an fcc structure and the basis is made of two silicon atoms. If we anchor one atom at the fcc lattice point, for example

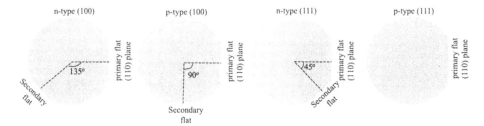

Figure 3.4 Semiconductor wafers are sold with major crystallographic indices and dopant type denoted by primary and secondary flats. These are common notations for 4 in. wafers.

at $(0, 0, 0)$, the other silicon atom is then at $(a/4, a/4, a/4)$ as shown in figure 3.5(a). In a conventional fcc unit cell, there are four lattice points and eight silicon atoms. In a primitive unit cell, there are two silicon atoms. The silicon crystal structure is called diamond structure, which is shared by several other crystals such as germanium and diamond. The zinc blende structure [Figure 3.5(b)], such as that of gallium arsenide (GaAs), is similar to the diamond structure but the basis is made of two different atoms, one Ga and one As atom for GaAs. If we take the Ga atom at $(0,0,0)$, then the As atom is at $(a/4, a/4, a/4)$. Since the Ga and the As atoms are different from one another, the zinc blende crystal structure has fewer symmetry operations than the diamond structure. Graphite has a close-packed hexagonal structure, as shown in figure 3.5(c).

Example 3.1 *Density of Si crystals*

Silicon is an fcc lattice with a lattice constant of 5.43 Å and two atoms per lattice site. Determine the density of Si crystals.

Solution: An fcc lattice has four lattice points. Since there are two Si atoms at each lattice point, there is a total of eight Si atoms per conventional fcc cell. Each atom weighs $28 \times 1.67 \times 10^{-27}$ kg, where 28 is the number of protons and neutrons,

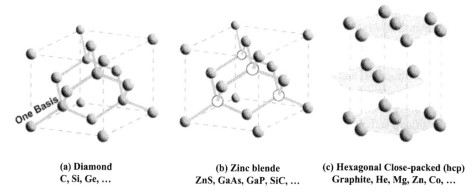

Figure 3.5 Three-types of common crystal structure: (a) diamond, (b) zinc blende, (c) close-packed hexagonal.

and 1.67×10^{-27} kg is the weight of a proton or neutron. The density of Si crystal is thus

$$\rho = \frac{8 \times 28 \times 1.67 \times 10^{-27} \text{kg}}{(5.43 \times 10^{-10} \text{ m})^3} = 2.34 \times 10^3 \text{ kg m}^{-3}$$

Real crystals also have *defects*, at which the periodicity of the crystal is disturbed. The defects can be divided into the following three types: points, lines, and planes. Examples of *point defects* are vacancies, where the atoms at the lattice points do not exist, and impurities, where the original atoms are substituted by different atoms. Another form of a point defect is an interstitial defect, where an additional atom is inserted in the space that does not belong to any allowed atomic site in the original lattice. Examples of *line defects* are *dislocations*. The two simplest types of dislocation are the *edge dislocation* and *screw dislocation*, as shown in figures 3.6(a) and (b). The edge dislocation can be constructed by inserting an extra plane of atoms in the upper half of the crystal. The screw dislocation can be thought of as the result of cutting the crystal partway through with a knife and shearing it parallel to the edge of the cut by one lattice vector. Dislocations can only form loops or be terminated at the surfaces (or interfaces), as is vortex in a fluid flow. The number of dislocations is measured by the dislocation density (dislocations/cm^2), which is the number of dislocation lines that intersect a unit area in the crystal. Typical values for the dislocation density are 10^8 cm^{-2}, and they vary significantly depending on how the materials are made. Because the dislocated regions are highly stressed, it takes a relatively small disturbance (shear stress) to move the location of the dislocation to the next lattice plane along the directions drawn in Figure 3.6. Thus crystals with dislocations have a lower plastic deformation limit. But this is not the whole story. In highly dislocated crystals, because there are many defects and dislocations along the path of the dislocation motion, the work needed to move these dislocations over other defects is much higher. Thus, poorly prepared crystals that have many dislocations can be harder than relatively good but yet not perfect (because the existence of dislocations) crystals. Examples of *planar defects* are the grain boundaries between two small crystalline regions inside a crystal, or polycrystals.

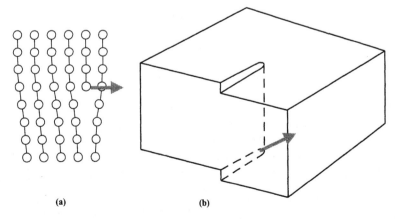

Figure 3.6 Illustrations of (a) edge and (b) screw dislocations. The arrows mark the direction of motion of the dislocation.

3.1.3 Crystal Bonding Potential

What holds the atoms together in a crystal? Fundamentally, it is the electron–electron and electron–nucleus interactions between atoms that hold them together. We touched on the topic of interatomic potential in chapter 1 and will now give a more detailed discussion, focusing on solids. More discussion on the interatomic potential will be given in chapter 10, for molecular dynamic simulations.

The force interaction between atoms *always* consists of a long-range attractive force and a short-range repulsive force. The short-range repulsive force is effective, due to the Pauli exclusion principle, when the inner-shell electrons or the nuclei of the atoms begin to overlap. Two often-used empirical expressions for the repulsive potential between the atoms separated by a distance r are

$$U_R(r) = \frac{B}{r^{12}} \quad \text{(Lennard–Jones)} \tag{3.3}$$

and

$$U_R(r) = U_0 e^{-r/\zeta} \quad \text{(Born–Mayer)} \tag{3.4}$$

where B, ζ, and U_0 are empirical constants determined from experimental data, such as the interatomic spacing and the binding energy. The differences between various types of crystals, however, are mainly due to the long-range attractive interaction, which will be discussed below.

Molecular crystals are characterized by the dipole–dipole interaction between atoms. An isolated atom is not polarized, but, when another atom is close by, the electrical field of electrons from the neighboring atom distorts the positions of the electrons and the nucleus of the current atom, creating an induced dipole. The attractive potential between the induced dipole of two atoms is given by

$$U_A = -\frac{A}{r^6} \tag{3.5}$$

Combining this attractive potential (van der Waals potential) with the Lennard–Jones potential for the repulsive force, we obtain the Lennard–Jones interaction potential between a pair of atoms i and j in a crystal as

$$U_{ij} = \frac{B}{r_{ij}^{12}} - \frac{A}{r_{ij}^{6}} \tag{3.6}$$

This potential is most appropriate for crystals of inert atoms (such as argon atoms which form a crystal at low temperatures) that have a neutral, spherically symmetric charge. Such a potential, however, can also be used to describe the interactions between similar atoms or molecules in liquidus or gaseous phases. A crystal structure is stable when the total potential energy of the system

$$U = \frac{1}{2}\sum_{i \neq j}\left(\frac{B}{r_{ij}^{12}} - \frac{A}{r_{ij}^{6}}\right) = \frac{1}{2}\sum_{i \neq j} 4\varepsilon \left(\left(\frac{\sigma}{r_{ij}}\right)^{12} - \left(\frac{\sigma}{r_{ij}}\right)^{6}\right) \tag{3.7}$$

reaches a minimum, as required by the second law of thermodynamics for a stable system. In eq. (3.7), $\sigma = (B/A)^{1/6}$ and $\varepsilon = A^2/(4B)$, and the factor of one-half is there because the potential is shared between two atoms and the summation double-counts the potential. Values of Lennard–Jones parameters for noble-gas crystals are given in table 3.1.

Table 3.1 Lennard–Jones potential parameters for noble-gas crystals

	Neon	Argon	Krypton	Xenon
Crystal structure	fcc	fcc	fcc	fcc
Lattice constant (Å)	4.46	5.31	5.64	6.13
ε (10^{-20} J)	0.050	0.167	0.225	0.320
ε (eV)	0.031	0.0104	0.014	0.0200
σ (Å)	2.74	3.40	3.65	3.98

Source: Ashcroft and Mermin, 1976.

Example 3.2 *Lattice constant*

Determine the lattice constant of an fcc crystal described by the Lennard–Jones potential in terms of σ in eq. (3.7).

Solution: We assume that the nearest neighbor distance is R. We first compute the potential energy for any one atom i interacting with the rest of the atoms in the crystal. The total energy of the crystal with N atoms is thus $N/2$ times this energy. From eq. (3.7), we have

$$U_{\text{tot}} = \frac{(4\varepsilon)N}{2} \sum_j \left(\left(\frac{\sigma}{Rp_{ij}}\right)^{12} - \left(\frac{\sigma}{Rp_{ij}}\right)^{6} \right) \quad (E3.2.1)$$

where p_{ij} is the interatomic distance in terms of the neighbor distance R. For an fcc crystal, we can deduce that

$$\sum_j p_{ij}^{-12} = 12.13188 \qquad \sum_j p_{ij}^{-6} = 14.45392 \quad (E3.2.2)$$

The lattice constant is the point when U_{tot} is minimum. Thus,

$$\left[\frac{dU_{\text{tot}}}{dR}\right]_{R=R_0} = -2N\varepsilon \left[12 \times 12.13 \frac{\sigma^{12}}{R_0^{13}} - 6 \times 14.45 \frac{\sigma^6}{R_0^7} \right] = 0 \quad (E3.2.3)$$

which leads to

$$R_0/\sigma = 1.09 \quad (E3.2.4)$$

The observed values for Ne, Ar, Kr, Xe are $R_0/\sigma = $ 1.14, 1.11, 1.10, 1.09, very close to the calculation.

In *ionic crystals*, such as NaCl, the single valence electron in the sodium atom moves to the chlorine atom so that both Na$^+$ and Cl$^-$ have closed electron shells but, meanwhile, become charged. The Coulomb potential among the ions becomes the major attractive force. The potential energy of any ion i in the presence of other ions j is then

$$U_{i,A} = \sum_{i \neq j} \frac{\pm q^2}{4\pi \varepsilon_0 r_{ij}} = -\frac{\alpha q^2}{4\pi \varepsilon_0 r_0} \tag{3.8}$$

where q is the charge per ion, ε_0 the dielectric permittivity of free space, and r_0 the nearest-neighbor separation. The parameter α is called the Madelung constant and is related to the crystal structure. This attractive potential, combined with an appropriate repulsive potential, constitutes the total potential energy in ionic crystals.

Covalent bonds are formed when electrons from neighboring atoms share common orbitals, rather than being attached to individual ions as in ionic crystals. Biological molecules are often formed through covalent bonding. Many inorganic systems are also covalently bonded, such as the H$_2$ molecule. The electron in each hydrogen atom of an H$_2$ molecule shares a common orbital (one spin-up and the other spin down) with the other electron in the other H atom. The covalent bond is strongly directional. In the case of the H$_2$ molecule, the bond is oriented along the line of the two nuclei. Diamond, silicon, and germanium are all *covalent crystals*. Each atom has four electrons in the outer shell and forms a tetrahedral system of covalent bonds with four neighboring atoms, as indicated in figure 3.5(a). In certain crystals, such as GaAs, both covalent and ionic bonding are important. Fundamentally, the covalent bonding force is also due to charge interaction. However, unlike the van der Waals force in molecular crystals or the electrostatic force in ionic crystals, it is more difficult to construct simple expressions for covalent crystals. Empirical potentials have been developed, such as the Stillinger–Weber potential for silicon (Stillinger and Weber, 1985). Expressions for various empirical potentials will be presented in chapter 10.

In covalent bonds, electrons are preferentially concentrated in regions connecting the nuclei, leaving some regions in the crystal with low charge concentration, as illustrated in figure 3.7(a). Metals and their associated *metallic bonds* can be considered an extreme case of covalent bonds, in which the bonds begin to overlap and all regions of the crystal become filled up with charges [figure 3.7(b)]. In the case of total filling of the empty space, it becomes impossible to tell which electron belongs to which atom. One can imagine the entire crystal as one big molecule with the electrons shared amongst all

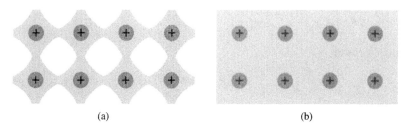

(a)   (b)

Figure 3.7 Distribution of electrons (gray area) in (a) a covalent bonding crystal and (b) a metallic bonding crystal (after Ashcroft and Mermin, 1970).

the atoms. In this case we say the electrons are delocalized and can wander throughout the crystal; in other words, they become free electrons. Later, we will see how the delocalization of electrons can be explained by solving the Schrödinger equation.

3.1.4 Reciprocal Lattice

We know that any periodic function can be expanded as a Fourier series. For example, if a time-dependent function $f(t)$ is periodic with a period of T, it can be expanded into a Fourier series as

$$f(t) = \sum_{n=-\infty}^{\infty} \left(a_n \sin\left(\frac{2\pi n}{T}t\right) + b_n \cos\left(\frac{2\pi n}{T}t\right) \right)$$
$$= \sum_{n=-\infty}^{\infty} (a'_n e^{in\omega t} + b'_n e^{-in\omega t}) \qquad (3.9)$$

where a_n and b_n are the coefficients of the Fourier series and a'_n and b'_n in the second step can be obtained by expressing sine and cosine functions as complex exponential functions. The angular frequency $\omega = 2\pi/T$ is the Fourier conjugate of the temporal periodicity such that $e^{i\omega T} = 1$, which ensures that $f(t)$ is periodic for every $\Delta t = T$; that is, $f(t + T) = f(t)$.

In a crystal, the atoms are periodic in space and, consequently, we expect that some atom-related functions are periodic. The simplest example is the potential energy, which should have a periodicity corresponding to the unit cell. Let's first consider a one-dimensional lattice, with a lattice constant of a. A spatial-dependent function $f(x)$, with a periodicity a, $f(x) = f(x + a)$, can be similarly expanded into a Fourier series,

$$f(x) = \sum_{n=-\infty}^{\infty} (a'_n e^{ink_x x} + b'_n e^{-ink_x x}) \qquad (3.10)$$

where the wavevector, $k_x = 2\pi/a$, is the Fourier conjugate to spatial periodicity a. Using eq. (3.10), one can easily show that $f(x)$ repeats for every $\Delta x = a$; that is, $f(x + a) = f(x)$.

The above example is for a one-dimensional lattice. But crystals are three dimensional. How can we expand a function that is periodic over the three-dimensional crystal into a Fourier series? We will use the charge distribution in the crystal, $n(\mathbf{r})$, as an example. It should be invariant with any translational lattice vector $\mathbf{R}$; that is, $n(\mathbf{r} + \mathbf{R}) = n(\mathbf{r})$. We will first give the following answer and then show that the given Fourier expansion indeed satisfies the required periodicity,

$$n(\mathbf{r}) = \sum_{\mathbf{G}} n_{\mathbf{G}} e^{i\mathbf{r}\cdot\mathbf{G}} \qquad (3.11)$$

where $\mathbf{G}$ and the inverse transformation are given by

$$\mathbf{G} = m_1 \mathbf{b}_1 + m_2 \mathbf{b}_2 + m_3 \mathbf{b}_3 \qquad (m_1, m_2, m_3 \text{ are integers}) \qquad (3.12)$$

$$n_{\mathbf{G}} = \frac{1}{V} \int_{\text{unit cell}} n(\mathbf{r}) e^{-i\mathbf{r}\cdot\mathbf{G}} dV \qquad (3.13)$$

and $(\mathbf{b}_1, \mathbf{b}_2, \mathbf{b}_3)$ are conjugated to the primitive lattice vectors $(\mathbf{a}_1, \mathbf{a}_2, \mathbf{a}_3)$ through

$$\mathbf{b}_1 = 2\pi(\mathbf{a}_2 \times \mathbf{a}_3)/V, \quad \mathbf{b}_2 = 2\pi(\mathbf{a}_3 \times \mathbf{a}_1)/V, \quad \mathbf{b}_3 = 2\pi(\mathbf{a}_1 \times \mathbf{a}_2)/V \qquad (3.14)$$

where $V = \mathbf{a}_1 \bullet (\mathbf{a}_2 \times \mathbf{a}_3)$ is the volume of the primitive unit cell in real space. One can easily prove the following relations,

$$\mathbf{a_i} \bullet \mathbf{b_j} = 2\pi \delta_{ij} \text{ where } \delta_{ij} = \begin{cases} 1 & i = j \\ 0 & i \neq j \end{cases} \qquad (3.15)$$

and δ_{ij} is the Kronecker delta. With the above definitions, we show now that $n(\mathbf{r})$ is indeed invariant with any translational lattice vector in the real space $\mathbf{R}(= n_1\mathbf{a}_1 + n_2\mathbf{a}_2 + n_3\mathbf{a}_3)$, where n_1, n_2, and n_3 are integers:

$$n(\mathbf{r} + \mathbf{R}) = \sum_G n_G e^{i(\mathbf{r}+\mathbf{R})\bullet\mathbf{G}} = \sum_G n_G e^{i\mathbf{r}\bullet\mathbf{G}+i\mathbf{R}\bullet\mathbf{G}}$$
$$= \sum_G n_G e^{i\mathbf{r}\bullet\mathbf{G}+i2\pi(n_1m_1+n_2m_2+n_3m_3)} = \sum_G n_G e^{i\mathbf{r}\bullet\mathbf{G}} = n(\mathbf{r})$$

We used eq. (3.15) in the third step and the fact that m_i and n_i are integers in the fourth step. Thus we see that the new set of vectors introduced, $(\mathbf{b}_1, \mathbf{b}_2, \mathbf{b}_3)$, which has a unit of m^{-1}, is the corresponding Fourier conjugate to the real space lattice vector $(\mathbf{a}_1, \mathbf{a}_2, \mathbf{a}_3)$. We can use $(\mathbf{b}_1, \mathbf{b}_2, \mathbf{b}_3)$ and eq. (3.12) to construct a new lattice called the *reciprocal lattice*, with **G** as the reciprocal lattice vector. Previous definitions on real space lattices, such as unit cells and the Wigner–Seitz primitive unit cell, are equally applicable to such a reciprocal lattice. This reciprocal space is the Fourier conjugate of the real space. Figures 3.8(a) and (b) show a primitive unit cell of an fcc lattice in real space and its corresponding reciprocal space, based on the Wigner–Seitz construction. The Wigner–Seitz primitive unit cell in reciprocal space is also called the first Brillouin zone, which will be used extensively later to represent the electron and phonon energy levels in solids. Although the representation of crystal properties in the reciprocal lattice may be unfamiliar to some readers, they undoubtedly have seen the spectrum representation of time-varying

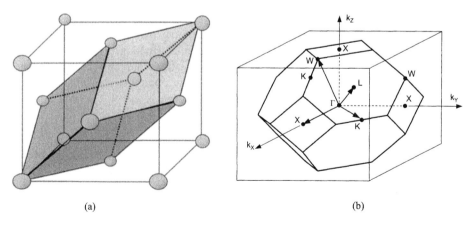

Figure 3.8 Conventional and primitive unit cells in real and reciprocal unit cells of an fcc lattice (a) in real space, (b) in reciprocal space.

ENERGY STATES IN SOLIDS 89

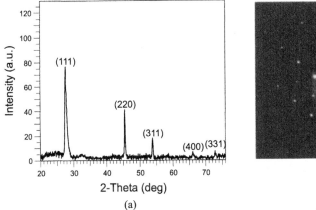

Figure 3.9 (a) X-ray 2-theta scan of a germanium crystal and (b) an electron diffraction pattern for a bismuth nanowire (Courtesy of Dr. Z.F. Ren and Dr. M.S. Dresselhaus).

signals in electrical engineering, or the spectral-dependent radiative properties of matter, which conveniently express time-dependent electrical signals or electromagnetic fields into stationary spectral properties through Fourier transformations. The representation of properties of crystals in the reciprocal space assumes a similar role. Specific symbols have been designated for different directions of the reciprocal lattice. For example, in figure 3.8(b), the Γ point is the center of the Brillouin zone. The Γ–X direction represents the [100] direction of the real lattice, and Γ–L the [111] direction of the real lattice.

Although a very abstract concept, the reciprocal lattice can actually be easily mapped out with diffraction experiments. When electron waves or X-rays (electromagnetic waves) with the proper energy are directed onto a crystal, the reflection or transmission occurs only along specific directions, as shown in figure 3.9(a) (X-ray reflection) and figure 3.9(b) (electron transmission). Such phenomena, known as diffraction, can be attributed to the superposition of the incident waves and scattered waves. Consider an incident wave (an electron beam or an X-ray beam) from the source along direction $\mathbf{k}$. At any point in the crystal, the magnitude of the wave is proportional to $e^{i\mathbf{k}\cdot(\mathbf{r}-\mathbf{r}_s)}$, where $\mathbf{r}_s$ is the location of the source relative to the origin of coordinates, as shown in figure 3.10(a). The wave scattered into the detector is then proportional to $n(\mathbf{r})e^{i\mathbf{k}'\cdot(\mathbf{r}_d-\mathbf{r})}$, where $\mathbf{k}'$ is the propagation direction of the scattered wave and $\mathbf{r}_d$ is the position of the detector. Because each atom scatters the incident wave, the total magnitude of the wave, S, at the detector is

$$S \propto \int_{\text{whole crystal}} e^{i\mathbf{k}\cdot(\mathbf{r}-\mathbf{r}_s)} n(\mathbf{r}) e^{i\mathbf{k}'\cdot(\mathbf{r}_d-\mathbf{r})} dV = e^{i(\mathbf{k}'\cdot\mathbf{r}_d - \mathbf{k}\cdot\mathbf{r}_s)} \int_{\text{whole crystal}} n(\mathbf{r}) e^{i(\mathbf{k}-\mathbf{k}')\cdot\mathbf{r}} dV$$

$$= \sum_{\mathbf{G}} e^{i(\mathbf{k}'\cdot\mathbf{r}_d - \mathbf{k}\cdot\mathbf{r}_s)} \int_{\text{whole crystal}} n_{\mathbf{G}} e^{i(\mathbf{G}+\mathbf{k}-\mathbf{k}')\cdot\mathbf{r}} dV \qquad (3.16)$$

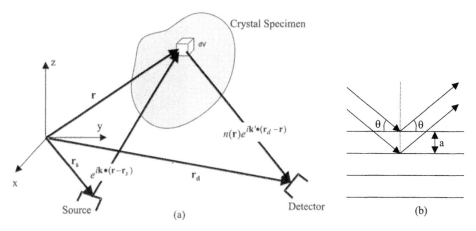

Figure 3.10 Derivation of the Bragg diffraction condition for (a) the general case and (b) the one-dimensional case.

where we have used eq. (3.11) to expand $n(\mathbf{r})$ into a Fourier series. Because the exponential function $e^{i(\mathbf{G}+\mathbf{k}-\mathbf{k}')\cdot\mathbf{r}}$ is a rapidly varying function in the crystal with both negative and positive values, the above integral will be close to zero except when the exponent vanishes, that is, when

$$\mathbf{G} + \mathbf{k} - \mathbf{k}' = 0 \qquad (3.17)$$

Equation (3.17) is called the Bragg condition for diffraction and determines the relationship between incident ($\mathbf{k}$) and diffracted ($\mathbf{k}'$) waves. The wavevectors $\mathbf{k}$ and $\mathbf{k}'$ are determined by the relative positions of the source, the sample, and the detector, and therefore the reciprocal lattice vectors $\mathbf{G}$, and thus the crystal structure, can be determined from diffraction experiments.

The Bragg condition, eq. (3.17), was derived by considering the scattering of individual atoms. We can treat each crystal plane as a continuous sheet and establish an equivalent condition based on the interference of the waves reflected from all the parallel crystal planes in a specific direction. Consider the special set of crystal planes separated by a distance a, as shown in figure 3.10(b), and an incident wave (photon or electron) of wavelength λ at an angle θ. Constructive interference between waves reflected from crystal planes occurs when the phase difference of the waves scattered between two consecutive planes is πn. From figure 3.10(b), we see that the path difference is $2a \sin\theta$. Thus diffraction peaks will be observed when the path difference is multiples of the wavelength,

$$2a \sin\theta = n\lambda \qquad (3.18)$$

It can be shown that eq. (3.18) is a special case of eq. (3.17). The proof is left as an exercise.

Example 3.3 *X-ray diffraction*

One way of using X-ray diffraction to determine the crystal structure is the rotating crystal method. In this method, an X-ray of fixed wavelength λ is directed onto the sample at a fixed direction, and the diffracted X-ray is measured by a fixed detector; refer to figure 3.10(a). The crystal is rotated to change the angle of incidence θ [figure 3.10(b)] with respect to a special crystal plane. When the Bragg condition is not satisfied, the detector will register very little signal. But when crystals are rotated to the positions where the Bragg condition, eq. (3.18), is satisfied, the detector will register a peak. A typical scan curve is shown in figure 3.9(a). Different peaks correspond to different crystal planes. For an X-ray of wavelength 1 Å and a first-order diffraction peak at $\theta = 30°$, the corresponding spacing between the two crystal planes is

$$a = \frac{1\text{Å}}{2 \sin 30°} = 1\text{Å}$$

3.2 Electron Energy States in Crystals

In the previous chapter, we discussed the energy levels of single atoms and harmonic oscillators. These energy levels are typically discrete. In solids, the wavefunctions of closely spaced atoms begin to overlap and form new wavefunctions and, correspondingly, new energy levels. We will see that the energy levels become more continuous than those of individual atoms. This trend can be thought of as the result of the broadening of the energy levels of individual atoms to avoid the overlapping of wavefunctions because, according to the Pauli exclusion principle, each quantum state can have only a maximum of one electron. In crystals, the most fundamental characteristic is the periodicity of the lattice. Such periodicity brings in many new features to the allowable energy levels of electrons as well as phonons. In this section, we will start from a simple one-dimensional model to examine the effect of periodicity on the electronic energy levels and then extend the discussion to three-dimensional crystals.

3.2.1 One-Dimensional Periodic Potential (Kronig–Penney Model)

Let us consider a simple one-dimensional lattice. The potential field is a periodic function, as sketched in figure 3.11(a). At the location of each ion, the electrons are attracted by the ion and have the lowest potential. As an approximation to the actual atomic potential distribution in a crystal as in figure 3.11(a), we consider a square periodic potential as shown in figure 3.11(b) and want to find out the energy levels, assuming there is only one electron inside such a periodic potential. As in the case that the hydrogen energy level can be used to explain the periodic table, the existence of many electrons in a

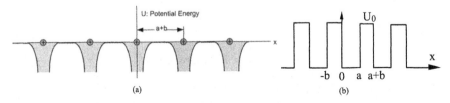

Figure 3.11 One-dimensional periodic potential model: (a) sketch of atomic potential; (b) Kronig-Penney model.

crystal does not change the main picture obtained from the one-electron assumption. This one-electron, rectangular periodic potential model is called the Kronig–Penney model. The Schrödinger equation is then

$$-\frac{\hbar^2}{2m}\frac{d^2\Psi}{dx^2} + (U - E)\Psi = 0 \qquad (3.19)$$

The potential distribution $U(x)$ is given by

$$U(x) = \begin{cases} 0 & 0 < x \leq a \\ U_0 & -b < x \leq 0 \end{cases} \qquad (3.20)$$

subject to the following periodicity requirement

$$U(x + a + b) = U(x) \qquad (3.21)$$

Solutions for eq. (3.19) are

$$\Psi = Ae^{iKx} + Be^{-iKx} \quad (0 < x \leq a) \qquad (3.22)$$

$$\Psi = Ce^{Qx} + De^{-Qx} \quad (-b \leq x \leq 0) \qquad (3.23)$$

where

$$E = \frac{\hbar^2 K^2}{2m} \text{ and } U_0 - E = \frac{\hbar^2 Q^2}{2m} \qquad (3.24)$$

and K and Q are to be determined, from which the eigen energy E of the electron inside such a periodic potential is to be extracted.

Four boundary conditions are needed to determine the unknown coefficients A, B, C, and D. We can use the continuity of the wavefunction and its derivative at $x = 0$, which gives

$$A + B = C + D \qquad (3.25)$$

$$iK(A - B) = Q(C - D) \qquad (3.26)$$

Two more boundary conditions are necessary to determine the four unknown coefficients. We can consider the continuity of the wavefunction and its derivative at $x = a$, but this requires that we know the wavefunction in $a < x \leq a + b$. The wavefunction in this region can be related to that in the region $-b < x < a$ because the potential is periodic. Due to the periodicity in the potential, the wavefunction at any two points separated by a lattice vector is related through the *Bloch theorem*,

$$\Psi(\mathbf{r} + \mathbf{R}) = \Psi(\mathbf{r}) \exp(i\mathbf{k} \bullet \mathbf{R}) \qquad (3.27)$$

where $\mathbf{R}$ is a lattice vector and $\mathbf{k}$ is the wavevector of the crystal. The Bloch theorem implies that the wavefunction values at two equivalent points ($\mathbf{r}$ and $\mathbf{r} + \mathbf{R}$) inside a crystal differ by only a phase factor $\exp(i\mathbf{k} \bullet \mathbf{R})$ and that we need to know only the wavefunction inside one unit cell. For the one-dimensional problem being considered, denoting k as the magnitude of **k** in the x-direction, eq. (3.27) is

$$\Psi[x + (a + b)] = \Psi(x) \exp[ik(a + b)] \quad (3.28)$$

We should distinguish the wavevector k from the propagation vector of the solution, K in eq. (3.22). The latter contains the energy of the electrons that we want to find. We will explain later, in more detail, the meaning of wavevector k. We want to find a relation between k and E, which is equivalent to a relation between k and K.

From Bloch's theorem, we know that if the wavefunction for $-b < x < 0$ is given by eq. (3.23), the wavefunction for $a < x < a+b$ is then given by eq. (3.23) multiplied by $\exp[ik(a+b)]$. The continuity requirements for the wavefunction and its derivative at $x = a$ are then

$$Ae^{iKa} + Be^{-iKa} = (Ce^{-Qb} + De^{Qb}) \exp[ik(a+b)] \quad (3.29)$$

$$iK(Ae^{iKa} - Be^{-iKa}) = Q(Ce^{-Qb} - De^{Qb}) \exp[ik(a+b)] \quad (3.30)$$

Now we have four equations, eqs. (3.25), (3.26), (3.29), (3.30), and four unknowns, A, B, C, D. Examining these equations indicates that this is a set of linear homogeneous equations and is thus again an eigenvalue problem, and a solution exists only when the determinant of the coefficients A, B, C, and D equals zero. From this condition, we arrive at the following equation

$$\frac{Q^2 - K^2}{2KQ} \sinh(Qb)\sin(Ka) + \cosh(Qb)\cos(Ka) = \cos[k(a+b)] \quad (3.31)$$

where "$\sinh(x)$" and "$\cosh(x)$" are hyperbolic sine and cosine functions. For a given wavevector k, the only unknown in the above equation is the electron energy E, which is embedded in both K and Q. Thus the above equation can be used to determine a relationship between E and k. To get a better idea of what the solution looks like, let's assume $b \to 0$ and $U_0 \to \infty$, but keep $Q^2 ba/2 (= P)$ equal to a constant. Under this approximation, $\sinh(Qb) \approx Qb$, and $\cosh(Qb) \approx 1$. Equation (3.31) reduces to

$$\frac{P}{Ka} \sin Ka + \cos Ka = \cos ka \quad (3.32)$$

We can solve the above equation for (Ka) as a function of (ka) and use eq. (3.24) to find out allowable energy E from K. One important observation is that the magnitude of the left-hand side of eq. (3.32) can be larger than 1 whereas the right-hand side cannot. Therefore, the equation has no solution for those values of K (and thus energy) where the absolute values of the left-hand side are larger than one. A graphical representation of the left-hand side is shown in figure 3.12, where the right-hand side is bounded within $[-1, 1]$. In the shaded region, there is no solution for K, and thus no electrons with energies corresponding to such K values exist. We can convert the solution for K into energy, and redraw the graph as a function of ka as shown in figure 3.13(a). The figure shows that, for each wavevector k, there are multiple values for the electron energy E.

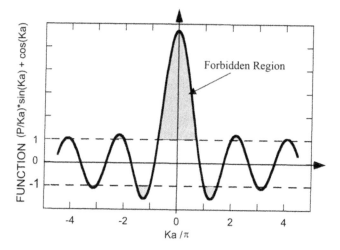

Figure 3.12 Left-hand side of eq. (3.32) as a function of Ka/π. Because the right-hand side is always less than or equal to one, there are regions (the shaded area) where no solution for Ka/π exists, and thus no electrons exist with energy corresponding to the values of K in these regions.

The electron energy forms quasi-continuous bands (because k itself is quasi-continuous) separated from each other by a minimum gap that occurs at $ka = s\pi$ ($s = 0, \pm1, \pm2, \ldots$), or $k = s\pi/a$, at which the right-hand side of eq. (3.32) is ±1. Figure 3.13(a) implies that there are multiple values of k for each energy E. However, the Bloch theorem, eq. (3.28), says that wavefunctions correspond to the wavevectors k separated by $m(2\pi/a)$ (since $b = 0$) are identical, they are the same quantum state and should be counted only once. Thus, rather than plotting the energy eigenvalues for all the wavevectors, we can plot them in one period, as shown in figure 3.13(b). This

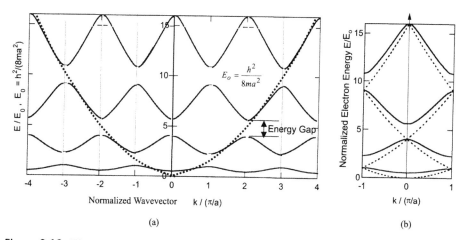

Figure 3.13 Electron energy as a function of its wavevector: (a) extended zone representation; (b) reduced zone representation. Dashed lines are free electron energy levels. Solid lines from Kronig–Penney model.

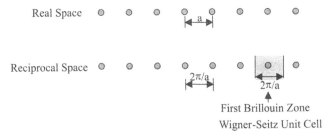

Figure 3.14 The Brillouin zone and Wigner–Seitz unit cell of a one-dimensional lattice.

way of representation is called the *reduced-zone* representation. Often, only half of the band, $[0, \pi/a]$, needs to be drawn because the band is symmetric for both positive and negative wavevector values. The relationship between the energy and the wavevector, as exemplified in figure 3.13(b), is the dispersion relation.

We see that the minimum separation between two energy bands occurs at $k_m (= s\pi/a, s = 0, \pm 1, \pm 2, \ldots)$. What do these k_m stand for and why do the minimum separations occur at k_m? For the one-dimensional lattice being considered with a lattice constant equal to a, its reciprocal lattice is also one-dimensional with a lattice constant equal to $2\pi/a$. The Wigner–Seitz cell in the reciprocal lattice, which is the first Brillouin zone as we explained before, is shown in figure 3.14. The boundaries of this primitive unit cell in the reciprocal space are at $\pm \pi/a$. Thus k_m represents the lattice vectors constructed using the primitive lattice vector of the Wigner–Seitz cell in the reciprocal space for the one-dimensional lattice. When we generalize to three-dimensional crystals, k_m will be replaced by the reciprocal lattice vector **G**. In most cases, the energy gap occurs at the Brillouin zone boundaries, that is, when **k** = **G**. This is not a coincidence since it results from the interference effects of electrons in periodic structures. This mechanism is not very different from the observation of diffraction peaks by X-ray and electron beams that we discussed in section 3.1.4. We also plotted the energy dispersion of a free electron in the reduced-zone representation in figure 3.13, which does not show an energy jump at k_m but is otherwise similar to that of an electron inside the periodic potential. The main effect of the periodic potential is to modify the band structure near k_m, as a result of the diffraction of the electron waves. More discussion on wave interference will be given in chapter 5.

We now determine the value of the wavevector k in the Bloch theorem, using the *Born–von Karman* periodic boundary condition. This boundary condition deals with the end points of a crystal. Ordinarily, we would think that the two end points are different from the internal points. For many applications, however, it is not necessary to distinguish the boundary points from the internal points, because a crystal usually has a tremendous number of lattice points (this is not true for quantum wells, quantum wires, and quantum dots). The Born–von Karman boundary condition requires that the wave functions at the two end points be equal to each other; that is, the two end points [points 1 and $N+1$ in figure 3.15(a)] are overlapped to form a lattice loop as shown in figure 3.15(b),

$$\Psi[x + N(a+b)] = \Psi(x) \qquad (3.33)$$

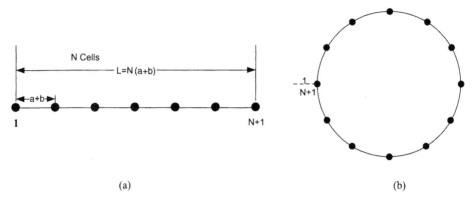

Figure 3.15 The von Karman boundary condition joins the two boundary points in (a) into a periodic loop in (b).

Using Bloch's theorem, eq. (3.23) can be written as

$$\Psi(x) = \Psi(x)\exp[ikN(a+b)] \tag{3.34}$$

The above equation imposes conditions on the allowable wavevectors of the Bloch wave inside the periodic potential

$$k = \frac{2\pi n}{N(a+b)} = \frac{2\pi n}{L} \quad (n = 0, \pm 1, \pm 2, \ldots) \tag{3.35}$$

where L is the length of the crystal.

What is the meaning of the wavevector k in the Bloch theorem? The exponential factor e^{ikx} in eq. (3.28) resembles that for a free electron, as we discussed in section 2.3.1 and eq. (2.34). In the latter case, $\hbar k$ represents the momentum of the free electron. The momentum of an electron in the crystal, however, should be calculated from the wavefunction using the momentum operator $-i\hbar\nabla\Psi$. Such a calculation would show that $\hbar k$ is not the momentum of the electron in a crystal. Nevertheless, in many ways $\hbar k$ for a periodic potential behaves as the momentum of a free electron and thus it is called the *crystal momentum*. The momentum conservation rule during the collision of several particles (now electrons, and later to be generalized to photons and phonons) is

$$\sum_i \hbar \mathbf{k}_i = \sum_f \hbar(\mathbf{k}_f + \mathbf{G}) \tag{3.36}$$

where the indices i and f mean the states before and after the collision, respectively. The addition of the reciprocal lattice vector maintains both $\mathbf{k}_i$ and $\mathbf{k}_f$ into the first Brillouin zone, and is a consequence of the identical wavefunctions and energy eigenvalues for waves with wavevector $\mathbf{k}$ and $\mathbf{k} + \mathbf{G}$.

The simple model of one electron inside a periodic potential carries many important messages which we will discuss below. This model shows that the electron does not belong to an individual atom. Its wavefunction extends over the whole crystal; therefore, it can be considered a free electron. In reality, the scattering due to the distortion of potential from the perfect periodicity reduces the spatial extension of the wavefunction.

ENERGY STATES IN SOLIDS   97

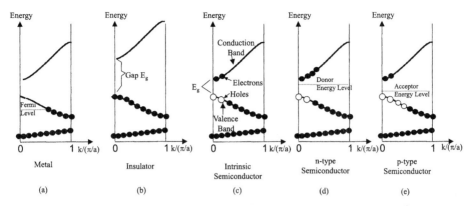

Figure 3.16 Explanation of metals, insulators, and semiconductors based on the one-dimensional band structure. (a) Electrons in metal partially fill a band. The top-most level (E_f) is called the Fermi level. (b) Electrons fill to the top of the band. When the energy gap E_g is large, no electrons can be excited to the next higher energy band and the material is an electrical insulator. (c) When the energy gap E_g is relatively small, some electrons can be thermally excited to the next higher energy band (called the conduction band), leaving the same number of empty states (holes) in the valence band. The material is an intrinsic semiconductor. (d) Impurities (more commonly called dopants) may have an energy level close to that of the conduction band. Electrons can be excited from the impurities and fall into the conduction band, resulting in more electrons than holes. Such a semiconductor is called an n-type semiconductor and the dopants are called donors. (e) When the impurity energy levels are close to the valence band, electrons are excited from the valence band into the impurity level, leaving more holes behind. Such semiconductors are called p-type and the impurities are called acceptors.

Yet still the mean free path of an electron can be as long as thousands of angstroms, and the number of atoms in a cube on the order of one mean free path is enormous, $\sim 10^6$ to $\sim 10^8$ atoms. It is amazing that an electron can zigzag through these atoms without getting scattered. Because of this behavior, we often treat electrons as a gas and neglect the ions completely, except when considering their occasional scattering effect.

Although the above solution is valid only for one electron, the existence of multiple electrons does not affect the qualitative picture of the energy bands, as long as the Coulomb potential between electrons is small compared to the potentials between electrons and ions. With such a simple picture of the energy bands, we can begin to understand the difference between insulators, metals, and semiconductors. In the first Brillouin zone, there are N allowable wavevectors for a lattice chain with N lattice points. Because each wavevector represents a wavefunction and each wavefunction can have a maximum of two electrons with different spins, each band can have a maximum of $2N$ electrons for a one-dimensional lattice. At zero temperature, the filling rule for the electrons is that they always fill the lowest energy level first, as required by thermodynamics. For alkali metals and noble metals that have one valence (free) electron per primitive cell, the band is only half filled since there are only N valence electrons in this case, as shown in figure 3.16(a). The topmost energy level that is filled with electrons at zero kelvin is called the *Fermi level*. The electron energy and momentum can be changed (almost) continuously within the same band because the separation between successive energy levels is small. Thus, the electrons in these metals can flow freely, making the materials good electrical conductors. If the valence electrons exactly fill one or more bands, leaving others empty [figure 3.16(b)], the crystal will be an insulator at zero kelvin and can be an insulator or a semiconductor at other temperatures, depending on the value of the minimum energy gap between the filled and the empty band. If a filled band is separated by a large energy gap (>3 eV) from the next higher unfilled band, one cannot change the energy and the momentum of an electron in the filled band easily; that is, these electrons cannot move freely and the material is an insulator [figure 3.16(b)]. A semiconductor is essentially similar to an insulator in terms of bandgap. The difference between them is that for a semiconductor, the gap between the filled and the empty bands is not so large (<3 eV), such that some electrons have enough thermal energy to jump across the gap to the empty band above (called the conduction band) [figure 3.16(c)], and these electrons can conduct electricity (these materials are called intrinsic semiconductors). The unoccupied states left behind also leave room for the electrons in the original band (called the valence band) to move. It turns out that the motion of these electrons in the valence band is equivalent to vacant states moving as positive electrons, or holes. The energy of these holes is opposite to that of electrons; it is a minimum at the peak in the valence band and increases as the electron energy becomes smaller. Impurities are added to most semiconductors and these impurities have energy levels somewhere within the bandgap; some are close to the bottom edge of the conduction band or the top edge of the valence band (or band edge). The electrons in the impurity levels can be thermally excited to the conduction band if their level is close to the bottom of the conduction band, thus creating more electrons than holes in the semiconductor. Such semiconductors are called n-type and the impurities are called donors [figure 3.16(d)]. Similarly, if the impurity energy level is close to the valence band edge, electrons in the valence band can be excited to the impurity levels, leaving more empty states or holes behind. Such semiconductors are called p-type and the impurities are called acceptors. Apparently, the number of electrons that are excited to the conduction band depends on the thermal energy of these electrons. This is why semiconductors are especially sensitive to temperature variation. Refer to Appendix B for more discussion on semiconductors.

3.2.2 Electron Energy Bands in Real Crystals

The energy band structures for real materials are more complex but have similarities to the one-dimensional potential model. In three-dimensional crystals, there are different crystallographic directions. Each of the directions has a different periodicity and a correspondingly different potential profile. Taking a cubic conventional cell as an example, the period is equal to the lattice parameter a in the $\langle 100 \rangle$ direction, $\sqrt{2}a$ in the $\langle 110 \rangle$ direction, and $\sqrt{3}a$ in the $\langle 111 \rangle$ direction. In each direction, due to the difference in the potential and periodicity, the energy bands will be different. The energy bands can be plotted along each of these directions and, particularly, along the major crystallographic directions, as in figure 3.17 for Cu, Si, and GaAs, the latter two being common semiconductors. The special points on the first Brillouin zone surface, such as X and L, which are indicated in figure 3.8(b), represent the major crystallographic directions. In copper, the Fermi level at zero kelvin falls inside a band. Therefore, electrons close to the Fermi level are free to move with minimum thermal energy disturbance. For Si and GaAs, the Fermi level is

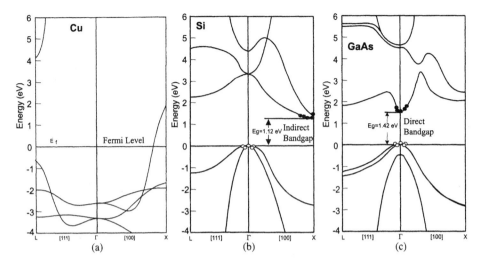

Figure 3.17 Electron band structures of (a) copper (after Mattheiss, 1964), (b) silicon, and (c) GaAs (after Chelikowsky and Cohen, 1976). Copper is a metal because the Fermi level falls inside the bands. The Fermi level for Si and GaAs at zero temperature is at the top of the valence band ($E = 0$). Silicon is an indirect gap semiconductor since the minimum of the conduction band and that of the valence band are not at the same wavevector location. GaAs is a direct gap semiconductor because the minima occur at the same wavevector ($k = 0$ for this case). All bandgap values are those at 300 K.

at $E = 0$ so that all bands below this level are filled. Above the filled bands, an energy gap exists in which no electrons are allowed at $T = 0$ K. The values and locations of the energy gap are different for dissimilar crystallographic directions, and the absolute minimum gap is called the *bandgap*. GaAs is a direct gap semiconductor because the minima of the conduction and the valence bands occur at the same wavevector. Si is an indirect gap semiconductor because the two minima do not occur at the same wavevector. Direct and indirect gap semiconductors have major differences in their optical properties. Direct bandgap semiconductors are more efficient photon emitters, semiconductor lasers are made of direct gap semiconductors such as GaAs, whereas most electronic devices are built on silicon technology.

For semiconductors, since most electrons are close to the minimum of the conduction band and holes are close to the minimum of the valence band, it is convenient to express the band structure near the minima in analytical form. Since the minima typically mean that the first-order derivative, $\partial E/\partial k$, is zero (as long as the first-order derivative exists), the second-order terms often are used. For the conduction band, the expansion of the electron allowable energy level near the minimum is often written in the form

$$E - E_c = \frac{\hbar^2}{2}\left(\frac{k_x^2}{m_{11}} + \frac{k_y^2}{m_{22}} + \frac{k_z^2}{m_{33}}\right) \quad (3.37)$$

where

$$m_{ij} = \frac{\hbar^2}{(\partial^2 E/\partial k_i \partial k_j)} \quad (3.38)$$

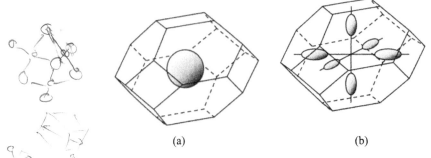

Figure 3.18 Constant electron energy surfaces in reciprocal space (or k-space): (a) a spherical band such as in GaAs; (b) an elliptical band such as in Si (after Shur, 1990).

is representative of the local curvature of the E–$\mathbf{k}$ relation and is called the effective mass, and E_c is the bottom of the conduction band, or band edge. The effective mass derives its name from the fact that the free electron energy can be expressed as $E = \hbar^2 k^2/2m$. The effective mass is a second-order tensor, according to eq. (3.38). Equation (3.37) is valid only when x, y, and z are the principal directions of the effective mass tensor and E_c is at $\mathbf{k} = 0$. For a spherical band, $m_{11} = m_{22} = m_{33}$ and the energy–wavevector relation, that is, the dispersion relation, reduces to the free electron form, except that the free electron mass is replaced by the effective mass. Using this effective mass, we can treat the motion of electrons in the conduction band (or the motion of holes in the valence band, which depends on the hole effective mass) as free electrons (or holes).

By setting the energy in eq. (3.37) to a constant, we can plot the constant energy surface in k-space, that is, inside the first Brillouin zone. Examples of constant energy surfaces are given in figure 3.18.

3.3 Lattice Vibration and Phonons

The previous section deals with the electron energy levels in solids. We now turn our attention to the vibrational energy levels of atoms, or the lattice vibration. Here, the term "lattice" really means the lattice together with its basis (we recall that the lattice in crystallography is a mathematical abstraction of periodic points in space). Similar to what we have done for the electronic energy levels, a simple one-dimensional model will render the fundamental characteristics of lattice vibrations. Therefore, we will start our discussion with a one-dimensional model and then move on to more complicated cases.

3.3.1 One-Dimensional Monatomic Lattice Chains

Let us consider first a one-dimensional monatomic chain, as shown in figure 3.19. By limiting our consideration to monatomic chains, we have made the following simplifications: (1) the mass at each lattice point is the same; (2) the separation between adjacent atoms is the same; and (3) the force interactions between adjacent atoms are the same. Such simplifications are no longer valid when the basis comprises more than one atom.

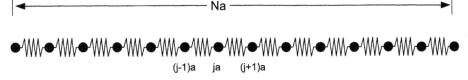

Figure 3.19 One-dimensional monatomic lattice chain model.

We will make the following assumptions in the analysis. First, we consider the force interaction between the nearest neighbors only. Second, the interaction force between atoms is assumed to be a harmonic force (which obeys Hooke's law) with spring constant K. This assumption can be justified in a similar fashion to that done for harmonic oscillators. Now consider a typical atom j. The displacement of atom j from its equilibrium position x_j^0 is

$$u_j = x_j - x_j^0 \tag{3.39}$$

The force acting on atom j comes from two components. One is due to the relative displacement between atom $(j-1)$ and atom j, and the other is due to the relative displacement between atom j and $(j+1)$. The net force is then

$$F_j = K(u_{j+1} - u_j) - K(u_j - u_{j-1}) \tag{3.40}$$

At this point, we have two choices for continuing this discussion. The first choice includes writing out the potential and the Schrödinger equation, and then solving for the allowable energy. But for our understanding here, a classical approach is easier to grasp and the quantum effect can be added into the classical solution later. Let's apply Newton's second law to atom j to obtain

$$m\frac{d^2 u_j}{dt^2} = K(u_{j+1} - u_j) - K(u_j - u_{j-1}) \tag{3.41}$$

The above equation is a special form of the differential wave equation

$$m\frac{\partial^2 u}{\partial t^2} = Ka^2 \frac{\partial^2 u}{\partial x^2} \tag{3.42}$$

which has a solution of the form $u \propto e^{-i(\omega t - kx)}$. Such a similarity suggests a wave type of solution for Eq. (3.41),

$$u_j = A \exp[-i(\omega t - kja)] \tag{3.43}$$

where ja is the discrete equilibrium location of atom j, ω is the frequency, and k is the wavevector. The major difference between this "guessed" solution and the conventional continuous wave, $\exp[-i(\omega t - kx)]$, lies in the use of "ja" as a discrete lattice coordinate rather than an infinitely divisible continuum coordinate "x" used in a continuous medium, because talking about the vibration at locations other than the atomic sites is meaningless.

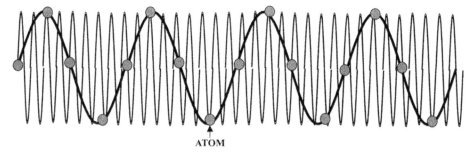

Figure 3.20 Snapshots of the atom displacements represented by two different wavevectors that differ by $2\pi/a$. The displacements of the atoms are the same for the two given wavevectors drawn in the figure. The short wavelength mode means that atom displacements would exist in the empty space between two closest atoms, which is not possible. Thus, the allowable wavevectors for phonons are limited to the first Brillouin zone.

Substituting the guessed solution (3.43) into eq. (3.41), we get

$$-m\omega^2 = K[e^{ika} + e^{-ika} - 2] \qquad (3.44)$$

or

$$\omega = 2\sqrt{\frac{K}{m}} \left| \sin\frac{ka}{2} \right| \qquad (3.45)$$

The allowable wavevector can similarly be determined from the von Karman periodic boundary condition, as we did for electrons,

$$k = \frac{2\pi n}{Na} \qquad (n = 0, \pm 1, \pm 2, \ldots) \qquad (3.46)$$

When treating the electronic energy levels, we assert that the states for k and $[k + m(2\pi/a)]$ are identical, as shown in figure 3.13(a). For the lattice vibration, we can argue physically that the allowable values of n for lattice vibrations lie between $-N/2$ and $N/2$ and the extension of k beyond the first Brillouin zone is meaningless. From $k = 2\pi/\lambda$, where λ is the wavelength, we observe that if $n > N/2$, λ will be less than twice the atom spacing; this means that there are no atoms between one period, as shown in figure 3.20. Talking about atomic displacement in empty spaces with no atoms is meaningless. Thus, the allowable wavevector for a lattice vibration is naturally confined to the first Brillouin zone.

Equation (3.46), together with the limitation on n, states that for a monatomic, one-dimensional lattice chain with N atoms there are N allowable wavevectors. Each of these wavevectors corresponds to one mode of the lattice vibration. Each mode is mathematically described by Eq. (3.43) and is called a *normal mode*. Normal modes are a familiar concept in the vibration analysis of structures, drums, and musical instruments.

Figure 3.21(a) shows the dispersion relationship between the vibrational frequency and the wavevector for a monatomic lattice chain. Although this relationship is nonlinear, very often the Debye approximation is used, which assumes a linear dispersion relation

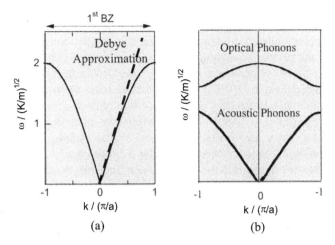

Figure 3.21 Phonon dispersion of a one-dimensional (a) monatomic lattice chain and (b) diatomic lattice chain. The Debye approximation use a linear relationship between the frequency and the wavevector. BZ stands for Brillouin zone.

between the frequency and the wavevector. This approximation is valid at low frequencies but is not a good approximation at high frequencies. In the very low frequency region the lattice vibration carries the sound wave. We will return to a discussion of the Debye approximation in the next section.

3.3.2 Energy Quantization and Phonons

The above treatment is based on classical mechanics. In the classical solution, the vibrational energy at each frequency is determined by the amplitude A in eq. (3.43). If we use the Schrödinger equation to solve the same problem, we will witness that the dispersion relation is the same. The difference is that the energy of the lattice vibration is quantized. This result is similar to the case of a simple harmonic oscillator that we dealt with in chapter 2, for which the vibrational frequency is the same as that of the classical mass–spring system [eq. (2.56)], but the energy of the oscillator is quantized. At each frequency determined by eq. (3.45), the allowable energy levels are

$$E_n = h\nu \left(n + \frac{1}{2}\right) \qquad (n = 0, 1, 2, 3, \ldots) \qquad (3.47)$$

This quantization is identical to that of a single harmonic oscillator, eq. (2.54), and to Planck's relation between photon energy and its frequency. The similarity leads to the concept of phonons, parallel to the concept of photons, as the minimum quantum of lattice vibration. Each phonon has an energy of $h\nu$. Its momentum is, according to the de Broglie relation, $h/\lambda = \hbar k$. Under the phonon picture, we can forget about atoms and consider the lattice waves or the phonon particles in a crystal just as we treat the electromagnetic waves and photons in a box. For many materials (nonmetals), heat is conducted by lattice vibrations. From the analogy between photons and phonons, it is logical to expect that lattice heat conduction should obey a law similar to the

Stefan–Boltzmann law, which is proportional to the fourth power of the temperature. This temperature dependence is not evident in most of (if not all) the heat conduction calculations that we learned in an undergraduate heat transfer course. The reason is that in most cases the phonon scattering is so strong that these waves cannot travel a long distance. At very low temperatures, in fact, we can calculate heat conduction just as we do with radiation (Casimir, 1938). A similar situation arises when dealing with very thin films, for which internal scattering does not occur as often as in bulk materials.

Although eq. (3.47) is identical to the expression of a single spring-mass harmonic oscillator, there are indeed differences. For an isolated harmonic oscillator only one vibrational frequency exists, while for a lattice chain N wavevectors (where N is the total number of atoms in the chain) present, and thus N frequencies are possible. Each of these allowable frequencies supports an energy that must be a multiple of $h\nu$. How many phonons actually occupy each state at a specific frequency is a point that we will address in the next chapter.

We should emphasize that although photons and phonons can be thought of as particles with an energy $h\nu$, these particles are fundamentally different from electrons. Some of the differences are discussed here. Unlike electrons, phonons and photons at rest do not have mass; and are fictitious particles since they are the quantization of the normal mode of a field. Electrons obey the Pauli exclusion principle, which says that each quantum state can have only one electron at most. Photons and phonons are not limited by the Pauli exclusion principle. Each quantum state, which corresponds to one set of wavevectors, can have many phonons and photons, as is indicated by eq. (3.47). The differences in behavior between the various quantum species will later be reflected in their statistical behavior, which we will discuss in more detail in chapter 4. Due to the differences in their statistical behavior, electrons are called *fermions*, while phonons and photons are called *bosons*.

3.3.3 One-Dimensional Diatomic and Polyatomic Lattice Chains

If there is more than one atom at each lattice point, several changes should be made to the above monatomic model. First, the masses of two adjacent atoms may be different (such as in a GaAs crystal) or the same (such as the two silicon atoms in the basis of a silicon crystal). Second, the distances between two adjacent atoms may be different. And third, the spring constants between adjacent atoms can also be different. Taking these differences into consideration and following a similar analysis as for the one-dimensional lattice chain, one can derive the dispersion relation between the vibrational frequency and the wavevector. For a diatomic lattice chain, a typical solution looks like figure 3.21(b). The lower branch is similar to that of a monatomic lattice chain, and the upper branch is similar to the folded representation of the electronic energy levels. Note that a gap exists between the two branches. The upper branch is due to the additional degrees of vibrational freedom caused by the additional atom in a unit cell. In the long wavelength limit, the vibration of the two atoms at each lattice point can be either in-phase or out-of phase, as illustrated in figure 3.22. Clearly, the out-of-phase modes require more energy. The lower frequency branch is called the acoustic branch, and the higher frequency one is called the optical branch, because the high-frequency phonons in the optical branch can interact with electromagnetic waves more easily. In general, for a one-dimensional lattice chain with m atoms in a basis and N lattice points

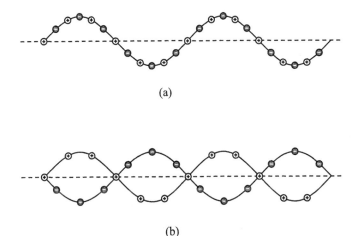

Figure 3.22 In the long wavelength limit, the two atoms in the acoustic phonon branch vibrate in phase (a) but are out of phase in the optical branch (b).

in the chain, there exists one acoustic branch with N acoustic modes and $(m-1)$ optical branches with $(m-1)N$ optic modes.

3.3.4 Phonons in Three-Dimensional Crystals

In a one-dimensional lattice, the phonon waves are longitudinal waves. In a three-dimensional crystal, the atoms can vibrate in three dimensions. Thus, we will have three vibrational branches for the acoustic modes—one longitudinal and two transverse branches. Furthermore, if there are m atoms per lattice point, then $3(m-1)$ optical branches exist, of which $2(m-1)$ are transverse optical phonons and the rest are longitudinal optical phonons. In a transverse wave, the atomic displacement direction is perpendicular to the wave propagation direction. The two transverse branches will coincide with each other if the two vibrational directions are symmetric. As with electrons, phonon dispersions along different crystallographic directions are different. Figure 3.23(a) shows the phonon dispersion relations for lead, which has an fcc structure and one atom as a basis, and thus three acoustic branches along each crystallographic direction. Figure 3.23(b) shows the phonon dispersion for Si, which has an fcc structure but two atoms as the basis. Thus, Si has three acoustic branches and three optical branches in each direction. In some high-symmetry directions, such as the [100] directions, the two transverse phonon branches collapse onto one curve.

3.4 Density of States

In the previous chapter, we introduced the concept of degeneracy for quantum states that have the same energy levels. From the one-dimensional electron model, we see that each wavevector k corresponds to one wavefunction and thus two quantum states after we include the electron spin. Because the dispersion relation is symmetric for

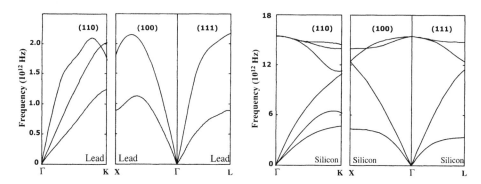

Figure 3.23 Phonon dispersion relations for lead and silicon; both have an fcc lattice. Lead has one atom as the basis, so only acoustic phonons are present. Silicon has two atoms as a basis, and thus there are three acoustic phonons and three optical phonons (lead after Brockhouse et al., 1962, and silicon after Giannozzi et al., 1991).

both positive and negative k values, the degeneracy for electrons at each energy level can be regarded as 4. In three-dimensional crystals, however, the dispersion picture is totally different because each crystallographic direction has its own dispersion relation between the energy and the wavevector. There exist potentially many combinations of wavevectors that have the same energy. As shown in figure 3.24, for the constant energy surface in the wavevector space for a spherical band, that is, equal effective mass [eq. (3.37)] but in two dimensions. Clearly, there can be many wavevectors on each constant energy surface. Because the energy levels in solids are quasi-continuous, we use the concept of density of states to describe the energy degeneracy. We will discuss next how the density of states is defined, and derive expressions for the density of states for electrons, phonons, and photons, using simplified dispersion relations derived earlier.

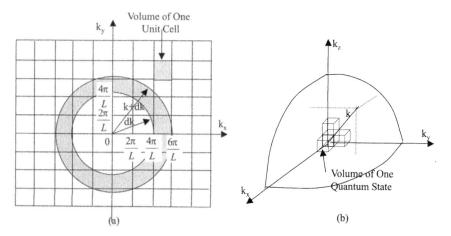

Figure 3.24 Constant energy surface in k-space for a spherical band, and volume of one electron state in the k-space: (a) two-dimensional projection; (b) three-dimensional view.

3.4.1 Electron Density of States

Consider a spherical parabolic band with the following relationship between the energy and the wavevector,

$$E - E_c = \frac{\hbar^2}{2m^*}(k_x^2 + k_y^2 + k_z^2)$$

$$= \frac{\hbar^2 k^2}{2m^*} \left(k_x, k_y, k_z = 0, \pm\frac{2\pi}{Na}, \pm\frac{4\pi}{Na}, \ldots, \pm\frac{\pi}{a} \right) \quad (3.48)$$

where k is the magnitude of the wavevector,

$$k^2 = k_x^2 + k_y^2 + k_z^2 \quad (3.49)$$

For each constant E value, there can be potentially many combinations of k_x, k_y, k_z satisfying eq. (3.48). To find the density of states, we refer to figure 3.24. The allowable wavevectors for k_x, k_y, and k_z are integral multiples of $2\pi/L$, where $L(= Na)$ is the length of the crystal along the x, y, and z directions. The volume of each quantum mechanical state in three-dimensional k-space is $(2\pi/L)^3$. The number of states between k and $k + dk$ in three-dimensional space is then

$$\text{\# states} = 2 \times \frac{4\pi k^2 dk}{(2\pi/L)^3} = \frac{V k^2 dk}{\pi^2} \quad (3.50)$$

where the factor of two accounts for the electron spin and $V(= L^3)$ is the crystal volume. In the above treatment, we have implicitly assumed that k and E are continuous functions. This should be valid as long as the number of atoms in the system is large enough (N is large).

On the basis of eq. (3.50), we can define the density of states as the number of quantum states per unit interval of the wavevector and per unit volume,

$$D(k) = \frac{\text{\# states between } k \text{ and } k + dk}{V dk} = \frac{k^2}{\pi^2} \quad (3.51)$$

We can also define the *density of states* as the number of states per unit volume and per unit energy interval

$$D(E) = \frac{\text{\# states between } E \text{ and } E + dE}{V dE} = \frac{1}{2\pi^2} \left(\frac{2m^*}{\hbar^2}\right)^{3/2} (E - E_c)^{1/2} \quad (3.52)$$

where we have used eq. (3.48) to replace k and dk in terms of E and dE. Sometimes, we also define the density of states on the basis of a frequency interval (as we will do for phonons). For electrons, eq. (3.52) is used most often. A schematic of eq. (3.52) is shown in figure 3.25.

The density of states is a purely mathematical convenience, but nevertheless, it is central for correctly counting the number of electrons and the energy (or charge and momentum) that they carry. As a simple example of how the density of states is needed, let's evaluate the electron energy of the topmost level at $T = 0$ K, that is, the Fermi level E_f. At 0 K, the filling of electron quantum states starts from the lowest energy

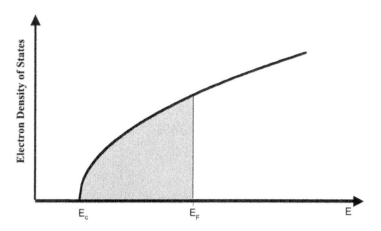

Figure 3.25 Density of states of electrons in a bulk crystal.

level and moves up from one energy level to the next until all electrons are placed into distinct quantum states. The number of electrons per unit volume at $T = 0$ K is

$$n = \int_{E_c}^{E_f} D(E)dE = \frac{1}{3\pi^2}\left(\frac{2m^*}{\hbar^2}\right)^{3/2}(E_f - E_c)^{3/2} \qquad (3.53)$$

Often, E_c is taken as a reference point and set to zero. From the above relation, we can calculate easily the Fermi level from the given electron number density and the effective mass.

Example 3.4 *Fermi level*

A gold crystal has an fcc lattice with one gold atom per lattice point and a lattice constant of 4.08 Å. Every gold atom has one valence electron. Estimate the electron Fermi level in a gold crystal.

Solution: We assume that the electron effective mass in gold is identical to that of free electrons. For an fcc lattice, there are four lattice points and thus four valence electrons. Consequently, the electron number density is

$$n = 4/(4.08 \times 10^{-10})^3 = 5.89 \times 10^{28} \text{ m}^{-3} = 5.89 \times 10^{22} \text{ cm}^{-3}$$

From eq. (3.53), the Fermi level at 0 K is

$$E_f = \frac{\hbar^2}{2m}(3\pi^2 n)^{2/3} = 8.66 \times 10^{-19} \text{ J} = 5.4 \text{ eV}$$

Comment: We give the number density in terms of cm^{-3} because such a unit is commonly used in semiconductors. Typical dopant levels for silicon-based semiconductor devices are $\sim 10^{17}$–10^{19} cm^{-3}, which is much smaller than those in metals.

3.4.2 Phonon Density of States

The phonon dispersion relation differs considerably from the electron dispersion relation, as is seen by comparing eq. (3.45) with eq. (3.37). Rather than having spins, phonons have different polarizations (branches) in crystals. One common simplification to the phonon dispersion is the Debye approximation, which assumes a linear relation between the frequency and the wavevector,

$$\omega = v_D |\mathbf{k}| = v_D k \tag{3.54}$$

Following a similar procedure to our previous treatment of electrons, we calculate the density of states of phonons. The volume of one phonon state in **k**-space is $(2\pi/L)^3 = (2\pi)^3/V$. Under the Debye approximation, the density of states for phonons per unit volume and per unit frequency interval is then

$$D(\omega) = \frac{dN}{Vd\omega} = 3\frac{4\pi k^2 dk/(2\pi/L)^3}{Vd\omega} = \frac{3\omega^2}{2\pi^2 v_D^3} \tag{3.55}$$

where a factor of 3 has been added to account for the three polarizations of phonons.

The Debye approximation, as represented by eq. (3.54), implies that all phonon branches have the same speed in all directions; in other words, the medium is isotropic. In reality, even a cubic crystal does not have the same velocity in different crystallographic directions. This isotropic medium will therefore have a different lattice constant a_D compared to real crystals. To calculate the equivalent lattice constant of a Debye crystal, we require that the total number of states in this isotropic crystal equal that of a real crystal. We assume that k_D is the wavevector at the boundary of the Brillouin zone, that is, $k_D = \pi/a_D$ in the Debye model. The number of states existing in a real crystal having N ions is equal to $3N$.* A Debye crystal should contain the same number of states,

$$3N = 3 \times \frac{4\pi k_D^3/3}{(2\pi)^3/V} \tag{3.56}$$

which gives

$$k_D = \left(\frac{6\pi^2 N}{V}\right)^{1/3} \quad \text{or} \quad a_D = \left(\frac{\pi V}{6N}\right)^{1/3} \tag{3.57}$$

where k_D is called the Debye cutoff wavevector. Correspondingly, the Debye frequency is $\omega_D = v_D k_D$.

It should be noted that the Debye approximation does not represent the reality at the Brillouin zone boundary, where the dispersion is flat, as shown in figures 3.21 and 3.23, as well as eq. (3.45). It is also not suitable for optical phonons. For the latter, a much better approximation is to set all the frequencies to the same value, that is, $\omega = \omega_E$ (for each branch). If there are N' lattice points in the crystal, there should be a total of N' modes for each branch, with the degeneracy equal to N'. This approximation was first used by Einstein to explain the specific heat, and the resultant theory is called the Einstein model. A more thorough discussion regarding the Einstein model will be given in the next chapter.

---

*This includes both acoustic and optical modes. In a Debye model, however, all the modes are approximated by the three identical acoustic branches.

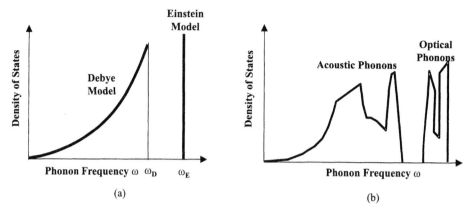

Figure 3.26 Phonon density of states, (a) approximated by the Debye and Einstein models and (b) in real crystals.

The densities of states for the Debye and Einstein models are illustrated in figure 3.26. The Debye model gives $D \propto \omega^2$, while the Einstein model gives a spike at ω_E. The densities of states in real crystals can be quite different from the predictions of these simple models, as illustrated in figure 3.26(b). At each frequency that the phonons intersect the zone boundary, a singularity, called the van Hove singularity, appears in the density of states because the dispersion curve is perpendicular to the zone boundary.

3.4.3 Photon Density of States

Photons also have a linear dispersion between frequency and its wavevector, $\omega = ck$, which is identical to that of phonons under the Debye approximation. Consider an electromagnetic wave in a cubic box of length L. The electromagnetic fields can be decomposed into normal modes using Fourier series, as we did for phonons. The allowable wavevectors are then

$$k_x, k_y, k_z = 0, \pm 2\pi/L, \pm 4\pi/L, \ldots \qquad (3.58)$$

Hence, as before, photons share much commonality with phonons. However, significant differences exist: unlike phonon waves in a crystal, which have a minimum wavelength as imposed by the interatomic distance, no such a limit presents on the wavevector for photons. Following a derivation similar to phonons, we can obtain the density of states for an electromagnetic wave as

$$D(\omega) = \frac{dN}{V d\omega} = \frac{\omega^2}{\pi^2 c^3} \qquad (3.59)$$

One difference in the above equation from eq. (3.55) is that a factor of two rather than three is used to reflect the fact that electromagnetic waves (photons) have two transverse polarizations, whereas phonons can be longitudinally polarized as well. The other difference is that while the phonon density of states has a cut-off frequency given by the Debye frequency, photons do not have such a cutoff frequency.

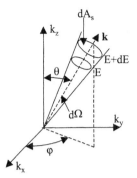

Figure 3.27 Differential density of states and solid angle.

3.4.4 Differential Density of States and Solid Angle

Although the density of states is usually defined on the basis of the magnitude of wavevector k or energy, we found that it is useful in the study of transport process to define a differential density of states. We first define the solid angle Ω in k-space as (figure 3.27)

$$d\Omega = \frac{dA_s}{k^2} = \sin\theta\, d\theta\, d\varphi \tag{3.60}$$

where dA_s is a differential area perpendicular to the **k** direction and θ and ϕ are polar and azimuthal angles, defined in figure 3.27. With this definition, it is easy to show that the solid angle over the entire space is 4π. The differential density of states, along a specific wave vector direction **k** is defined as

$$dD(E, \mathbf{k}) = \frac{\text{No. of states within}(E, E+dE) \text{ and } d\Omega}{V dE d\Omega} = \frac{D(E)}{4\pi} \tag{3.61}$$

where the second equality applies to isotropic dispersions only.

3.5 Energy Levels in Artificial Structures

We touched upon quantum wells and quantum dots in the previous chapters. These structures can be made by various synthesis routes such as molecular beam epitaxy and self-assembly. The energy states of electrons, phonons, and photons in such structures are often different from those in their bulk counterparts. Many of the novel properties of these artificial structures originate from the different energy states and, consequently, different densities of states. These artificial structures can be categorized into two groups. One imposes new boundary conditions, as in quantum wells and quantum dots, and the other creates new periodicity, as in superlattices. We will briefly illustrate some examples in this section.

3.5.1 Quantum Wells, Wires, Dots, and Carbon Nanotubes

A quantum well can be formed by sandwiching a thin film between two other materials. For example, a thin layer of GaAs (typically < 200 Å can be sandwiched between

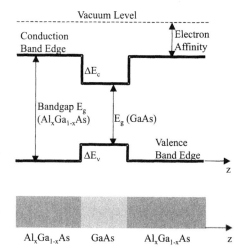

Figure 3.28 A quantum well can be formed by sandwiching one material (GaAs) between two materials (Al$_x$Ga$_{1-x}$As). The top figure shows the band-edge alignment. The band-edge offset provides the potential barrier to confine electrons in the GaAs layer.

two Al$_x$Ga$_{1-x}$As layers, where the AlAs volume fraction x can be controlled to a high precision between 0 and 1. Both GaAs and AlAs are semiconductors. AlAs (bandgap 2.17 eV) and its alloy with GaAs, Al$_x$Ga$_{1-x}$As, have larger bandgaps than GaAs (bandgap 1.42 eV). When two materials form an interface, a general rule for the alignment of the bands is that the vacuum level must be the same. The energy required to bring an electron from the conduction band edge to the vacuum level, that is, the energy needed to take an electron out of the conduction band to vacuum, is called the electron affinity [figure 3.28], which is different for different materials. For the Al$_x$Ga$_{1-x}$As/GaAs/Al$_x$Ga$_{1-x}$As sandwich structure, the final band-edge alignment is shown in figure 3.28. A potential difference exists at the interface in both the conduction band and the valence band, called the band-edge offset. For electrons in the conduction band, the conduction band-edge offset ΔE_c provides the potential barrier to form a quantum well. Similarly, the valence band-edge offset provides a potential well for holes. In chapter 2, we solved the energy levels for a one-dimensional quantum well. In a realistic quantum well constructed of a thin film, the electrons are not constrained in the x–y plane (the film plane) and the potential barrier is not infinitely high. Hence, the solution is more complicated and will not be pursued here (see exercise 2.12). Instead, we assume for simplicity an infinite potential barrier height in the z-direction. We then can obtain the following energy–wavevector relation from the Schrödinger equation,

$$E(k_x, k_y, n) = \frac{\hbar^2}{2m^*}(k_x^2 + k_y^2) + n^2 \frac{\hbar^2 \pi^2}{2m^* d^2} \quad (n = 1, 2, \ldots, N) \quad (3.62)$$

where d is the width of the quantum well and m^* the electron effective mass. In the above relation, the dispersion relations in the k_x and k_y directions are the same as in the bulk material, but in the z-direction, the energy becomes discrete as given by the one-dimensional particle-in-a-box model presented in chapter 2.

Example 3.5 *Quantum well density of states*

For the energy dispersion relation given by eq. (3.62), determine the corresponding electron density of states.

ENERGY STATES IN SOLIDS   113

Solution: The allowable wavevectors for k_x and k_y are

$$k_x, k_y = \pm \frac{2\pi n}{Na} \quad (n = 0, 1, 2, \ldots) \quad \text{(E3.5.1)}$$

where a is the lattice constant. Thus, the area per state in the k_x, k_y plane is $(2\pi/L)^2$. We can rewrite eq. (3.62) as

$$E(k_x, k_y, n) = \frac{\hbar^2 k_{xy}^2}{2m^*} + n^2 \frac{\hbar^2 \pi^2}{2m^* d^2} \quad \text{(E3.5.2)}$$

where $k_{xy}^2 = k_x^2 + k_y^2$. Examining eq. (E3.5.2), we see that for energy E larger than $E_n = n^2 \hbar^2 \pi^2/(2m^* d^2)$ but smaller than $E_{(n+1)}$, there exist n series of k_{xy} values that can satisfy eq. (E3.5.2) because the value n in the last term can be any integer between 1 and n. For each of these n series, the number of states between k_{xy} and $k_{xy} + dk_{xy}$ is

$$N = \text{No. of states} = \frac{4\pi k_{xy} dk_{xy}}{(2\pi/L)^2} = \frac{A k_{xy} dk_{xy}}{\pi} \quad \text{(for each allowable series of } k_{xy})$$

where $A = L^2$ is the area along the x-y plane. From eq. (E3.5.2) we get

$$dE = \frac{\hbar^2}{m^*} k_{xy} dk_{xy} \quad \text{(for each allowable series of } k_{xy}) \quad \text{(E3.5.3)}$$

So the electron density of states per energy interval and per unit area of film for each allowable k_{xy} series is

$$D_1(E) = \frac{N}{AdE} = \frac{m^*}{\pi \hbar^2} \quad \text{(E3.5.4)}$$

For an energy state $E_n < E < E_{(n+1)}$, the total number of states is $D(E) = nD_1(E)$. The electron density of states for such a dispersion relation is a staircase, as illustrated in figure E3.5.

As with electrons, phonon energy states in quantum structures are also altered because of new boundary conditions. Figure 3.29 compares the phonon dispersion and the phonon density of states in a freestanding thin film (Yang and Chen, 2000). The phonon spectrum

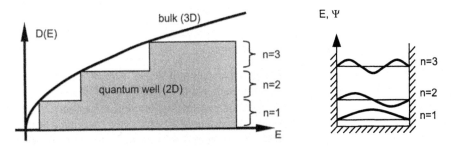

Figure E3.5 Density of states of electrons in a quantum well.

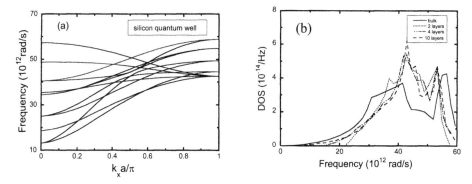

Figure 3.29 (a) Phonon dispersion and (b) density of states in quantum wells (Yang and Chen, 2000).

change can be seen experimentally through, for example, Raman spectroscopy, which probes the phonons through the frequency shift of a photon that interacts with a phonon (Weisbuch and Vinter, 1991). Numerous studies have been devoted to the effects of phonon confinement in quantum structures (Bannov et al., 1995). Recent applications include the use of phonon confinement to reduce thermal conductivity and thus, increase the thermoelectric energy conversion efficiency (Chen, 2001).

The quantum effects for nanometer-scale wires (quantum wires) and nanometer-scale dots (quantum dots) are expected to be even stronger than in quantum wells because of the additional boundary conditions on the electron or phonon motion in one or two more directions. A recent discovery is that of nanoscale tubular structures, particularly carbon nanotubes (Iijima, 1991). A carbon nanotube can be considered as the rolling of an atomic sheet (or several atomic sheets) of graphite carbon (Dresselhaus et al., 2001). Graphite has a close-packed hexagonal structure, as shown in figure 3.5(c). The bonding between different layers is through the van der Waals bond, which is weaker than the covalent bonds within each layer. If only one atomic layer rolls up, the nanotube thus formed is called a single-walled carbon nanotube. If several layers roll up, the nanotube formed is called multiwalled. Depending on the nanotube diameter and the orientation of the major crystallographic directions with the nanotube axis, the nanotube can be a semiconductor or a metal, due to quantum size effects. The electron and phonon energy states in carbon nanotubes are very different from those in their bulk materials, leading to some special properties. The mechanical strength and thermal conductivity of these tubes are expected to be very high (Kim et al., 2001). Research is actively exploring various properties and applications of carbon nanotubes (Dresselhaus et al., 2001).

3.5.2 Artificial Periodic Structures

We have observed that the periodicity that naturally exists in bulk crystals plays a crucial rule in determining the electron and phonon energy levels. Natural systems are three dimensional, with a periodicity determined by the lattice constants. One can also create artificial periodic structures, for example, by repeatedly growing a thin layer of GaAs and a thin layer of AlAs on the same substate. In fact, artificial periodic structures have been used widely in optical coatings, such as in the making of optical

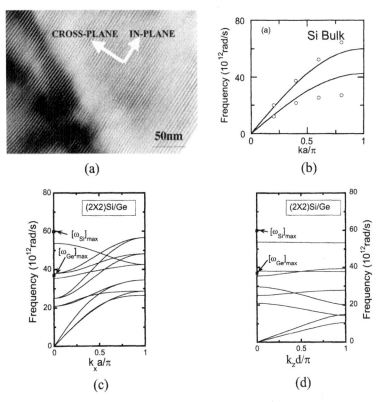

Figure 3.30 (a) A Si/Ge superlattice (Borca-Tasciuc et al., 2000); (b) acoustic phonons in bulk silicon; and phonon dispersion (c) along and (d) perpendicular to the superlattice plane (Yang and Chen, 2004).

interference filters that consist of alternating layers of quarter-wavelength thin films (Heavens, 1965). An optical wave can be totally reflected when the wavelength matches the period thickness, as the Bragg condition in figure 3.10(b) and eq. (3.18) dictates. We will discuss optical interference filters further in chapter 5. In an analogy to the optical interference filters, Esaki and Tsu (1970) proposed superlattices, which are periodic structures with the thickness of each layer less than the electron or phonon mean free path. Since the electron and phonon mean free path and wavelength are very short, such an analogy requires much thinner films than those used in optical interference filters. Consequently, the concept was demonstrated only after the invention of advanced thin film deposition techniques such as molecular beam epitaxy (Chang et al., 1973). Electron wave propagation and energy states in superlattices can be modeled using the Kronig–Penney model, leading to drastically different electrical and optical properties from those of their constituent materials (Weisbuch and Vinter, 1991). Phonons can also exhibit similar behavior, forming bandgaps and new energy states (Narayanamurti et al., 1979). Figure 3.30 shows an example of a model Si/Ge superlattice and the calculated phonon spectra for acoustic phonons in a model Si/Ge-like superlattice, together with the acoustic phonon dispersion relations in bulk Si (Yang and Chen, 2001). We can

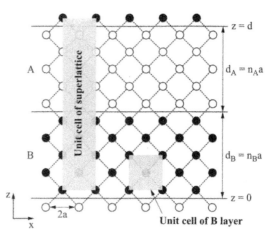

Figure 3.31 Unit cell of a superlattice made of two cubic crystals, being a tetrahedron with a much larger number of atoms as the basis.

see that, along the superlattice film plane, the period is still a and thus the maximum reciprocal lattice vector in the first Brillouin zone is π/a. In the cross-plane direction, the lattice constant of the original bulk lattice no longer represents the true periodicity in this direction. It is the total thickness of one period, d, that represents the lattice periodicity in this direction. Thus the maximum width of the first Brillouin zone is π/d rather than π/a. The superlattice clearly has very different dispersion relations to those of the bulk materials. In the cross-plane direction, for example, a small gap, called a minigap, forms at every $k_z = \pi/d$ in the phonon spectrum, similar to the electronic gap formation in a one-dimensional Kronig–Penney model. The dispersion of high-frequency acoustic phonons in bulk silicon becomes very flat inside a superlattice because there are no corresponding phonons in adjacent germanium layers; these phonons are confined inside the silicon layer. Another way to think of these confined "acoustic" phonons is that the unit cell of a Si/Ge superlattice is no longer a cube as in bulk silicon or germanium, but a tetrahedron as shown in figure 3.31. It is a new material with a new unit cell that has more than one atom as the basis (at least one silicon and one germanium atom). However, we should remind ourselves that to form the new phonon or electron spectra, as theoretically predicted using idealized models (such as Kronig–Penney or harmonic lattice dynamics), the mean free path must be much longer than one period thickness. We will discuss this point in more detail in chapter 5.

Superlattice structures made of alternating layers of thin films have artificial periodicity only in one direction. By periodically arranging quantum wires and quantum dots, one can also make quantum wire and quantum dot superlattices that have artificial periodicity in two or three directions.

The periodicity in naturally existing crystals creates electronic bandgaps and phonon branches, partially due to the fact that electrons and phonons can sample and feel individual atoms and potential barriers and thus experience diffraction and interference effects. This is not the case for optical waves (except X-rays) in naturally existing crystals, since visible light usually has a quite long wavelength and thus averages over a large volume of the crystal. The artificial optical interference filters that are made of alternating quarter-wavelength layers create frequency ranges (called stop bands) along certain directions that can completely reflect all incident photons. This phenomenon corresponds to the photon bandgap, similar to that in the Kronig–Penney model for electrons. By extending such a concept to three dimensions to make three-dimensional periodic structures with periods comparable to optical wavelength, Yablonovitch (1986) proposed the concept of three-dimensional photonic bandgap structures. These photonic crystals have become a very active research field and have potential applications in lasers, telecommunication, and optical coatings (Joannopoulos et al., 1995, 1997).

3.6 Summary of Chapter 3

The contents of this chapter are often covered, in a solid-state physics course, in at least three individual chapters: crystal structure, electronic energy states, and phonon energy states (Kittel, 1996; Ashcroft and Mermin, 1976). This condensed chapter introduces the terminology and often-used methodologies for the analysis of energy states in crystalline structures.

The most important characteristic of crystals is their periodicity, which is described by a lattice. Real crystals are obtained by attaching a basis to each lattice point. The basis can consist of one atom or a cluster of atoms. Lattices are described by the primitive lattice vectors. A primitive unit cell contains one lattice point, but a conventional unit cell can have more than one lattice point. One way to construct a primitive unit cell unambiguously is to form the Wigner–Seitz cell. In three-dimensional space, a total of 14 lattice types exists. The Miller index method is commonly used to denote crystal planes and directions.

A lattice is periodic in real space, and we often express a periodic function in terms of its Fourier transformation. The Fourier conjugate of real space is called the reciprocal space. The primitive lattice vectors in reciprocal space can be calculated from the primitive lattice vectors in real space. Diffraction experiments provide an image of reciprocal space. A Wigner–Seitz cell in reciprocal space is called the first Brillouin zone. Later, we express the energy dispersion of electrons and phonons in the frist Brillouin zone.

In a periodic structure, the electronic energy levels form energy bands. The band formation is demonstrated by the solution of the Schrödinger equation based on the Kronig–Penney model. In real crystals, each crystallographic direction has its own dispersion relation. The electronic band structure determines whether a material is metal, semiconductor, or insulator. In metals, an electronic band is only partially filled and electrons can move to the empty quantum states within the same band. The topmost electron energy level at 0 K is called the Fermi level. If a band is totally filled and the next band has an energy gap from this band, electrons cannot move within the band. Whether the electron can go to the next energy band depends on the magnitude of the bandgap compared to the thermal energy which is 26 meV at 300 K. A material can be a semiconductor if the bandgap is relatively small such that there exist some electrons with high enough energy to jump to the conduction band, leaving some vacant quantum states behind. If the bandgap is very large, no electrons can jump to the conduction band and the material is an insulator.

In a semiconductor, the motion of electrons in the valence band can be described by the motion of equivalent positive charges, called holes, that occupy the empty states in the band. Semiconductors can be intrinsic or extrinsic. Extrinsic semiconductors are obtained by adding impurities that have an energy level close to the conduction or valence band such that they can contribute electrons to the conduction band or holes to the valence band. Semiconductors can also be divided into direct gap or indirect gap, depending on whether the minima of the conduction band and the valence band have identical or different wavevectors, respectively. The motion of electrons or holes in a semiconductor is similar to the motion of free electrons, except that their masses should be replaced by their effective masses.

The vibration of atoms in a crystal was investigated using Newtonian mechanics, assuming harmonic force interaction between nearest neighbors. The vibration can be decomposed into normal modes, each having a specific wavevector and frequency. The dispersion relation obtained between the wavevector and the frequency is identical to the result from quantum mechanics. Quantum mechanics requires, however, that each normal mode be quantized and that the minimum quantum has an energy of $h\nu$. This minimum energy quantum is called a phonon, similar to a photon which is the quantized normal mode of an electromagnetic wave. Phonons can exist only in the first Brillouin zone since larger wavevectors are meaningless in terms of atomic displacement. In three-dimensional crystals, one longitudinal acoustic mode and two transverse acoustic modes exist along each crystallographic direction. If there are m atoms in a basis and N lattice points, $3N$ acoustic modes and $3(m-1)N$ optical modes exist.

An important counting method for the degeneracy of energy levels is the density of states. The density of states can be defined on the basis of per unit magnitude of the wavevector, per unit frequency interval, or per unit energy interval. The most often used definition is based on the per unit energy interval. We have shown how to compute the density of states for electrons, phonons, and photons, for three-dimensional structures as well as structures of lower dimensions.

Artificial nanostructures, such as quantum wells, superlattices, carbon nanotubes, and photonic crystals, can have artificial energy levels and densities of states. These artificial structures break down the analysis for bulk materials by imposing new boundary conditions and creating new periodicities that do not exist naturally in bulk materials. The new energy states and densities of states lead to novel properties that are often hard to find in bulk materials. To form the new energy spectra, the mean free path of the carriers (electrons, phonons, or photons) usually should be much larger than the characteristic length; thus the most interesting regime is often at nanoscale, when these conditions are relatively easily satisfied, which speaks strongly for the current interest in nanotechnology.

3.7 Nomenclature for Chapter 3

a lattice constant or length in figure 3.11, m
a primitive or conventional lattice vector, m
A constant; area, m^2
b length in figure 3.11, m
b reciprocal space primitive lattice vector, m^{-1}
B constant
d thickness of quantum well or period of superlattice, m
D density of states per unit volume, m^{-3}
E energy, J
E_c conduction band edge, J
f arbitrary function, periodic in time or space

F	interatomic force, N	Γ	center of the first Brillouin zone for an fcc lattice
$\mathbf{G}$	reciprocal space lattice vector, m^{-1}		
h	Planck constant, J s	δ	delta function, eq. (3.15)
$\hbar$	Planck constant divided by 2π, J s	ΔE	band edge offset, J
k	magnitude of wavevector, m^{-1}	ε	parameter in Lennard–Jones potential, J
$\mathbf{k}$	wavevector, m^{-1}		
K	spring constant between atoms, N m^{-1}; quantity defined by eq. (3.24)	ε_0	electrical permittivity in vacuum, C^2 m^{-2} N^{-1}
L	L-point of the Brillouin zone of an fcc lattice, [111] direction; length of crystal, m	ζ	parameter in Born–Meyer repulsive potential, m, eq. (3.4)
		θ	polar angle
m	mass, kg	κ_B	Boltzmann constant, J K^{-1}
m_{ij}	effective mass tensor, kg	λ	wavelength, m
m^*	effective mass, kg	ν	frequency of phonons and photons, Hz
n	integer; local electron density, m^{-3}		
		ρ	density, kg m^{-3}
N	total number of atoms in the crystal; number of states	σ	parameter in Lennard–Jones potential, m
P	constant in eq. (3.32)	ϕ	azimuthal angle
q	charge per ion, C	Ψ	wavefunction
Q	quantity defined by eq. (3.24)	ω	angular frequency, rad.Hz
r	separation between atoms	Ω	solid angle, srad
r_0	nearest neighbor separation, m	( )	Miller index of a crystal plane
$\mathbf{r}$	atom position	{ }	Miller index for identical crystal planes
$\mathbf{R}$	translational vector		
s	integer	[ ]	vector along a crystallographic direction
S	electron or photon amplitude at detector		

Subscripts

t	time, s
T	period in time, s
u	atom displacement from its equilibrium position
U	interatomic potential, J
v	velocity, m s^{-1}
V	volume of primitive unit cell, m^3; volume of crystal
x	atom coordinate
X	X-point of fcc reciprocal lattice, [100] direction
α	Madelung constant for ionic crystals

A	attractive
c	conduction band
D	Debye
f	Fermi level
G	bandgap
ij	between atom i and atom j
R	repulsive
x, y, z	Cartesian coordinate direction
v	valence band

Superscript

*	complex conjugate

3.8 References

Ashcroft, N.W., and Mermin, N.D., 1976, *Solid State Physics*, Saunders College, Philadelphia.

Bannov, N., Aristov, V., Mitin, V., and Stroscio, M.A., 1995, "Electron Relaxation Time due to the Deformation-Potential. Interaction of Electron with Confined Acoustic Phonons in a Free-Standing Quantum Well," *Physical Review B*, vol. 51, pp. 9930–9942.

Borca-Tasciuc, T., Liu, W.L., Zeng, T., Song, D.W., Moore, C.D., Chen, G., Wang, K.L., Goorsky, M.S., Radetic, T., Gronsky, R., Koga, T., and Dresselhaus, M.S., 2000, "Thermal Conductivity of Symmetrically Strained Si/Ge Superlattices," *Superlattices and Microstructures*, vol. 28, pp. 119–206.

Brockhouse, B.N., Arase, T., Caglioti, G., Rao, K.R., and Woods, D.B., 1962, "Crystal Dynamics of Lead. I. Dispersion Curves at 100°K," *Physical Review*, vol. 128, pp. 1099–1111.

Casimir, H.B.G., 1938, "Note on the Conduction of Heat in Crystals" *Physica*, vol. 5, pp. 495–500.

Chang, L.L., Esaki, L., Howard, W.E., and Ludeke, R., 1973, "The Growth of GaAs–GaAlAs Superlattice," *Journal of Vacuum Science and Technology*, vol. 10, pp. 11–16.

Chelikowsky, J.R., and Cohen, M.L., 1976, "Nonlocal Pseudopotential Calculations for the Electronic Structure of Eleven Zinc-Blende Structures," *Physical Review B*, vol. 14, pp. 556–582.

Chen, G., 2001, "Phonon Heat Conduction in Low-Dimensional Structures," in *Semiconductors and Semimetals, Recent Trends in Thermoelectric Materials Research III*, vol. 71, pp. 203–259, Academic Press, San Diego.

Dresselhaus, M.S., Dresselhaus, G., and Phaedon, A., 2001, *Carbon Nanotubes: Synthesis, Structure, Properties, and Applications*, Springer, New York.

Esaki, L., and Tsu, R., 1970, "Superlattice and Negative Differential Conductivity in Semiconductors," *IBM Journal of Research and Development*, vol. 14, pp. 61–65.

Giannozzi, P., de Cironocoli, S., Pavone, P., and Baroni, S., 1991, "Ab Initio Calculation of Phonon Dispersions in Semiconductors," *Physical Review B*, vol. 43, pp. 7231–7242.

Heavens, O.S., 1965, *Optical Properties of Thin Solid Films*, Dover Publications, New York.

Iijima, S., 1991, "Helical Microtubules of Graphitic Carbon," *Nature*, vol. 354, pp. 56–58.

Joannopoulos, J.D., Meade, R.D., and Winn, J., 1995, *Photonic Crystals*, Princeton University Press, Princeton, NJ.

Joannopoulos, J.D., Villeneuve, P.R., and Fan, S., 1997, "Photonic Crystals: Putting a New Twist on Light," *Nature*, vol. 386, pp. 143–149.

Kim, P., Shi, L., Majumdar, A., and McEuen, P.L., 2001, "Thermal Transport Measurements of Individual Multiwalled Nanotubes," *Physical Review Letters*, vol. 87, pp. 215502/1–4.

Kittel, C., 1996, *Introduction to Solid State Physics*, 7th ed., Wiley, New York.

Mattheiss, L.F., 1964, "Energy Bands for Iron Transition Series," *Physical Review*, vol. 134, pp. A970–A973.

Narayanamurti, V., Stormer, H.L., Chin, M.A., Gossard, A.C., and Wiegmann, W., 1979, "Selective Transmission of High-Frequency Phonons by a Superlattice: The 'Dielectric' Phonon Filter," *Physical Review Letters*, vol. 43, pp. 2012–2016.

Shur, M., 1990, *Physics of Semiconductor Devices*, Prentice Hall, Englewood Cliffs, NJ.

Stillinger, F.H., and Weber, T., 1985, "Computer Simulation of Local Order in Condensed Phases of Silicon," *Physical Review B*, vol. 31, pp. 5262–5271.

Weisbuch, C., and Vinter, B., 1991, *Quantum Semiconductor Structures*, Academic Press, Boston.

Yablonovitch, E., 1986, "Inhibited Spontaneous Emission in Solid-State Physics and Electronics," *Physics Review Letters*, vol. 58, pp. 2059–2062.

Yang, B., and Chen, G., 2000, "Lattice Dynamics Study of Phonon Heat Conduction in Quantum Wells," *Physics of Low-Dimensional Structures Journal*, vol. 5/6, pp. 37–48.

Yang, B., and Chen, G., 2001, "Anisotropy of Heat Conduction in Superlattices," *Microscale Thermophysical Engineering*, vol. 5, pp. 107–116.

3.9 Exercises

3.1 *Number of atoms.* How many silicon atoms are there in a cube of 100 Å, 1000 Å, and 1 μm?

3.2 *Density of crystals.* The lattice constants of germanium and GaAs are 5.66 Å and 5.65 Å, respectively. Ge has a diamond structure and GaAs has a zinc-blende structure. Calculate their density.

ENERGY STATES IN SOLIDS   121

3.3 *Unit cell in real and reciprocal space.* A body-centered cubic lattice has the following primitive translation vector:

$$\mathbf{a}_1 = \frac{1}{2}a(-\hat{\mathbf{x}} + \hat{\mathbf{y}} + \hat{\mathbf{z}}); \quad \mathbf{a}_2 = \frac{1}{2}a(\hat{\mathbf{x}} - \hat{\mathbf{y}} + \hat{\mathbf{z}}); \quad \mathbf{a}_3 = \frac{1}{2}a(\hat{\mathbf{x}} + \hat{\mathbf{y}} - \hat{\mathbf{z}})$$

(a) Construct the Wigner–Seitz cell in real space.
(b) Find out the corresponding primitive translation vector in the reciprocal space and prove that the reciprocal lattice is an fcc structure.
(c) Sketch the Wigner–Seitz cell in the reciprocal space, that is, the first Brillouin zone.

3.4 *Lennard–Jones potential.* The values of the Lennard–Jones potential for noble gas crystals are given in table 3.1. For argon crystal,
(a) Calculate the interatomic distance.
(b) Calculate the energy at the minimum (called cohesive energy).
(c) Calculate the effective spring constant between two argon atoms.

3.5 *Miller index.* Index the following planes in a silicon crystal: (100), (110), (111), (121).

3.6 *X-ray diffraction.* In an X-ray diffraction experiment a, the angle formed between the incident ray and the detector (2θ) is $90°$ and first-order diffraction peaks are observed with the X-ray wavelength at 1 Å. Determine the distance of the crystal planes in this specific direction.

3.7 *Kronig–Penney model.* For $P = 5$, find out possible solutions for Ka in eq. (3.32). Convert the solution into a relationship between wavevector and energy and plot the solution in
(a) extended zone representation,
(b) reduced zone representation.

3.8 *Phonon spectra of a diatomic lattice chain.* Consider a diatomic chain of atoms as shown in figure P3.8. The masses of the two atoms are different but the spacing and the spring constant between them are the same. Derive the following given expression for the phonon dispersion in this diatomic lattice chain. Schematically draw the phonon dispersion you obtained

$$\omega^2 = K\left(\frac{M_1 + M_2}{M_1 M_2}\right) \pm \left[K^2 \left(\frac{M_1 + M_2}{M_1 M_2}\right)^2 - \frac{4K^2}{M_1 M_2}\sin^2\left(\frac{1}{2}ka\right)\right]^{1/2}$$

where K is the spring constant and k the wavevector with the following values

$$k = \pm\frac{2\pi}{Na}, \pm\frac{4\pi}{Na}, \ldots, \frac{\pi}{a}$$

and N is the total number of lattice points in the chain.

3.9 *Electron density of states.* For an elliptical electronic band given by eq. (3.37), derive an expression for the electron density of states.

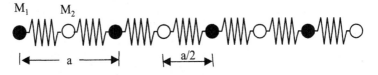

Figure P3.8 Figure for problem 3.8.

3.10 *Electron density of states inside quantum wires.* The electron energy dispersion in an infinite potential barrier quantum wire can be expressed as

$$E(k_x, \ell, n) = \frac{\hbar^2 k_x^2}{2m^*} + \frac{\hbar^2 \pi^2}{2m^*}\left[\left(\frac{\ell}{L_y}\right)^2 + \left(\frac{n}{L_z}\right)^2\right]$$

where ℓ, n can take integer values $1, 2, \ldots$. Derive an expression for the electron density of states and plot this expression for $L_y = L_z = 50$ Å.

3.11 *Electron density of states inside quantum dots.* Determine the electron density of states of a cubic quantum dot with side length $d = 20$ Å, assuming that the electron effective mass equals the free electron mass.

3.12 *Phonon density of states.* Assuming that phonons of a three-dimensional crystal obey the following isotropic dispersion relation,

$$\omega = 2\sqrt{\frac{K}{m}}\left|\sin\frac{ka}{2}\right|$$

where a is the lattice constant, derive an expression for the phonon density of states.

3.13 *Debye approximation.* Derive an expression for sound velocity from eq. (3.45). Calculate the sound velocity for a monatomic fcc crystal along (100) and (111) directions, using this simplified expression. Assume that the mass of the atom is 9.32×10^{-23} kg, the lattice constant of the conventional fcc unit cell is 5.54×10^{-10} m, and the spring constant is 7600 N m^{-1}.

3.14 *Transverse and longitudinal phonons.* Consider three separate acoustic phonons in a three-dimensional isotropic medium with an effective lattice constant of 2.5 Å. The dispersion for each branch is $\omega_L = v_L k$, $\omega_t = v_t k$ (degenerate). For $v_L = 8000$ m s^{-1} and $v_t = 5000$ m s^{-1}, plot the density of states as a function of frequency.

3.15 *Size effects on density of states.* The density of states expressions we derived are valid when the separations between states are small and the number of states is large, such that we can calculate the number of states by eq. (3.50). In small geometries, the energy separations between different states can be large and the number of states at each energy level can be small, so that eq. (3.50) is no longer valid. As an example, consider a cubic cavity of $(2 \,\mu\text{m})^3$ size. Find out how many states are allowed to exist inside the cavity for electromagnetic waves with a wavelength in the range of 0.5–1 μm, using the following two methods:

  (a) by finding out how many sets of (k_x, k_y, k_z) are allowed in this cavity that fall into the given wavelength range;
  (b) by integrating eq. (3.59) over the given wavelength range.

4

Statistical Thermodynamics and Thermal Energy Storage

The quantum mechanics principles covered in the previous two chapters give the allowable energy states of matter. The number of allowable states in typical macroscopic matter is usually very large, and at any instant, the matter can be at any one of these states. Although our mathematical treatments in the previous two chapters were based on solving the steady-state Schrödinger equation for the energy states and the wavefunctions, matter will not stay at one particular quantum state (a microscopic state) for long because of the interactions among particles (atoms, molecules, electrons, and phonons) in the matter. For example, we assumed a harmonic potential between atoms to obtain the phonon dispersion relation. In reality, the interatomic potential is not harmonic, as one can easily infer from examining the Lennard–Jones potential. When the anharmonicity (the deviation from the harmonic potential) is small, the solutions of the Schrödinger equation for the quantum states based on the harmonic potential are approximately correct. Yet a small degree of anharmonicity can cause a rapid ($\sim 10^{-9}$–10^{-13} s) change of the matter from one quantum state to another. Due to the large number of quantum states available in matter, it is impractical to follow the real time evolution of matter among its allowable quantum states. The bridge connecting the allowable quantum states to the macroscopic behavior is provided by statistical thermodynamics, which determines the probability that matter will be at a particular quantum state when it is at equilibrium. Through statistical thermodynamics, temperature enters into the picture of energy storage and transport.

In this chapter, we focus on the equilibrium state of a system and discuss different probability functions for systems under different constraints, such as an isolated system or a system at constant temperature. From the probability distribution functions, we will show how to calculate the internal energy and specific heat of a system, including nanostructures, using what we learned in previous chapters about the energy levels, degeneracy, and the density of states.

4.1 Ensembles and Statistical Distribution Functions

In an actual experiment, we often follow the time history of a system and observe its time-averaged behavior. In analysis, however, following the time history requires the solution of master equations that govern the motion of a large number of particles, such as the Newton equations of motion and the time-dependent Schrödinger equation. Although, with increasing computational power, such computation is becoming feasible for limited situations, as in the molecular dynamics simulations to be introduced in chapter 10, for most applications direct computation of the time history is impractical. Statistical thermodynamics avoids the time averaging by introducing *ensembles*, which are large collections of systems, each representing a microscopic state that satisfies the macroscopic constraints. Examples of the macroscopic constraints of a system are its total energy, temperature, and volume. The quantities to be measured are averaged over the ensemble at a fixed time, rather than, as in an experimental situation, over a time period of a single system. A fundamental assumption made in statistical mechanics is that the ensemble average of an observed quantity is equal to the time average of the same quantity. This assumption is called the *ergodic hypothesis*. The study of conditions necessary for this hypothesis to be valid is an ongoing research area (Kubo et al., 1998). Our analysis will assume that all systems are ergodic. Depending on the macroscopic constraints, various ensembles are developed, each has a probability distribution for the microscopic states in the ensemble that differs from other ensembles. In the following, we will discuss three ensembles: microcanonical, canonical, and grand canonical ensembles.

4.1.1 Microcanonical Ensemble and Entropy

Unlike classical thermodynamics, which completely neglects the microscopic processes in a system, statistical thermodynamics builds the system properties from its microscopic states (Kittel and Kroemer, 1980; Callen, 1985). We consider an *isolated* macroscopic system with a volume V, a total number of particles N, and a total energy of U (macroscopic constraints). The quantum states of the system that satisfy these macroscopic constraints are called the accessible quantum states (or simply accessible states). Given these macroscopic constraints, together with detailed information about the interatomic potentials between the particles in the system and the initial conditions, one could in principle solve the Schrödinger equation to follow the temporal evolution of the system among the accessible quantum states. The macroscopic properties of the system, such as temperature and pressure, are a measure of the average corresponding microscopic properties over a certain amount of time.

How can we calculate the average values of this system? If we performed an experiment, we would measure these values as a function of time and carry out a time average. In statistical mechanics, instead of tracing the time evolution of the system, we focus on the probability of a system being at a specific accessible quantum state. A *fundamental*

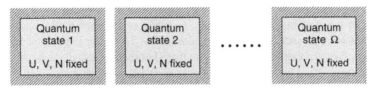

Figure 4.1 A microcanonical ensemble is made of isolated systems with fixed U, V, and N. Each system corresponds to one accessible quantum state of the original system.

postulate in statistical mechanics is that an isolated macroscopic system samples every accessible quantum state with equal probability. This postulate is also called the *principle of equal probability*. If Ω is the total number of accessible quantum states, the probability of each accessible quantum state, denoted by s, being sampled is

$$P(s) = 1/\Omega \tag{4.1}$$

Once the probability of each accessible quantum state is known, we can construct a way to calculate a desired macroscopic quantity $\langle X \rangle$ of a macroscopic system. We first evaluate the corresponding property X (such as temperature, pressure) for each accessible quantum state, and then calculate the average according to

$$\langle X \rangle = \sum_{s=1}^{\Omega} P(s) X(s) \tag{4.2}$$

Note that the summation is over all accessible quantum states.

Because each accessible quantum state is a state of the system that satisfies the macroscopic constraints and each one has equal probability to be sampled, we are effectively dealing with a collection of Ω stationary systems, as shown in figure 4.1. These systems are identical from the macroscopic points of view; that is, they have the same U, V, and N and are all isolated from their surroundings. This collection of Ω systems is called an *ensemble*. A fixed U, V, N ensemble is called a *microcanonical ensemble*. The principle of equal probability is valid only for each system in a microcanonical ensemble. Later, we will introduce canonical and grand canonical ensembles and derive the probability of each system in such ensembles on the basis of results we obtained from the microcanonical ensemble. Equation (4.2) means that each of the stationary systems in the ensemble is sampled once in the computing of the average. Such an average is called the ensemble average. For a microcanonical ensemble, $P(s) = 1/\Omega$, thus

$$\langle X \rangle = \sum_{s=1}^{\Omega} X(s)/\Omega \tag{4.3}$$

The idea of equal probability for each accessible quantum state in an isolated system may seem unreasonable for some readers. For example, for a system of 10^{23} particles, one accessible quantum state might be that one particle has energy U and the rest have zero energy. This distribution of energy among N particles seems to be a quite unlikely event but the principle of equal probability states that such a state is just as probable as any other quantum states, for example, a state in which each particle has an energy U/N (assuming that such a state is also accessible). The latter is perceived to be more likely. This concern can be resolved by noting that there is usually a large number of accessible states close to the latter case (large degeneracy) so that an actual observation would most likely sample one of these high-degeneracy states.

Suppose that we have determined the number of accessible states Ω of an isolated system with fixed U, V, and N. How can we relate Ω to macroscopic thermodynamic quantities? For a microcanonical ensemble, we rarely use eq. (4.3) to calculate the average quantities. Rather, we use a crucial link established by Ludwig Boltzmann (1844–1906), who showed that Ω is directly related to the entropy of the system through

$$S = \kappa_B \ln \Omega \qquad (4.4)$$

where κ_B is the Boltzmann constant ($= 1.38 \times 10^{-23}$ J K^{-1}). Equation (4.4) is called the *Boltzmann principle*. This relation between entropy and accessible states is consistent with the typical interpretation in classical thermodynamics that entropy is a measure of the randomness of the system. The larger the number of accessible states, the more freedom a macroscopic system has, and thus the more randomness it has. The Boltzmann constant (and its units) is a conversion factor that makes the classical definition of entropy consistent with its microscopic definition while the logarithm satisfies the additiveness of the entropy used in classical thermodynamics. For example, if the system can be divided into two subsystems, each having Ω_1 and Ω_2 accessible states, the total number of accessible states of the system is then $\Omega_1 \times \Omega_2$. Equation (4.4) leads to $S = S_1 + S_2$, satisfying the additiveness of entropy.

Because Ω is constrained by U, V, and N, we anticipate that it is a function of these variables, and thus so is entropy,

$$S = S(U, V, N) \qquad (4.5)$$

If we know the above function, we can construct all other thermodynamic properties of the system. For example, eq. (4.5) can be written as

$$dS = \left(\frac{\partial S}{\partial U}\right)_{V,N} dU + \left(\frac{\partial S}{\partial V}\right)_{U,N} dV + \left(\frac{\partial S}{\partial N}\right)_{U,V} dN$$

The above equation, when combined with the more familiar form of $dU = T\,dS - p\,dV + \mu\,dN$ from classical thermodynamics, immediately leads to

$$\frac{1}{T} = \left(\frac{\partial S}{\partial U}\right)_{V,N}, \frac{p}{T} = \left(\frac{\partial S}{\partial V}\right)_{U,N}, -\frac{\mu}{T} = \left(\frac{\partial S}{\partial N}\right)_{U,V} \qquad (4.6)$$

where μ is the chemical potential. Equation (4.6) means that if we know the function $S(U,V,N)$ [or $U(S,V,N)$], we can determine all other thermodynamic state variables such as temperature, pressure, and chemical potential. The function $S(U,V,N)$ is called a *thermodynamic potential*.

It is important to realize that a thermodynamic potential must be expressed in terms of its corresponding variables U, V, and N. Not any arbitrary combination of macroscopic thermodynamic properties can be the variables of a thermodynamic potential. For example, a known function of entropy $S(T, p, N)$ with T, p, N as variables does not give all the other thermodynamic properties of the system and is thus not a thermodynamic potential. In the next section, we will examine two other types of ensemble, each leads to the construction of a thermodynamic potential that gives a full description of the macroscopic thermodynamic properties of the system.

4.1.2 Canonical and Grand Canonical Ensembles

In practice, we are often more interested in determining how system properties change with temperature. The microcanonical ensemble is not convenient for this purpose since one of its natural thermodynamic variables is internal energy rather than temperature. Instead of considering an isolated system of fixed energy, let's now consider a system of given temperature T, fixed volume V, and fixed number of particles N. To maintain the system at a fixed temperature, we assume that it is in contact with a thermal reservoir at the same temperature. A thermal reservoir is a very large object, such that its temperature does not change even if there is energy transfer between the system and the reservoir. Because of this energy exchange, the system energy is not fixed. Our goal is to obtain the probability of finding the system at a specific accessible quantum state with energy E_i, satisfying the macroscopic constraints of constant T, V, and N. We will see that, because of the change of constraint from fixed U to fixed T, the probability of observing each accessible quantum state is no longer identical as in a microcanonical ensemble.

We start from the microcanonical ensemble statistics established in the previous section by considering that the system (and let's call it the original system) and the thermal reservoir form a large isolated system (we will call it the combined system) with fixed U_t, V_t, and N_t. The combined system has a total number of Ω_t quantum states. We can construct a microcanonical ensemble of the combined system as shown in figure 4.2. In such an ensemble, because the reservoir is very large, the total number of accessible quantum states of the combined system Ω_t is much larger than the number of accessible quantum states Ω_s of the original system. It is very likely that corresponding to a specific accessible quantum state of the original system there are many accessible quantum states in the reservoir. Rather than examining a microcanonical ensemble of the combined system as shown in figure 4.2(a), we construct a new ensemble made of Ω_s systems. Each of the systems in the new ensemble is an accessible quantum state of the original system with macroscopic constraints of constant T, V, and N, as shown in figure 4.2(b), and is in thermal equilibrium with a common reservoir. Because each system in the new ensemble may have a different number of accessible states in the reservoir, the chance that we observe each system in the ensemble is no longer identical to each other as in a microcanonical ensemble. Consequently, we call this ensemble a *canonical ensemble*. We would like to determine the probability of finding a system at a specific accessible quantum state with an energy E_i. Let's assume that there are $\Omega_r(U_t - E_i)$ accessible states in the reservoir corresponding to this specific system state. The probability of observing the system is then

$$P(E_i) = \frac{\Omega_r(U_t - E_i)}{\Omega_t(E_t)} \tag{4.7}$$

128 NANOSCALE ENERGY TRANSPORT AND CONVERSION

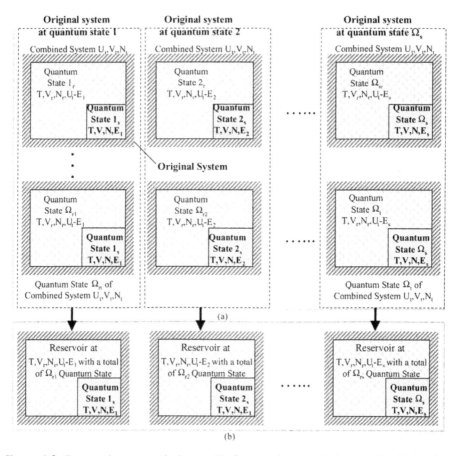

Figure 4.2 Constructing a canonical ensemble from a microcanonical ensemble. (a) A microcanonical ensemble for the combined system comprises the reservoir and the original system, with a total of Ω_t systems, each with equal probability. Each column represents microcanonical ensembles with the original system in one quantum state, while the reservoirs may have many quantum states. (b) A canonical ensemble has Ω_s systems with fixed T, V, and N. Each system has a number of quantum states in the reservoir and thus not every system has an equal probability of being observed.

To find Ω_r in the above expression, we consider another combined system made of one quantum state of the original system at energy E_i and Ω_r quantum states of the reservoir with an energy $(U_t - E_i)$. This "new" combined system occupies only one column in figure 4.2(a) of the previously established "old" combined microcanonical ensemble and is also a microcanonical ensemble because of its fixed energy. Assuming that the entropy of this new combined system is S_r and the entropy of the old microcanonical ensemble in figure 4.2(a) is $S_t(U_t)$, we can use eq. (4.4) to rewrite eq. (4.7) as

$$P(E_i) = \frac{\exp[S_r(U_t - E_i)/\kappa_B]}{\exp[S_t(U_t)/\kappa_B]} \qquad (4.8)$$

If U is the average energy of the original system, then the total entropy of the old combined system is

$$S_t(U_t) = S_r(U_t - U) + S(U) \tag{4.9}$$

By a Taylor expansion, we can express $S_r(U_t - E_i)$ as

$$\begin{aligned} S_r(U_t - E_i) &= S_r[(U_t - U) + (U - E_i)] \\ &= S_r(U_t - U) + \left.\frac{\partial S_r}{\partial E_r}\right|_{U_t - U} (U - E_i) \\ &= S_r(U_t - U) + \frac{U - E_i}{T} \end{aligned} \tag{4.10}$$

where we have applied eq. (4.6) to the new combined microcanonical ensemble,

$$\frac{1}{T} = \left[\frac{\partial S_r}{\partial E_r}\right]_{U_t - U} \tag{4.11}$$

Substituting eq. (4.10) into eq. (4.8) leads to

$$P(E_i) = \exp\left[\frac{U - TS}{\kappa_B T}\right] \exp\left[-\frac{E_i}{\kappa_B T}\right] \tag{4.12}$$

where we have used $S(U) = S_t(U_t) - S_r(U_t - U)$, that is, eq. (4.9), to eliminate the entropy of the thermal reservoir. We recognize that $F = U - TS$ is the *Helmholtz free energy* of the system—a thermodynamic potential with natural variables of T, V, and N. Given $F(T, V, N)$, we can calculate all other thermodynamic properties of the system. We use the probability normalization requirement to find F,

$$\sum_i P(E_i) = \exp\left[\frac{F}{\kappa_B T}\right] \sum_i \exp\left[-\frac{E_i}{\kappa_B T}\right] = 1 \tag{4.13}$$

where the summation is over all the accessible quantum states of the system. Equation (4.13) gives

$$F(T, V, N) = -\kappa_B T \ln Z \tag{4.14}$$

where Z is called the *canonical partition function*

$$Z = \sum_i \exp\left[-\frac{E_i}{\kappa_B T}\right] \tag{4.15}$$

Substituting eq. (4.14) into eq. (4.12) leads to the probability of a fixed V, N, T system at a quantum state having energy E_i as

$$P(E_i) = e^{-E_i/k_B T}/Z \tag{4.16}$$

The factor $\exp(-E_i/k_B T)$ is called the *Boltzmann factor*. It is widely seen in different disciplines of science. You may have encountered this exponential form somewhere else.

For example, the Arrhenius law governing chemical reactions is a manifestation of the Boltzmann factor.

A similar analysis can be extended to consider a system of fixed volume that exchanges both energy and particles with a larger reservoir. From classical thermodynamics, we know that the driving force for particle flow is the chemical potential that is defined by eq. (4.6). An ensemble of such a system in equilibrium with the reservoir is called a *grand canonical ensemble*. The variables for such an ensemble are T, V, and μ, and the corresponding thermodynamic potential is called the *grand canonical potential* $G(T, V, \mu)$. Following a similar analysis to that of a canonical ensemble, the probability of a quantum state having energy E_i and number of particles N_i is given by

$$P(E_i, N_i) = \frac{\exp\left[\frac{N_i\mu - E_i}{\kappa_B T}\right]}{\Im} \tag{4.17}$$

The exponential factor is called the *Gibbs factor* and the denominator is the *grand canonical partition function*, given by

$$\Im(T, V, \mu) = \sum_{N_i}\sum_{E_i} \exp\left[\frac{N_i\mu - E_i}{\kappa_B T}\right] = \sum_{N_i} \gamma^{N_i} Z_i \tag{4.18}$$

where $\gamma = \exp(\mu/\kappa_B T)$ and the double summation is over all accessible energy states and number of the particles of the system. The grand canonical potential is

$$G(T, V, \mu) = U - TS - \mu N = -\kappa_B T \ln \Im \tag{4.19}$$

where U and N are the average energy and number of particles of the system, respectively.

4.1.3 Molecular Partition Functions

The above discussion shows that if the partition function is known, the thermodynamic potential of a system can be determined, and consequently all other thermodynamic quantities are also known. We discuss below the partition function of gas molecules.

Let's start with the partition function of a single molecule. To find the accessible quantum states, we can extend the solution for the energy levels of a particle-in-a-potential-well model to the three-dimensional case and arrive at the following form of the quantized energy levels for the translational motion of a gas molecule in a cubic box of length L [see eq. (E2.1.6)],

$$E = \frac{\pi^2 \hbar^2}{2mL^2}(n_x^2 + n_y^2 + n_z^2) \text{ where } n_x, n_y, n_z = 1, 2, 3, \ldots \tag{4.20}$$

and the density of states per unit volume is

$$D(E) = \frac{1}{4\pi^2}(2m/\hbar^2)^{3/2} E^{1/2} \tag{4.21}$$

Substituting eq. (4.20) into eq. (4.15) leads to the canonical partition function for the translational motion of one gas molecule,

$$Z = \sum_{n_x=1}^{\infty} \sum_{n_y=1}^{\infty} \sum_{n_z=1}^{\infty} \exp\left[-\frac{\pi^2 \hbar^2 (n_x^2 + n_y^2 + n_z^2)}{2mL^2 \kappa_B T}\right] \quad (4.22)$$

To evaluate the above triple summation, we first notice that $\lfloor \pi^2 \hbar^2 / (2mL^2 \kappa_B T) \rfloor$ in the exponent is a very small number such that the exponential function is slowly varying. Second, n_x, n_y, and n_z are integers spaced by Δn_x (or Δn_y, Δn_z) = 1. Due to the above two reasons, the summation can be well approximated by integration,

$$Z = \int_0^{\infty} \int_0^{\infty} \int_0^{\infty} \exp\left[-\frac{\pi^2 \hbar^2 (n_x^2 + n_y^2 + n_z^2)}{2mL^2 \kappa_B T}\right] dn_x dn_y dn_z$$

$$= V \left(\frac{2\pi m \kappa_B T}{h^2}\right)^{3/2} = \frac{V}{\lambda^3} \quad (4.23)$$

where $V = L^3$ and

$$\lambda = \frac{h}{\sqrt{2\pi m \kappa_B T}} \quad (4.24)$$

is called the *thermal de Broglie wavelength*.

Another way to arrive at the same answer as eq. (4.23) is to realize that the triple summation in eq. (4.22) is essentially sampling all the quantum states. With a slowly varying exponential, we can convert this summation over the quantum states into an integration over allowable energy levels, using the density-of-states,

$$Z = V \int_0^{\infty} \exp\left[-\frac{E}{\kappa_B T}\right] D(E) dE$$

$$= V \int_0^{\infty} \frac{1}{4\pi^2} \left(\frac{2m}{\hbar^2}\right)^{3/2} E^{1/2} \exp\left[-\frac{E}{\kappa_B T}\right] dE = \frac{V}{\lambda^3} \quad (4.25)$$

which is identical to eq. (4.23).

For a dilute gaseous system with N molecules, the energy eigenvalues from a single particle-in-a-box model apply to every particle because the potential interactions among the particles are weak. The total energy of the system equals the summation of the energy of all particles $E_t = E_1 + E_2 + \cdots + E_N$, where E_i represents the possible energy values of particle i and is given by eq. (4.20) without any additional constraints for the indices n_{xi}, n_{yi}, n_{zi}, because the total energy of a canonical system is not a prior constraint of the system and thus can take any value. From eq. (4.15), we can write the canonical particle function for this N-molecule system as

$$Z_N = \sum_{n_{x1}, n_{y1}, n_{z1}} \sum_{n_{x2}, n_{y2}, n_{z2}} \cdots \sum_{n_{xN}, n_{yN}, n_{zN}} \exp\left(-\frac{E_1 + E_2 + \cdots + E_N}{\kappa_B T}\right) \quad (4.26)$$

where each summation is over all possible states of molecule i as determined by indices (n_{xi}, n_{yi}, n_{zi}) with specified values as in eq. (4.22). If the molecules were distinguishable, that is, if every quantum state of particle i were different from that of particle j, eq. (4.22) could be factorized into Z^N. Real molecules, however, are identical to each other and are thus indistinguishable. This indinstinguishability affects how we perform the summations in eq. (4.26) because one accessible quantum state should be counted only once in the summation. The indistinguishability of molecules means that if two molecules i and j are at an identical quantum state, for example, with $n_{xi} = n_{xj} = 10, n_{yi} = n_{yj} = 10$, and $n_{zi} = n_{zj} = 10$, they should be counted in the summation of eq. (4.26) once only rather than twice because no way exists to distinguish the two molecules in the system. In general, counting such indistinguishable cases is difficult, but can be done when the accessible quantum states of one molecule as given by eq. (4.20) are much larger than the total number of molecules such that no two molecules occupy the same energy states at the same time. Under this dilute-gas limit, we overcount the indistinguishable molecules by $N!$ when treating them as distinguishable, where $N! = N(N-1)(N-2)\ldots 1$ is the factorial of N. Thus in the dilute gas limit, the canonical partition function, eq. (4.26), can be simplified to

$$Z_N = \frac{Z^N}{N!} = \frac{1}{N!}\left(\frac{V}{\lambda^3}\right)^N \tag{4.27}$$

From the canonical partition function Z_N, we can calculate the Helmholtz energy of the gas system:

$$F(T, V, N) = -\kappa_B T \ln Z_N = -N\kappa_B T \left[\ln V - \frac{3}{2}\ln\left(\frac{h^2}{2\pi m \kappa_B T}\right)\right]$$
$$+ \kappa_B T(N \ln N - N) \tag{4.28}$$

where we have used the Stirling approximation: $\ln N! \approx N \ln N - N$. This approximation is valid when N is large, which is typically the case. With $F(T, V, N)$ known, all other thermodynamic quantities of the system can be obtained. For example, from $dF = -S\,dT - p\,dV + \mu\,dN$, we can calculate the pressure as

$$p = -\left(\frac{\partial F}{\partial V}\right)_{T,N} = \frac{\kappa_B T N}{V} \tag{4.29}$$

and the internal energy can be calculated from

$$U = \sum_i E_i P(E_i) = \sum_i \frac{E_i \exp(-E_i/\kappa_B T)}{Z_N}$$
$$= \kappa_B T^2 \frac{\partial \ln Z_N}{\partial T} = \frac{3N\kappa_B T}{2} \tag{4.30}$$

The last two equations should be familiar. Equation (4.29) is the ideal gas law that applies to dilute gas. Equation (4.30) is equivalent to eq. (1.28) and is a special case of the *equipartition theorem*, which says that, at high temperature, every degree of freedom with a quadratic energy term contributes $\kappa_B T/2$ to the average energy of the system.

A monatomic molecule has three degrees of translational freedom only and the energy is in quadratic form, as eq. (4.20) shows. Each quadratic term in energy contributes $\kappa_B T/2$, and thus, we have an average energy of $3\kappa_B T/2$ for each molecule.

For a polyatomic gas, the energy level of each molecule can be separated into translational, vibrational, rotational, and electronic components:

$$E = E_t + E_v + E_r + E_e \tag{4.31}$$

and the corresponding partition function is

$$\begin{aligned} Z &= \sum e^{-E/\kappa_B T} \\ &= \sum e^{-E_t/\kappa_B T} \sum e^{-E_v/\kappa_B T} \sum e^{-E_r/\kappa_B T} \sum e^{-E_e/\kappa_B T} \\ &= Z_t Z_v Z_r Z_e \end{aligned} \tag{4.32}$$

where Z_t, Z_v, Z_r, and Z_e are the canonical partition functions for each energy component of the molecule as represented by the corresponding subscripts. Once the partition function for one molecule is known, the canonical partition function for a dilute system of N molecules can be calculated from eq. (4.27).

Before concluding this section, we present a criterion that determines when the dilute gas limit, which leads to the factorial $N!$ in eq. (4.27), is valid. The requirement for this limit is that the number of quantum states for one molecule is much larger than the number of molecules in the box. Thus if the number of quantum states between zero energy and the average molecular energy is much larger than the number of molecules per unit volume, then the dilute gas assumption should be valid; in other words,

$$\int_0^{3\kappa_B T/2} D(E)dE \gg \frac{N}{V} \tag{4.33}$$

where the left-hand side is the number of quantum states with an energy between zero and $3\kappa_B T/2$. Substituting eq. (4.21) into eq. (4.33) and carrying out the integration, we obtain a criterion for the dilute gas assumption to be valid as

$$\frac{6N}{\pi V}\left(\frac{h^2}{12 m \kappa_B T}\right)^{3/2} \ll 1 \text{ or } \frac{N}{V}\left(\frac{\pi}{6}\right)^{1/2} \lambda^3 \ll 1 \tag{4.34}$$

Because the intermolecular distance is of the order of $(V/N)^{1/3}$, eq. (4.34) also means that the thermal de Broglie wavelength must be much smaller than the intermolecular distance.

Example 4.1 *Canonical partition function*

Derive an expression for the canonical partition function of the rotational modes of a H_2 molecule in a box of H_2 gas.

Solution: We have obtained in chapter 2 the energy, eq. (2.65), and degeneracy, eq. (2.66), of a rigid rotor as

$$E_\ell = \frac{\hbar^2}{2I}\ell(\ell+1) = B\ell(\ell+1) \quad (\ell = 0, 1, 2, \ldots, |m| \leq \ell) \tag{E4.1.1}$$

$$g(\ell) = 2\ell + 1 \tag{E4.1.2}$$

where ℓ and m are the two quantum numbers of rational wavefunctions, and B is the rotational constant. The canonical partition function for the rotational modes is

$$\begin{aligned}Z_r &= \sum_{\ell,m} \exp\left(-\frac{E_\ell}{\kappa_B T}\right) = \sum_{\ell=0}^{\infty} g(\ell) \exp\left(-\frac{E_\ell}{\kappa_B T}\right) \\ &= \int_0^{\infty} (2\ell+1) \exp\left[-\frac{B\hbar\ell(\ell+1)}{\kappa_B T}\right] d\ell = \frac{8\pi^2 I \kappa_B T}{h^2} = \frac{T}{\theta_r}\end{aligned} \tag{E4.1.3}$$

where θ_r is called the rotational temperature

$$\theta_r = \frac{hB}{\kappa_B} = \frac{h^2}{8\pi^2 \kappa_B I} \tag{E4.1.4}$$

In eq. (E.4.1.3), the first summation over all ℓ and m is over all quantum states and the second summation over ℓ is over all energy levels. Similarly to eq. (4.23), we have converted the summation into an integral.

Comments. For hydrogen, $B = 1.8 \times 10^{12}$ Hz and $\theta_r = 85.3$ K. The transformation in eq. (E.4.1.3) from the summation into the integral is valid only when T is much larger than θ_r, that is, when changing ℓ by 1 does not change the exponential rapidly. So eq. (E.4.1.4) is valid only for $T \gg \theta_r$. In the limit, when T is comparable to θ_r or smaller, we can take the first few terms of the summation to get

$$Z_r = 1 + 3\exp\left(-\frac{3\theta_r}{T}\right) + \cdots \tag{E4.1.5}$$

4.1.4 Fermi–Dirac, Bose–Einstein, and Boltzmann Distributions

Let's now consider the probability of electrons occupying a specific quantum state. We assume that we have determined the accessible quantum states for electrons in a given system. From the Pauli exclusion principle, each quantum state can have a maximum of one electron. If the system is at equilibrium with a temperature T, we wish to determine the probability of one quantum state having energy E being empty or occupied by one electron. We take this specific quantum state as our system, and the rest of the accessible quantum states of the original system are grouped into the reservoir. There can be energy and particle exchanges between the new system and its reservoir because an electron can fluctuate randomly between this quantum state and other quantum states. Thus the appropriate ensemble for the new system is the grand canonical ensemble. The grand canonical partition function for the new system can be evaluated from eq. (4.18),

$$\Im(T, V, \mu) = \sum_{N_i=0}^{1} \gamma^{N_i} Z_i = 1 + \exp\left(\frac{\mu - E}{\kappa_B T}\right) \quad (4.35)$$

where $N_i = 0$ means that the quantum state is unoccupied, with system energy at zero, and $N_i = 1$ means that the state is occupied, with system energy at E. According to eq. (4.17), the probability that this quantum state is empty or occupied is, respectively,

$$P(E_i = 0, N_i = 0) = P_0 = \frac{1}{1 + \exp\left(\frac{\mu-E}{\kappa_B T}\right)} \quad \text{(empty)} \quad (4.36)$$

and

$$P(E_i = E, N_i = 1) = P_1 = \frac{\exp\left(\frac{\mu-E}{\kappa_B T}\right)}{1 + \exp\left(\frac{\mu-E}{\kappa_B T}\right)} \quad \text{(occupied)} \quad (4.37)$$

The average number of occupancy of this quantum state is thus

$$\langle n \rangle \equiv f(E) = 0 \times P(E_i = 0, N_i = 0) + 1 \times P(E_i = E, N_i = 1)$$

$$= \frac{1}{\exp\left(\frac{E-\mu}{\kappa_B T}\right) + 1} \quad (4.38)$$

and the average energy of this quantum state is

$$\langle E \rangle = 0 \times P(E_i = 0, N_i = 0) + E \times P(E_i = E, N_i = 1)$$

$$= \frac{E}{\exp\left(\frac{E-\mu}{\kappa_B T}\right) + 1} = Ef(E) \quad (4.39)$$

$\langle n \rangle$, or in a more popular symbol f, is called the *Fermi–Dirac distribution function*. Electrons and other particles that obey the Fermi–Dirac distributions are called *fermions*. Figure 4.3 illustrates this distribution function. Recall that μ is the chemical potential. When the energy is a few times of $\kappa_B T$ smaller than the chemical potential, the distribution is close to one, indicating that most of the energy states below the chemical potential are occupied. When the energy is a few times of $\kappa_B T$ larger than the chemical potential, the distribution function is close to zero, indicating that most states above the chemical potential are empty. Because the motion of electrons means that there must be unoccupied states for the electrons to fill, only the electrons close to the chemical potential are active in carrying the charge. At zero temperature, the chemical potential equals the Fermi level. In some fields, however, particularly electrical engineering, the chemical potential and the Fermi level are used interchangeably.

Next, let's consider the probability of phonons or photons occupying an accessible quantum state of the system. Unlike electrons, the number of phonons or photons in a system is not conserved. Thus N is not a thermodynamic variable for the system,

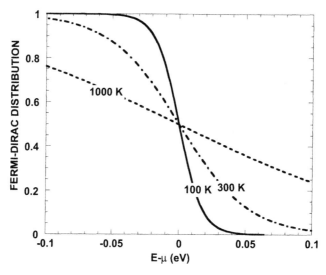

Figure 4.3 Fermi–Dirac distribution as a function of the electron energy relative to the chemical potential.

and, correspondingly, neither is the chemical potential. We know that for an accessible quantum state of the system, with frequency v, there can be an arbitrary number n of photons or phonons such that the total energy of this state is $E = (n + 1/2)hv$ ($n = 0, 1, 2, \ldots$). Following a similar argument as for electrons, we take this quantum state to be our new system and the remaining quantum states to be the reservoir. Since neither the chemical potential nor the particle number is a thermodynamic variable, the new system is best described by a canonical ensemble with the canonical partition function

$$Z(v) = \sum_{n=0}^{\infty} \exp\left(-\frac{(n+1/2)hv}{\kappa_B T}\right) = \frac{\exp\left(-\frac{hv}{2\kappa_B T}\right)}{1 - \exp\left(-\frac{hv}{\kappa_B T}\right)} \quad (4.40)$$

The probability that the quantum state (the new system) has n particles (photons or phonons) is thus

$$P(v, n) = \frac{\exp\left(-\frac{(n+1/2)hv}{\kappa_B T}\right)}{Z} = \exp\left(-\frac{nhv}{\kappa_B T}\right)\left[1 - \exp\left(-\frac{hv}{\kappa_B T}\right)\right] \quad (4.41)$$

and the average number of the particles, or the occupancy of the quantum state, is

$$\langle n \rangle \equiv f(v) = \sum_{n=0}^{\infty} n P(v, n) = \frac{1}{\exp\left(\frac{hv}{\kappa_B T}\right) - 1} \quad (4.42)$$

This equation is the *Bose–Einstein distribution function*, and the particles obeying this distribution are called *bosons*. Figure 4.4 shows the Bose–Einstein distribution. Because each particle has energy hv, the average energy of the quantum state is

$$\langle E \rangle = hv f(v) \quad (4.43)$$

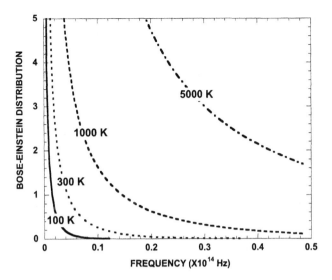

Figure 4.4 Bose–Einstein distribution as a function of the frequency of the carriers (phonons and photons).

where we have neglected the zero-point energy, which does not participate in heat transfer processes.

Other boson systems, such as gas molecules, can have a fixed number of particles. For such bosons, we should use the grand canonical ensemble as for fermions, and the general Bose–Einstein distribution can be written as

$$\langle n \rangle \equiv f(E) = \frac{1}{\exp\left(\frac{E-\mu}{\kappa_B T}\right) - 1} \quad (4.44)$$

where μ is again the chemical potential of the boson gas.

The Bose–Einstein distribution changes the "plus one" in the denominator of the Fermi–Dirac distribution into minus one. In the limit of low occupancy (high energy and high temperature), both Bose–Einstein and Fermi–Dirac distributions reduce to the Boltzmann distribution function

$$f(E, T, \mu) = \exp\left(-\frac{E-\mu}{\kappa_B T}\right) \text{ or } f(E) = \exp\left(-\frac{E}{\kappa_B T}\right) \quad (4.45)$$

This distribution function is considered as "classical", while the Fermi–Dirac and Bose–Einstein distributions are "quantum." Thus, for the statistical distributions, difference between "classical" and "quantum" statistics lies merely in the "one" of the denominator!

4.2 Internal Energy and Specific Heat

The statistical distribution functions establish a link with temperature between the quantum state and its energy level. With the distribution functions, we can investigate the properties of matter at finite temperatures. In this section we consider the internal energy and specific heat. Recall that the constant volume specific heat per unit volume, C_V [J m^{-3} K^{-1}], is defined as

$$C_V = \frac{1}{V}\left(\frac{\partial U}{\partial T}\right)_V \tag{4.46}$$

where U is the average internal energy of the system. We will consider the internal energy and specific heat of various energy carriers in this section.

4.2.1 Gases

For a dilute monatomic gas, the total internal energy is given by eq. (4.30). Consequently, the volumetric specific heat is

$$C_V = \frac{1}{V}\frac{3}{2}\kappa_B N \tag{4.47}$$

Since the number of molecules per mole equals Avogadro's constant $N_A = 6.02 \times 10^{23}$ mol^{-1}, the specific heat per mole for a monatomic gas is

$$c_V = \frac{3}{2}\kappa_B N_A = \frac{3}{2}R \tag{4.48}$$

where $R(=\kappa_B N_A = 8.314$ J K^{-1} mol$^{-1})$ is the universal gas constant.

For a diatomic gas, we should consider the contributions from the rotational and vibrational states. We already have from eq. (E.4.1.3) the rotational partition function of one diatomic molecule,

$$Z_r = \sum_{l=0}^{\infty}(2\ell+1)\exp\left[-\frac{\theta_r \ell(\ell+1)}{T}\right] \tag{4.49}$$

We will consider next the vibrational partition function. The vibrational energy of a harmonic oscillator was derived in chapter 2 as

$$E = h\nu\left(n+\frac{1}{2}\right) \quad (n=0,1,2,\ldots) \tag{4.50}$$

The vibrational partition function is thus

$$Z_v = \sum_{n=0}^{\infty}\exp\left(-\frac{h\nu(n+1/2)}{\kappa_B T}\right) = \frac{\exp\left(-\frac{\theta_v}{2T}\right)}{\exp\left(\frac{\theta_v}{T}\right)-1} \tag{4.51}$$

where $\theta_v = h\nu/\kappa_B$ is called the vibrational temperature.

In addition, the molecule also has electronic energy states. From the solution of the electronic energy levels in chapter 2 for a hydrogen atom, we know that the electronic energy levels are high and that their separations are large. So we can take the first term only of the electronic partition function

$$Z_e = g_{e1} \exp\left[-\frac{E_{e1}}{\kappa_B T}\right] + g_{e2} \exp\left[-\frac{E_{e2}}{\kappa_B T}\right] + \cdots \approx g_{e1} \exp\left[-\frac{E_{e1}}{\kappa_B T}\right] \quad (4.52)$$

where E_{ei} is the ith electronic energy level and g_{ei} is the degeneracy for that energy level. From eqs. (4.27) and (4.32), the canonical partition function for N molecules is

$$Z_N = \frac{(Z_t Z_r Z_v Z_e)^N}{N!} \quad (4.53)$$

and the average internal energy of the molecule, according to eq. (4.30), is thus

$$U = \kappa_B T^2 \frac{\partial}{\partial T} \left\{ \ln\left(\frac{(Z_t Z_r Z_v Z_e)^N}{N!}\right) \right\}$$

$$= \kappa_B T^2 N \left\{ \frac{\partial}{\partial T}(\ln Z_t) + \frac{\partial}{\partial T}(\ln Z_r) + \frac{\partial}{\partial T}(\ln Z_v) \right.$$

$$\left. + \frac{\partial}{\partial T}(\ln Z_e) - (\ln N - 1) \right\} \quad (4.54)$$

The volumetric specific heat can be obtained by taking the derivative of U with respect to T at constant V. The translational energy contribution to the specific heat is given by eq. (4.47). The electronic energy level contribution to the specific heat is

$$C_{V,e} = \frac{N}{V} \frac{\partial}{\partial T} \left[\kappa_B T^2 \frac{\partial}{\partial T} \ln Z_e \right] = 0 \quad (4.55)$$

This result is because the electrons are only sitting in the first energy states and their contribution to the total system energy does not change with temperature. The contribution of rotational energy states to specific heat is

$$C_{V,r} = \frac{N}{V} \frac{\partial}{\partial T} \left[\kappa_B T^2 \frac{\partial}{\partial T} \ln \left(\sum_{\ell=0}^{\infty} (2\ell+1) \exp\left[-\frac{\theta_r \ell(\ell+1)}{T}\right] \right) \right] \quad (4.56)$$

We know, from eq. (E4.1.3), that the summation in the above equation is proportional to T/θ_r at high temperatures. In this limit, the contribution of the rotational energy level to the specific heat is

$$C_{V,r} = N\kappa_B/V \text{(at high temperature)} \quad (4.57)$$

This result is again a manifestation of the equipartition theorem. A diatomic molecule has two degrees of rotational freedom. So at high temperatures, when the rotational levels are fully excited, each molecule contributes $2 \times \kappa_B T/2 = \kappa_B T$ to the average energy. At low temperatures, the rotational specific heat must be calculated from the full rotational partition function in the summation format, eq. (4.56). Similarly, the contribution of the vibrational energy state to the specific heat is

$$C_{V,v} = \frac{\kappa_B N}{V} \frac{\theta_v^2}{T^2} \frac{e^{\theta_v/T}}{(e^{\theta_v/T} - 1)^2} \quad (4.58)$$

At high temperatures, the above formula leads to

$$C_{V,v} \approx \frac{\kappa_B N}{V} \quad (4.59)$$

which is again a manifestation of the equipartition theorem. After obtaining the contributions from all the energy modes, we calculate the total specific heat of a diatomic molecule by summing each of the contributing terms: $C_V = C_{V,t} + C_{V,r} + C_{V,v} + C_{V,e}$. The following example shows more numerical details.

Example 4.2 *Specific heat of* H_2

The rotational temperature of a hydrogen molecule is 85.3 K and its vibrational temperature is 6332 K. Plot the specific heat of hydrogen gas as a function of temperature.

Solution: From eqs. (4.48), (4.56), and (4.58), we can write the total specific heat per mole of a diatomic gas as

$$\frac{c_V}{R} = \frac{3}{2} + \left(\frac{\theta_v}{T}\right)^2 \frac{e^{\theta_v/T}}{(e^{\theta_v/T} - 1)^2}$$

$$+ \frac{\partial}{\partial T}\left[T^2 \frac{\partial}{\partial T} \ln\left(\sum_\ell (2\ell + 1) \exp\left[-\frac{\theta_r \ell(\ell+1)}{T}\right]\right)\right] \quad (E4.2.1)$$

The last term in the above equation can be written as

$$\frac{c_{V,r}}{R} = \left(\frac{\theta_r}{T}\right)^2 \times$$

$$\frac{Z_r \sum_\ell (2\ell+1)\ell^2(\ell+1)^2 \exp\left[-\frac{\theta_r\ell(\ell+1)}{T}\right] - \left\{\sum_\ell (2\ell+1)\ell(\ell+1)\exp\left[-\frac{\theta_r\ell(\ell+1)}{T}\right]\right\}^2}{Z_r^2}$$

(E4.2.2)

A computer program is used to carry out the above summation. Figure E4.2 plots the variation of c_V/R with temperature. At low temperatures, only the translational energy levels are fully excited and the specific heat is $3R/2$. As the temperature increases, the rotational energy levels become excited and contribute to the specific heat up to a maximum of R so that the total specific heat reaches $5R/2$. At even higher temperatures, the vibrational energy levels start contributing to the specific heat, which approaches a final value of $7R/2$.

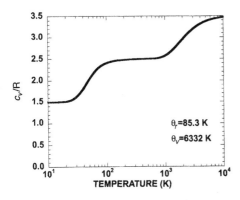

Figure E4.2 Specific heat of H_2 gas as a function of temperature.

4.2.2 Electrons in Crystals

Now we investigate the specific heat of electrons in a crystal. We assume that the electrons have a parabolic band with an isotropic effective mass

$$E - E_c = \frac{\hbar^2}{2m^*}(k_x^2 + k_y^2 + k_z^2) \qquad (4.60)$$

We obtained the density of states in chapter 3, eq. (3.52),

$$D(E) = \frac{1}{2\pi^2}\left(\frac{2m^*}{\hbar^2}\right)^{3/2}(E - E_c)^{1/2} \qquad (4.61)$$

The total number of electrons per unit volume is thus

$$n = \int_0^\infty f(E, T, \mu) D(E)\, dE \qquad (4.62)$$

From eq. (4.62), the chemical potential as a function of temperature can be determined for a given n. For $T = 0$, the above relation leads to

$$n = \int_{E_c}^{\mu} D(E)\, dE = \frac{1}{3\pi^2}\left(\frac{2m^*}{\hbar^2}\right)^{3/2}(\mu - E_c)^{3/2} \qquad (4.63)$$

We have already obtained this relation, eq. (3.53), in chapter 3. The chemical potential μ at $T = 0$ is called the Fermi level, E_f.* At other temperatures, eq. (4.62) cannot be explicitly integrated. However, when $(E - \mu)/k_B T \gg 1$, which is the classical limit, we can use the Boltzmann distribution as an approximation of the Fermi–Dirac distribution. Equation (4.62) can be integrated explicitly,

$$n = \int_{E_c}^{\infty} \exp\left(\frac{-E + \mu}{\kappa_B T}\right) \frac{1}{2\pi^2}\left(\frac{2m^*}{\hbar^2}\right)^{3/2}(E - E_c)^{1/2}\, dE = N_c \exp\left(-\frac{E_c - \mu}{\kappa_B T}\right) \qquad (4.64)$$

---

*In electronics, however, E_f is often used to represent the chemical potential at all temperatures.

with

$$N_c = 2\left(\frac{2\pi m^* \kappa_B T}{h^2}\right)^{3/2} \quad (4.65)$$

Equation (4.64) is often used to determine the chemical potential level in doped semiconductors, as will be seen from the following example.

Example 4.3 *Chemical potential level in doped semiconductors*

Silicon is a widely used semiconductor material, and it is often doped with phosphorus to form an n-type semiconductor. Determine the chemical potential of an n-type semiconductor doped with phosphorus with a concentration of 10^{17} cm^{-3} at 300 K, assuming that every phosphorus atom contributes one free electron to the conduction band and neglecting thermally excited electrons from the valence band. Although the silicon conduction bands are not spherical [figure 3.18(b)], they can be approximated by an isotropic band with an effective mass equal to $0.33m$, where m is the free electron mass.

Solution: Silicon has six identical conduction bands [figure 3.18(b)]. When counting all six bands, eq. (4.64) should be written as

$$n = 12\left(\frac{2\pi m * \kappa_B T}{h^2}\right)^{3/2} \exp\left(-\frac{E_c - \mu}{\kappa_B T}\right) \quad (E4.3.1)$$

Taking $n = 10^{17}$ cm^{-3}, we can find the chemical potential as

$$\frac{\mu - E_c}{\kappa_B T} = \ln\left[\frac{n}{12}\left(\frac{2\pi m^* \kappa_B T}{h^2}\right)^{-3/2}\right]$$

$$= \ln\left[\frac{10^{23}}{12}\left(\frac{2\pi \times 0.33 \times 9.1 \times 10^{-31} \times 1.38 \times 10^{-23} \times 300}{6.6^2 \times 10^{-68}}\right)^{-3/2}\right]$$

$$= -5.65 \quad (E4.3.2)$$

Thus

$$\mu - E_c = -5.65 \times 26 \text{ meV} = -147 \text{ meV} \quad (E4.3.3)$$

Comments. 1. The negative sign means that the chemical potential is below the conduction band edge. The silicon bandgap at room temperature is 1.12 eV. Thus the chemical potential level is within the bandgap. In fact, only in this case, the Boltzmann approximation we used in eq. (4.64) is applicable because the electron energy inside the conduction band, minus the chemical potential, is much larger than $\kappa_B T$. If the chemical potential is close to the band edge or falls inside the conduction band, which is the case when the semiconductors are heavily doped, we need to carry out numerical integration with the Fermi–Dirac statistical distribution.

2. The value of the chemical potential needs a reference point. Equation (4.64) suggests that it is the relative difference between μ and E_c that determines the electron number density, and thus this difference is the value of the chemical potential. In chapter 6 (figure 6.9), we will give a more detailed discussion on the reference point issue.

To calculate the specific heat of electrons, we first formulate the internal energy of electrons as

$$U(T) = \int_{E_c}^{\infty} E F(E, T, \mu) D(E) dE \qquad (4.66)$$

For convenience, we limit our discussion to metals so that the number of electrons per unit volume n_e is fixed. We further take $E_c = 0$ as reference, eq. (4.62) becomes

$$n_e = \int_0^{\infty} f(E, T, \mu) D(E) dE = \text{constant} \qquad (4.67)$$

We can use eq. (4.67) to rewrite eq. (4.66) as

$$U(T) = \int_0^{\infty} (E - E_f) f(E, T, \mu) D(E) dE + E_f n_e \qquad (4.68)$$

where E_f is the Fermi level (μ at $T = 0$ K). In eq. (4.68), since only f is temperature dependent, we obtain the heat capacity of the electron system as

$$C_e = \int_0^{\infty} (E - E_f) \frac{df(E, T, \mu)}{dT} D(E) dE \qquad (4.69)$$

Typically, df/dT is nonzero only in the region close to the chemical potential. If the density of states does not vary rapidly around μ, we can use its value at $E = \mu$ and pull $D(\mu)$ out of the integration. In addition, the change of μ with temperature in metal is very small because E_f is very large. We can thus neglect the temperature dependence of μ and set $\mu \approx E_f$. Under these approximations, eq. (4.69) becomes

$$C_e \approx D(\mu) \int_0^{\infty} (E - E_f) \frac{df(E, T, \mu)}{dT} dE$$

$$= D(\mu) \int_0^{\infty} \frac{(E - E_f)(E - \mu)}{\kappa_B T^2} \frac{\exp\left(\frac{E-\mu}{\kappa_B T}\right)}{\left[\exp\left(\frac{E-\mu}{\kappa_B T}\right) + 1\right]^2} dE$$

$$\approx \kappa_B^2 T D(E_f) \int_{-E_f/\kappa_B T}^{\infty} \frac{x^2 e^x}{(e^x + 1)^2} dx \qquad (4.70)$$

Since $E_f/\kappa_B T$ is very large, the above integral can be evaluated by setting the lower limit to $-\infty$, leading to the following expression for the specific heat

$$C_e = \frac{1}{2}\pi^2 n_e \kappa_B T/T_f \qquad (4.71)$$

where $T_f = E_f/\kappa_B$ is called the *Fermi temperature*. In deriving eq. (4.71), we used the relationship $n_e = 2E_f D(E_f)/3$, which can be obtained from eqs. (3.52) and (3.53). Thus the specific heat of electrons is linearly dependent on temperature.

4.2.3 Phonons

4.2.3.1 Debye Model

In chapter 3, we obtained the phonon density of states per unit volume under the Debye approximation when the three acoustic phonon polarizations are identical [eq. (3.55)],

$$D(\omega) = \frac{dN}{V\,d\omega} = 3 \times \frac{\omega^2}{2\pi^2 v_D^3} \qquad (4.72)$$

The total energy of phonons per unit volume is

$$U = \int_0^{\omega_D} \hbar\omega f(T,\omega) D(\omega)\,d\omega = \frac{3}{2\pi^2 v_D^3} \int_0^{\omega_D} \frac{\hbar\omega^3\,d\omega}{\exp(\hbar\omega/\kappa_B T) - 1} \qquad (4.73)$$

and the volumetric specific heat of phonons can be calculated from

$$C = \frac{\partial U}{\partial T} = \frac{3\hbar^2}{2\pi^2 v_D^3 \kappa_B T^2} \int_0^{\omega_D} \frac{\omega^4 \exp(\hbar\omega/\kappa_B T)}{[\exp(\hbar\omega/\kappa_B T) - 1]^2}\,d\omega \qquad (4.74)$$

From eqs. (3.56) and (3.57), the Debye frequency ω_D, Debye velocity v_D, and Debye temperature θ_D are related through

$$\omega_D = \frac{\pi v_D}{a_D} = \frac{\kappa_B \theta_D}{\hbar} \qquad (4.75)$$

where a_D is the effective lattice constant under the Debye model. Substituting eq. (4.75) into eq. (4.74), we get

$$C = \frac{3\hbar^2}{2\pi^2 (a_D \omega_D/\pi)^3 \kappa_B T^2} \int_0^{\omega_D} \frac{\omega^4 \exp(\hbar\omega/\kappa_B T)}{[\exp(\hbar\omega/\kappa_B T) - 1]^2}\,d\omega$$

$$= \frac{3\pi \kappa_B}{2 a_D^3}\left(\frac{T}{\theta_D}\right)^3 \int_0^{\theta_D/T} \frac{x^4 e^x\,dx}{(e^x - 1)^2} \qquad (4.76)$$

Using eq. (3.57), the specific heat can be further written as

$$C = 9\kappa_B \left(\frac{N}{V}\right)\left(\frac{T}{\theta_D}\right)^3 \int_0^{\theta_D/T} \frac{x^4 e^x dx}{(e^x - 1)^2} \qquad (4.77)$$

where N/V is the number of atoms per unit volume. At low temperatures, the integration limit can be set to infinity, leading to the familiar T^3 law,

$$C(T) = \frac{36\pi^4 \kappa_B}{15}\left(\frac{N}{V}\right)\left(\frac{T}{\theta_D}\right)^3 \propto T^3 \qquad (4.78)$$

Generally, the Debye temperature is unknown and the above expression is used to calculate the Debye temperature from experimentally measured values of specific heat. If the Debye model is accurate, a single value of the Debye temperature should be able to fit all of the temperature-dependent specific heat data. Such a situation happens rarely, however, and the Debye temperature is sometimes given as a function of temperature. This temperature-dependent Debye temperature is because the Debye model assumes a linear dispersion, which is not valid for phonons close to the boundary of the first Brillouin zone. In particular, it is completely wrong for optical phonons, for which the Einstein model is more appropriate, as we discuss below.

4.2.3.2 Einstein Model

Einstein's model assumes that all phonons have the same frequency ω_E and is thus more appropriate for optical phonons. We assume that there are N' states; that is, N' is the number of lattice points or primitive cells for each optical phonon polarization.* The total energy of the crystal per unit volume due to the contribution of the optical phonons with a frequency ω_E is then

$$U = N_p \frac{N' f(T, \omega_E) \hbar \omega_E}{V} = \frac{N_p N' \hbar \omega_E}{V[\exp(\hbar \omega_E / \kappa_B T) - 1]} \qquad (4.79)$$

where the factor N_p accounts for the number of polarizations of optical phonons at this frequency. The specific heat per unit volume is then

$$C = \frac{\partial U}{\partial T} = N_p \kappa_B \frac{N'}{V} \frac{(\hbar \omega_E / \kappa_B T)^2 \exp(\hbar \omega_E / \kappa_B T)}{[\exp(\hbar \omega_E / \kappa_B T_B) - 1]^2} \qquad (4.80)$$

The contributions of other optical phonons at a different frequency can be similarly calculated. At high temperature, both the Debye model and the Einstein model lead to the same result, as required by the equipartition theorem because the oscillator has three directions and each direction has two degrees of freedom (kinetic energy plus potential energy).

Clearly, the Debye model will be more appropriate for acoustic phonons and the Einstein model for optical phonons. At low temperatures, acoustic phonons are normally

---

*Notice that this N' is different from N in the Debye model, in which all phonon modes (including the optical modes) are lumped as three identical acoustic modes.

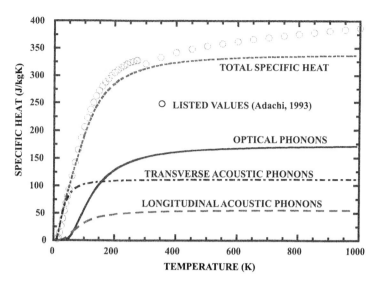

Figure 4.5 Estimated contribution of different phonon branches to the specific heat of GaAs (Chen, 1997).

excited, so the Debye approximation is more appropriate. At room and higher temperatures, both acoustic and optical phonons are excited and a combination of the two models is more appropriate. Figure 4.5 shows the estimated contributions of different phonon polarizations to the specific heat of GaAs (Chen, 1997). In this figure, a sine-function was assumed for the acoustic phonon dispersion.

4.2.4 Photons

Photons are bosons and obey the Bose–Einstein distribution. We have obtained the photon density of states in a three-dimensional cavity in eq. (3.59),

$$D(\omega) = \frac{dN}{V\,d\omega} = \frac{\omega^2}{\pi^2 c^3} \tag{4.81}$$

From eq. (4.81), the photon energy density per unit volume per unit angular frequency interval is

$$U_\omega = f(\omega, T)\hbar\omega D(\omega) = \frac{\hbar}{\pi^2 c^3} \frac{\omega^3}{[\exp(\hbar\omega/\kappa_B T) - 1]} \tag{4.82}$$

Since a photon propagates in all directions at the speed of light c, the intensity is then*

$$I_\omega = \frac{cU_\omega}{4\pi} = \frac{\hbar}{4\pi^3 c^2} \frac{\omega^3}{[\exp(\hbar\omega/\kappa_B T) - 1]} \tag{4.83}$$

---

*See section 6.1.3 for a more detailed explanation of intensity.

STATISTICAL THERMODYNAMICS AND THERMAL ENERGY STORAGE   147

Equation (4.83) is the Planck blackbody radiation law, expressed in terms of per angular frequency interval. In terms of wavelength, we have

$$I_\lambda = I_\omega \left|\frac{d\omega}{d\lambda}\right| = \frac{C_1/\pi}{\lambda^5[\exp(C_2/\lambda T) - 1]} \quad (4.84)$$

where $C_1 = 2\pi hc^2$ and $C_2 = hc/\kappa_B$. The blackbody emissivity power that is given in eq. (1.9) can be obtained easily from the above expression for intensity through $e_\lambda = \pi I_\lambda$. Integration of eq. (4.82) for frequencies ranging from 0 to ∞ leads to the total photon energy density

$$U = \frac{4}{c}\sigma T^4 \quad (4.85)$$

where $\sigma(= 5.67 \times 10^{-8} \text{ W m}^{-2} \text{ K}^{-4})$ is the Stefan–Boltzmann constant. The total intensity is

$$I = \frac{\sigma T^4}{\pi} \quad (4.86)$$

and the blackbody emissive power is thus

$$e_b = \pi I = \sigma T^4 \quad (4.87)$$

Although the concept of specific heat is seldom used in radiation, we can follow the previous treatment for electrons and phonons and calculate the photon specific heat,

$$C = \frac{16\sigma T^3}{c} \quad (4.88)$$

which has the same temperature dependence as the specific heat of phonons at low temperatures [eq. (4.78)].

Example 4.4 *Electron and phonon contributions to specific heat*

The Debye temperature of gold is 170 K and its Fermi level is 5.53 eV. Compute the specific heat of phonons and electrons in the temperature range of 0–1000 K.

Solution: The phonon and electron contributions to specific heat are given by

$$\text{Phonon: } C = 9\kappa_B \left(\frac{N}{V}\right)\left(\frac{T}{\theta_D}\right)^3 \int_0^{\theta_D/T} \frac{x^4 e^x dx}{(e^x - 1)^2}$$

$$\text{Electron: } C_e = \frac{1}{2}\pi^2 n_e \kappa_B T/T_f$$

Gold has an fcc structure with a lattice constant of 4.08 Å, and the number of atoms per unit cell is 4. Each atom contributes one valence electron. We have $n_e = N/V = 4/(4.08)^3 \times 10^{-30} \text{ m}^{-3}$. The Fermi temperature $T_f = E_f/k_B = 64,115 \text{ K}$. Substituting these numbers into the above expressions, we obtain the phonon

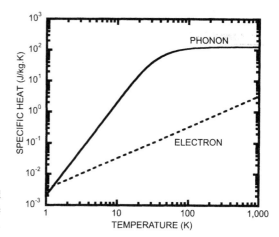

Figure E4.4 Phonon and electron contributions to the specific heat of gold.

and electron specific heats. The volumetric specific heats are converted into mass specific heat ($c = C/\rho$) and plotted in figure E4.4. We observe that the electron specific heat is typically much smaller than the phonon specific heat, except at very low temperatures.

4.3 Size Effects on Internal Energy and Specific Heat

In nanostructures, we expect that the internal energy and specific heat will be different from what we have given in the preceding sections. The differences come from two sources: one is physical and the other is mathematical. On the physics side, the energy levels, and their associated densities of states, will differ from those in bulk materials, as we have seen in chapter 3. On the mathematical side, for bulk materials we have replaced the summation over all energy states by integration in calculating the total energy. For nanostructures, this approximation may no longer be accurate.

A few experimental and theoretical studies exist about the size effects on the phonon specific heat of nanostructures. For systems of dimensionality d of 1 or higher ($d = 1$ for nanowires, $d = 2$ for films, and $d = 3$ for bulk structures), the Debye model predicts that at low temperatures the lattice specific heat should be proportional to T^d. The most common criterion for dimensional crossover is to compare the average phonon wavelength λ to the length scale of the structure. The average phonon wavelength can be estimated from the spectral-dependent phonon internal energy [the integrand of eq. (4.73)], similar to that obtaining the Wien's displacement law, eq. (1.11), from the Planck law. Well below the Debye temperature, this wavelength is given by $\lambda T \approx 50$ nm K for sound velocity v_s (5000 m s^{-1}). For example, a 3 nm Si thin film would be expected to exhibit $C \sim T^3$ behavior at 50 K ($\lambda \approx 1$ nm), but $C \sim T^2$ behavior at 5 K ($\lambda \approx 10$ nm). Some questions exist about whether the resulting low-dimensional specific heat should be larger or smaller than the corresponding bulk value. A simple model (Dames et al., 2004) summing over all of the normal modes of an elastic continuum with free boundaries predicts that the low-temperature specific heat of

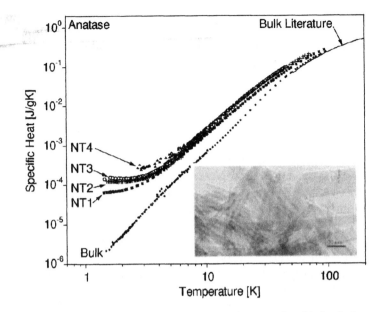

Figure 4.6 Experimental specific heat of anatase TiO_2 nanotubes (four kinds of tubes synthesized under different conditions) and that of bulk TiO_2 (Dames et al., 2004). Insert shows transmission electron micrographs of the nanotubes.

1- or 2-dimensional systems should exceed the bulk value, but several other calculations predict the low-temperature, low-dimensional specific heat should be reduced compared to bulk (Prasher and Phelan, 1998; Yang and Chen, 2000), slightly larger than the bulk value (Grille et al., 1996), or varying from both below and above bulk (Hotz and Siems, 1987; Tosic et al., 1992).

Limited experimental data are available to test these theories. The first studies were on zero-dimensional (0D) metallic nanoparticles of ~2–10 nm diameter, where experiments (Novotny and Meincke, 1973; Chen et al., 1995) show a specific heat enhanced by 50–100% at temperatures where the average phonon wavelength is comparable to the diameter of the nanoparticles, an exponential decay at lower temperatures, and an asymptotic return to bulk values at higher temperatures. These results have been successfully explained by theories that sum over all of the normal modes of an elastic sphere with free boundaries (Baltes and Hilf, 1973; Lautenschlager, 1975; Nonnenmacher, 1975). For anatase nanoparticles, Wu et al. (2001) reported enhancement by 20% for particles of about 15 nm diameter between 78 K and 370 K. In figure 4.6, we compare the experimental data on the specific heat of compacted titanium dioxide (TiO_2) nanotubes with that of bulk TiO_2, and show that the nanotubes have higher specific heat at low temperatures (Dames et al., 2004).

Considerable effort has been devoted to studying the specific heat of carbon nanotubes (CNT). Yi et al. (1999) observed a linear temperature dependence down to 10 K in multi-walled (MW) CNT, in close agreement with isolated sheets of graphene. In contrast, another MWCNT experiment by Mizel et al. (1999) showed a much steeper decay, with temperatures of about $T^{2.5}$ down to ~1 K, a better match to graphite. Bundles of single-walled (SW) CNT were studied by both Hone et al. (2000) and Mizel et al. (1999), and exhibited a linear or slightly superlinear temperature dependence from ~100 K down to ~2–4 K. At lower temperatures, Lasjaunias et al. (2003) reported a transition to T^3 attributed to the filling up of inter-tube modes, plus a surprising additional term proportional to $T^{0.34}$ or $T^{0.62}$ below ~1 K that was qualitatively attributed to localized excitations of atomic rearrangement as in glasses and amorphous materials. In all of these CNT, the specific heat is bounded between that of graphite and graphene. Various theoretical efforts have had mixed success at explaining these MWCNT and SWCNT measurements by extending isolated tube models to include the effects of interlayer coupling (in MWCNT) and intertube coupling (Mizel et al., 1999; Hone et al., 2000; Zhang et al., 2003). Overall, more work is needed to reconcile the diverse experimental results with theory (Dresselhaus and Eklund, 2000).

In comparison with phonons, we anticipate that the specific heat of electrons will have a stronger size dependence, due to the following factors: (1) the energy quantization of electrons is more dramatic than that of phonons; (2) the specific heat also depends on the Fermi level, particularly the rate of change of the density of states at the Fermi level. In our derivation of the electron specific heat in metals, we assumed that the density of states does not change much near the Fermi level. For nanostructures, the sharp features in the electronic density of states suggest that this assumption may not be valid. Indeed, existing studies show that the specific heat is a strong function of the size (Ghatak and Biswas, 1994; Lin and Shung, 1996).

For photons, we are not interested in the specific heat but rather in the energy density or emission spectrum from small objects. Since thermal radiation can have relatively long wavelengths, the issue of size effects on the energy density of the emission spectrum from a small object has been studied for various geometries (Rytov, 1959; Rytov et al., 1989). One interesting question is whether the thermal emission from any structure at any specific wavelength can exceed the blackbody radiation given by the Planck law. For example, the density of states in photonic crystals can be very different from that in free space. It can be inferred that in the frequency region where the photon density of states of the photonic crystal is larger than that in its parent crystal, the energy density of the thermal radiation inside the photonic crystal can exceed that in its parent crystal. However, not all of the energy can be emitted into free space since the density of states in free space is limited by eq. (4.81), and thus the maximum emissive power in an open free space is the blackbody radiation. There are, however, some recent experimental reports of the far-field thermal emission from photonic crystals being larger than that of the blackbody, although the physics is not clear (Lin et al., 2003). At small scales, however, radiative heat exchange can exceed that between two blackbodies due to the tunneling of evanescent and surface waves (Polder and Van Hove, 1971; Tien and Cunnington, 1973; Pendry, 1999; Mulet et al.; 2002; Narayanaswamy and Chen, 2003), which we will discuss in more detail in the next chapter. Another example is that the emissivity of particles with a diameter comparable to the wavelength can exceed 1 because of the diffraction effect (Bohren and Huffman, 1983).

4.4 Summary of Chapter 4

Through statistical mechanics, this chapter establishes the link between the energy states and temperature for a system in equilibrium. A system in equilibrium makes rapid transitions among accessible quantum states as a function of time. A fundamental assumption in statistical mechanics is that, in an isolated system, every accessible quantum state has an equal probability of being sampled by the system. Because it is hard to follow the time evolution of the system, we use an ensemble average to replace the time average for quantities of interest and assume that the ensemble average equals the time average. This assumption is called the egordicity assumption. An ensemble is made from a collection of systems, each of which is one accessible quantum state of the original system (but is stationary). Depending on the macroscopic constraints for the original system, we can establish different ensembles. We discussed the following three ensembles.

A *microcanonical ensemble* corresponds to an original system that is isolated with fixed U, V, and N. Each system in the ensemble represents one accessible quantum state in the original system. The most important relation for such an ensemble is the Boltzmann principle, which relates the total number of quantum states Ω of the original system (and thus the number of systems in the ensemble) to the entropy of the system,

$$S(U, V, N) = \kappa_B \ln \Omega \qquad (4.89)$$

Through the Boltzmann principle and the first law of thermodynamics, we can obtain other thermodynamic properties of the system,

$$\frac{1}{T} = \left(\frac{\partial S}{\partial U}\right)_{V,N}, \frac{p}{T} = \left(\frac{\partial S}{\partial V}\right)_{U,N}, -\frac{\mu}{T} = \left(\frac{\partial S}{\partial N}\right)_{U,V} \qquad (4.90)$$

Thus S is the thermodynamic potential of a microcanonical ensemble and its natural variables are U, V, and N.

A *canonical ensemble* is more appropriate for a system with fixed T, V, and N. In this case, the system can exchange energy with its surroundings and we assumed that the surroundings are a thermal reservoir. Although we can combine the system and the reservoir to establish a microcanonical ensemble for the combined system, it is more convenient to establish a canonical ensemble in which each of the systems is an accessible quantum state of the original system and each can exchange energy with the reservoir. Since a number of accessible quantum states exists in the reservoir corresponding to one specific accessible quantum state of the system, the probability of observing any one system in the canonical ensemble is no longer equal, as in a microcanonical system. The probability of finding a quantum state with energy E_i is

$$P(E_i) = e^{-E_i/\kappa_B T}/Z \text{ where } Z = \sum_i \exp\left[-\frac{E_i}{\kappa_B T}\right] \qquad (4.91)$$

In eq. (4.91), $e^{-E_i/\kappa_B T}$ is the familiar Boltzmann factor and Z is called the canonical partition function. The canonical partition function is related to the Helmholtz free energy, a thermodynamic potential with natural variables as T, V, and N, through

$$F(T, V, N) = U - TS = -\kappa_B T \ln Z \qquad (4.92)$$

If, rather than having N fixed, we consider a system with fixed T, V, and μ, such a system exchanges not only energy but also particles with its reservoir. We can construct an the ensemble that consists of a number of systems, each corresponds to one accessible quantum states in the original system and can exchange energy and particles with the reservoir. This ensemble is called the *grand canonical ensemble*, and the probability of finding a particular system with energy E_i and number of particles N_i is given by

$$P(E_i, N_i) = \frac{\exp\left[\frac{N_i\mu - E_i}{\kappa_B T}\right]}{\Im} \tag{4.93}$$

where the exponential is called the Gibbs factor and

$$\Im(T, V, \mu) = \sum_{N_i}\sum_{E_i} \exp\left[\frac{N_i\mu - E_i}{\kappa_B T}\right] = \sum_{N_i} \lambda^{N_i} Z_i \tag{4.94}$$

is the grand canonical partition function.

After establishing the probabilities and partition functions for different ensembles, we applied them to different particles. The partition function for a dilute gas made of N indistinguishable molecules is given by

$$Z_N = \frac{(Z_t Z_r Z_v Z_e)^N}{N!} \tag{4.95}$$

where Z_t, Z_r, Z_v, and Z_e are the partition functions for the translational, the rotational, the vibrational, and the electronic states, respectively.

For electrons and other fermions, the average number of particles in a specific accessible quantum state with an energy E is given by the Fermi–Dirac distribution function,

$$f(E, T, \mu) = \frac{1}{\exp\left(\frac{E-\mu}{\kappa_B T}\right) + 1} \tag{4.96}$$

On the basis of f, we can calculate the amount of energy this quantum state contributes to the total energy as $\langle E \rangle = E \times f$.

For phonons, photons, and other bosons, the average number of particles in a specific quantum state with frequency v is given by the Bose–Einstein distribution

$$f(v, T) = \frac{1}{\exp\left(\frac{hv}{\kappa_B T}\right) - 1} \tag{4.97}$$

Given this average number of particles, each having an energy hv, we can calculate the contribution of this specific accessible quantum state to the total system energy as $\langle E \rangle = hvf$.

In the classical limit, when the exponential in these distributions is much larger than one, both Bose–Einstein and Fermi–Dirac distributions reduces to the classical Boltzmann distribution

$$f(E, T, \mu) = \exp\left(-\frac{E - \mu}{\kappa_B T}\right) \tag{4.98}$$

With the distribution functions, we are in a position to count the energy of the system, from which other properties related to the energy, such as the specific heat for electrons and phonons and the emissive power for blackbody radiation, can be determined. Since, at each allowable energy level, the system can be degenerate as measured by the density of states $D(E)$, the total energy at this level is $U(E) = E \times f \times D(E)$. If we want the total energy of the system, we should sum $U(E)$ over all energy levels. In a three-dimensional space (bulk materials), the summation is often replaced by integration since the separation between energy levels is usually very small. This procedure leads to the following results for electrons, phonons, and photons:

Electron specific heat: $C_V \propto T$ (4.99)

Phonon specific heat: $C_V \propto T^3$ (at low temperature) (4.100)

$C_V =$ constant (at high temperatures) (4.101)

Photon emissive power: $I_\omega = \dfrac{\hbar}{4\pi^3 c^2} \dfrac{\omega^3}{[\exp(\hbar\omega/\kappa_B T) - 1]}$ (Planck's law) (4.102)

For nanostructures, the statistical distributions are still valid as long as the systems are in thermal equilibrium. However, there are several reasons that could invalidate the derivations of the specific heat and emissive power for bulk materials. One is that the energy levels are different in nanostructures from these in macrostructures, which will change the density of states. Second is that the energy separation is usually large and the replacement of the summation by integration over energy is no longer valid as it was for bulk materials.

With the contents of this chapter and the previous two chapters, the readers are encouraged to read through current literature. With persistency and patience, readers may find that they begin to understand (or partially understand) some of the nanoscience and nanotechnology research topics.

4.5 Nomenclature for Chapter 4

a	lattice constant, m	h	Planck constant, J s
c	speed of light	$\hbar$	Planck constant divided by 2π, J s
c_v	specific heat per mole, J K^{-1} mol^{-1}	I	intensity, W m^{-2} srad^{-1}
C	volumetric specific heat, J m^{-3} K^{-1}	k	magnitude of wave vector, m^{-1}
D	density of states per unit volume, m^{-3}	L	length of box, m
		m	mass, kg
e_b	blackbody radiation emissive power, W m^{-2}	m^*	effective mass, kg
		n	quantum number; electron number density, m^{-3}
E	energy, J		
E_c	conduction band edge, J	N	total number of particles in the system
E_f	Fermi level, J		
f	distribution function	N_A	Avogadro constant, mol^{-1}
F	Helmholz free energy, J	N_p	number of polarization of optical phonons
G	grand canonical potential, J		

R	universal gas constant, $J\,K^{-1}\,mol^{-1}$	ω	angular frequency, rad.Hz
s	accessible quantum state	Ω	number of accessible states in a microcanonical system
S	entropy, $J\,K^{-1}$		
T	temperature, K	$\Im$	grand canonical partition function
U	system energy, J		
v	speed, $m\,s^{-1}$	$\langle\rangle$	ensemble average
V	system volume, m^3		
x	integration variable		Subscripts
Z	canonical partition function		
θ	temperature, K	D	Debye
κ_B	Boltzmann constant, $J\,K^{-1}$	e	electronic
λ	thermal de Broglie wavelength, m	f	at Fermi level
		i	ith energy level
μ	chemical potential, J	r	reservoir; rotational
ν	frequency of phonons or photons, Hz	t	total
		v	vibrational
ρ	density, $kg\,m^{-3}$	V	constant volume
σ	Stefan–Boltzmann constant, $W\,m^{-2}\,K^{-4}$	x, y, z	Cartesian coordinate direction

4.6 References

Adachi, S., ed., 1993, *Properties of Aluminum Gallium Arsenide*, INSPEC, London.
Baltes, H.P., and Hilf, E.R., 1973, "Specific Heat of Lead Grains," *Solid State Communications*, vol. 12, pp. 369–373.
Bohren, C.F., and Huffman, D.R., 1983, *Absorption and Scattering of Light by Small Particles*, Wiley, New York.
Callen, H.B., 1985, *Thermodynamics and an Introduction to Thermostatistics*, Wiley, New York.
Chen, Y.Y., Yao, Y.D., Hsiao, S.S., Jen, S.U., Lin, B.T., Lin, H.M., and Tung, C.Y., 1995, "Specific Heat Study of Nanocrystalline Palladium," *Physical Review B*, vol. 52, pp. 9364–9369.
Chen, G., 1997, "Size and Interface Effects on Thermal Conductivity of Superlattices and Periodic Thin-Film Structures," *Journal of Heat Transfer*, vol. 119, pp. 220–229.
Dames, C., Poudel, B., Wang, W.Z., Huang, J.Y., Ren, Z.F., Sun, Y., Oh, J.I., Opeil, C., Naughton, S.J., Naughton, M.J., and Chen, G., "Low-dimensional phonon specific heat of titanium dioxide nanotubes," submitted for publication.
Dresselhaus, M.S., and Eklund, P.C., 2000, "Phonons in Carbon Nanotubes," *Advances in Physics*, vol. 49, pp. 705–814.
Ghatak, K., and Biswas, S., 1994, "Influence of Quantum Confinement on the Heat Capacity of Graphite," *Fizika A*, vol. 3, pp. 7–24.
Grille, H., Karch, K., and Bechstedt, F., 1996, "Thermal Properties of $(GaAs)_M(Ga_{1-x}Al_xAs)_N(001)$ Superlattices," *Physica B*, vol. 119–120, pp. 690–692.
Hone, J., Batlogg, B., Benes, Z., Johnson, A.T., and Fischer, J.E., 2000, "Quantized Phonon Spectrum of Single-Wall Carbon Nanotubes," *Science*, vol. 289, pp. 1730–1733.
Hotz, R., and Siems, R., 1987, "Density of States and Specific Heat of Elastic Vibrations in Layer Structures," *Superlattices and Microstructures*, vol. 3, pp. 445–454.
Kittel, C., and Kroemer, H., 1980, *Thermal Physics*, 2nd ed., Freeman, New York.
Kubo, R., Toda, M., and Hashitsume, N., 1998, *Statistical Physics II*, 2nd ed., Springer, Berlin.

Lasjaunias, J.C., Biljakovic, K., Monceau, P., and Sauvajol, J.L., 2003, "Low-Energy Vibrational Excitations in Carbon Nanotubes Studied by Heat Capacity," *Nanotechnology*, vol. 14, pp. 998–1003.

Lautenschlager, R., 1975, "Improved Theory of the Vibrational Specific Heat of Lead Grains," *Solid State Communications*, vol. 16, pp. 1331–1334.

Lin, M.F., and Shung, W.W.K., 1996, "Electronic Specific Heat of Single-Walled Carbon Nanotubes," *Physical Review B*, vol. 54, pp. 2896–2900.

Lin, S.Y., Moreno, J., and Fleming, J.G., 2003, "Three-Dimensional Photonic-Crystal Emitters for Thermal Photovoltaic Power Generation," *Applied Physics Letters*, vol. 83, pp. 380–381.

Mizel, A., Benedict, L.X., Cohen, M.L., Louie, S.G., et al., 1999, "Analysis of the Low-Temperature Specific Heat of Multiwalled Carbon Nanotubes and Carbon Nanotube Ropes," *Physical Review B*, vol. 60, pp. 3264–3270.

Mulet, J.-P., Joulain, K.L., Carminati, R., and Greffet, J.-J., 2002, "Enhanced Radiative Heat Transfer at Nanometric Distances," *Microscale Thermophysical Engineering*, vol. 6, pp. 209–222.

Narayanaswamy, A., and Chen, G., 2003, "Surface Modes for Near-Field Thermophotovoltaics," *Applied Physics Letters*, vol. 82, pp. 3544–3546.

Nonnenmacher, Th.F., 1975 "Quantum Size Effect on the Specific Heat of Small Particles," *Physics Letters A*, vol. 51, pp. 213–214.

Novotny, V., and Meincke, P.P.M., 1973, "Thermodynamic Lattice and Electronic Properties of Small Particles," *Physical Review B*, vol. 8, pp. 4186–4199.

Pendry, J.B., 1999, "Radiative Exchange of Heat between Nanostructures," *Journal of Physics, Condensed Matter*, vol. 11, pp. 6621–6633.

Polder, D., and Van Hove, M., 1971, "Theory of Radiative Heat Transfer between Closely Spaced Bodies," *Physical Review B*, vol. 4, pp. 3303–3314.

Prasher, R.S., and Phelan, P.E., 1998, "Size Effects on the Thermodynamic Properties of Thin Solid Films," *Journal of Heat Transfer*, vol. 120, pp. 1078–1086.

Rytov, S.M., 1959, *Theory of Electric Fluctuations and Thermal Radiation*, Air Force Cambridge Research Center, Bedford, MA, Report AFCRC-TR-59-162.

Rytov, S.M., Kravtsov, Y.A., and Tatarskii, V.I., 1989, *Principles of Statistical Radiophysics*, vol. 3, Springer-Verlag, Berlin.

Tien, C.L., and Cunnington, G.R., 1973, "Cryogenic Insulation Heat Transfer," *Advances in Heat Transfer*, vol. 9, pp. 349–417.

Tosic, B.S., Setrajcic, J.P., Mirjanic, D.L., and Bundalo, Z.V., 1992, "Low-Temperature Properties of Thin Films," *Physica A*, vol. 184, pp. 354–366.

Wu, X.M., Wang, L., Tan, Z.C., Li, G.H., and Qu, S.S., 2001, "Preparation, Characterization, and Low-Temperature Heat Capacities of Nanocrystalline TiO_2 Ultrafine Powder," *Journal of Solid State Chemistry*, vol. 156, pp. 220–223.

Yang, B., and Chen, G., 2000, "Lattice Dynamics Study of Phonon Heat Conduction in Quantum Wells," *Physics of Low-Dimensional Structures*, vol. 5/6, pp. 37–48.

Yi, W., Lu, L., Zhang, D.-L., Pan, Z.W., and Xie, S.S., 1999, "Linear Specific Heat of Carbon Nanotubes," *Physical Review B*, vol. 59, pp. R9015–R9018.

Zhang, S., Xia, M., Zhao, S., Xu, T., and Zhang, E., 2003, "Specific Heat of Single-Walled Carbon Nanotubes," *Physical Review B*, vol. 68, 075415/1–7.

4.7 Exercises

4.1 *Grand canonical ensemble.* Establish a grand canonical ensemble and derive the probability distribution for the ensemble, that is, eq. (4.17).

4.2 *Thermal de Broglie wavelength.* Calculate the thermal de Broglie wavelength of a He molecule at 300 K and show that the dilute gas condition, eq. (4.34), is satisfied at 1 atm and 300 K.

4.3 *Specific heat of monatomic gas.* Derive an expression for the specific heat of a box of He gas and plot it as a function of temperature.

4.4 *Entropy of mixing.* There are two tanks of gas. Both tanks have N molecules and a volume V, and are at the same temperature and pressure. The two tanks are connected by a pipe with a valve. After the valve is opened, the gases in both tanks eventually mix into a homogeneous mixture. Show the following:

(a) If the two gases are identical, there is no change in entropy due to the mixing.

(b) If the two gases are different, the mixing causes an entropy production of $2N \ln 2$.

The difference in the results is called the Gibbs paradox and comes from the distinguishability of the molecules.

4.5 *Bose–Einstein distribution.* Plot the Bose–Einstein distribution as a function of frequency for $T = 100$ K, 300 K, and 1000 K. Compare with the Boltzmann distribution at the same temperatures.

4.6 *Electrons in semiconductors.* A semiconductor has a parabolic band structure

$$E - E_c = \frac{\hbar^2}{2m^*}(k_x^2 + k_y^2 + k_z^2)$$

The Fermi level in the semiconductor could be above or below the conduction band edge. Take the electron effective mass as the free electron mass. For $\mu - E_c = 0.05$ eV and $T = 300$ K, do the following in the range 0.0 eV $< E - E_c < 0.1$ eV:

(a) Plot the Fermi–Dirac distribution as a function of E,

(b) Plot the density of states as a function of E,

(c) Calculate the product of $f(E,T)D(E)$, which means the average number of electrons at each E, and plot the product as a function of E,

(d) Calculate the product of $E f(E,T)D(E)$, which means the actual energy at each allowable energy level, and plot the product as a function of E.

Repeat the questions for $\mu - E_c = -0.05$ eV.

4.7 *Chemical potential.* The number of electrons in the conduction band can be assumed to be equal to the dopant concentration. Calculate the chemical potential levels relative to the band edge for the dopant concentrations of 10^{18} cm^{-3} and 10^{19} cm^{-3}, assuming free electron mass and $T = 300$ K.

4.8 *Debye crystal.* A crystal has a Debye velocity of 5000 m s^{-1}, and a Debye temperature of 500 K. For $T = 300$ K,

(a) Plot the Bose–Einstein distribution as a function of ω,

(b) Plot the density of states as a function ω using the Debye model.

(c) Plot fD as a function of frequency ω.

(d) Plot $\hbar \omega f D$ as a function of ω.

(e) Compute the specific heat of the crystal as a function of temperature for $1 < T < 1000$ K.

4.9 *Blackbody radiation.* Consider the blackbody radiation at $T = 300$ K.

(a) Plot the Bose–Einstein distribution as a function of angular frequency ω.

(b) Plot the density of states as a function of ω, using the Debye model.

(c) Plot fD as a function of ω.

(d) Plot $\hbar \omega f D$ as a function of ω.

(e) Compute the emissive power as a function of temperature and the corresponding specific heat.

(f) Also compare (a)–(e) with corresponding questions for phonons in problem 4.8.

4.10 *Specific heat of diatomic molecules.* A diatomic molecule has one rotational energy state at 100 meV and one vibrational energy state at 1 eV. Plot the contribution of this molecule to the specific heat of a box of such molecules as a function of temperature.

4.11 *Electron specific heat of a quantum well.* Derive and plot the electron internal energy and specific heat for an infinite-barrier-height quantum well, $L_z = 20$ Å and 100 Å, as a function of temperature. Take the electron effective mass equal to the free electron mass and an electron density $n_e = 2 \times 10^{28}$ m^{-3}.

4.12 *Electron specific heat of quantum dots.* Derive and plot the electron internal energy and specific heat for a cubic quantum dot with infinite potential barrier height with $L = 20$ Å or 100 Å as a function of temperature. Take the electron effective mass equal to the free electron mass and an electron density $n_e = 2 \times 10^{28}$ m^{-3}.

4.13 *Phonon specific heat.* Assuming that phonons obey the following dispersion relation (three-dimensional isotropic medium)

$$\omega = 2\sqrt{\frac{K}{m}} \left| \sin \frac{|\mathbf{k}|a}{2} \right|$$

where a is the lattice constant, K the spring constant, and $\mathbf{k}$ the wavevector. Derive an expression for the phonon internal energy and specific heat.

4.14 *Fermi level and specific heat in Au.* The valence electron concentration in gold is 5.9×10^{22} cm^{-3}.

(a) Calculate the Fermi level in gold at zero temperature.

(b) What is the corresponding Fermi temperature?

(c) Estimate the electronic contribution to the specific heat of gold at 300 K.

(d) Calculate the Fermi level at 300 K.

4.15 *Phonon specific heat in Ge.* Germanium has an fcc structure with two Ge atoms per basis and a lattice constant of 5.66 Å. On the basis of the equipartition theorem, estimate the phonon specific heat per unit mass in germanium at high temperatures and compare it with the experimental specific heat value at 300 K.

4.16 *Phonon high temperature specific heat—Debye model.* Prove that at high temperatures the Debye model leads to a specific heat of $3k_B N$, where N is the number of atoms in the crystal.

4.17 *Diamond specific heat.* The Debye temperature of diamond is 1320 K. Calculate the specific heat of diamond at 300 K and compare it with the literature value (the lattice constant of diamond is 3.567 Å).

4.18 *Phonon specific heat in a quantum dot.* A bulk crystal has a Debye velocity of 5000 m s^{-1} and a Debye temperature of 300 K. Assuming that phonons in a quantum dot obey the same dispersion relation as those in the bulk material, but considering the discrete nature of the wavevectors, compute the specific heat of a cubic quantum dot with the following lengths: 10 Å, 20 Å, and compare it with the specific heat of the bulk crystal.

4.19 *Blackbody radiation in a small cavity.* Consider thermal radiation in equilibrium inside such a cubic cavity. Compute the radiation energy density in a cubic cavity of length $L = 1\ \mu\text{m}$ at $T = 400$ K and compare it with the Planck distribution obtained by assuming that the cavity is very large compared to the wavelength.

4.20 *Entropy of one phonon state.* From eqs. (4.14) and (4.40), show that the entropy, s, of one phonon state having a frequency ω obeys the following relationship:

$$\frac{\hbar\omega}{T} f_0(1+f_0) = -\frac{\kappa_B T}{\hbar}\frac{\partial s}{\partial \omega}$$

Where f_0 is the Bose–Einstein distribution.

5

Energy Transfer by Waves

The wave–particle duality of matter from quantum mechanics implies that energy carriers have both wave and particle characteristics. One way to think about this duality is that material waves are granular rather than continuous. For example, a phonon wave at frequency v contains a discrete number of identical waves, each having an energy hv. A fundamental property of waves is their phase information. A coherent wave has a fixed relationship between two points in space or at two different times. Due to the fixed phase relationship, the superposition of waves from the same source creates interference and diffraction phenomena that are familiar in optics.

The wave characteristics of matter (electrons, phonons, and photons) are important for transport processes at interfaces and in nanostructures. We have seen in previous chapters that the size effects on energy quantization can be considered as a result of the formation of standing waves. In this chapter, we will discuss the reflection of waves at a single interface, and interference and tunneling phenomena in thin films and multilayers. We will make parallel presentations for three major energy carriers: electrons, photons, and phonons. We have discussed rather extensively in chapters 2 and 3 the electron waves based on the Schrödinger equation. Optical wave effects are readily observable and can be understood from classical electrodynamics based on the Maxwell equations, which will be reviewed in this chapter. For phonons, we will adapt a continuum approach based on the acoustic waves, rather than on the discrete lattice dynamics method we used in chapter 3. The acoustic-wave-based approach allows us to treat phonons in parallel with electrons and photons. We will see that wave reflection, interference, and tunneling phenomena can occur for all three types of carriers and the descriptions of these phenomena are also similar (table 1.4), despite the differences in their statistical behavior, dispersion, and origin (table 1.3), as we discussed in previous chapters. Readers familiar with any of these waves can use the analogy for understanding the other waves.

For macroscale transport processes, however, we seldom consider the phase of material waves. Rather, we treat the entities as particles. Why and when can we do so? Section 5.6 answers these questions and briefly discusses transport in the partially coherent regime.

5.1 Plane Waves

When throwing a stone into water, one can observe a concentric wave propagating outward. Television antennas emit electromagnetic waves that are approximately spherical. Rather than considering these nonplanar waves, we will carry out most of our discussion in this chapter on the basis of plane waves, although the phenomena to be discussed also exist for other forms of waves such as the cylindrical or spherical waves. A *plane wave* is one that has a constant amplitude at any plane perpendicular to the direction of propagation at any fixed time. These waves must satisfy the equation governing their motion. Later, we will discuss these governing equations, such as the Maxwell equations for electromagnetic waves. Before getting into these details, let's first examine some common forms of plane waves. For example, in chapter 2, we showed that the wavefunction of a free electron is [eq. (2.34)]

$$\Psi(x,t) = A_1 e^{-i(\omega t - kx)} + A_2 e^{-i(\omega t + kx)} \tag{5.1}$$

where the first term represents a plane wave traveling in the positive x-direction and the second term in the negative x-direction. These are scalar plane waves because the wavefunction is a scalar. Other waves, such as the electromagnetic field, are vector waves because the electric/magnetic fields have directions. We can express a harmonic, vector plane wave propagating in three-dimensional space as

$$\mathbf{F}(t, \mathbf{r}) = \mathbf{A} \sin(\omega t - \mathbf{k} \bullet \mathbf{r}) \tag{5.2}$$

where $\mathbf{A}$ represents the amplitude and direction of the field (e.g., electric, magnetic, or atomic displacement), ω is the angular frequency, $\mathbf{k}$ [with components (k_x, k_y, k_z)] is the wavevector representing the direction of wave propagation and its spatial periodicity ($|\mathbf{k}| = 2\pi/\lambda$), and $\mathbf{r}$ is the spatial coordinate. Equation (5.2) is a plane wave because all the points $\mathbf{r}$ satisfying $\mathbf{k} \bullet \mathbf{r} = constant$ form a plane perpendicular to the wavevector $\mathbf{k}$, and the field $\mathbf{F}$ is a constant on this plane at any given time.

Very often, it is much more convenient to use the complex representation rather than the sine and cosine representation for the waves. For example, instead of eq. (5.2), we can write $\mathbf{F}$ as

$$\mathbf{F}(t, \mathbf{r}) = \mathbf{A} \exp[-i(\omega t - \mathbf{k} \bullet \mathbf{r})] \tag{5.3}$$

When using such a complex representation, we resort to either the real part or the imaginary part of the final solution as the solution of the problem, in accordance with whether the initial or boundary conditions are represented in terms of cosine (real part) or sine (imaginary part) functions. For example, the imaginary part of $\mathbf{F}$ in eq. (5.3) is identical to eq. (5.2) and we thus expect that the solution to a physical problem will be the imaginary part of the complex variables used in solving the governing equations.

In the following sections, we will examine three types of waves: the electron wave as a material waves, the electromagnetic wave governing the radiation transfer, and the acoustic wave representing lattice vibration.

5.1.1 Plane Electron Waves

In chapter 2, we dealt extensively with electron waves in planar geometries, such as free electrons and electrons in a potential well. The wavefunction of a plane electron wave propagating along the positive x-direction is

$$\Psi(x, t) = A \exp[-i(\omega t - kx)] \tag{5.4}$$

From the Schrödinger equation, we obtained in chapter 2 the following dispersion relation between the electron energy E and wavevector k

$$k = \sqrt{\frac{2m(E - U)}{\hbar^2}} \tag{5.5}$$

where U is the electrostatic potential. The particle current (or flux) can be calculated from [eq. (2.31)]:

$$\mathbf{J} = \frac{i\hbar}{2m}(\Psi \nabla \Psi^* - \Psi^* \nabla \Psi) = \text{Re}\left[\frac{i\hbar}{m}\Psi \nabla \Psi^*\right] \tag{5.6}$$

As we will see later, this flux expression is similar to the Poynting vector that represents the energy flux of electromagnetic and acoustic waves.

5.1.2 Plane Electromagnetic Waves

In this section, we will introduce the Maxwell equations that govern the propagation of electromagnetic waves. We will show that a plane wave of the form of eq. (5.3) satisfies the Maxwell equations and discuss how to calculate the energy flux of the electromagnetic waves.

An electromagnetic wave in vacuum is characterized by an *electric field vector* **E** [N C^{-1} = V m^{-1}], and a *magnetic field vector* **H** [C m^{-1} s^{-1} = A m^{-1}]. When the electromagnetic field interacts with a medium, under the force of the electric and magnetic fields the electrons and ions of the atoms in the medium are set into motion. These electrons and ions generate their own electric and magnetic fields that influence each other and are superimposed onto the external fields. For example, the positive ions and negative electrons of an atom under an external field will be deformed from the original equilibrium condition, forming an electrical dipole.* A measure of the capability of the material to respond to the incoming electric field is the electric polarization per unit

---

*A dipole is a pair of positive charge Q and negative charge $-Q$, separated by a small distance a. The dipole moment of the pair of charges equals $p = Qa$.

volume, or the *dipole moment* per unit volume **P** [C m^{-2}], which is related to the electric field through the electric susceptibility χ,

$$\mathbf{P} = \varepsilon_0 \chi \mathbf{E} \tag{5.7}$$

where ε_0 is the vacuum permittivity, $\varepsilon_0 = 8.85 \times 10^{-12}$ [C^2 N^{-1} m^{-2} = F m^{-1}], and the electric susceptibility is nondimensional. The total field inside the medium is measured by the *electric displacement* **D** [C m^{-2}], which is a superposition of the contributions from the external electric field and the electric polarization,

$$\mathbf{D} = \varepsilon_0 \mathbf{E} + \mathbf{P} = \varepsilon_0 (1 + \chi) \mathbf{E} = \varepsilon \mathbf{E} \tag{5.8}$$

where ε is called the electrical *permittivity* of the medium.

The electron and ion motion in a medium also induces a magnetic field, which is superimposed onto the external magnetic field. A measure of the total magnetic field inside the medium is called the *magnetic induction* **B** [N s m^{-1} C^{-1} = N A^{-1} m^{-1}],

$$\mathbf{B} = \mu \mathbf{H} \tag{5.9}$$

where μ is the magnetic *permeability*. In vacuum and in most materials, $\mu = \mu_0 = 4\pi \times 10^{-7}$ N s^2 C^{-2} = H m^{-1}. If a material has μ larger than μ_0, it is called paramagnetic, and if $\mu < \mu_0$ it is diamagnetic. Many materials are nonmagnetic, with $\mu = \mu_0$.

The electric dipoles and electric polarization are most appropriate when describing the distortions of electrons bound to ions. The free electrons that are not bound to any atoms will also be set into motion by the external electric field. The motion of free electrons forms a current, which is related to the electric field through Ohm's law,

$$\mathbf{J}_e = \sigma_e \mathbf{E} \tag{5.10}$$

where σ_e [C^2 N^{-1} m^{-1} s^{-1} = Ω^{-1} m^{-1}] is the electrical *conductivity*.

The propagation of an electromagnetic wave is governed by the following *Maxwell equations*:

$$\nabla \times \mathbf{E} = -\frac{\partial \mathbf{B}}{\partial t} \tag{5.11}$$

$$\nabla \times \mathbf{H} = \frac{\partial \mathbf{D}}{\partial t} + \mathbf{J}_e \tag{5.12}$$

$$\nabla \bullet \mathbf{D} = \rho_e \tag{5.13}$$

$$\nabla \bullet \mathbf{B} = 0 \tag{5.14}$$

where ρ_e is the net charge density [C m^{-3}]. Equation (5.11) is the Faraday law, which states that a changing magnetic field induces an electric field. Without the first term on the right-hand side, eq. (5.12) is the Ampère law, which says an electric current induces a magnetic field. Maxwell's (1831–1879) ingenuity lies in the first term of the right-hand side of eq. (5.12), which represents the current due to electron oscillation around the ion even though the electrons are not free to move. This additional term, however, places the electric and the magnetic fields at similar positions with respect to time and space, and endows the electromagnetic field with a wave type of behavior, as we will see later. The Maxwell equations govern the propagation of all electromagnetic waves, including, for example, the signals of cellular phones, radios and televisions, lasers, light, thermal radiation, and X-rays, despite the fact that these waves are generated by different sources. The differences between these waves, in terms of propagation, are mainly the wavelength.

We will next demonstrate how to derive a wave type of equation for the electric field. By taking the curl of eq. (5.11), we get

$$\nabla \times \nabla \times \mathbf{E} = -\frac{\partial}{\partial t}(\nabla \times \mathbf{B}) \tag{5.15}$$

The left-hand side can be manipulated using the vector identity

$$\nabla \times \nabla \times \mathbf{E} \equiv \nabla(\nabla \bullet \mathbf{E}) - \nabla^2 \mathbf{E} \tag{5.16}$$

For a region free of electrical charge, eq. (5.13) leads to

$$\nabla \bullet \mathbf{D} = 0 \text{ and thus } \nabla \bullet \mathbf{E} = 0 \tag{5.17}$$

We should mention that the second of the above equations is based on the assumption that the dielectric constant ε is independent of space, which is not the case for the photonic crystal that we discussed in chapter 3. Substituting eqs. (5.16) and (5.17) into eq. (5.15) and using $\mathbf{B} = \mu \mathbf{H}$ [eq. (5.9)] yields

$$-\nabla^2 \mathbf{E} = -\mu \frac{\partial}{\partial t}(\nabla \times \mathbf{H}) \tag{5.18}$$

To eliminate $\mathbf{H}$, we substitute eq. (5.12) into the above equation and utilize eqs. (5.8) and (5.10),

$$\nabla^2 \mathbf{E} = \mu \varepsilon \frac{\partial^2 \mathbf{E}}{\partial t^2} + \mu \sigma_e \frac{\partial \mathbf{E}}{\partial t} \tag{5.19}$$

If $\sigma_e = 0$, eq. (5.19) is clearly a wave type of equation. The first-order derivative in eq. (5.19) introduces damping to the wave, because of the absorption by free electrons. A similar equation for the magnetic field can be derived. We seek a solution to eq. (5.19) in the form of a plane wave, eq. (5.3),

$$\mathbf{E}(\mathbf{r}, t) = \mathbf{E}_0 \exp[-i(\omega t - \mathbf{k} \bullet \mathbf{r})] \tag{5.20}$$

where $\mathbf{E}_0$ represents both the amplitude and direction of the electric field. Substituting eq. (5.20) into eq. (5.19), we obtain

$$\mathbf{k} \bullet \mathbf{k} = \mu \varepsilon \omega^2 + i\mu\sigma_e \omega = \mu\omega^2[\varepsilon_0(1+\chi) + i\sigma_e/\omega] = \mu\hat{\varepsilon}\omega^2 \tag{5.21}$$

or

$$|\mathbf{k}| = \frac{N\omega}{c_0} \tag{5.22}$$

with

$$c_0 = \frac{1}{\sqrt{\varepsilon_0 \mu_0}}, \hat{\varepsilon} = \varepsilon_0(1+\chi) + i\sigma_e/\omega, N = \sqrt{\frac{\mu\hat{\varepsilon}}{\mu_0\varepsilon_0}} = n + i\kappa \tag{5.23}$$

where c_0 is the speed of light in vacuum, $\hat{\varepsilon}$ is called the *complex permittivity*, and N is called the *complex refractive index* or *complex optical constant*.* The real part of N, n, is the usual refractive index of materials. The imaginary part of N, κ, is called the *extinction coefficient* and measures the damping of the electromagnetic field, which arises not only from the absorption of free electrons [the conductivity part in eq. (5.21)], but also from the dipole oscillation of bound electrons and other mechanisms [the susceptibility part of eq. (5.21)]. For nonmagnetic materials ($\mu = \mu_0$),

$$N = \sqrt{\frac{\hat{\varepsilon}}{\varepsilon_0}} = \sqrt{\varepsilon_r} = n + i\kappa \tag{5.24}$$

and $\varepsilon_r = \hat{\varepsilon}/\varepsilon_0$ is called the *dielectric constant* or dielectric function. Neither N nor ε_r is really a constant, as their names suggest, because they are dependent on wavelength. Studies on the wavelength dependence of the dielectric function can lead to insights into the material constituents and energy states. For example, some electron and phonon states can be identified from measuring the dielectric function or complex refractive indices. There exists a large library of complex refractive indices of materials (Palik, 1985).

Substituting eq. (5.22) into eq. (5.20), we see that the electric field of a harmonic electromagnetic wave can be expressed as

$$\mathbf{E}(\mathbf{r}, t) = \mathbf{E}_0 \exp\left[-i\omega\left(t - \frac{N}{c_0}\hat{\mathbf{k}} \bullet \mathbf{r}\right)\right] \tag{5.25}$$

where $\hat{\mathbf{k}}$ is the unit vector along the wavevector direction. With the electric field determined, the magnetic field can be computed according to eq. (5.11). One can further prove that the electromagnetic wave is a transverse wave, and that the electric and magnetic fields are perpendicular to each other:

$$\mathbf{E} \perp \mathbf{H} \perp \mathbf{k} \tag{5.26}$$

In the special case that a plane wave is traveling along the x-direction with the electric and magnetic fields pointing in the y- and z-directions, respectively, the electric and magnetic fields can be expressed as

$$\mathbf{E}_y = E_{ya} \exp\left[-i\omega\left(t \mp \frac{Nx}{c_0}\right)\right]\hat{\mathbf{y}} \tag{5.27}$$

$$\mathbf{H}_z = H_{za} \exp\left[-i\omega\left(t \mp \frac{Nx}{c_0}\right)\right]\hat{\mathbf{z}} \tag{5.28}$$

where the minus and plus signs represent waves propagating in the positive and negative x-direction, respectively. Substituting eqs. (5.27) and (5.28) back into the Maxwell equation (5.11), we can see that the electric and the magnetic fields are related:

$$H_{za} = \pm\frac{N}{\mu c_0}E_{ya} \tag{5.29}$$

where the plus and minus signs correspond to those in the exponential factor.

---

*N is sometimes expressed as $N = n - i\kappa$, coupled with a plane wave of the form $\mathbf{E}(\mathbf{r}, t) = \mathbf{E}_0 \exp[i(\omega t - \mathbf{k} \bullet \mathbf{r})]$ rather than eq. (5.20). As long as they are consistent, the end results will be the same.

Now let us see how we can calculate the energy flow associated with electromagnetic fields. We start by manipulating the Maxwell equations. Taking the dot product of **H** with eq. (5.11) and the dot product of **E** with eq. (5.12), then, subtracting the resulting two equations, we get

$$-\nabla \bullet (\mathbf{E} \times \mathbf{H}) = \frac{\partial}{\partial t}\left(\frac{1}{2}\mu \mathbf{H} \bullet \mathbf{H} + \frac{1}{2}\varepsilon \mathbf{E} \bullet \mathbf{E}\right) + \mathbf{E} \bullet \mathbf{J}_e \qquad (5.30)$$

where we have used the vector identity, $\nabla \bullet (\mathbf{E} \times \mathbf{H}) = \mathbf{H} \bullet (\nabla \times \mathbf{E}) - \mathbf{E} \bullet (\nabla \times \mathbf{H})$ We identify the meaning of each term on the right-hand side of eq. (5.30) as

Magnetic field energy density [J m^{-3}]: $\qquad \frac{1}{2}\mu \mathbf{H} \bullet \mathbf{H}$ $\qquad$ (5.31)

Electric field energy density [J m^{-3}]: $\qquad \frac{1}{2}\varepsilon \mathbf{E} \bullet \mathbf{E}$ $\qquad$ (5.32)

Joule heating [W m^{-3}]: $\qquad \mathbf{E} \bullet \mathbf{J}_e$ $\qquad$ (5.33)

To see what $\mathbf{E} \times \mathbf{H}$ means in eq. (5.30), we integrate the equation over a volume,

$$-\iiint \nabla \bullet (\mathbf{E} \times \mathbf{H})dV$$
$$= \iiint \left[\frac{\partial}{\partial t}\left(\frac{1}{2}\mu \mathbf{H} \bullet \mathbf{H} + \frac{1}{2}\varepsilon \mathbf{E} \bullet \mathbf{E}\right) + \mathbf{E} \bullet \mathbf{J}_e\right]dV \qquad (5.34)$$

and rewrite the left-hand side into a surface integral, using Gauss's theorem,

$$-\oiint (\mathbf{E} \times \mathbf{H}) \bullet \hat{\mathbf{n}} dA$$
$$= \iiint \left[\frac{\partial}{\partial t}\left(\frac{1}{2}\mu \mathbf{H} \bullet \mathbf{H} + \frac{1}{2}\varepsilon \mathbf{E} \bullet \mathbf{E}\right) + \mathbf{E} \bullet \mathbf{J}_e\right]dV \qquad (5.35)$$

where the surface integration is carried out over the surface enclosing the volume and $\hat{\mathbf{n}}$ is the local normal of the surface, pointing outward. The right-hand side of eq. (5.35) represents the rate of change of the stored electromagnetic energy inside the volume plus the heat generated. The first law of thermodynamics requires that this must be supplied by the energy flow into the volume across the boundaries, hence the left-hand side must be this energy flow. Because of the negative sign and the fact that $\hat{\mathbf{n}}$ points outward, the cross product $\mathbf{E} \times \mathbf{H}$ must mean the rate of energy flow into the volume. We call this product the *Poynting vector* **S** [W m^{-2}]

$$\mathbf{S} = \mathbf{E} \times \mathbf{H} \qquad (5.36)$$

The Poynting vector represents the instantaneous energy flux. It oscillates at twice the frequency of the electromagnetic field. Currently, no electronic devices can measure such a fast signal. What can be measured is the time-averaged Poynting vector

$$\langle \mathbf{S} \rangle = \frac{1}{T}\int_{t}^{t+T} \mathbf{S} dt' \qquad (5.37)$$

If the complex representation of **E** and **H** is used, as is the case most of the time, it can be shown that the time-averaged Poynting vector can be calculated from

$$\langle \mathbf{S} \rangle = \frac{1}{2}\text{Re}(\mathbf{E}_c \times \mathbf{H}_c^*) \tag{5.38}$$

where the subscript c is used, only in this equation, to emphasize the complex representation of **E** and **H**, and the superscript * means the complex conjugate. Because we use the complex representation most of the time, we will drop the subscript c whenever it is clear, as we have been doing so far. The time-averaged Poynting vector expression, eq. (5.38), is similar to the particle flux expression, eq. (5.6), for quantum mechanical waves. As we move on, we will see more similarities between these waves.

As an example, we consider a plane wave propagating in the positive x-direction as given by eqs. (5.27) and (5.28). The corresponding Poynting vector is

$$\begin{aligned}\langle \mathbf{S} \rangle &= \frac{1}{2}\text{Re}(\mathbf{E}_c \times \mathbf{H}_c^*) \\ &= \frac{1}{2}\text{Re}\left\{E_{ya}H_{za}^*\exp\left[-i\omega\left(t - \frac{Nx}{c_0}\right)\right]\exp\left[i\omega\left(t - \frac{N^*x}{c_0}\right)\right]\right\}(\hat{\mathbf{y}} \times \hat{\mathbf{z}}) \\ &= \frac{1}{2}\text{Re}\left\{E_{ya}\frac{N^*}{\mu c_0}E_{ya}^*\exp\left[\frac{i\omega(N-N^*)x}{c_0}\right]\right\}\hat{\mathbf{x}} \\ &= \frac{1}{2}\text{Re}\left\{\frac{n - i\kappa}{\mu c_0}\exp\left[-\frac{2\omega\kappa x}{c_0}\right]\right\}E_{ya}^2\hat{\mathbf{x}} \\ &= \frac{1}{2}\frac{n}{\mu c_0}\exp\left[-\frac{4\pi\kappa x}{\lambda_0}\right]E_{ya}^2\hat{\mathbf{x}} \\ &= \frac{1}{2}\frac{n}{\mu c_0}e^{-\alpha x}E_{ya}^2\hat{\mathbf{x}}\end{aligned} \tag{5.39}$$

where λ_0 is the wavelength in vacuum and we have used eq. (5.29) to replace H_{za} by E_{ya}, and

$$\alpha = \frac{4\pi\kappa}{\lambda_0} \quad [m^{-1}] \tag{5.40}$$

is called the *absorption coefficient*. Its inverse, $\delta = 1/\alpha$, is called the *skin depth*. Equation (5.39) shows that as the electromagnetic field propagates, the energy decays exponentially. The skin depth is where the energy flux has dropped by e^{-1}. With a $\kappa \sim 0.1$, the skin depth is of the order of one wavelength λ_0. For optical fibers, a low absorption coefficient is essential. For a $\delta \sim 1000$ m and $\lambda_0 = 1.55$ μm, which is the wavelength used in long-distance optical communication, κ, must be less than 10^{-10}. For metals, κ is usually large in the range from visible to far infrared, and thus electromagnetic fields usually do not penetrate far into metal.

In closing this section, we comment further on the relationship between **E**, **H**, and **k**. Equation (5.26) says that they form an orthogonal set. Usually, it is further understood that this set follows the so-called right-hand rule: with the right hand fully extended and the thumb perpendicular to the four other fingers, close the hand by turning the four fingers, which originally pointed in the **E**-field direction, toward the **H**-field direction; the thumb then points in the **k**-direction. In artificial materials with both negative ε and negative μ, however, the Maxwell equations actually require that **E**, **H**, and **k** follow the left-hand rule (Veselago, 1968). Such left-handed materials may have interesting properties such as negative refractive index arising from taking the negative root of eq. (5.21), and could focus light to a spot much smaller than the wavelength (Pendry, 2000; Shelby et al., 2001).

5.1.3 Plane Acoustic Waves

Having explored quantum mechanical material waves and electromagnetic waves in the form of plane waves, we examine in this section the appropriate form of plane lattice waves. We discussed in chapter 3 lattice waves based on a simple one-dimensional lattice chain model, treating atoms as discrete points. In the long wavelength range, we can neglect the atomic structure and treat the medium as a continuum. The lattice wave in this range can be described by acoustic wave equations. Our discussion of lattice waves in this section will be based on acoustic wave equations, which resemble strongly the electromagnetic waves dealt with in the previous section.

In the continuum representation, the acoustic wave propagation can be described in terms of the local medium *displacement*, **u**, from its equilibrium position, or, more often, the velocity **v** of this displacement,

$$\mathbf{v} = \frac{d\mathbf{u}}{dt} \tag{5.41}$$

The displacement can be related to the *strain tensor* $\overline{\overline{S}}$, where the "=" above the symbol means that it is a second-rank tensor, which can be represented by a (3×3) matrix with 9 components, S_{ij} ($i, j = 1, 2, 3$), calculated from

$$S_{ij}(\mathbf{r}, t) = \frac{1}{2}\left(\frac{\partial u_i}{\partial x_j} + \frac{\partial u_j}{\partial x_i}\right) \tag{5.42}$$

The strain can be further related to the stress tensor $\overline{\overline{\sigma}}$, which is again a second-rank tensor, through the generalized Hooke law. The force acting on any surface with a norm $\hat{n}$ is $\mathbf{F} = \overline{\overline{\sigma}} \bullet \hat{n}$, where the product of a tensor with a vector is carried out using the following matrix product rule:

$$\begin{pmatrix} F_x \\ F_y \\ F_z \end{pmatrix} = \begin{pmatrix} \sigma_{xx} & \sigma_{xy} & \sigma_{xz} \\ \sigma_{yx} & \sigma_{yy} & \sigma_{yz} \\ \sigma_{zx} & \sigma_{zy} & \sigma_{zz} \end{pmatrix} \begin{pmatrix} n_x \\ n_y \\ n_z \end{pmatrix} \tag{5.43}$$

Using eqs. (5.41)–(5.43) and Newton's second law of motion, a relationship between the stress tensor and displacement can be obtained and finally expressed in terms of the displacement velocity. The acoustic wave equations thus obtained in their general form are very complicated in an anisotropic medium with damping (Auld, 1990), because the constitutive relationship between the stress and the strain involves the elastic stiffness tensor and the viscosity tensor (representing damping), both are fourth-rank tensors with

81 components, although symmetry requirements render many of the matrix components to be zero. We will not list the acoustic wave equations but focus only on the special case when the medium is isotropic with no damping, for which only two constants are needed for the stiffness tensor. These two constants, denoted here as λ_L and μ_L, are called the Lame constants. In this case, the acoustic wave equations are significantly simplified. In particular, if we assume a plane wave of the form $\mathbf{v}\exp[-i(\omega t - k\hat{\mathbf{k}}\bullet\mathbf{r})]$, where k is the magnitude of the wavevector and $\hat{\mathbf{k}}$ is the unit wavevector, the acoustic wave equations lead to the following eigenvalue equation (Auld, 1990)

$$k^2 \begin{bmatrix} c_{11}\hat{k}_x^2 + \mu_L(1-\hat{k}_x^2) & (\lambda_L+\mu_L)\hat{k}_x\hat{k}_y & (\lambda_L+\mu_L)\hat{k}_x\hat{k}_z \\ (\lambda_L+\mu_L)\hat{k}_y\hat{k}_x & c_{11}\hat{k}_y^2 + \mu_L(1-\hat{k}_y^2) & (\lambda_L+\mu_L)\hat{k}_y\hat{k}_z \\ (\lambda_L+\mu_L)\hat{k}_z\hat{k}_x & (\lambda_L+\mu_L)\hat{k}_z\hat{k}_y & c_{11}\hat{k}_z^2 + \mu_L(1-\hat{k}_z^2) \end{bmatrix} \begin{bmatrix} v_x \\ v_y \\ v_z \end{bmatrix} = \rho\omega^2 \begin{bmatrix} v_x \\ v_y \\ v_z \end{bmatrix} \quad (5.44)$$

where $c_{11} = \lambda_L + 2\mu_L$. The above eigenvalue equation determines the dispersion relation between ω and $\mathbf{k}$, and is called the *Christoffel* equation.

Consider a plane transverse wave propagating along the z-direction and vibrating in the x-direction,

$$\mathbf{v}_T = A_T e^{-i(\omega t - k_T z)}\hat{\mathbf{x}} \quad (5.45)$$

Substituting eq. (5.45) into (5.44) leads to the following solution

$$\mu_L k_T^2 = \rho\omega^2 \text{ or } \omega = v_T k_T \quad (5.46)$$

where $v_T = (\mu_L/\rho)^{1/2}$ is the velocity of the wave. There is an identical transverse wave vibrating in the y-direction since the medium is isotropic, as is the case when we consider phonons in three-dimensional crystals on the basis of lattice-dynamics calculations. A longitudinal wave that vibrates along the z-direction,

$$\mathbf{v}_L = A_L e^{-i(\omega t - k_L z)}\hat{\mathbf{z}} \quad (5.47)$$

also satisfies eq. (5.44). Substituting eq. (5.47) into (5.44) yields the following dispersion relation

$$k_L^2 c_{11} = \rho\omega^2 \text{ or } \omega = v_L k_L \quad (5.48)$$

where the longitudinal wave velocity $v_L = (c_{11}/\rho)^{1/2} = [(\lambda_L + 2\mu_L)/\rho]^{1/2}$. Thus, unlike electromagnetic waves, which are transverse and always have the electric field $\mathbf{E}$ and the magnetic field $\mathbf{H}$ perpendicular to the wave propagation direction $\hat{\mathbf{k}}$, acoustic waves can exist in both transverse and longitudinal forms. Sound propagation in gases is a longitudinal wave. For a plane wave propagating along an arbitrary direction $\hat{\mathbf{k}}$, we can express the transverse and longitudinal plane waves in the general form

$$\mathbf{v}_{T1} = \hat{\mathbf{a}} A_{T1} e^{-i(\omega t - k_T \hat{\mathbf{k}}\bullet\mathbf{r})}, \quad (\hat{\mathbf{a}} \bullet \hat{\mathbf{k}} = 0) \quad (5.49)$$

$$\mathbf{v}_{T2} = \hat{\mathbf{a}} \times \hat{\mathbf{k}} A_{T2} e^{-i(\omega t - k_T \hat{\mathbf{k}}\bullet\mathbf{r})} \quad (5.50)$$

$$\mathbf{v}_L = \hat{\mathbf{k}} A_L e^{-i(\omega t - k_L \hat{\mathbf{k}}\bullet\mathbf{r})} \quad (5.51)$$

where eq. (5.49) defines a transverse acoustic wave with displacement direction $\hat{\mathbf{a}}$ perpendicular to $\hat{\mathbf{k}}$, and eq. (5.50) denotes another transverse acoustic wave with displacement normal to both $\hat{\mathbf{a}}$ and $\hat{\mathbf{k}}$. Based on the displacement velocity, components of the stress tensor can be calculated from

$$\frac{\partial}{\partial t}\begin{bmatrix}\sigma_{xx}\\ \sigma_{yy}\\ \sigma_{zz}\\ \sigma_{yz}\\ \sigma_{xz}\\ \sigma_{xy}\end{bmatrix}=\begin{bmatrix}c_{11} & c_{12} & c_{12} & 0 & 0 & 0\\ c_{12} & c_{11} & c_{12} & 0 & 0 & 0\\ c_{12} & c_{12} & c_{11} & 0 & 0 & 0\\ 0 & 0 & 0 & c_{44} & 0 & 0\\ 0 & 0 & 0 & 0 & c_{44} & 0\\ 0 & 0 & 0 & 0 & 0 & c_{44}\end{bmatrix}\begin{bmatrix}\partial v_x/\partial x\\ \partial v_y/\partial y\\ \partial v_z/\partial z\\ \partial v_y/\partial z+\partial v_z/\partial y\\ \partial v_x/\partial z+\partial v_z/\partial x\\ \partial v_x/\partial y+\partial v_y/\partial x\end{bmatrix} \quad (5.52)$$

where $c_{12} = \lambda_L$ and $c_{44} = \mu_L$. Equation (5.52) is a form of the Hooke law for an isotropic medium without damping. The symmetry relations $\sigma_{xy} = \sigma_{yx}$, $\sigma_{yz} = \sigma_{zy}$, and $\sigma_{xz} = \sigma_{zx}$ can be used to obtain all nine components of the second-order stress tensor.

The time-averaged power carried by the acoustic wave can be calculated from the *acoustic Poynting vector* [W m^{-2}]

$$\mathbf{J}_a = -\frac{1}{2}\text{Re}(\mathbf{v}^* \bullet \bar{\bar{\boldsymbol{\sigma}}}) \quad (5.53)$$

which is again similar to that of an electromagnetic wave [eq. (5.38)] and of quantum mechanical material waves [eq. (5.6)].

The above discussion on acoustic waves and electromagnetic waves is clearly very sketchy and also mathematically involved. The main purpose is to get the reader familiar with the plane waves propagating along an arbitrary wavevector $\mathbf{k}$ direction, as represented by eq. (5.3) and the flux carried by the plane waves. In the next section, we will examine how these plane waves behave at an interface.

5.2 Interface Reflection and Refraction of a Plane Wave

When a wave, be it an electron, photon, or phonon wave, meets a boundary, it will be reflected and refracted. In this section, we will determine how much of the incoming wave is reflected and how much is refracted, by imposing boundary conditions for these waves. The reader will find that although the electron, photon, and phonon waves are quite different in nature, the expressions for their respective interface reflectivity and transmissivity, which we will define soon, are similar.

5.2.1 Electron Waves

At an interface between two materials, an electrical potential generally exists as shown in figure 3.28. We examine the transport of a plane electron wave across an interface represented by a step potential as shown in figure 5.1. Although the more general case of a wave with any arbitrary angle of incidence can be treated, we focus only on the normal incidence case, that is, when the wave is traveling along the z-direction.

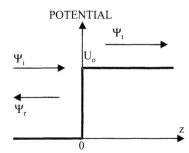

Figure 5.1 Reflection and transmission of an electron wave at an interface, caused by the potential barrier at the interface.

The wavefunctions of the incident, Ψ_i, the reflected, Ψ_r, and the transmitted waves Ψ_t, can be expressed as

$$\Psi_i = A_i e^{-i(\omega t - k_1 z)}, \quad \Psi_r = A_r e^{-i(\omega t + k_1 z)}, \quad \Psi_t = A_t e^{-i(\omega t - k_2 z)} \qquad (5.54)$$

where k_1 and k_2 are the electron wavevectors in the two media, respectively, and

$$\omega = \frac{E}{\hbar}, \quad k_1 = \sqrt{\frac{2mE}{\hbar^2}}, \quad k_2 = \sqrt{\frac{2m(E - U_0)}{\hbar^2}} \qquad (5.55)$$

These wavefunctions have been obtained in sections 2.3.1 and 3.2.1. Using the boundary conditions on the continuity of the wavefunction and its first-order derivative, similar to eqs. (3.25) and (3.26), we obtain the *reflection* and *transmission coefficients*

$$r = \frac{A_r}{A_i} = \frac{k_1 - k_2}{k_1 + k_2} \quad \text{and} \quad t = \frac{A_t}{A_i} = \frac{2k_1}{k_1 + k_2} \qquad (5.56)$$

The corresponding *reflectivity* and *transmissivity* of the particle flux are

$$\left. \begin{array}{l} R = \dfrac{J_r}{J_i} = \left|\dfrac{k_1 - k_2}{k_1 + k_2}\right|^2 \\[1em] \tau = \dfrac{J_t}{J_i} = \dfrac{\operatorname{Re}\left[k_2^* A_t A_t^*\right]}{\operatorname{Re}\left[k_1^* A_i A_i^*\right]} = \dfrac{4\operatorname{Re}(k_1 k_2^*)}{|k_1 + k_2|^2} = 1 - R \end{array} \right\} \qquad (5.57)$$

where the incident, reflected, and transmitted particle fluxes (J_i, J_r, and J_t) are calculated from eq. (5.6).

When $E < U_0$, it can be shown that $R = 1$ and $\tau = 0$, which means that all electrons are reflected (total reflection). In this case, however, the wavefunction Ψ_t is not zero, as one can easily show by substituting k_2 from eq. (5.55), which gives an imaginary k_2 when $E < U_0$, into eq. (5.54). The wavefunction Ψ_t thus obtained decays exponentially from the interface but is nonetheless not zero. This transmitted wave, however, does not carry a net particle flux across the boundary ($J_t = 0$), and it is called an *evanescent wave*, which will be discussed in section 5.4.

When $E > U_0$, electrons will be partially reflected and partially transmitted, whereas classical mechanics would lead to 100% transmission without reflection. Making a mundane daily analogy, this means that if one throws a stone at a wall, there is some probability that the stone will be reflected back even if it is thrown higher than the wall. Such a scenario is clearly contrary to our common experience but is possible in the quantum world for very small particles such as electrons. In fact, the reflection phenomenon occurs for all waves, as we will see later. Thus, wave mechanics differs significantly from particle mechanics at a single interface.

5.2.2 Electromagnetic Waves

The reflection and refraction of light at an interface is a more familiar process for many readers, and shares many similarities to the behavior of electron waves at an interface. Although we considered only the case of normal incidence for electron waves, we will treat here the more general case of oblique incidence of an electromagnetic wave onto an interface. As shown in figure 5.2, a plane electromagnetic wave propagates along direction $\mathbf{k}_i$ (wavevector direction) and meets an interface with norm $\hat{\mathbf{n}}$. The reflected wave and refracted wave propagate along the $\mathbf{k}_r$ and $\mathbf{k}_t$ directions, respectively. We call the plane formed by $\mathbf{k}_i$ and $\hat{\mathbf{n}}$ the plane of incidence, and the angle formed between $\hat{\mathbf{n}}$ and $\mathbf{k}_i$ the angle of incidence. The electromagnetic wave is a transverse wave, so the **E**-field and the **H**-field can have any orientation in the plane normal to **k**. We can decompose the electric field into two components, one parallel to the plane of incidence and the other perpendicular to the interface. When an electric field is parallel to the plane of incidence, as is the case of figure 5.2, it cannot be parallel to the interface unless the angle of incidence is zero. Its conjugate magnetic field component, in this case pointing out of the paper, is perpendicular to the plane of incidence and is thus always parallel to the interface. This wave is called a transverse magnetic wave, or TM-polarized wave. Sometimes the notations p (parallel polarized) and $/\!/$ (relative to E) are also used. If the electrical field component is perpendicular to the plane of incidence, the wave is called a transverse electric wave or TE-polarized. Notations TE, s (perpendicularly polarized), and $\perp$ are often used interchangeably. We will limit our discussion to positive media, that is, those with refractive indices of the form of eq. (5.24) for which **E**, **H**, and **k** obey the right-hand rule.

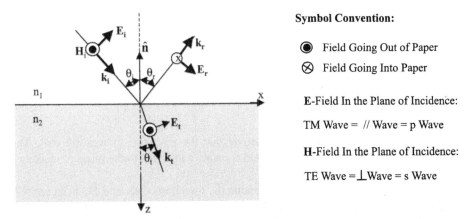

Figure 5.2 Reflection and refraction of an electromagnetic field at an interface.

We need to establish boundary conditions for the electric and magnetic fields to determine the reflection and transmission at the interface. By applying the Maxwell equations to a very thin control volume surrounding an interface, the following *boundary conditions* can be obtained (Born and Wolf, 1980)

$$\hat{n} \cdot (\mathbf{D}_2 - \mathbf{D}_1) = \rho_s \tag{5.58}$$

$$\hat{n} \times (\mathbf{E}_2 - \mathbf{E}_1) = 0 \tag{5.59}$$

$$\hat{n} \cdot (\mathbf{B}_2 - \mathbf{B}_1) = 0 \tag{5.60}$$

$$\hat{n} \times (\mathbf{H}_2 - \mathbf{H}_1) = J_s \tag{5.61}$$

where ρ_s [C m^{-2}] and $\mathbf{J}_s$ [A m^{-1}] are the net surface charge density and the surface current density, respectively, $\mathbf{E}_1$ and $\mathbf{E}_2$ are the total electric fields on the two sides of the interface, and similarly $\mathbf{H}_1$ and $\mathbf{H}_2$ are the total magnetic fields on the two sides of the interface. To obtain the "total" $\mathbf{E}$ and $\mathbf{H}$ for side 1 of figure 5.2, we need to sum up the incident and the reflected fields. Equation (5.58) means that the difference between the normal components of the electric displacements across the interface must be equal to interface charge density, while eq. (5.59) means that the tangential components of the electric field must be continuous. Equation (5.60) says that the normal components of the magnetic induction must be continuous, while eq. (5.61) means that the difference of the tangential components of the magnetic field across a surface equals the surface current density.

With the above boundary conditions, we can determine the amount of reflection and transmission of an incident electromagnetic wave onto a surface. We consider a plane TM wave incident onto a surface at an incident angle θ_i. The wavevector directions of the incident, reflected, and transmitted waves are $(\sin\theta_i, 0, \cos\theta_i)$, $(\sin\theta_r, 0, -\cos\theta_r)$, and $(\sin\theta_t, 0, \cos\theta_t)$, respectively. Using a plane wave of the form of eq. (5.25), the incident, reflected, and transmitted electric fields can be expressed as

$$\mathbf{E}_{//i} \exp\left[-i\omega\left(t - \frac{n_1 x \sin\theta_i + n_1 z \cos\theta_i}{c_0}\right)\right] \tag{5.62}$$

$$\mathbf{E}_{//r} \exp\left[-i\omega\left(t - \frac{n_1 x \sin\theta_r - n_1 z \cos\theta_r}{c_0}\right)\right] \tag{5.63}$$

$$\mathbf{E}_{//t} \exp\left[-i\omega\left(t - \frac{n_2 x \sin\theta_t - n_2 z \cos\theta_t}{c_0}\right)\right] \tag{5.64}$$

respectively. Here, we temporarily assume that the refractive indices are real. The subscript "//" means that the electric field is polarized parallel to the plane of incidence (TM wave as shown in figure 5.2).

Some readers may ask how to determine the direction of $\mathbf{E}_r$ and $\mathbf{H}_r$ in figure 5.2. The answer is that a correct assumption of the direction is not important as long as both $\mathbf{E}_r$ and $\mathbf{H}_r$ follow the right-hand rule. The signs in the final results will take care of the directions. Notice the sign change in eq. (5.63) before z in the exponent due to the change in the wave propagation direction for the reflected wave. Based on a similar derivation of eq. (5.29), the magnitude of the magnetic field, which is pointing out of the plane of the paper, is related to the electric field by

$$H_y = \frac{n}{\mu c_0} E_{//} \text{(forward)}, \quad H_y = -\frac{n}{\mu c_0} E_{//} \text{(backward)} \quad (5.65)$$

where the "forward" denotes waves propagating along the positive z-direction (incident and refracted waves) and "backward" applies to the reflected wave propagating along the negative z-direction, the subscript "y" of H denotes that H points perpendicular to the plane of incidence, in the y-direction, and $E_{//}$ is the magnitude of vector $\mathbf{E}_{//}$. In figure 5.2 we show the reflected magnetic fields pointing into the paper because of the negative sign in eq. (5.65) for the reflected wave. In reality, the actual sign change may be in the reflected electric field rather than the magnetic field. As long as one is consistent with the mathematical operations, the end result will give the correct sign.

We consider a surface free of net charge and current, and take this surface as $z = 0$. To determine the magnitudes of the reflected and refracted fields, we need consider only the continuity of the tangential components. The boundary conditions on the normal components will be automatically satisfied. Applying the continuity of the electric field, and noting that the electric field of a TM wave is not perpendicular to the surface, we use the component along the x-direction,

$$\cos\theta_i E_{//i} \exp\left[i\omega\frac{n_1 x \sin\theta_i}{c_0}\right] + \cos\theta_r E_{//r} \exp\left[i\omega\frac{n_1 x \sin\theta_r}{c_0}\right]$$
$$= \cos\theta_t E_{//t} \exp\left[i\omega\frac{n_2 x \sin\theta_t}{c_0}\right] \quad (5.66)$$

where we have dropped the factor of $e^{-i\omega t}$ since it is contained in all the terms and cancels out. Since x can take any value, the above equation is valid only when the exponents are equal. This gives

$$n_1 \sin\theta_i = n_1 \sin\theta_r = n_2 \sin\theta_t \quad (5.67)$$

which leads to the *Snell law* for reflection and refraction

$$\theta_i = \theta_r \text{ and } n_1 \sin\theta_i = n_2 \sin\theta_t \quad (5.68)$$

Substituting eqs. (5.67) and (5.68) back into eq. (5.66) leads to

$$\cos\theta_i E_{//i} + \cos\theta_i E_{//r} = \cos\theta_t E_{//t} \quad (5.69)$$

which gives one equation relating $E_{//i}$, $E_{//r}$, and $E_{//t}$. Another relation can be obtained on the basis of the continuity of the tangential components of the magnetic field at the interface, since there is no surface current. On the basis of eq. (5.65), we can write the continuity of the tangential component of the magnetic field, eq. (5.61), as

$$n_1 E_{//i} - n_1 E_{//r} = n_2 E_{//t} \quad (5.70)$$

Solving eqs. (5.69) and (5.70), we obtain the reflection coefficient $r_{//}$ and transmission coefficient $t_{//}$ for a TM wave as

$$r_{//} = \frac{E_{//r}}{E_{//i}} = \frac{-n_2 \cos\theta_i + n_1 \cos\theta_t}{n_2 \cos\theta_i + n_1 \cos\theta_t} \tag{5.71}$$

$$t_{//} = \frac{E_{//t}}{E_{//i}} = \frac{2n_1 \cos\theta_i}{n_2 \cos\theta_i + n_1 \cos\theta_t} \tag{5.72}$$

Similarly, for a TE wave,

$$r_\perp = \frac{n_1 \cos\theta_i - n_2 \cos\theta_t}{n_1 \cos\theta_i + n_2 \cos\theta_t} \tag{5.73}$$

$$t_\perp = \frac{2n_1 \cos\theta_i}{n_1 \cos\theta_i + n_2 \cos\theta_t} \tag{5.74}$$

Equations (5.71)–(5.74) are called the *Fresnel coefficients* of reflection and transmission.

The above discussion assumes real refractive indices for both media. If any of the media is absorbing, one can prove that the expressions for the Snell law and the Fresnel coefficients remain the same except that we should replace n by the complex refractive index N. This statement can raise questions. For example, the Snell law becomes $n_1 \sin\theta_i = N_2 \sin\theta_t$. If N_2 is complex, the angle of refraction θ_t is also complex. What does a complex angle mean? To answer this question, one can substitute the complex angle into the transmitted wave expression, eq. (5.64), and see that in this case the constant amplitude surface does not coincide with the constant phase surface. Such a wave with different constant amplitude and phase surfaces is called an inhomogeneous wave. The proof is left as an exercise.

The Fresnel coefficients give the magnitudes of reflected and transmitted fields. To calculate the energy flux going across the interface, we need to examine the Poynting vector. For a TM wave, we have

$$\langle\mathbf{S}\rangle = \frac{1}{2}\text{Re}(\mathbf{E}\times\mathbf{H}^*) = -\frac{1}{2}\text{Re}(E_z H_y^*)\hat{\mathbf{x}} + \frac{1}{2}\text{Re}(E_x H_y^*)\hat{\mathbf{z}}$$

$$= \frac{n}{2\mu c_0} E_{//}^2 \sin\theta\,\hat{\mathbf{x}} + \frac{n}{2\mu c_0} E_{//}^2 \cos\theta\,\hat{\mathbf{z}}$$

$$= \langle S_x\rangle\hat{\mathbf{x}} + \langle S_z\rangle\hat{\mathbf{z}}$$

Since only the component in the z-direction goes across the interface, we define the reflectivity and transmissivity, based on the power arriving per unit area normal to the interface, $S_{i,z}$, as

$$\text{Reflectivity:} \quad \left.\begin{array}{l} R_{//} = \frac{S_{//r,z}}{S_{//i,z}} = \frac{S_{//r}}{S_{//i}} = |r_{//}|^2 \\ R_\perp = |r_\perp|^2 \end{array}\right\} \tag{5.75}$$

$$\text{Transmissivity: } \tau_{//} = \frac{S_{//t,z}}{S_{//i,z}} = \frac{S_{//t}}{S_{//i}} = \frac{\text{Re}(N_2^* \cos\theta_t)}{\text{Re}(N_1^* \cos\theta_i)} |t_{//}|^2 \Bigg\} \quad (5.76)$$

$$\tau_\perp = \frac{\text{Re}(N_2 \cos\theta_t)}{\text{Re}(N_1 \cos\theta_i)} |t_\perp|^2$$

Equations (5.75) and (5.76) apply to the cases when either or both of the two media are absorbing. When medium 1 is non-absorbing, it can be shown that

$$R + \tau = 1 \quad (5.77)$$

for both TM and TE waves. However, when medium 1 is also absorbing, it can be shown that the above intuitive expression is no longer valid. This is because of the interference of the incident and reflected waves (Knittl, 1976).

At normal incidence, the reflectivity can be simplified to

$$R = R_{//} = R_\perp = \left|\frac{k_1 - k_2}{k_1 + k_2}\right|^2 = \left|\frac{N_2 - n_1}{N_2 + n_1}\right|^2 = \frac{(n_2 - n_1)^2 + (\kappa_2 - \kappa_1)^2}{(n_2 + n_1)^2 + (\kappa_2 + \kappa_1)^2} \quad (5.78)$$

which is identical to eq. (5.57) for the reflectivity of an electron wave. For an air/glass interface, where $n = 1$ for air and $n \approx 1.45$ for glass between $\lambda = 0.5$ and $0.6\,\mu\text{m}$, the reflectivity at normal incidence is $\sim 3.4\%$. For an air/silicon interface (n for Si ≈ 4 between $\lambda = 0.5$ and $0.6\,\mu\text{m}$), the reflectivity is $\sim 36\%$ at normal incidence.

In figure 5.3, we show the reflectivity for typical dielectric and metallic materials as a function of the angle of incidence. The figure shows that the reflectivity depends on the polarization of the incident radiation. The reflectivity for a TM wave can be zero for dielectric materials. From eq. (5.71), this happens when the numerator equals zero, that

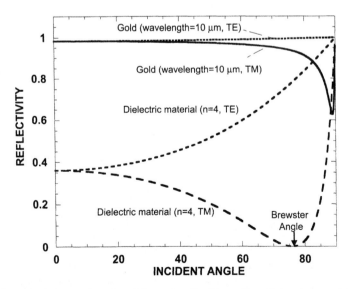

Figure 5.3 Reflectivity as a function of the angle of incidence for a dielectric material with $n = 4$ and for gold with $N = 10.8 + i51.6$.

is, when $\sin(2\theta_i) = \sin(2\theta_t)$ or $\theta_t = \pi/2 - \theta_i$. Substituting this condition into the Snell law, eq. (5.67), leads to the following incidence angle, θ_B, at which the reflectivity of the TM mode is zero,

$$\tan\theta_B = n_2/n_1 \tag{5.79}$$

This angle is called the *Brewster angle*. At this angle, only TE waves are reflected. This phenomenon can be exploited to control the polarization of white light and is also the cause of shiny dark (seemingly wet) surfaces on the freeway on a sunny day.

Another interesting situation is when the refractive index of medium 1 (incident side) is larger than that of medium 2. Because the maximum angle of the refracted wave is $\theta_t = 90°$, there exists an angle of incidence above which no real solution for θ_t exists. This *critical angle* happens when, according to the Snell law,

$$n_1 \sin\theta_c = n_2 \sin 90° \text{ or } \theta_c = \arcsin(n_2/n_1) \tag{5.80}$$

Above this angle, the reflectivity equals one; that is, all the incident energy is reflected. This is called total internal reflection and is the basis of waveguides that confine the photon waves laterally, as in an optical fiber and a semiconductor laser. An optical fiber has a core region and a cladding layer [figure 5.4(a)]. The refractive index in the core region is higher than in the cladding layer. If light is launched into the fiber at an angle of incidence (relative to the core/cladding interface) larger than the critical angle, the light will be bounced inside the core only without leakage, thus traveling a long distance along the fiber if the absorption coefficient of the core is small. However, if the angle of incidence is smaller than the critical angle, the light can escape the fiber core. In a semiconductor laser, light is emitted through electron–hole recombination inside the active region. The emitted light spreads over the core region and is confined by cladding layers that have a lower refractive index than the core [figure 5.4(b)]. We mention here, however, that even though the reflectivity is one and transmissivity is zero for a wave incident above the critical angle, there is still a nonzero electromagnetic

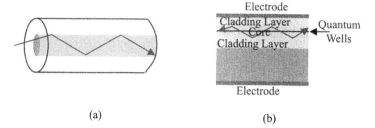

Figure 5.4 Examples of waveguides based on total internal reflection. (a) An optical fiber is made of a core (a few microns wide) and a cladding layer in concentric cylindrical configuration. (b) An edge-emitting semiconductor layer has a planar geometry. Light is emitted inside the quantum wells (a few tens of angstroms) through electron–hole recombination. The emitted light is confined inside the core ($\sim$ microns in total thickness) by total internal reflection effects. Some of the emitted light comes out from the edges because the side surfaces (mirrors or facets) has a nonzero transmissivity.

wave in medium 2. This is called the *evanescent wave*, similar to the evanescent electron wave we mentioned in the previous section. We will discuss the evanescent wave in more detail in section 5.4.

Example 5.1

A 0.5 W laser with a beam diameter of 1 mm and wavelength 0.5 μm is directed at normal incidence to a piece of aluminum. The complex refractive index of aluminum at 0.5 μm is $0.769 + i6.08$. Determine the distribution of heat generation inside the aluminum.

Solution: From eq. (5.39), we know that the Poynting vector, thus the intensity of the laser beam, decreases exponentially inside the film. The distribution of the laser intensity can be expressed as

$$I = (1 - R)I_i e^{-\alpha x} \text{ [W m}^{-2}\text{]} \tag{E5.1.1}$$

where x is the coordinate perpendicular to the surface, R is the reflectivity, and α the absorption coefficient. The decrease in intensity is converted into heat. So, the heat generation distribution is

$$\dot{q} = -\frac{dI}{dx} = (1 - R)\alpha I_i e^{-\alpha x} \text{ [W m}^{-3}\text{]} \tag{E5.1.2}$$

We can calculate R and α as

$$R = \left|\frac{1-N}{1+N}\right|^2 = \left|\frac{0.231 - 6.08i}{1.769 + 6.08i}\right|^2 = 0.923 \tag{E5.1.3}$$

$$\alpha = \frac{4\pi \kappa}{\lambda} = \frac{4\pi \times 6.08}{0.5 \times 10^{-6} \text{ m}} = 1.53 \times 10^8 \text{ m}^{-1} \tag{E5.1.4}$$

Substituting these values and $I_i = 0.5 \text{ W}/(\pi \times 0.001^2/4) = 6.34 \times 10^5$ W m^{-2} into the heat generation distribution expression leads to the final answer as

$$q = 7.5 \times 10^{12} e^{-\alpha x} \text{ [W m}^{-3}\text{]} \tag{E5.1.5}$$

Comment. Because α is very large, heat absorption occurs only in the region near the surface.

With the expression for the surface reflectivity and transmissivity, the emissivity of the surface at the same wavelength can be readily calculated. If medium 2 is semi-infinite, the energy transmitted into the medium will eventually be absorbed and thus the absorptivity equals the transmissivity at the interface. From Kirchoff's law, the emissivity equals the absorptivity at the same incident direction and the same wavelength.

5.2.3 Acoustic Waves

The reflection and refraction of acoustic waves can be treated similarly to the electromagnetic waves. The continuity requirements for acoustic waves are that the displacement velocity and the force at the interface must be continuous:

$$\sum \mathbf{v}_1 = \sum \mathbf{v}_2 \quad \text{and} \quad \sum \bar{\bar{\sigma}}_1 \cdot \hat{\mathbf{n}} = \sum \bar{\bar{\sigma}}_2 \cdot \hat{\mathbf{n}} \tag{5.81}$$

where $\hat{\mathbf{n}}$ is the norm of the interface and $\bar{\bar{\sigma}} \cdot \hat{\mathbf{n}}$ is the force acting on the interface, as expressed by eq. (5.43), and the summation is over all the fields (for example, the incident and reflected) on each side. While continuity of force is always required, even in a lattice dynamics simulation, continuity of displacement velocity is not always true at the atomic scale. For example, the two atoms at the two sides of an interface can have different displacement velocities. In the long wavelength limit, however, the continuum assumption is reasonable and eq. (5.81) is valid.

Using the boundary conditions and plane acoustic waves of the form of eqs. (5.49)–(5.51), one can again derive expressions for the reflectivity and transmissivity of acoustic waves at an interface as for electromagnetic and electron waves. The derivation, however, is more complex because acoustic waves have three polarizations and one must consider the possibility of coupling among these polarizations—for example, whether a longitudinal wave can excite a transverse component in the reflected and transmitted waves (Auld, 1990). The simplest case is when the medium is isotropic and the incident wave is a transverse wave with displacement polarized in the direction perpendicular to the plane of incidence (called a horizontally polarized shear wave or SH wave). In this case, only one SH reflected wave and one SH transmitted wave are excited. From the boundary conditions and following a similar procedure as for optical waves, one can derive the reflection and transmission coefficients for an SH wave as

$$\left. \begin{array}{l} r_s = \dfrac{v_r}{v_i} = \dfrac{Z_1 \cos \theta_i - Z_2 \cos \theta_t}{Z_1 \cos \theta_i + Z_2 \cos \theta_t} \\[2ex] t_s = \dfrac{v_t}{v_i} = \dfrac{2 Z_1 \cos \theta_i}{Z_1 \cos \theta_i + Z_2 \cos \theta_t} \end{array} \right\} \tag{5.82}$$

where $Z = (\rho c_{44})^{1/2} = \rho v_T$ is the *acoustic impedance*, which plays a similar role to the optical refractive index. These equations are identical to the Fresnel coefficients of a TE wave [eqs. (5.73) and (5.74)] with the acoustic impedances replacing the refractive indices. The relation between the incident angle θ_i and the refraction angle θ_t is given by the Snell law, which assumes the following form for an SH incident wave:

$$\frac{\sin \theta_i}{v_{T1}} = \frac{\sin \theta_t}{v_{T2}} \tag{5.83}$$

On the basis of the reflection and tranmission coefficients and the acoustic Poynting vector, eq. (5.53), one can calculate the energy reflectivity and transmissivity for acoustic waves. At normal incidence, the acoustic reflectivity for an SH wave is

$$R_s = \left| \frac{Z_1 - Z_2}{Z_1 + Z_2} \right|^2 \tag{5.84}$$

ENERGY TRANSFER BY WAVES   179

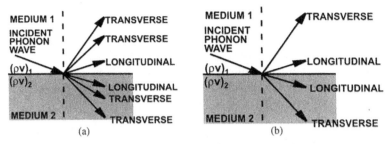

Figure 5.5 Reflection and refraction of acoustic waves in (a) anisotropic and (b) isotropic media, for longitudinally polarized waves (L waves) or transversely polarized waves with the displacement vector in the plane of incidence (vertically polarized shearwave or SV waves).

For a transverse wave polarized in the plane of incidence (vertically polarized shear wave or SV wave) and for a longitudinally polarized incident wave (L wave), coupling of different polarizations can occur. In general, an incident wave can excite three reflected waves and three transmitted waves, as shown in figure 5.5(a). For isotropic media, the two transverse waves are degenerate, so the picture reduces to 5.5(b). Correspondingly, the Snell law looks slightly different,

$$\frac{\sin\theta_i}{v_i} = \frac{\sin\theta_{rL}}{v_{L1}} = \frac{\sin\theta_{rT}}{v_{T1}} = \frac{\sin\theta_{tL}}{v_{L2}} = \frac{\sin\theta_{tT}}{v_{T2}} \quad (5.85)$$

where subscripts r and t represent the reflected and transmitted components, and L and T represent longitudinal (L) and transverse modes (SH and SV).

For isotropic media, the reflection and transmission coefficients can be found by solving a 4×4 matrix equation (Auld, 1990). An example of phonon reflectivity and transmissivity at an interface between two isotropic media with acoustic properties similar to those of Si and Ge is shown in figure 5.6 (Chen, 1999). It shows

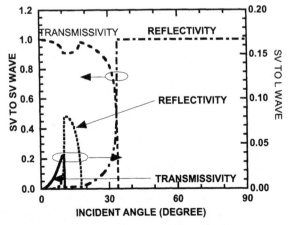

Figure 5.6 Phonon transmissivity and reflectivity at an interface similar to Si and Ge for an acoustic wave (incident from the Ge side) with the displacement vector polarized in the plane of incidence (Chen, 1999), that is, an SV incident wave. Note that L waves are also excited.

that part of the transverse incident wave is converted into a longitudinal wave. This is called mode conversion. Total reflection of acoustic waves can also occur and several critical angles can exist, as one can infer from eq. (5.85), for refracted waves of different polarizations. In figure 5.6, the total reflection of an SV wave from a germanium-like medium into a silicon-like medium occurs at 33°. Above this angle, an evanescent wave exists in the silicon side, similar to the evanescent electron and photon waves.

5.2.4 Thermal Boundary Resistance

The reflection of waves reduces the number of forward-going carriers (electrons, phonons, and photons) and is thus a source of resistance to the carrier flow. For the current flow, electron interface reflection creates an additional electrical resistance. For photons, highly reflective materials are used as radiation shields for thermal insulation, as exemplified in multilayer thermal insulation materials (Tien and Cunnington, 1973). The reflection of acoustic waves results in a resistance to heat flow, called the thermal boundary resistance. This is well known in cryogenics, where thermal boundary resistance was first discovered to exist between liquid helium and solid walls and is called the Kapitza resistance (Kapitza, 1941; Swartz and Pohl, 1989). The same phenomenon also occurs for two solid interfaces and was first treated theoretically by Little (1959).

We take figure 5.7(a) as a model system, where a heat current flows across the interface from medium 1 into medium 2, the media being maintained at two different temperatures. We neglect the temperature drop inside each material due to phonon scattering, but will come back to this issue in chapter 7. We first count the phonon heat flux going from medium 1 into medium 2, $q_{1 \to 2}$ [W m^{-2}], which can be written as

$$q_{1 \to 2} = \sum_{p=1}^{3m} \left[\frac{1}{V_1} \sum_{k_x = -k_{max}}^{k_{max}} \sum_{k_y = -k_{max}}^{k_{max}} \sum_{k_z = 0}^{k_{max}} v_{z1} \hbar \omega \tau_{12} f(\omega, T_{e1}) \right] \quad (5.86)$$

where T_{e1} represents the temperature of the phonons coming toward the interface and $f(\omega, T_{e1})$ is the Bose–Einstein distribution for phonons at T_{e1}, and τ_{12} is the phonon transmissivity from medium 1 into medium 2. The summations are over all the wavevectors of k_x, k_y, and positive wavevector of k_z, so that only phonons coming toward the interface are counted. The maximum wavevector is π/a; that is, the boundaries of the first Brillouin zone. Equation (5.86) is a mixture of the phonon concept we learnt in chapter 3 and the acoustic wave propagation discussed in this chapter. The transmissivity can, for example, be calculated on the basis of the discussion in the previous section. The term $\hbar \omega f(\omega, T_{e1})$ inside eq. (5.86) represents the average energy per quantum state and the summation over the wavevectors denotes the allowable quantum states of medium 1 within a volume V_1. The division by the volume is to obtain the energy density per unit volume. These phonons are propagating toward the interface at a speed v_{z1}—the phonon velocity along the z-direction. The summation over p denotes the phonon polarizations, that is, $3m$ if there are m atoms per basis. To carry out the summation over wavevectors

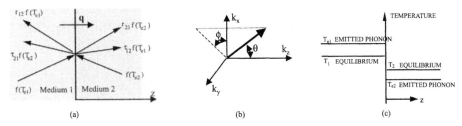

Figure 5.7 (a) Transmission and reflection of phonons at an interface. (b) Coordinate system for thermal boundary resistance evaluation. (c) Difference between the emitted phonon temperature and the local equivalent equilibrium temperature.

in eq. (5.86), we will convert it into an integration using the differential density-of-states concept outlined in section 3.4.4. For an isotropic medium, we have

$$q_{1\to 2} = \iint_{\Omega_1 > 2\pi} \left[\int_0^{\omega_{D1}} \tau_{12} v_{z1} \hbar \omega f(\omega, T_{e1}) \frac{D_1(\omega)}{4\pi} d\omega \right] d\Omega_1$$

$$= \frac{1}{4\pi} \int_0^{2\pi} d\phi_1 \int_0^{\pi/2} \sin\theta_1 d\theta_1 \left[\int_0^{\omega_{D1}} v_1 \cos\theta_1 \tau_{12}(\omega, \phi_1, \theta_1) \hbar \omega f(\omega, T_{e1}) D_1(\omega) d\omega \right]$$

(5.87)

where ω_{D1} is the Debye frequency of medium 1 and the angles are denoted in figure 5.7(b). We use the notation $\Omega_1 > 2\pi$ to represent the half-space solid angle toward the interface from medium 1. We should point out, however, that the Debye assumption is not really necessary for the derivation. Similarly, there exists a phonon heat flux from medium 2 to medium 1. The net heat flux from material 1 into material 2 is the difference of the two,

$$q = \iint_{\Omega_1 \geq 2\pi} \left[\int v_1 \cos\theta_1 \hbar \omega f(\omega, T_{e1}) \tau_{12}(\omega, \phi_1, \theta_1) D_1(\omega)/4\pi \, d\omega \right] d\Omega_1$$
$$- \iint_{\Omega_2 \leq 2\pi} \left[\int v_2 \cos\theta_2 \hbar \omega f(\omega, T_{e2}) \tau_{21}(\omega, \phi_2, \theta_2) D_2(\omega)/4\pi \, d\omega \right] d\Omega_2 \quad (5.88)$$

where we have suppressed the integration limits and used $d\Omega = \sin\theta \, d\theta \, d\phi$ for simplicity in notation; details can be worked out by following eq. (5.87). At thermal equilibrium, that is, when $T_{e1} = T_{e2}$, the net heat flux is zero, so that

$$\int_{\Omega_1 \geq 2\pi} \left[\int v_1 \cos\theta_1 \hbar \omega f(\omega, T_{e1}) \tau_{12}(\omega, \phi_1, \theta_1) D_1(\omega) d\omega/4\pi \right] d\Omega_1$$
$$= \int_{\Omega_2 \leq 2\pi} \left[\int v_2 \cos\theta_2 \hbar \omega f(\omega, T_{e1}) \tau_{21}(\omega, \phi_2, \theta_2) D_2(\omega) d\omega/4\pi \right] d\Omega_2 \quad (5.89)$$

Equation (5.89) is an example of the *principle of detailed balance*, which requires that no net flux of any kind (heat, particle, or charge) crosses an interface if the system is at equilibrium. Using eq. (5.89), we can write eq. (5.88) as

$$q = \int_{\Omega_1 \geq 2\pi} \left[\int v_1 \cos\theta_1 \hbar\omega [f(\omega, T_{e1}) - f(\omega, T_{e2})] \tau_{12}(\omega, \phi_1, \theta_1) D_1(\omega)/4\pi d\omega \right] d\Omega_1 \quad (5.90)$$

Thus, with the help of the principle of detailed balance, we can express the flux with properties in one of the two media plus, of course, the transmissivity. This greatly simplifies the calculations. When the difference between T_{e1} and T_{e2} is small, eq. (5.90) can be further written as

$$q = \frac{T_{e1} - T_{e2}}{\mathfrak{R}_e} \quad (5.91)$$

where $\mathfrak{R}_e$ is the *specific thermal boundary resistance* [K m^2 W^{-1}],

$$\frac{1}{\mathfrak{R}_e} = \frac{1}{4\pi} \int_0^{2\pi} d\phi_1 \int_0^{\pi/2} d\theta_1 \int_0^{\omega_{D1}} v_1 \cos\theta_1 \sin\theta_1 \hbar\omega \frac{\partial f(\omega, T)}{\partial T} \tau_{12}(\omega, \phi_1, \theta_1) D_1(\omega) d\omega$$

$$= \frac{1}{2} \int_0^1 \left[\int_0^{\omega_{D1}} v_1 C_1(\omega) \tau_{12}(\omega, \mu_1) d\omega \right] \mu_1 d\mu_1 \quad (5.92)$$

where $\mu = \cos\theta$ is the directional cosine and $C_1(\omega)[= \hbar\omega D(\omega)\partial f/\partial T]$ is the spectral (mode) specific heat. The second equation in (5.92) used the fact that for an isotropic medium τ_{12} is independent of Φ. When the temperature is low and τ is independent of frequency, it can be shown from the above equation that

$$\mathfrak{R}_e \propto T_e^{-3} \quad (5.93)$$

because of the T^3 dependence of the specific heat. The agreement of eq. (5.93) with experimental results depends on how the transmissivity τ is calculated. Modeling results based on acoustic wave relations such as eq. (5.82) generally agree well with experimental results at very low temperatures, as shown in figure 5.8, and such a model is called the acoustic mismatch model (Little, 1959). At higher temperatures, however, the experimental results deviate from the acoustic mismatch model. This is because the thermal boundary resistance, calculated on the basis of the phonon transmissivity discussed in section 5.2.3, requires that phonon scattering at the interface be specular and elastic; that is, the phonon frequency does not change and the Snell law is obeyed. These conditions are satisfied at very low temperatures when the phonon wavelengths are long, but are more difficult to satisfy at higher temperatures when the phonon wavelengths contributing dominantly to heat transfer are comparable or shorter than the surface roughness. At room temperature, the average phonon wavelength is ∼ 10–20 Å in most

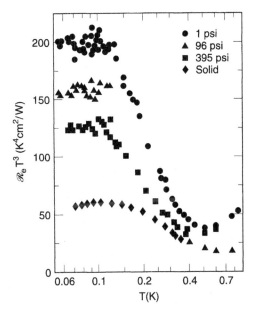

Figure 5.8 Thermal boundary resistance at low temperature (Swartz and Pohl, 1989; courtesy of APS Associate Publisher).

materials. A few atomic layers mixing will create a rough interface that no longer scatters phonons specularly.

Despite intense research, there is no generally accepted way to calculate the thermal boundary resistance at higher temperatures. One rather crude model is called the diffuse mismatch model (Swartz and Pohl, 1989), which assumes that phonons emerging from the interface do not really bear any relationship with their origin; in other words, one cannot tell which side they come from. This assumption implies that

$$R_{12} = \tau_{21} \text{ or } 1 - \tau_{12} = \tau_{21} \tag{5.94}$$

where the second equation comes from the energy conservation identity $R_{12} + \tau_{12} = 1$. We recall again that subscript 12 means from medium 1 into medium 2 and vice versa. By substituting the above relationship into eq. (5.89) and further assuming a linear dispersion for each acoustic wave polarization, Swartz and Pohl (1989) obtained

$$\tau_{d12} = \frac{1/v_2^2}{1/v_1^2 + 1/v_2^2} \tag{5.95}$$

where $1/v^2$ comes from the product of the density of states [eq. (3.55)] and the velocity. This relation is valid at low temperatures. At higher temperatures, a similar treatment leads to (Dames and Chen, 2004),

$$\tau_{d12}(T_e) = \frac{v_2 U_2(T_e)}{v_1 U_1(T_e) + v_2 U_2(T_e)} = \frac{1}{1 + [v_1 U_1/v_2 U_2]} \tag{5.96}$$

where U is the volumetric internal energy [eq. (4.73)],

$$U = \int_0^{\omega_D} \hbar \omega f(\omega, T_e) D(\omega) d\omega = \int_0^{T_e} C(T) dT \tag{5.97}$$

Chen (1998) assumed that the specific heat is independent of temperature and further expressed eq. (5.96) in terms of the specific heat. Equation (5.96), however, is more accurate and includes all other simplified cases, including eq. (5.95). Using the Debye approximation, the ratio in the denominator of eq. (5.96) can be expressed as

$$\frac{v_1 U_1}{v_2 U_2} = \frac{1/v_1^2 \int_0^{\omega_{D1}} \omega^3 f(\omega, T_e) d\omega}{1/v_1^2 \int_0^{\omega_{D2}} \omega^3 f(\omega, T_e) d\omega} \tag{5.98}$$

We should caution, however, that the diffuse mismatch model discussed above is a very crude approximation and is clearly not valid when the two materials are very similar. In this case, the transmissivity should approach one but eq. (5.96) predicts a transmissivity approaching 0.5.

The case when the two materials are similar, and in the limit of an identical material with an imaginary interface, poses another dilemma for the usual thermal boundary resistance expression defined on the basis of eq. (5.91). In this limit, the thermal boundary resistance should clearly be zero but eq. (5.92) gives a finite value, even if one sets $\tau_{12} = 1$. This dilemma (Little, 1959) arises from the temperature definition used in eq. (5.91). So far, we have been careful in saying that T_e represents the temperature of the phonons coming toward the interface, and avoided discussing what are the real temperatures of medium 1 and medium 2 at the interface. This is because phonon interface transport is a highly nonequilibrium process and it is hard to define a temperature. Referring to figure 5.7(a), we see that on each side of the interface there are three groups of phonons. For example, on the side of medium 1, one group is the incoming phonons with a temperature T_{e1}, the other group is the reflected phonons leaving the interface, which has an energy distribution determined by the convolution of the incoming phonon at T_{e1} and the interface reflectivity. The third group comes from the transmission of side 2, with an energy distribution determined by the convolution of phonons at temperature T_{e2} and the interface transmissivity. The local phonon energy spectra at the interface are thus very different from that of the incoming phonons and cannot be represented by an equilibrium, or close to equilibrium, distribution with a single equilibrium temperature (Katerberg et al., 1977). However, if these phonons were to adiabatically approach an equilibrium, we could obtain the final equilibrium temperature of this adiabatic system and will call this temperature the *equivalent equilibrium temperature*. This equivalent equilibrium temperature is really just a measure of the local energy density rather than one that represents the spectral characteristics of the energy distribution, but it is consistent with the local equilibrium approximation used in heat transfer calculations such as the Fourier law. Figure 5.7(c) illustrates the difference between the incoming phonon temperature and the equivalent equilibrium temperature. It can be shown that the equivalent temperature can be related to the incoming phonon temperature by

$$T_1 = T_{e1} - (T_{e1} - T_{e2}) \int_0^1 \tau_{12}(\mu_1) d\mu_1 / 2 \tag{5.99}$$

$$T_2 = T_{e2} + (T_{e1} - T_{e2}) \int_0^1 \tau_{21}(\mu_2) d\mu_2 / 2 \tag{5.100}$$

where $\mu = \cos\theta$ is the directional cosine and we have assumed that transmissivity is independent of angle Φ. On the basis of this equivalent temperature and the consideration of the deviation of the phonon density of states from the Debye model, Chen and Zeng (2001) arrived at the following expression for the thermal boundary resistance:

$$\Re = \frac{T_1 - T_2}{q} = \frac{2[1 - \langle \int_0^1 \tau_{12}(\mu_1) d\mu_1 + \int_0^1 \tau_{21}(\mu_2) d\mu_2 \rangle /2]}{\int_0^1 [\int \tau_{12}(\mu_1) v_1 C_1(\omega) d\omega] \mu_1 d\mu_1} \quad (5.101)$$

Note that here we have dropped the subscript "e" because the temperatures are defined as a measure of the local energy density, not on the properties of the phonons coming toward the interface, as in eq. (5.91). Strictly speaking, the spectral specific heat should be evaluated at T_e but, under the assumption of a small temperature difference, it can be evaluated as the average of the equivalent equilibrium temperatures on the two sides. Equation (5.101) leads to zero thermal boundary resistance when the transmissivity from both sides is equal to one, that is, when no interface exists. This resolves the dilemma in eq. (5.92) that gives a nonzero thermal boundary resistance even when the transmissivity equals one (Little, 1959). When measuring thermal boundary resistance at low temperatures, it is possible to anchor temperature sensors so that T_{e1} and T_{e2} are measured, so that eq. (5.91) is a valid definition. Most high-temperature measurements of thermal boundary resistance, however, cannot determine T_e and the thermal boundary resistance values are obtained on the basis of the *equivalent equilibrium T*. Consequently, one should pay attention to using correct models to explain the experimental data.

The thermal boundary resistance discussed here exists even if the interface is perfect, as long as there exists phonon reflection at the interface. The order of magnitude of such thermal boundary resistance is, according to eq. (5.101), $\Re \sim 1/(Cv)$, where C is the volumetric specific heat and v is the phonon speed. Taking $C \approx 10^6$ J m^{-3} K^{-1} at room temperature and $v \approx 1000$ m s^{-1}, the thermal boundary resistance is then $\sim 10^{-9}$–10^{-8} K m^2 W^{-1}, consistent with experimental data for nearly perfect interfaces (Costescu et al., 2003). Less ideal interfaces have higher thermal boundary resistances (Stoner and Maris, 1993). Although the value of thermal boundary resistance for ideal interfaces seems exceedingly small, it becomes dominant for nanoscale systems with a large number of interfaces. For example, the thermal conductivity of superlattices in the direction perpendicular to the film plane is found to be dominated by the thermal boundary resistance. It should be pointed out, however, that the value of the thermal boundary resistance in a multilayer structure can differ from that of a single interface (Chen, 1998). In macroscopic structures, the thermal boundary resistance at the interfaces can be much larger because the two materials are not in perfect contact. We will not discuss these cases here.

5.3 Wave Propagation in Thin Films

In thin films, there are multiple interfaces. We should first emphasize that these thin films do not have to be an actual material. A thin vacuum space between two parallel plates can be considered a thin film. These interfaces will cause the reflection of the incident waves. The reflected waves can be superimposed on the incoming wave to cause interference effects that lead to the thickness dependence of reflectivity and transmissivity. One other new phenomenon that may occur in thin films is tunneling, which makes the total reflection phenomenon that occurs at one interface disappear. These interference and tunneling processes can occur for photons, phonons, and electrons. In this section, we will first examine the interference phenomenon. The formulation established can also be applied to tunneling processes, which will be discussed in section 5.4.

5.3.1 Propagation of EM Waves

There are three ways to derive an expression for the radiative properties (reflectivity and transmissivity) of thin films: the field-tracing method, the resultant wave method, and the transfer matrix method, as explained in figure 5.9. The field-tracing method, figure 5.9(a), follows the trajectory of the wave and counts each reflection and transmission when the wave meets an interface (Born and Wolf, 1980), using the Fresnel reflection and transmission coefficients. This method is intuitive but cumbersome. Because all the forwarding waves in the same medium have the same exponential factor, we can sum them up into one wave with a undetermined amplitude and call this wave the resultant wave [figure 5.9(b)]. Similarly, all the backward propagating waves in the same medium can be summed into a resultant wave. There are then four resultant waves in the single layer thin film situation, one reflected, two inside the film (forward and backward), and one transmitted, as shown in figure 5.9(b). The amplitude of each resultant wave will be determined by applying the boundary conditions at the two interfaces. The transfer matrix method combines all the waves (both forward and backward) in each medium into one wave, and uses a matrix to relate the electric and magnetic fields between two different points inside a medium, as shown in figure 5.9(c). Because the tangential components of the electric and magnetic fields are continuous across the interface when no interface charge or interface current exists, the transfer matrix method can be easily extended to multilayers. We will therefore focus on the transfer matrix method.

Consider a TM wave, for example, the x-component of the electric field and the y-component of the magnetic field inside the film, as a function of location z:

$$E_x(z) = \cos\theta_2 E^+ e^{i\varphi(z)} + \cos\theta_2 E^- e^{-i\varphi(z)} \qquad (5.102)$$

$$H_y(z) = \frac{n_2}{\mu c_0}[E^+ e^{i\varphi(z)} - E^- e^{i\varphi(z)}] \qquad (5.103)$$

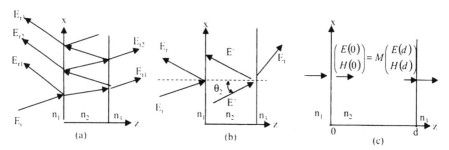

Figure 5.9 Three methods of treating reflection and transmission of electromagnetic fields through a thin film: (a) the field tracing method; (b) the resultant wave method; (c) the transfer matrix method.

where E^+ and E^- are the amplitudes of the resultant forward and backward propagating waves inside the film and

$$\varphi_{(z)} = \frac{\omega n_2 z \cos \theta_2}{c_0} \tag{5.104}$$

where θ_2 is the angle formed between wavevector direction and z. Again, if n_2 is complex, this angle is also complex, and can be calculated according to the Snell law. In the above equations, we have dropped the terms $\exp(-i\omega t)$ and $\exp(-k_x x)$ because all terms have these factors and eventually cancel.

We want to relate the electric and magnetic fields at any location z inside the film to these fields at the interface $z = 0$. This can be realized by first taking $z = 0$ in eq. (5.102) and (5.103) and then eliminating E^+ and E^- in these equations,

$$E_x(z) = E_x(0) \cos \varphi(z) + ip_2 H(0) \sin \varphi(z) \tag{5.105}$$

$$H_y(z) = \frac{i}{p_2} E_x(0) \sin \varphi(z) + H_y(0) \cos \varphi(z) \tag{5.106}$$

where $p_2 = [\cos \theta_2 / (n_2 / \mu c_o)]$ is the *surface impedance* for a TM wave. The above equations can be written in matrix form

$$\begin{pmatrix} E_x(z) \\ H_y(z) \end{pmatrix} = \begin{pmatrix} \cos \varphi(z) & ip_2 \sin \varphi(z) \\ \frac{i}{p_2} \sin \varphi(z) & \cos \varphi(z) \end{pmatrix} \begin{pmatrix} E_x(0) \\ H_y(0) \end{pmatrix} \tag{5.107}$$

Taking $z = d$ and inverting the above matrix, we get

$$\begin{pmatrix} E_x(0) \\ H_x(0) \end{pmatrix} = \begin{pmatrix} \cos \varphi_2 & -ip_2 \sin \varphi_2 \\ -\frac{i}{p_2} \sin \varphi_2 & \cos \varphi_2 \end{pmatrix} \begin{pmatrix} E_x(d) \\ H_y(d) \end{pmatrix} = M \begin{pmatrix} E_x(d) \\ H_y(d) \end{pmatrix} \tag{5.108}$$

where $\varphi_2 = \varphi(d)$ and M is the second-order matrix in the above equation. We call M the interference matrix. It is easy to show that $|M| = 1$.

Equation (5.108) relates the electric and magnetic fields inside the film at $z = d$ to their values at the boundary $z = 0$. To find the reflectivity or transmissivity, we need to further relate them to the fields outside the film through the boundary conditions. For a boundary free of charge and current, eqs. (5.58) and (5.61) dictate that the electric and magnetic fields are continuous, which means that at $z = 0$,

$$E_x(0) = E_i \cos \theta_i + E_r \cos \theta_r = E_{ix} + E_{rx} \tag{5.109}$$

$$H_y(0) = \frac{n_1}{\mu c_0}(E_i - E_r) = \frac{1}{p_1}(E_{ix} - E_{rx}) \tag{5.110}$$

and at $z = d$, only the transmitted wave exists,

$$E_x(d) = E_t \cos \theta_t = E_{tx} \tag{5.111}$$

$$H_y(d) = \frac{n_3}{\mu c_0} E_t = \frac{1}{p_3} E_{tx} \tag{5.112}$$

where $p_1 = \cos\theta_i/(n_1/\mu c_0)$ and $p_3 = \cos\theta_t/(n_3/\mu c_0)$, and we have assumed that μ is the same for all layers because most materials are diamagnetic in the infrared to visible frequency range. We can again write the above equations in matrix form,

$$\begin{pmatrix} E_x(0) \\ H_y(0) \end{pmatrix} = \begin{pmatrix} 1 & 1 \\ \frac{1}{p_1} & -\frac{1}{p_1} \end{pmatrix} \begin{pmatrix} E_{ix} \\ E_{rx} \end{pmatrix} \tag{5.113}$$

$$\begin{pmatrix} E_x(d) \\ H_y(d) \end{pmatrix} = \begin{pmatrix} 1 \\ \frac{1}{p_3} \end{pmatrix} E_{tx} \tag{5.114}$$

We now combine eqs. (5.113), (5.114), and (5.108), using the continuity of E_x and H_y at the interfaces, to get

$$\begin{pmatrix} 1 & 1 \\ \frac{1}{p_1} & -\frac{1}{p_1} \end{pmatrix} \begin{pmatrix} E_{ix} \\ E_{rx} \end{pmatrix} = \begin{pmatrix} m_{11} & m_{12} \\ m_{21} & m_{22} \end{pmatrix} \begin{pmatrix} 1 \\ \frac{1}{p_3} \end{pmatrix} E_{tx} \tag{5.115}$$

where m_{ij} are the elements of the *interference matrix M*. Inverting the matrix of the left-hand side and multiplying out the three matrices, we obtain

$$\begin{pmatrix} E_{ix} \\ E_{rx} \end{pmatrix} = \frac{1}{2} \begin{pmatrix} (m_{11} + \frac{1}{p_3} m_{12}) + (m_{21} + \frac{1}{p_3} m_{22}) p_1 \\ (m_{11} + \frac{1}{p_3} m_{12}) - (m_{21} + \frac{1}{p_3} m_{22}) p_1 \end{pmatrix} E_{tx} \tag{5.116}$$

From the above matrix, we get the reflection and transmission coefficients through the film as

$$r = \frac{E_r}{E_i} = \frac{E_{rx}}{E_{ix}} = \frac{(m_{11} + \frac{1}{p_3} m_{12}) - (m_{21} + \frac{1}{p_3} m_{22}) p_1}{(m_{11} + \frac{1}{p_3} m_{12}) + (m_{21} + \frac{1}{p_3} m_{22}) p_1} \tag{5.117}$$

and

$$t = \frac{E_t}{E_i} = \frac{E_{tx}/\cos\theta_t}{E_{ix}/\cos\theta_i} = \frac{2 c_{tm}}{(m_{11} + \frac{1}{p_3} m_{12}) + (m_{21} + \frac{1}{p_3} m_{22}) p_1} \tag{5.118}$$

where $c_{tm} = \cos\theta_i/\cos\theta_t$. For a TE wave, the above expressions are still valid if p and c_{tm} are replaced by

$$p = -\frac{n\cos\theta}{\mu c_0} \quad \text{and} \quad c_{te} = 1 \tag{5.119}$$

With the reflection and transmission coefficients known, we can calculate the reflectivity and transmissivity according to eqs. (5.75) and (5.76). For absorbing films, the above formulation is still valid by if n is replaced with the complex refractive index N.

The power of the matrix method can be best appreciated when dealing with multilayers of thin films. In this case, we can relate the electric and magnetic field inside the ith layer at both interfaces by the interference matrix M_i for that layer. Since the transverse components of the electric and magnetic fields are continuous at each interface that is free of net charge and current, the total interference matrix of the whole multilayer structure is

$$M = M_1 M_2 M_3 \ldots M_n \tag{5.120}$$

Thus, with such a simple substitution, all previous expressions for the single-layer film are still valid.

For a single layer of film, eqs. (5.117) and (5.118) can be written as

$$r = \frac{r_{12} + r_{23}e^{2i\varphi_2}}{1 + r_{12}r_{23}e^{2i\varphi_2}} \qquad (5.121)$$

and

$$t = \frac{t_{12}t_{23}e^{i\varphi_2}}{1 + r_{12}r_{23}e^{2i\varphi_2}} \qquad (5.122)$$

where r_{12}, r_{23} and t_{12}, t_{23} are the Fresnel reflection and transmission coefficients from medium 1 into medium 2 or from medium 2 to medium 3. The above formula is valid for both TM and TE waves.

On the basis of these expressions, we can calculate the reflectivity and transmissivity of the film. For a nonabsorbing film,

$$R = |r|^2 = \frac{r_{12}^2 + r_{23}^2 + 2r_{12}r_{23}\cos 2\varphi_2}{1 + 2r_{12}r_{23}\cos 2\varphi_2 + r_{12}^2 r_{23}^2} \qquad (5.123)$$

$$\tau = \frac{n_3 \cos\theta_t}{n_1 \cos\theta_i}|t|^2 = \frac{(1-r_{12}^2)(1-r_{23}^2)}{1 + 2r_{12}r_{23}\cos 2\varphi_2 + r_{12}^2 r_{23}^2} \qquad (5.124)$$

If the optical constants of any media are complex, we should use eq. (5.76) to calculate the transmissivity, and carry out complex number operation, $R = rr^*$ and $\tau = tt^*$.

The cosine function in eqs. (5.123) and (5.124) suggests that the reflectivity and transmissivity vary as a function of thickness, and when there is no absorption the variation is periodic, as shown in figure 5.10. This periodic variation in reflectivity and

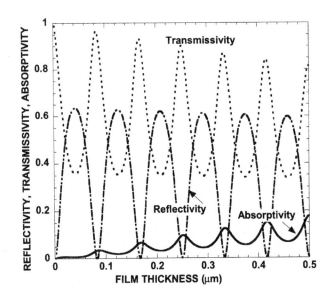

Figure 5.10 Reflectivity, transmissivity, and absorptivity of a thin film as a function of the film thickness, assuming vacuum on both sides.

transmissivity is the *interference phenomenon*, caused by the constructive or destructive superposition of the reflected and the incident waves. The maximum or minimum in the reflectivity can be found by setting $dR/d\varphi_2 = 0$, which leads to

$$\sin 2\varphi_2 = 0 \tag{5.125}$$

or

$$\frac{4\pi n_2 L \cos\theta_2}{\lambda_0} = m\pi \tag{5.126}$$

$$d = \frac{m\lambda_0}{4n_2 \cos\theta_2} \tag{5.127}$$

Under the above condition, eq. (5.123) becomes

$$R = \left(\frac{r_{12} - r_{23}}{1 - r_{12}r_{23}}\right)^2 = \left(\frac{n_1 n_3 - n_2^2}{n_1 n_3 + n_2^2}\right)^2 \quad \text{(for odd } m = 2l+1\text{)} \tag{5.128}$$

$$R = \left(\frac{r_{12} + r_{23}}{1 + r_{12}r_{23}}\right)^2 = \left(\frac{n_1 - n_3}{n_1 + n_3}\right)^2 \quad \text{(for even } m = 2l\text{)} \tag{5.129}$$

where the first equality in the above two equations is valid for an arbitrary angle of incidence while the second is for normal incidence only. When the film thickness is $(2\ell+1)\lambda_0/(4n_2\cos\theta_2)$, the reflectivity R can be a maximum ($n_2 < n_3$) or a minimum ($n_2 < n_3$). Zero reflection occurs when the film has a refractive index $\sqrt{n_1 n_3}$ and its thickness satisfies eq. (5.127) for odd m. Such interference phenomena are the basis for *antireflection coatings*. When the film thickness is $\ell\lambda_0/(2n_2\cos\theta_2)$, the reflectivity does not depend on the second layer.

The reflectivity and transmissivity of multilayer thin films can be calculated using the transfer matrix method. In practice, the reflectivity and transmissivity of multilayers can be controlled quite accurately with various thin-film deposition techniques and the possibility of controlling spectral and directional properties is large. One special example is the *Bragg reflector*, which is made from two alternating layers of thin films, figure 5.11(a). Each layer has a thickness equal to one-quarter of the light wavelength inside the film. Although, at one interface, the reflectivity between the two materials may be small, the coherent superposition of the reflected fields can create a reflectivity that is close to 100%. Such Bragg reflectors are used as coatings for mirrors that are highly reflective at a specific required wavelength, such as for lasers and X-rays. Figure 5.11(b) gives an example of the reflectivity of a quarter-wavelength mirror, similar to those used in special semiconductor laser structures called vertical-cavity surface-emitting lasers (Koyama et al., 1989; Walker, 1993). The reflectivity in certain spectral regions can reach 100%, meaning that no electromagnetic fields in that wavelength regime exist inside the reflector. These spectral regions, called stop bands, occur when the round-trip phase difference through one period (two layers) equals $2\ell\pi$, that is, when the forward and backward propagating fields inside the films cancel each other,

$$\frac{4\pi n_1 d_1 \cos\theta_1}{\lambda_0} + \frac{4\pi n_2 d_2 \cos\theta_2}{\lambda_0} = 2\ell\pi \tag{5.130}$$

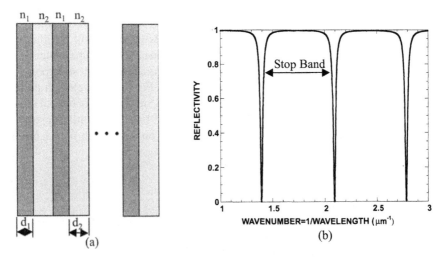

Figure 5.11 (a) A Bragg reflector is a periodic thin-film structure. (b) Calculated reflectivity of a Bragg reflector as a function of the incident photon wavelength for a reflector with refractive indices of 3 and 3.5 and a corresponding thickness of 417 Å and 352 Å for each layer.

where the subscripts 1 and 2 denote layer 1 and layer 2 respectively. Denoting $a = n_1 d_1 \cos\theta_1 + n_2 d_2 \cos\theta_2$ as the optical thickness of one period, the above equation can be written as (Knittl, 1976)

$$ka = \ell\pi \qquad (5.131)$$

where $k(= 2\pi/\lambda_0)$ is the wavevector in vacuum. Equation (5.131) is identical to the condition of the electron bandgap formation discussed in chapter 3, which was obtained by solving the Schrödinger equation. We have said before that the formation of the electron bandgap is due to the cancellation of the electron waves inside the crystal. The discussion on the photon stop bands reinforces this picture. The similarities of these different waves, including electrons, photons, and phonons, have, in the past, been explored extensively to develop new concepts. For example, the *phonon interference filters* (Narayanamurti et al., 1979) and the electron minigaps (Esaki and Tsu, 1970), based on superlattices, benefited from the analogy of photon stop bands in interference filters. In return, it was exactly on the basis of the analogy of three-dimensional band structure in naturally existing crystals for electrons and phonons that the concept of three-dimensional *photonic crystals* was proposed (Yablonovitch, 1986), although one can also argue that this concept is an extension of the thin-film Bragg reflectors to three dimensions. Not only are these concepts very similar to each other; the mathematical techniques are also often interchangeable. For example, one popular approach for calculating the band structures of three-dimensional photonic crystals is based on a generalized transfer matrix method (Pendry, 1996).

5.3.2 Phonons and Acoustic Waves

In chapter 3, we considered phonon waves in a periodic lattice chain and discussed phonons in superlattices. The periodicity in naturally existing crystal lattices leads to the representation of phonons in the first Brillouin zone. The periodicity of superlattices adds an additional restriction to the phonon wavevector and leads to the folded zone representation and the formation of phonon minibands [figure 3.30]. Similar to the photon stop bands, the phonon minigaps formed in the dispersion of superlattices can be thought of as stop bands generated by multiple reflections and coherent superposition of the lattice waves, as for photons in periodic structures. For long-wavelength phonons, that is, acoustic waves, one can also use the transfer matrix method as for optical waves to calculate the transmission of lattice waves through single-layer and multilayer structures (Nayfeh, 1995). The reflectivity r and transmissivity t of an SH wave through a film with thickness d can be calculated from the following matrix

$$\begin{pmatrix} 1 \\ r \end{pmatrix} = A_i^{-1} M A_t \begin{pmatrix} t \\ 0 \end{pmatrix} \tag{5.132}$$

where the interference matrix is similar to that of an electromagnetic wave

$$M = \begin{pmatrix} \cos\varphi_{T2} & i\sin\varphi_{T2}/Y_2 \\ iY_2 \sin\varphi_{T2} & \cos\varphi_{T2} \end{pmatrix} \tag{5.133}$$

$$A_i = \begin{pmatrix} 1 & 1 \\ -Z_{Ti}\cos\theta_{Ti} & Z_{Ti}\cos\theta_{Ti} \end{pmatrix} \tag{5.134}$$

where $\varphi_{T2} = \omega d \cos\theta_2/v_{T2}$, $Y_2 = -Z_{T2}\cos\theta_2$, and A_t is obtained by replacing the subscript i in eq. (5.134) by t. The subscript T is used to represent properties of the transverse waves and, in this case, a transverse wave polarized perpendicular to the plane of incidence. The reflection and transmission coefficients are defined as

$$r = v_r(0)/v_i(0) \quad t = v_t(d)/v_i(0) \tag{5.135}$$

The matrix formulation for SH acoustic waves is clearly similar to that for optical waves. Multilayers can again be treated by simply replacing the interference matrix M with the product $M_1 M_2 \ldots M_{2n+1}$. The order of the matrices is the same as the sequence of the layers. For longitudinal waves (L) and vertically polarized transverse waves (SV) with the displacement polarized in the plane of incidence, the relationship between the incident, reflected, and transmitted wave velocity components of isotropic media is

$$\begin{pmatrix} v_{Ti}(0) \\ v_{Li}(0) \\ v_{Tr}(0) \\ v_{Lr}(0) \end{pmatrix} = B_i^{-1} M B_t \begin{pmatrix} v_{Tt}(d) \\ v_{Lt}(d) \\ 0 \\ 0 \end{pmatrix} \tag{5.136}$$

where v_{Ti} and v_{Li} are the amplitudes of the displacement velocities of the incident transverse and longitudinal waves, respectively, and subscripts r and t represent the reflected and transmitted waves, as usual. Matrix B_i is a 4×4 matrix given by

$$B_i = \begin{pmatrix} -\sin\theta_{Ti} & \cos\theta_{Li} & \sin\theta_{Ti} & -\cos\theta_{Li} \\ \cos\theta_{Ti} & \sin\theta_{Li} & \cos\theta_{Ti} & \sin\theta_{Li} \\ -\mu_1 k_{Ti}\sin 2\theta_{Ti} & (\lambda_1 + 2\mu_1\cos^2\theta_{Li})k_{Li} & -\mu_1 k_{Ti}\sin 2\theta_{Ti} & (\lambda_1 + 2\mu_1\cos^2\theta_{Li})k_{Li} \\ \mu_1 k_{Ti}\cos 2\theta_{Ti} & \mu_1 k_{Li}\sin 2\theta_{Li} & -\mu_1 k_{Ti}\cos 2\theta_{Ti} & -\mu_1 k_{Li}\sin 2\theta_{Li} \end{pmatrix}$$

$$\tag{5.137}$$

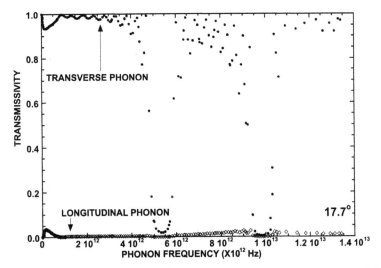

Figure 5.12 Transmissivity of a transverse acoustic wave polarized in the plane of incidence through a Si/Ge-like superlattice as a function of frequency with an incident angle of 17.7° (Chen, 1999).

In the above expressions, $k_i (= \omega/v_i)$ is the magnitude of the wavevector of the incident waves (SV or L, as distinguished by subscripts T and L). B_t is obtained by replacing subscript i with t, that is, from incident to transmitted waves. The interference matrix of the layer (with index 2) in eq. (5.136) is obtained from $M = B_2^{-1} N_2 B_2$, where B_2 is obtained by replacing i in eq. (5.137) by 2, and N_2 is given by

$$N_2 = \begin{pmatrix} e^{i\varphi_{T2}} & 0 & 0 & 0 \\ 0 & e^{i\varphi_{L2}} & 0 & 0 \\ 0 & 0 & e^{-i\varphi_{T2}} & 0 \\ 0 & 0 & 0 & e^{-i\varphi_{L2}} \end{pmatrix} \quad (5.138)$$

The transfer matrix is 4×4 because, as shown in eq. (5.136), the longitudinal and transverse waves are coupled and the conversion between these two waves is possible at the interface. With eq. (5.136), the reflectivity and transmissivity for an incident field (either v_{Ti} or v_{Li}) can be calculated.

Figure 5.12 shows an example of phonon transmissivity through a Si/Ge-like superlattice obtained by the transfer matrix method (Chen, 1999), for a transverse wave polarized in the plane of incidence at an angle of incidence of 17.7°. The stop bands in transmissivity (zero transmissivity) correspond to the minigaps obtained from lattice dynamics simulation (figure 3.30) (Yang and Chen, 2001). The figure also shows that some transverse incident waves are converted into longitudinal waves.

5.3.3 Electron Waves

The study of electron wave propagation in layered media started with the investigation on superlattices (Esaki and Tsu, 1970). The most popular approach has been based on solving the Schrödinger equation using the Kronig–Penney model. There are, however, also a few approaches based on the transfer matrix method for electron transport in superlattices (Tsu and Esaki, 1973; Huang and Wu, 1992). Because we have dealt extensively with the Kronig–Penney model in chapter 3 and described the transfer matrix method above, we will not present any details on the applications of the transfer matrix method to electron waves here.

5.4 Evanescent Waves and Tunneling

In section 5.2, we saw that when total internal reflection occurs for each of the three types of waves, an evanescent wave exists on the other side of the interface. The fields or wavefunctions of the evanescent wave decay exponentially from the interface. The time averaged net energy or particle flux carried by the evanescent wave is zero. However, if a third medium is brought close to the interface before the evanescent wave dies down completely, the evanescent wave can refract into this third medium. If this refracted wave is a propagating mode in the third medium, the evanescent wave becomes "revitalized" and a net energy or particle flux "tunnels" through the small region between the incident medium and the third medium. The reflection will no longer be total. In fact, one can even reach 100% transmission under appropriate conditions. The descriptions of these evanescent waves and the tunneling phenomena are based on the same mathematical expressions as we have obtained in the previous section.

5.4.1 Evanescent Waves

For electron reflection at a step potential, as shown, in figure 5.1, total reflection occurs when the electron energy E is smaller than the potential height U_0. In this case, eqs. (5.54)–(5.56) lead to

$$\Psi_t = \frac{2i|k_2|e^{-|k_2|z}}{k_1 + i|k_2|}\Psi_i \tag{5.139}$$

where

$$|k_2| = \sqrt{\frac{2m(U_0 - E)}{\hbar^2}} \tag{5.140}$$

Thus the penetration depth of the evanescent wave, which we define as the depth at which the wavefunction decays to e^{-1} of its boundary value, is

$$\delta = \frac{1}{|k_2|} \tag{5.141}$$

Taking $U_0 - E = 1$ eV, we get $\delta \approx 2$ Å, which is a very short distance.

For an electromagnetic wave incident above the critical angle, the Snell law gives

$$\sin\theta_t = \frac{n_1 \sin\theta_i}{n_2} > 1 \tag{5.142}$$

and thus

$$\cos\theta_t = \sqrt{1 - \sin^2\theta_t} = i\sqrt{\left(\frac{n_1 \sin\theta_i}{n_2}\right)^2 - 1} = i|\cos\theta_t| \quad (5.143)$$

Substituting the above expression into eq. (5.72) and then into eq. (5.64) gives the evanescent electric field for a TM wave as

$$\frac{2n_1 \cos\theta_i (\hat{\mathbf{x}}\cos\theta_t - \hat{\mathbf{z}}\sin\theta_t)}{n_2 \cos\theta_i + n_1|\cos\theta_t|i} |\mathbf{E}_{i/\!/}| \exp\left[-i\omega\left(t - \frac{n_2 x \sin\theta_t}{c_0}\right)\right] \exp\left(-z\frac{n_2\omega}{c_0}|\cos\theta_t|\right) \quad (5.144)$$

where $\hat{\mathbf{x}}$ and $\hat{\mathbf{z}}$ are is the unit vectors along the x and z coordinate directions. The above equation demonstrates that the evanescent field decays exponentially with a penetration depth

$$\delta = \frac{\lambda_0}{2\pi n_2 |\cos\theta_t|} \quad (5.145)$$

which is roughly of the same order as the wavelength inside the medium. Using eq. (5.144) and the corresponding expression for the magnetic field, it is also easy to show that the z-component time-averaged Poynting vector of the evanescent electromagnetic field is zero,

$$\langle S_z \rangle = \frac{1}{2}\text{Re}(\mathbf{E}\times\mathbf{H}^*)_z = \frac{1}{2}\text{Re}(E_x H_y)\hat{\mathbf{z}} = 0 \quad (5.146)$$

that is, no net energy flows across the interface. However, if the instantaneous Poynting vector is examined, it can be seen that there is instantaneous energy flow into and out of the second medium carried by the evanescent field. The net energy flow in and out averaged over time, however, is zero.

The above discussion shows the similarities between evanescent electron waves and electromagnetic waves. Evanescent acoustic waves can be similarly analyzed. We will not go into the details.

5.4.2 Tunneling

Tunneling of the evanescent waves may occur if a third medium is brought close to the first interface such that the exponentially decaying evanescent wave has finite magnitude at the interface between the second medium and the third medium. If the wave refracted into the third medium is propagating, a net flow of particles or energy from the first to the third medium occurs. In figure 5.13, we illustrate the tunneling of electromagnetic and electron waves. The analysis of the tunneling process can be based on the same methods that are used for treating interference phenomena. For electromagnetic waves passing through one layer of a thin film, for example, the transmissivity is given by eq. (5.122). Substituting relations, eqs. (5.104) and (5.143), into eq. (5.122), we get

$$t = \frac{t_{12}t_{23}\exp\left[-\frac{2\pi n_2 d|\cos\theta_2|}{\lambda_0}\right]}{1 + r_{12}r_{23}\exp\left[-\frac{4\pi n_2 d|\cos\theta_2|}{\lambda_0}\right]} \quad (5.147)$$

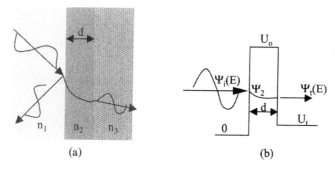

Figure 5.13 Tunneling of totally reflected waves, (a) for an electromagnetic wave and (b) for an electron wave.

where we have used θ_2 rather than θ_t to denote that the angle is for the wave inside the second medium, not medium 3. The transmissivity due to tunneling for a TM wave is

$$\tau = \frac{\text{Re}(n_3 \cos \theta_t)}{n_1 \cos \theta_1} |t|^2 \tag{5.148}$$

If the refractive index n_3 is larger than n_2, there are certain incident angles of θ_i that allow a real solution for θ_t, while θ_2 is an imaginary angle. From the Snell law, this occurs when the incident angle falls in the range

$$\sin^{-1}\left(\frac{n_2}{n_1}\right) \leq \theta_i \leq \sin^{-1}\left(\frac{n_3}{n_1}\right) \tag{5.149}$$

When tunneling happens, it goes without saying that the reflectivity is no longer 100% as for the case of total reflection. If the medium through which the wave tunnels is nonabsorbing, energy or particle conservation gives $R = 1 - \tau$.

For electrons, we can solve the Schrödinger equation for a barrier structure as shown in figure 5.13. The solution follows closely the method we used in section 3.2.1, which also resembles the derivation of the transfer matrix method for electromagnetic waves. The tunneling transmissivity through a potential barrier of height U_0 and width d is (Cohen-Tannoudji et al., 1997)

$$\tau = \frac{4E(U_0 - E)}{4E(U_0 - E) + U_0^2 \sinh^2[\sqrt{2m(U_0 - E)}d/\hbar]} \tag{5.150}$$

When the argument of the hyperbolic sine function is large, the above expression can be approximated as

$$\tau \approx \frac{16E(U_0 - E)}{U_0^2} \exp[-2\sqrt{2m(U_0 - E)}d/\hbar] = \frac{16E(U_0 - E)}{U_0^2} e^{-2|k_2|d} \tag{5.151}$$

The same tunneling phenomenon can also occur for phonons. Figure 5.14 shows the transmissivity of an acoustic wave through a superlattice, calculated on the basis of the transfer matrix method as a function of frequency and angle of incidence (Chen, 1999). At a low angle of incidence, the transmission behaves as normal and has several stop

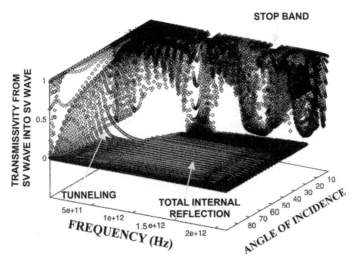

Figure 5.14 Phonon transmissivity through a Si/Ge-like superlattice, each layer 5 Å thick, showing the stop bands, the total reflection region, and the tunneling region (Chen, 1999).

bands. At a large angle of incidence, for which total reflection occurs, the transmissivity across the superlattice is not zero but decreases exponentially as the frequency increases, due to tunneling of acoustic waves.

Tunneling phenomena are the basis of several inventions that led to Nobel prizes, including the tunneling diode by Esaki (1958) and the scanning tunneling electron microscope (STM) (Binnig and Rhorer, 1982). The principle of an STM is shown in figure 5.15(a). A sharp tip is brought into close proximity with a conducting surface but without contacting the surface. Under an applied voltage, electrons tunnel through the vacuum gap and create a current in the loop. The current is extremely sensitive to the separation (sub-angstrom) between the tip and the surface, as one can easily

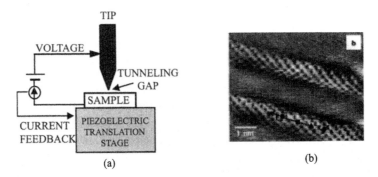

Figure 5.15 (a) The scanning tunneling microscope is based on the sub-angstrom level sensitivity of the tunneling current between a conducting tip to a conducting sample as a feedback to control the piezoelectric translation stage, which is also capable of sub-angstrom motion precision, to fly the tip over the sample and to obtain information on the topographical and electron structure of the sample surface. (b) STM image of two single-walled carbon nanotubes (Odom et al., 1998; courtesy of Nature Publishing Group).

see from eq. (5.151) since k_2 is of the order of ~ 1 Å^{-1}. As the tip is scanned over the sample, different regions have different potential barriers of different heights. By using the current as a feedback signal to control the tip–sample separation, one can map the electronic wavefunction surrounding individual atoms or the surface roughness. Figure 5.15(b) shows the STM images of two single-walled carbon nanotubes (Odom et al., 1998).

Since the invention of the STM, a host of other types of microscope have been invented, including the atomic force microscope (Binnig et al., 1986), photon scanning tunneling microscope (Reddick et al., 1989), scanning thermal microscope (Majumdar et al., 1995), and others. The photon scanning tunneling microscope is also based on the evanescent electromagnetic wave hovering above a surface. The atomic force microscope, however, is based on an even simpler principle: the effective spring constant between atoms can be quite large—much larger, for example, than that of a Si cantilever 3 μm (thickness) × 100 μm (length) × 10 μm (width). When such a cantilever is in contact with a sample via a sharp tip, the atoms of the sample will not be scratched off. Rather, the cantilever will be displaced. Angstrom-level displacement can be easily measured with either the STM (Binnig et al., 1986) or through laser deflection, making it possible to use such a device to measure the angstrom-level topography of surfaces, particularly for dielectric surfaces that cannot be characterized with an STM because the sample is nonconducting.

5.5 Energy Transfer in Nanostructures: Landauer Formalism

Knowing the transmissivity from one point to another in a system, one can easily estimate various fluxes associated with the carriers (charge, momentum, energy, etc.). Consider for example, the heat transfer between two reservoirs at temperatures T_1 and T_2, as shown in figure 5.16. The heat flux from reservoir 1 to 2 is

$$q_{1 \to 2} = \sum_p \left[\frac{1}{V_1} \sum_{k_{x1}=-k_{\max}}^{k_{\max}} \sum_{k_{y1}=-k_{\max}}^{k_{\max}} \sum_{k_{z1}=0}^{k_{\max}} v_{z1} E \tau_{12} f(E, T_1) \right] \quad (5.152)$$

where E is the energy of one carrier and τ_{12} is the transmissivity from point 1 to point 2 for the carrier with energy E, v_{z1} is the velocity of the carrier, the index p represents summation over all the polarizations of the carriers, and the wavevector summation indices are over all values of k_x and k_y and positive values of k_z. Equation (5.152) is a recasting of eq. (5.86) and is valid for electrons and photons as well as phonons. We can similarly write the reverse heat flux from point 2 to point 1. The difference between these heat fluxes gives the net heat flux between point 1 and point 2, similar to eq. (5.88). The

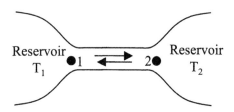

Figure 5.16 The Landauer formulation of the net (energy, charge, particle) flux between two points is based on the carrier transmissivity between the two points.

principle of detailed balance can once again be applied to obtain a relationship between the transmissivity from point 1 to point 2 and the reverse direction, as expressed in eq. (5.89). With such an approach, and when the difference between T_1 and T_2 is not large, the heat flux between point 1 and point 2 can be expressed as

$$q = (T_1 - T_2) \sum_p \int_{\Omega \geq 2\pi} \left[\int v_1 \cos\theta E \frac{\partial f(E,T)}{\partial T} \tau_{12}(E,\phi,\theta) D_1(E) dE/4\pi \right] d\Omega$$
$$= K \Delta T \qquad (5.153)$$

where K is the thermal conductance with units of $W\,m^{-1}\,K^{-1}$, Ω is the solid angle defined in figure 3.27, and T is the average temperature. One can write down similar expressions for the current density and particle flux. These types of expressions for flux are called the *Landauer formalism*, which views the transport as a transmission process (Imry and Landauer, 1999).

The key for applying the Landauer formalism is the calculation of the transmissivity. When scattering exists, the calculation of the transmissivity is more difficult and the Landauer formalism is less useful. When no internal scattering exists, which is also called ballistic transport, the transmissivity can be calculated relatively easily, as we have done for a single interface and multilayers, and the Landauer formalism is very convenient to use.

The effects of interference and tunneling in thin films, and more generally in nanostructures, on the transport processes can be studied from the Landauer formalism, using appropriately calculated transmissivity between two points. As an example, we consider radiative heat transfer between two parallel plates, paying special attention to the case when the spacing between the plates is small. Quite a few studies have been devoted to radiative heat transfer across small gaps. In a series of studies, Tien and co-workers (Cravolho et al., 1967; Domoto et al., 1970) investigated the effects of tunneling and interference on radiative heat transfer between small vacuum gaps, which are used in low-temperature thermal insulation materials (Tien and Cunnington, 1973). More extensive experiments were performed by Hargreaves (1969). The approach championed by Tien and co-workers was equivalent to the Landauer formalism. Polder and van Hove (1971) established a direct approach that considered the emission processes based on Rytov's electromagnetic field fluctuation theory (Rytov et al., 1987; Narayanaswamy and Chen, 2004). Pendry (1999) provided a slightly different point of view on radiative heat transfer in small gaps, based on the Landauer formalism. Figures 5.17(a) and (b) show the modeling and experimental data for radiative heat transfer between small gaps. Due to tunneling, the radiation flux increases as the vacuum gap decreases and a radiative heat flux much higher than that between two blackbodies can be realized, as shown in figure 5.17(c). Some recent applications of the tunneling phenomena include the scanning tunneling microscope (Xu et al., 1994) and thermophotovoltaics (DiMatteo et al., 2001; Whale and Cravalho, 2002). When the two objects are identical, the maximum radiation heat transfer can be increased by n^2 times the blackbody radiation heat transfer between two surfaces through tunneling of the internally reflected wave. The n^2 limit is the blackbody emissive power inside an object, which can be derived following similar steps as we arrived at eq. (4.83). In addition to the tunneling of evanescent waves, the tunneling of surface waves that decay exponentially on both sides of the interface

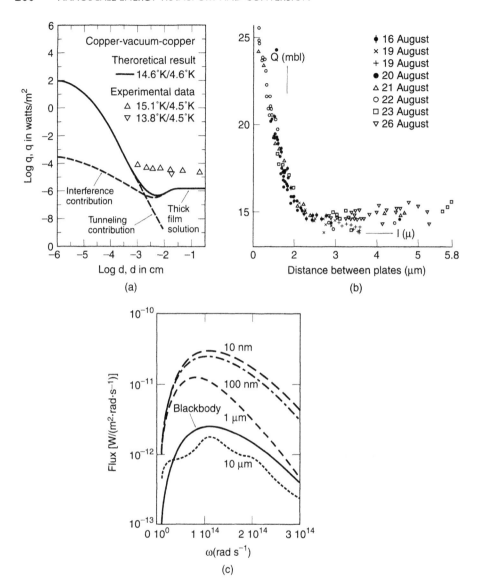

Figure 5.17 Size effects on radiation heat transfer between two parallel plates (a) at low temperature (Domoto et al., 1970; courtesy of ASME), (b) at room temperature (Hargreaves, 1969; courtesy of Elsevier). (c) Radiative heat flux as a function of frequency for radiation heat transfer between a plate at 300 K and another at 0 K, demonstrating that the heat flux can exceed the exchange between two blackbodies.

can also occur (Mulet et al., 2002; Narayanaswamy and Chen, 2003). The surface waves exist when the dielectric constants of the two sides of the interface are equal in magnitude but are of opposite signs (Raether, 1987). If one side of the interface is vacuum, the other side should have a negative dielectric constant close to unit, which can occur when the electrons or phonons are in resonance with electromagnetic waves

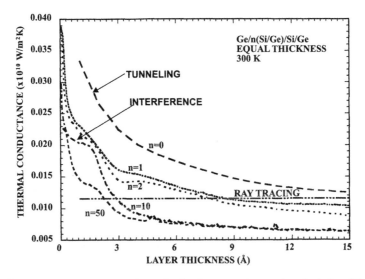

Figure 5.18 Thermal conductance of superlattice from transfer matrix calculation (Chen, 1999).

in the medium. The resonant modes of electromagnetic waves with optical phonons are called *phonon-polaritons* and those modes with electrons are called *plasmons*. Surface phonon-polaritons and surface plasmons have high energy density near the surface but decay rapidly away from the surface. Radiative heat flux through the tunneling of surface waves can significantly exceed the n^2 limit of evanescent waves.

The same interference and tunneling phenomena also affect the radiative properties of thin films grown on substrates. The emissivity of the surface will change with film thickness (Wong et al., 1995; Chen, 1996; Zhang et al., 2003). This affects, for example, the temperature of semiconductor wafers during thin film growth. The uncertainty in temperature measurement caused by the emissivity change is a significant factor in the design of semiconductor equipment used for rapid thermal processing (Nulman, 1989).

For phonons in thin films, interference and tunneling phenomena may also affect heat conduction in extremely thin films such as superlattices with very short spatial periods. Chen (1999) evaluated thermal conductance in the limit of no scattering of thin films and superlattices, as shown in figure 5.18. Generally, when the film thickness is less than a few monolayers, tunneling can increase the conductance. Lattice dynamics simulations lead to similar conclusions (Tamura et al., 1999; Simkin and Mahan, 2000; Yang and Chen, 2001, 2003). So far, there have been some experimental data that suggest this phenomenon, but they are not very conclusive (Capinski et al., 1999; Venkatasubramanian, 2000).

Example 5.2 *Universal quantum thermal conductance*

Develop a model for the thermal conductance of a square nanowire between two thermal reservoirs, neglecting the internal scattering and assuming the phonon transmissivity for each allowable mode is one.

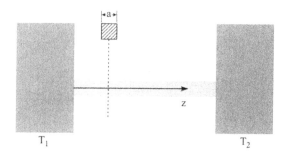

Figure E.5.2 Nanowire connecting two thermal reservoirs.

Solution: Consider a square nanowire as shown in figure E.5.2. In the cross-sectional direction, standing waves must be formed so that the allowable wavevectors in the x and y directions are

$$k_x = 2\pi \frac{m}{2a} = \frac{\pi m}{a}, \quad k_y = \frac{n\pi}{a} \quad (m, n = \pm 1, \pm 2, \ldots)$$

We assume that the linear dispersion relation as used in the Debye model is still valid. The allowable modes inside the nanowire are

$$\omega = ck = c\sqrt{\left(\frac{m\pi}{a}\right)^2 + \left(\frac{n\pi}{a}\right)^2 + k_z^2} \qquad (E5.2.1)$$

for each polarization. With a phonon transmissivity of one (which requires that the materials of the reservoirs and the nanowire are the same, and also a tapered joint between the wire and the reservoir, similar to that drawn in figure 5.16), we can use the Landauer formalism to express the heat transfer through the nanowire as

$$q_{12} = q_{1\to 2} - q_{2\to 1} = \frac{1}{2\pi} \sum_{m,n} \int v_z \hbar\omega \left[f(\omega, T_1) - f(\omega, T_2)\right] dk_z \qquad (E5.2.2)$$

where the factor $(1/2\pi)$ arises from the conversion of the summation over the quantum state determined by k_z in eq. (5.152) into an integration over k_z, because the separation between two consecutive k_z is $2\pi/L$, where L is the length of the wire. We also used direct summation for the modes in the x and y directions because the separations between two consecutive wavevectors in these directions are large when a is small. We will see in the next section that the phonon velocity v_z used for energy transport calculations should be the group velocity

$$v_z = d\omega/dk_z \qquad (E5.2.3)$$

Thus, eq. (E5.2.2) can be converted into integration over frequency,

$$q_{12} = \frac{1}{2\pi} \sum_{m,n} \int \hbar\omega \left[f(\omega, T_1) - f(\omega, T_2)\right] d\omega \qquad (E5.2.4)$$

If we further assume that $T_1 - T_2$ is small, the thermal conductance of the nanowire, eq. (E5.2.2), can be expressed as

$$K = \frac{q_{12}}{T_1 - T_2} = \frac{3}{2\pi} \sum_{m,n} \int_{\omega_{mn}}^{\omega_{max}} \hbar\omega \frac{df}{dT} d\omega \quad \text{(E5.2.5)}$$

where the factor of 3 represents the three phonon polarizations, ω_{max} is the maximum phonon frequency, and

$$\omega_{mn} = c\sqrt{\left(\frac{m\pi}{a}\right)^2 + \left(\frac{n\pi}{a}\right)^2} \quad \text{(E5.2.6)}$$

For the first few quantized modes (m, n are small, so that ω_{mn} is small) and at low temperature, eq. (E5.2.5) can be simplified to

$$K = \frac{3\kappa_B^2 T}{h} \sum_{m,n} \int_{x_{mn}}^{\infty} \frac{x^2 e^x}{(e^x - 1)^2} dx \quad \text{(E5.2.7)}$$

where $x_{mn} = \hbar\omega_{mn}/\kappa_B T$. When m and n are small, such that the lower limit can be extended to zero, the integral value is $\pi^2/3$. In this limit, the thermal conductance of each mode is

$$K_1 = \frac{\pi^2 \kappa_B^2 T}{3h} \quad \text{(E5.2.8)}$$

Comment. This thermal conductance expression does not depend on the material properties and thus is the same for all materials. Such universal conductance behavior also happens for electrons. Quantum size effects on thermal conductance have been observed experimentally (Angelescu et al., 1998; Schwab et al., 2000).

In closing this section, we would like to point out that the Landauer formalism, as expressed in eqs. (5.152) and (5.153), is based on the assumption that the carrier distributions going from point 1 to point 2 are determined by the equilibrium reservoir temperatures (and chemical potentials). In reality, the local carriers at point 1 are not at equilibrium because there are also carriers coming from point 2 with a characteristic temperature of T_2. So T_1 and T_2 actually do not represent the local temperatures at points 1 and 2. We have discussed this point carefully in connection with the treatment of the thermal boundary resistance as represented by figures 5.7(a)–(c). Whether one should use eq. (5.153) or the local equivalent equilibrium temperature (or chemical potential) depends on how experiments are conducted or how models are laid out. So far, most experiments are done in electron systems with the electrochemical potentials of the reservoirs measured, and thus the Landauer formulation is directly applicable. If one treats the transport inside the reservoirs concurrently with the ballistic transport between point 1 and point 2, the consistency of the definitions in each region must be considered carefully (Chen, 2003).

5.6 Transition to Particle Description

Our discussion in this chapter so far has centered on the wave nature of the energy carriers by considering their phase. From quantum mechanics, we know that energy carriers have both wave and particle characteristics. When the energy carriers are treated as waves, their particle characteristics are included, for example, by considering the energy of one phonon as $h\nu$. In macroscale, we often ignore the phase and treat the energy carriers as particles, either quantum particles such as photons or phonons, or simply as classical particles with no energy quantization. The question is then: when must we consider the energy carriers as waves, and when can we ignore the phase information and treat them simply as pure particles, either quantized particles or classical particles? We will attempt to answer these questions in this section.

5.6.1 Wave Packets and Group Velocity

In our previous discussion on energy propagation, such as eq. (5.152), we did not give much consideration to the meaning of the velocity. This velocity should represent the speed and direction of energy propagation and is usually the group velocity. To understand the *group velocity*, we first consider a plane wave traveling along the x-direction

$$\mathbf{A} e^{-i(\omega t - kx)} \qquad (5.154)$$

Its phase velocity is [eq. (2.4)]

$$v_{p,x} = \frac{dx}{dt} = \frac{\omega}{k} \qquad (5.155)$$

Is this velocity the speed of signal or energy propagation? Generally, the answer is no. We see that the plane wave represented by eq. (5.154) extends from minus infinity to plus infinity in both time and space. It has no start or finish, and does not represent any meaningful signal. In practice, a signal has a starting point and an ending point in time. Let's suppose that a harmonic signal at frequency ω_0 is generated during a time period $[0, t_0]$, as shown in figure 5.19(a). Such a finite-time harmonic signal can be decomposed through a Fourier series into the summation of true plane waves with time extending from minus infinity to plus infinity, as shown in figure 5.19(b). The frequencies of these plane waves are centered around ω_0 and their amplitudes decay as the frequency moves away from ω_0, as illustrated in figure 5.19(c). One can better understand these pictures by actually carrying out the Fourier expansion. Because each of the plane waves in such a series expansion is at a frequency slightly different from the central frequency ω_0, it also has a corresponding wavevector that is different from k_0, as determined by the dispersion relation between ω and k. The subsequent propagation of the signal can be obtained from tracing the spatial evolution of all these Fourier components as a function of time.

For simplicity, let's consider that the signal is an electromagnetic wave with the electric field points to y-direction. We pick only two Fourier components of equal amplitude and consider their superposition, one at frequency $\omega_0 - \Delta\omega/2$ and another at

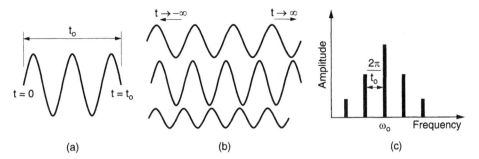

Figure 5.19 (a) A finite-time period signal generated between time period $(0, t_o)$ contains a spectrum of plane waves extending infinitely in time (b), with a power spectrum shown in (c). The propagation of these plane waves evolves into a wave packet.

frequency $\omega_0 + \Delta\omega/2$, propagating along the positive x-direction [Figure 5.20(a)]. The superposition of these two waves gives the electric field as

$$E_y(x, t) = a \cos\left[\left(\omega_0 - \frac{\Delta\omega}{2}\right)t - \left(k_0 - \frac{\Delta k}{2}\right)x\right]$$
$$+ a \cos\left[\left(\omega_0 + \frac{\Delta\omega}{2}\right)t - \left(k_0 - \frac{\Delta k}{2}\right)x\right]$$
$$= 2a \cos(\Delta\omega t - \Delta k x) \cos(\omega_0 t - k_0 x) \qquad (5.156)$$

The above electric field is shown schematically in figure 5.20(b). There appear to be two waves: one is the carrier wave at central frequency ω_0, another is the modulation of the carrier wave by a wave at frequency $\Delta\omega$. If the frequency $\Delta\omega$ is much smaller than ω_0, we can calculate the Poynting vector time-averaged over a period much shorter than

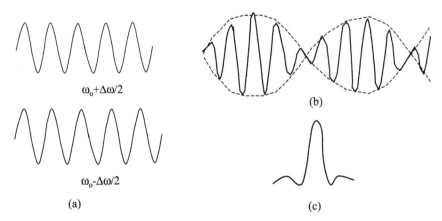

Figure 5.20 Example of the superposition of two plane waves (a) into wave packets (b). If there are many different frequency components, the superposition leads to a narrow wave packet (c).

$1/\Delta\omega$ but much longer than $1/\omega_0$ according to eq. (5.37), to obtain the average energy flux as

$$S(x,t) = \frac{4a^2 n}{c_0 \mu} \cos^2(t\,\Delta\omega - x\Delta k) \qquad (5.157)$$

which is another wave propagating at the speed

$$v_{g,x} = \frac{\Delta\omega}{\Delta k} \qquad (5.158)$$

This means that the energy is propagating at the speed of $v_{g,x}$ rather than the phase velocity. This $\mathbf{v}_g$ is called the group velocity. In the more general case of the existence of a spectrum of frequencies, the superposition of waves leads to a narrow *wave packet* as sketched in figure 5.20(c). The group velocity can be calculated from

$$\mathbf{v}_g = \nabla_{\mathbf{k}}\omega = \frac{\partial\omega}{\partial k_x}\hat{\mathbf{k}}_x + \frac{\partial\omega}{\partial k_y}\hat{\mathbf{k}}_x + \frac{\partial\omega}{\partial k_z}\hat{\mathbf{k}}_z \qquad (5.159)$$

The above derivation is by no means rigorous, but the concept of wave packets and group velocity is generally applicable to all waves. In the following, we will discuss two points related to the group velocity. One is whether the group velocity is always the velocity of energy flow. The other is the difference between the momentum of a wave packet and the crystal momentum.

The group velocity is usually considered to be the velocity of energy propagation of all carriers. We have shown this point in the above example for electromagnetic waves. The recognition of the significance of the group velocity actually started with the study of sound by Rayleigh (1945). It should also be pointed out, however, that the group velocity does not always represent the energy velocity. We can appreciate this from the step between eqs. (5.156) and (5.157), where we assumed that $\Delta\omega$ is much smaller than ω_0. In the case of a very large variation in the dispersion relation, the group velocity no longer represents the energy velocity. In figure 5.21, we show the real and the imaginary parts of the refractive index of silver. There is a region, called anomalous dispersion, in which the refractive index changes rapidly with wavelength. In this region, the real part of the refractive index is less than 1 and the phase velocity is larger than the speed of light. The group velocity is also larger than the speed of light. This does not mean a violation of the principle of relativity, however, which says that the maximum speed cannot be larger than the speed of light in free space. In this region, if we have a signal starting at $t = 0$ as shown in figure 5.19(a), one cannot superimpose all spectra of the light and still obtain a nice wave packet as sketched in figure 5.20(c). For such situations, the superimposed wave is more extended in space and the velocity of the majority of the energy propagation, called the signal velocity (Brillouin, 1960), is still less than the speed of light. A detailed discussion of this anomalous region is given by Bohren and Huffman (1983).

The group velocity of electrons is the velocity at which the electron wave packet moves in free space and inside a crystal. The energy dispersion of a free electron is

$$E = \frac{(\hbar\mathbf{k})^2}{2m} \qquad (5.160)$$

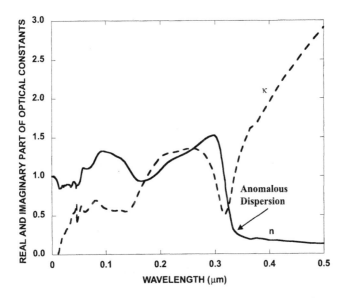

Figure 5.21 Refractive index of silver as a function of wavelength, showing the anomalous dispersion region in which both the phase velocity and the group velocity are larger than the speed of light. The signal velocity, however, is still smaller than the speed of light.

and thus the phase and group velocity are, respectively,

$$\mathbf{v} = \frac{E/\hbar}{k} = \frac{\hbar \mathbf{k}}{2m} \text{ and } \mathbf{v}_g = \frac{\partial (E/\hbar)}{\partial \mathbf{k}} = \frac{\hbar \mathbf{k}}{m} \qquad (5.161)$$

Clearly, the group velocity is consistent with the de Broglie relation $\mathbf{p} = \hbar \mathbf{k}$ and our classical relation $\mathbf{p} = m\mathbf{v}_g$, but not the phase velocity. When we deal with electron motion in crystals, however, $m\mathbf{v}_g$ does not normally equal $\hbar \mathbf{k}$, where $\mathbf{k}$ are the electron wavevectors determined from the von Karman boundary condition. We call $\hbar \mathbf{k}$ the crystal momentum and use it to satisfy the momentum conservation rules and to calculate the external force field, $\mathbf{F}_{ext}$

$$\mathbf{F}_{ext} = \frac{d(\hbar \mathbf{k})}{dt} \qquad (5.162)$$

The reason for doing so is that the periodic potential also exerts another force on the electrons. When the crystal momentum is used, one can carry out the calculations as if electrons are not subject to the internal field of the crystal (Aschroft and Mermin, 1976; Slater, 1967). For such calculations, however, one still should use the group velocity as defined by eq. (5.159) as the actual speed of motion of the electrons, while using the crystal momentum for the external force and the momentum conservation rules. Similar arguments hold for phonons inside crystals.

5.6.2 Coherence and Transition to Particle Description

When should we consider the phase of the waves and when not? The answer to this question is fundamental for the transport of all these carriers and has been studied in different disciplines. If the phase of the carriers must be considered, transport is coherent and the wave approaches illustrated in this chapter should be followed. In the other limit, the transport is incoherent. Transport in the incoherent regime will be treated in the following two chapters. In between the two limits is the partially coherent regime. Most engineering approaches for transport, built on diffusion equations, ignore the phase and treat carriers as incoherent particles. What are the conditions for these approaches to be valid?

Answers to these questions are by no means straightforward and vary with the types of carrier. For photons, the scattering is less frequent and mostly elastic; consequently, the discussion of coherence has been based more or less on the spectral purity (Born and Wolf, 1980). For electrons, inelastic scattering is strong and thus the discussion of coherence is closely related to scattering. There is less research on phonon coherence. Consequently, we will first discuss photon coherence and then electron coherence, followed by some discussion on phonons.

5.6.2.1 Coherence of Electromagnetic Waves

From eq. (5.156), we infer that the spatial spread of the wave packets in figure 5.20 is $\Delta x \, \Delta k \sim 2\pi$, or, denoting Δx as ℓ_c,

$$l_c \approx \frac{c}{\Delta \nu} \tag{5.163}$$

For electromagnetic wave propagation, this length is called the *coherence length* (Born and Wolf, 1980), which is inversely proportional to the effective bandwidth of the waves in the system. A ray of electromagnetic waves contains a series of wave packets fired by individual emitters in the source, as shown in figure 5.22(a). Each wave packet has a coherence length given by eq. (5.163). However, no phase relations exist between the wave packets.

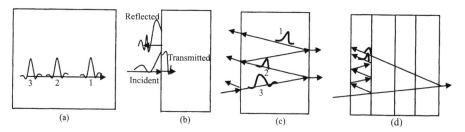

Figure 5.22 Traveling and interference of wave packets. (a) In a big domain, individual wave packets are uncorrelated and can be thought of as point particles. (b) At an interface, the tail of the wave packet and the reflected wave packet have a fixed relationship and thus can interfere with each other. (c) Inside a thick film, two wave packets can have transient interference but, since they do not have a fixed phase relationship, such transient interference can be anywhere inside the film. The end results are that no interference beats can be observed and thus geometrical optics should be used rather than wave optics. (d) In a periodic structure, however, the same wave packets are split many times at each interface and it is possible that the wave packets in a layer that are returned from other interfaces can overlap with the other wave packets inside the layer.

For blackbody thermal radiation, the energy uncertainty of the individual radiation emitters (atoms, electrons, or molecules) is of the order of $\kappa_B T$, due to the collision of the emitters with the reservoir, which also means that the effective bandwidth for thermal emission is $\kappa_B T/h$. Using this effective bandwidth and eq. (5.163), one can estimate that the coherence length is of the order of $hc/(\kappa_B T)$. A more detailed calculation of the coherence length leads to (Mehta, 1963)

$$\ell_c = 0.15 \frac{hc}{\kappa_B T} \tag{5.164}$$

This equation can be rewritten as

$$\ell_c T = 2167.8 \, \mu\text{m K} \tag{5.165}$$

For reasons to be explained later, eq. (5.164) will also be called *thermal length,* reflecting the origin of this coherence length. Compared with Wien's displacement law, the coherence length of a blackbody radiation field is very close to the wavelength corresponding to the peak radiation intensity.

The coherence length, taken as a measure of the wave packet size, gives an indication of whether the phase information needs to be considered for transport processes or not. If the size of the transport domain is much larger than the wave packets or the coherence length, then the wave packets can be treated as point-wise particles traveling through the domain, as shown in figure 5.22(a). When a wave packet meets a perfect interface, however, it will be reflected and refracted. The reflected wave packet has a fixed phase relationship with the incoming one and can thus interfere with the incoming wave packet [figure 5.22(b)]. This is why we always use the Fresnel formula—the wave solution of the Maxwell equations—to calculate the reflectivity and transmissivity of a perfect interface. If multiple interfaces exist, as in a film, the reflected wave packet in the domain can encounter another incoming wave packet [figure 5.22(c)]. Although these two wave packets can create a transient interference when they overlap, their overlapping locations are not fixed because no fixed phase relationship exists between the two wave packets coming from a random thermal source. Because the number of random wave packets inside a ray is large, on average, we can ignore the phase relationship inside the domain and treat the wave packets as particles. Transport in this regime is called incoherent.

Consider now a thin film with two interfaces. If the size of a wave packet is small compared to the film thickness, transport is incoherent. In this regime, we can neglect the interference phenomenon discussed in section 5.3 and use energy superposition to calculate the reflectivity and transmissivity of the film. The energy superposition, also called ray tracing, is based on tracing the trajectory of photons and their intensity, rather than the electromagnetic fields, thus completely neglecting the phase information of the electromagnetic fields. The next two chapters will discuss such particle transport in more detail. The horizontal line in figure 5.23 gives the transmissivity of a non-absorbing thin film calculated on the basis of the energy superposition method, which is independent of the film thickness. In the other limit, when the wave packet is large compared to the film thickness, the same wave packet can overlap after experiencing multiple reflections; that is, the tail of one wave packet is still entering the film while the head has gone through multiple reflections. In such a case, consideration of the phase of the waves becomes

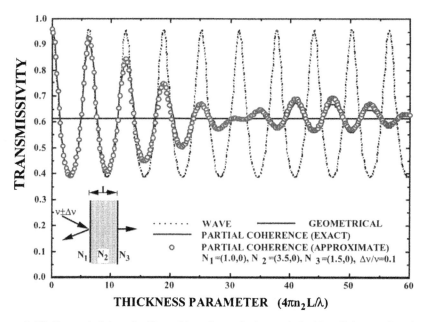

Figure 5.23 Transmissivity of a film subjected to polychromatic incident light as a function of the film thickness parameter calculated from different methods: wave approach, particle approach, and partial coherence theory. In the thin film limit, results from the partial coherence theory agree with wave approach. In the thick film limit, the theory agrees with the particle picture (Chen and Tien, 1992).

necessary and one can observe the interference phenomenon discussed in section 5.3 and shown in figure 5.23. If the number of overlapping reflections inside the film for the same wave packet is large enough, the solution obtained in section 5.3, which includes an infinite number of reflections for a single frequency (also called monochromatic), can be a good approximation, as shown in figure 5.23. In the intermediate case, when the wave packet is comparable to the film thickness, the same wave packet may overlap only partially inside the film or only within a few reflection cycles, and the transport falls into the partially coherent regime. In this regime, one can use the partial coherence theory for electromagnetic waves (Born and Wolf, 1980; Chen and Tien, 1992) to calculate the reflectivity and transmissivity. On the other hand, since a wave packet can be decomposed into the superposition of monochromatic waves, it is also permissible to calculate the reflectivity and transmissivity first for each frequency, using the wave formulation, and then to superimpose the results for all the frequencies to obtain the final results,

$$\bar{\tau} = \frac{J_t}{J_i} = \frac{\int_0^{\Delta\omega} \tau(\omega) J_i(\omega) d\omega}{\int_0^{\Delta\omega} J_i(\omega) d\omega} \text{ and } \bar{R} = \frac{J_t}{J_i} = \frac{\int_0^{\Delta\omega} R(\omega) J_i(\omega) d\omega}{\int_0^{\Delta\omega} J_i(\omega) d\omega} \qquad (5.166)$$

where J_i is the incident photon Poynting vector and $\Delta\omega$ is the spectral width of the incident photon. Figure 5.23 shows results obtained from the above spectral averaging (marked as exact) and from a partial coherence formulation (marked as approximate) (Chen and Tien, 1992). The two approaches lead to the same results, which shows that when the film is thin, the single frequency formulation is approximately correct, and

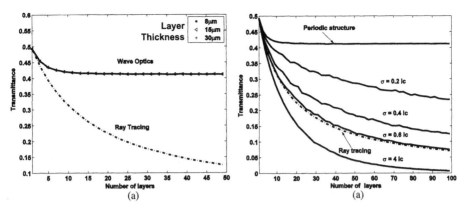

Figure 5.24 Directional and spectral averaged transmissivity of a Bragg reflector for a thermal radiation source at 1000 K. (a) Comparison between wave optics results and ray tracing results for perfect Bragg refectors at different period thickness, showing that for a thick period (thicker than the thermal length) the transmissivity does not depend on the period thickness, but the results from the two methods never agree with each other. (b) If the period varies randomly, the transmissivity decreases similarly to ray tracing but numerical values still do not always agree with those of the ray tracing method. Thus, random thickness variations do not justify the use of the particle picture. σ is standard deviation in thickness, and ℓ_c is the thermal wavelength given by eq. (5.165) (Hu et al., 2005).

when the film is thick, the particle picture should be used. For film thickness in between, in the partially coherent regime, either the spectral averaging method or a treatment based on partial coherence theory should be adopted.

The situation is more complicated if there are more interfaces. Consider a periodic structure such as a Bragg reflector, as shown in figure 5.22(d). A wave packet can experience multiple reflections in multiple layers and the coherence property of the wave packet can be altered. In a single-layer-thick film, for example, the multiple reflection of a narrow wave packet does not overlap with itself. In a multilayer structure, however, wave packets split at different interfaces have a chance of overlapping each other, as sketched in figure 5.22(d). Thus, in the case of Bragg reflectors, the spectral averaging of reflectivity and transmissivity based on wave optics and based on ray tracing lead to different results for a blackbody radiation source even with a period thickness much larger than the coherence length given by eq. (5.164). Figure 5.24(a) shows an example of transmissivity of a Bragg reflector for incident blackbody radiation at 1000 K, averaged over all frequencies and incident angles. The transmissivity, calculated from the wave method using the transfer matrix method, approaches a constant as the number of layers increases but is independent of the period thickness (for films with periods much larger than the blackbody coherence length). However, the results based on ray tracing continuously decrease with increasing number of periods. One can understand this phenomenon from the Kronig–Penney model for electron waves inside a periodic potential. In a periodic structure, the electron waves can extend over the whole structure, except in the bandgap region where no electrons exist at all. For these extended waves, the transmissivity is one even if the number of layers approaches infinity. On the other hand, if the phase of the waves is ignored, as in ray tracing, photons experience sequential reflection at every interface. As a result, the number of transmitted photons decrease with increasing number of periods, as shown in figure 5.24(a). In another demonstration, it has been shown that the blackbody radiation passing through two pinholes is partially coherent even for a pinhole separation as large as a few centimeters, much longer than the coherence length of the blackbody radiation source (Mandel and Wolf, 1995; James and Wolf, 1991; Santarsiero and Gori, 1992).

The above discussion throws us into trouble. The coherence length is not a proper measure for neglecting the phase information of waves, as shown in the case of Bragg reflectors. In practice, however, the modeling of radiation transport through thick multilayers, such as windows, is often done with the ray tracing method. How we can justify the use of the particle picture? There are three possible justifications: (1) surface roughness; (2) nonparallel surfaces; and (3) thickness variations. All these factors create a certain randomness in the phase of the reflected and refracted waves. However, randomness in a structure does not necessarily lead to the particle picture.

As an example, we consider that the thickness of a Bragg reflector has a certain randomness. For this case, the transfer matrix method is still applicable. Lu et al. (2005) computed the transmissivity of Bragg reflectors with different level of randomness in the film thickness, as shown in figure 5.24(b). With randomness in the period thickness, the transmissivity does decrease with increasing number of periods as with ray tracing. However, the wave approach still does not agree with the ray tracing method. Depending on the degree of randomness, the transmissivity from wave optics can be either larger or smaller than the ray tracing results.

It turns out that the decreasing transmissivity in the case of random thickness variations is due to another phenomenon—the *wave localization*. When the phases have enough randomness, the superposition of waves can create complete cancellation of waves at certain frequencies for the one-dimensional superlattice considered here (Sheng, 1990). Figure 5.25 shows the transmissivity as a function of wavelength. It can be seen that the transmissivity of high-frequency waves is nearly zero. These waves do not propagate through the structure because of destructive interference among the waves. This phenomenon is called localization. In a three-dimensional structure, localization implies that waves are localized in some region and do not propagate. If a wave is localized, the transmissivity decreases exponentially with thickness of the structure. Thermal radiation, however, contains waves of many frequencies, not all of which can be localized. Thus the transmissivity does not exactly follow an exponential behavior. The phenomenon of localization was first studied for electron waves by Anderson (1958), and similar phenomena have been found for all kinds of waves, including photon and phonon waves (Sheng, 1990). The investigation of localization phenomena is still a very active area of research. It has been found that localization can easily occur in one-dimensional and two-dimensional structures, and when the number of modes is small. However, it is much more difficult to create localization in three-dimensional structures, particularly for electromagnetic waves (Garcia-Martin et al., 2000). The example of Bragg reflectors with random thickness variations is a simple one-dimensional case and thus it is relatively easy to observe the localization phenomenon, as the computed transmissivity in figures 5.24(b) and 5.25 shows. Surface roughness, however, makes the wave three-dimensional. In these situations, it is likely that the phase of the waves can be ignored and the particle picture can be used, provided the coherence length is smaller than the characteristic length of the surface randomness.

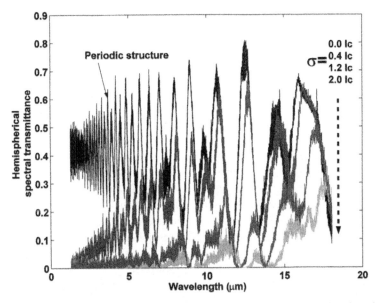

Figure 5.25 Photon transmissivity through a random Bragg reflection, showing that high-frequency photons are readily localized (Hu et al., 2005). The four curves have increasing thickness variations, σ, measured in terms of coherence length, as indicated by the dashed arrow.

5.6.2.2 Coherence of Electron Waves

The coherence of electron waves is an important issue for electron transport. One major difference between electrons and photons is the scattering mechanisms. Photon scattering is mostly elastic; that is, the wavelength and energy of the scattered photons are the same as that of the incoming photons and a fixed phase relationship between the incoming and scattered waves exists. The scattering of electrons can be *elastic*, in which the electrons merely change direction but have the same energy before and after the scattering, or *inelastic*, in which both the direction and the energy of the electrons are changed. The scattering of electrons by impurities and at the boundaries is elastic. The elastic scattering itself does not destroy the phase but the random locations of the impurities and the surface roughness may create enough randomness in the phase such that the particle approach is approximately valid. In other cases, the randomness can also create localization of the electron waves. On the other hand, inelastic scattering, such as electron–phonon scattering, randomizes the phases because the location and the phase of the electron–phonon scattering change all the time. These scattering events completely destroy the phase. If one uses the wave approach without proper consideration of the phase-destroying inelastic scattering processes, the results will be wrong.

Thus, electron transport has used a different set of definitions from those of photon transport (Ferry and Goodnick, 1997). Three major length scales that are often used in describing electron transport are the *mean free path*, the *phase coherence length*, and the *thermal length*. The mean free path is a measure of the average distance between successive scattering events and is $\Lambda \approx v_F \tau$, where v_F is the electron velocity at the Fermi level, or the Fermi velocity, and τ is the relaxation time, that is, the average time between successive collisions. The approximation sign is used because the Fermi velocity is only an approximation to the average electron velocity; this approximation works best in metals and in heavily doped semiconductors.* Not all the scattering events governing the mean free path are phase destroying. The inelastic scattering mean free path is $\Lambda_{in} = v_F \tau_\varphi$, where τ_φ is the inelastic relaxation time or phase-breaking time. The phase coherence length, also called the Thouless length, is defined as $\Lambda_\varphi = (a\tau_\varphi)^{1/2}$, where a is the electron diffusivity, which we will discuss more in the next chapter, and $a \approx v_F^2 \tau$. Because the relaxation time is used in the diffusivity, Λ_φ is slightly different from Λ_{in}. The use of diffusivity in the definition of phase coherence length implies that electrons may experience multiple elastic scattering, that is, diffusion, during the phase-breaking time. Typically, we have $\Lambda < \Lambda_\varphi < \Lambda_{in}$.

In addition to these length scales, there is also another length scale that is related to the thermal broadening of the energy levels of electrons. As in the discussion of the coherence length of blackbody radiation, the thermal broadening in energy is of the order of $\kappa_B T$. Thus, according to the Heisenburg uncertainty principle, the corresponding uncertainty in time is $h/(\kappa_B T)$. The thermal length is defined as $\Lambda_T = (a\hbar/\kappa_B T)^{1/2}$. Comparing this thermal length with the photon coherence length, eq. (5.164), the thermal length here is defined based on the diffusion transport, with the diffusion length given as $(a\tau_c)^{1/2}$, where τ_c is the characteristic time. The photon coherence length given by eq. (5.164) is based on the ballistic transport of photons of different energy spreading over $\kappa_B T$, with the transport length given by $v\tau_c$. Both lengths are a measure of the thermal spreading in the energy (wavelength) of the energy carriers, and thus are fundamentally similar concepts. This is the reason that we also call the coherence length given by eq. (5.164) the thermal length.

The phase coherence length Λ_φ and the thermal length Λ_T are usually used in judging whether transport is in the wave regime or the particle regime. If $\Lambda_T > \Lambda_\varphi$, the inelastic scattering is considered as the dominant phase-destroying process. Under this condition, if the structure characteristic length, such as the diameter of a nanowire or the width of a quantum well, is larger than Λ_φ, quantum states, as predicted by simple quantum well and quantum wire models in chapter 2, cannot be created because of the loss of phase correlation of the electron waves. The transport should be treated with the particle approach for such situations. If $\Lambda_\varphi > \Lambda_T$, thermal excitation is often considered as the dominant dephasing mechanism. Under this condition, if the structure characteristic length is much larger than Λ_T, it is often thought that the particle treatment leads to the same results as that of the wave approach. However, as pointed out before for photons, the wave and the particle approaches lead to the same results for simple geometries only. For periodic multilayer structures such as superlattices, the particle and the wave treatments do not lead to the same results, as explained in figures 5.22(c) and 5.24.

5.6.2.3 Coherence of Phonons

Phonon coherence from a transport point of view is the least considered one among electrons, photons, and phonons. The discussions on photons and electrons, however,

---

*For non-degenerate semiconductors, that is, semiconductors with the chemical potential lying inside the bandgap, v_F should be replaced by the thermal velocity $v_t = (3\kappa_B T/m^*)^{1/2}$, where m^* is the electron effective mass.

shed some light on the coherence issues of phonons. Scattering of phonons by boundaries and impurities is elastic and thus not phase destroying. Consequently, these scattering centers can either lead to localization or make the particle picture a good approximation. Because the dominant phonon wavelength is typically very short, ~10–20 Å at room temperature (Chen, 1997), interface roughness of the same order or much larger exists at most material interfaces or boundaries, and thus the particle-based treatment is likely to be valid for most practical situations. Phonon–phonon scattering, which is dominant in most materials at room temperature—a topic we will discuss in the next chapter—is inelastic. The mean free path of such scattering processes can be long, however. For example, in silicon, the estimated mean free path is 2500–3000 Å (Chen, 1998; Ju and Goodson, 1999). For structures much larger than the phonon–phonon scattering mean free path, the phase of phonons must be ignored. For structures smaller than the mean free path, rough interface scattering most likely can justify the particle treatment because of the short phonon wavelength, as mentioned before. The particle approach is particularly useful when the detailed interface structures are not clear and thus exclude a full-scale wave treatment.

One can also estimate the thermal coherence length on a basis similar to that for photons and electrons (Chen, 1997). Using an energy spread of $\kappa_B T$, we obtain the phonon thermal wavelength, defined as $v/\Delta v$, as $\sim vh/(\kappa_B T)$. At room temperature, taking a typical value of $v \sim 5000$ m s^{-1}, we get a thermal length of 10 Å. Such a thermal length, although useful as an indicator of width of the wave packet due to thermal spreading, cannot be applied to periodic structures such as superlattices (Chen, 1999) in deciding whether wave effect can be neglected, as in the case of the application of eq. (5.165) to Bragg reflectors.

The thermal conductivity of superlattices is a good example to illustrate the coherence issues related to phonon transport. It has been experimentally observed that the thermal conductivities of superlattices are significantly reduced in comparison with values obtained from the Fourier heat conduction law using the bulk properties of each layer (Yao, 1987; Chen et al., 1994; Lee et al., 1997; Capinski et al., 1999). The mechanisms of the thermal conductivity reduction, however, have been under debate (Chen, 2001; Yang and Chen, 2003). One approach is based on treating phonons as particles, with the phonons in each individual layer having their bulk properties but experiencing incoherent interface scattering. The other approach is based on treating superlattices as a new crystal structure with a unit cell spanning over one period of the superlattice, that is, treating phonons as coherent waves extending over the whole structure. The particle approach assumes that interface scattering destroys the phase. Particle-based model can fit experimental data based on the assumption of how many phonons are diffusely scattered at the interfaces. The coherent phonon wave approach, based on a pure harmonic lattice dynamics model, assumes that the reduction in thermal conductivity is caused by the phonon spectrum change and the associated reduction in group velocity, as shown in figure 3.30. The ideal lattice dynamics model, however, cannot predict the same order of thermal conductivity reduction as is experimentally observed. Recently, it has been shown that a lattice dynamics model based on damped lattice waves, that is, lattice waves that do not extend over the whole superlattice but can exist in one layer or over a few periods, can capture the same trends as experimental data, as shown in figure 5.26 (Yang and Chen, 2003). Such damped waves are created by introducing a complex wavevector into eq. (3.43) (Simkin and Mahan, 2000). The imaginary part of the wavevector, for example, can be due to the loss of coherence resulting from diffuse interface scattering

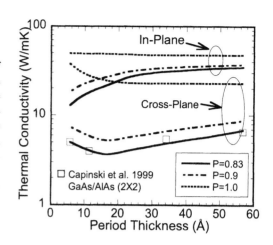

Figure 5.26 Thermal conductivity of superlattices obtained from a lattice dynamics model with damped lattice waves (Yang and Chen, 2003). The damping is determined by the interface specularity parameter p, representing the fraction of specularly scattered phonons. In the thin period limit, the results represent coherent transport whereas, in the thick period limit, the coherence is lost and the results represent the particle transport regime.

(Yang and Chen, 2003). In figure 5.26, the interface specularity parameter p represents the fraction of specularly scattered phonons, which are assumed to be coherent. The other phonons are diffusely scattered and are assumed to be incoherent. We first examine the case $p = 1$, that is, all phonons are specularly reflected and waves extend through the whole superlattice. In this case, the thermal conductivity is independent of the period thickness until the period is only 1–5 monolayers, in which regime the cross-plane thermal conductivity actually increases with decreasing film thickness. This recovery in thermal conductivity is due to phonon tunneling, as shown in figure 5.14. The fact that above about 10 Å the thermal conductivity does not change with thickness is related to thermal broadening. As we indicated earlier, the thermal length is also about 10 Å. However, in this case, the particle approach would lead to different results (Chen, 1999) and thus the wave and the particle approaches do not agree with each other. When p is less than one, the phonon waves are damped in the superlattice due to diffuse scattering. When the period thickness is large, the phonon waves are not coherent over many periods and thus the spectra calculated from lattice dynamics with imaginary wavevectors are close to those of bulk phonons. In this case, the thermal boundary resistances at interfaces dominate the thermal conductivity until the period becomes much larger than the mean free path in the bulk material, for which the thermal conductivity eventually approaches the predictions of the Fourier law (Chen, 1998). In the thin period limit, the superposition of coherent phonons extending over many periods leads to new phonon band structures and, correspondingly, wave phenomena such as stop bands, interference, and tunneling all contribute to the thermal conductivity behavior.

5.7 Summary of Chapter 5

This chapter discussed the wave picture of energy transport and the transition from the wave to the particle description. The purpose of section 5.1 was to familiarize the readers with various forms of waves including electromagnetic waves, acoustic waves, and material waves. The electromagnetic waves are governed by the Maxwell equations.

Acoustic waves, which are long-wavelength phonons, are described by the acoustic field equations or the Christoffel equation. Material waves are described by the Schrödinger equation. Solutions of the wave equations lead to the fields at each point as a function of time. The energy flux associated with each wave is usually a product of various fields, as given by the Poynting vector for electromagnetic and acoustic waves, and the particle flux expression for material waves. Although these waves are described by different governing equations, the key point is that all forms of waves share similar behavior, as is clearly demonstrated in the following four sections. Although the material presented in this chapter is diverse, some readers may be familiar with one or several forms of these waves and can understand other forms of waves by analogy.

At an interface, all waves experience the phenomena of reflection and refraction. A Snell-law type of relation governs the angles of incidence and transmission. The reflection and transmission coefficients, which are called Fresnel coefficients for electromagnetic waves, can be obtained by applying the appropriate boundary conditions for each type of wave. The expressions for these coefficients are quite similar among different types of waves. From the reflection and transmission coefficients, one can calculate the reflectivity and transmissivity of energy or particle flux. Several special cases for reflection and transmission of waves at one interface are of great importance. One example is total reflection, which occurs when the refractive index or the acoustic impedance of the medium at the incident side is larger than that at the transmission side for optical and acoustic waves, respectively, or the potential barrier is higher than the energy of the incident material waves. When total reflection occurs, an evanescent wave exists that extends into the second medium. The time-averaged energy or particle flux into the second medium carried by the evanescent wave is zero but the instantaneous field and energy are not zero. Thermal boundary resistance between two perfect solids is due to the reflection of phonons at the interface.

When multiple interfaces exist, superposition of waves due to reflection at multiple interfaces creates the familiar interference and tunneling phenomena in thin films. For multidimensional problems, which we did not discuss in this chapter here, the superposition of scattered waves leads to diffraction phenomena. We introduced the transfer matrix method for calculating the reflection and transmission coefficients of multilayers, which is valid for both interference and tunneling regimes. Interference gives the familiar oscillation of reflectivity and transmissivity of optical coatings as a function of the film thickness, and affects the thermal radiative properties of thin films and multilayers. In multilayer structures, particularly periodic structures, interference leads to the formation of stop-bands, which corresponds to the formation of gaps in the energy spectrum of electrons, phonons, and photons as discussed in chapter 3. Tunneling of evanescent waves that exist near the interface under appropriate conditions can occur when a third medium is brought close to the interface, before the evanescent wave significantly decays in the second medium, and when the third medium allows the propagation of the wave. The tunneling phenomenon is the basis of several recent inventions such as scanning tunneling microscopy for electrons and photons. It also occurs for acoustic waves and may affect heat conduction.

Given the transmissivity of heat carriers through two points of a system, we can calculate the net heat transfer (or other fluxes of interest) between the two points, using the Landauer formalism as manifested by eqs. (5.88) and (5.153). The Landauer formalism views transport as a transmission process. The net flux (energy or particle) between any two points A and B is the difference between the corresponding flux transmitted from A to B and that from B to A. The principle of detailed balance can be used to write the final flux in terms of the properties of one side (or one point) only, together with the transmissivity.

Calculation of energy or particle transport under the wave picture is often tedious and requires mathematical manipulation of the field quantities. In section 5.6, we discussed under what conditions we can neglect the phase information and treat energy carriers as particles. First, we demonstrated that the superposition of monochromatic waves leads to wave packets that propagate at the group velocity rather than the phase velocity. This group velocity is normally the velocity at which energy is propagating, but in a highly dispersive medium the group velocity is not necessarily the energy propagation velocity. The width of these wave packets is the coherence length, which is inversely proportional to the inverse effective spectrum width (or energy spread) of the carriers. If the coherence length is long compared to the structural characteristic length, the wave picture should be used. In the opposite limit, however, we should be more careful. We can treat the transport as particles as long as the wave packets split from the same original one, for example, through reflection at an interface, do not overlap at the same place and the same time. This often happens when the structural size is large compared to that of the wave packets. However, in periodic structures, such as Bragg reflectors and superlattices, the wave packets reflected at different layers can merge and still overlap. Consequently, the particle approach and wave approach do not agree with each other. Elastic scattering, caused by inhomogeneities such as impurities and interface roughness, does not destroy the coherence of the waves. The random elastic scattering can potentially lead to two effects. One is localization, for which the waves are localized and do not propagate. Localization is generally easier to observe in low-dimensional structures than in three-dimensional structures. The other effect is that random scattering and the subsequent superposition of scattered waves usually leads to results that are close to those obtained from the particle treatment. When the exact locations of the scattering centers and surface topology are not known, which is usually the case, the particle treatment leads to better agreement with experimental results. Inelastic scattering completely destroys the phase. When the structure characteristic length is much larger than the inelastic scattering mean free path, or, for electrons, the phase coherence length, the particle treatment is mandatory.

5.8 Nomenclature for Chapter 5

a	electron diffusivity, m s^{-2}	E	allowed energy level, J
A	amplitude and direction of field	E	electric field, N C^{-1} = V m^{-1}
B	magnetic induction, N s m^{-1} C^{-1}	f	probability distribution function
c_0	speed of light in vacuum, m s^{-1}	F	vector wave field; force, N
		h	Planck constant, J s
d	film thickness, m	$\hbar$	Planck constant divided by 2π, J s
D	electric displacement, C m^{-2}		

ENERGY TRANSFER BY WAVES

$\mathbf{H}$	magnetic field, $C\,m^{-1}\,s^{-1} = A\,m^{-1}$	$\mathbf{v}_g$	group velocity, $m\,s^{-1}$
i	imaginary number unit, $\sqrt{-1}$	Z	acoustic impedance, $kg\,m^{-2}\,s^{-1}$
$\mathbf{J}$	flux of particles, $s^{-1}\,m^{-2}$	α	absorption coefficient, m^{-1}
$\mathbf{J}_a$	acoustic wave power flux, $W\,m^{-2}$	δ	skin depth, m
$\mathbf{J}_e$	current density, $C\,s^{-1}\,m^{-2}$	$\Delta\nu$	spectral width, s^{-1}
$\mathbf{J}_s$	surface current density, $A\,m^{-1}$	ε	electric permittivity, $C^2\,N^{-1}\,m^{-2} = F\,m^{-1}$
k	magnitude of wavevector, m^{-1}	$\hat{\varepsilon}$	complex electric permittivity, $C^2\,N^{-1}\,m^{-2}$
$\mathbf{k}$	wavevector, m^{-1}	ε_0	electric permittivity of vacuum, $C^2\,N^{-1}\,m^{-2}$
$\hat{\mathbf{k}}$	unit vector along wavevector direction	ε_r	dielectric constant
l	integer	θ	angle, rad
ℓ_c	coherence length, m	κ	imaginary part of complex refractive index
m	mass, kg; integer	κ_B	Boltzmann constant, $J\,K^{-1}$
n	real part of complex refractive index N	λ	wavelength, m
N	complex refractive index or complex optical constant	λ_L	Lamb constant, $N\,m^{-2}$
p	surface impedance, $C^2\,J^{-1}\,s^{-1}$	Λ	mean free path, m
$\mathbf{P}$	polarization per unit volume, $C\,m^{-2}$	Λ_{in}	inelastic scattering mean free path, m
q	magnitude of heat flux, $W\,m^{-2}$	Λ_T	thermal wavelength, m
$\mathbf{q}$	heat flux vector, $W\,m^{-2}$	Λ_φ	phase coherence length, m
Q	charge, C	μ	magnetic permeability, $N\,s^2\,C^{-2}$
r	reflection coefficient	μ_L	Lamb constant, $N\,m^{-2}$
$\mathbf{r}$	position vector	ν	frequency of phonons and photons, s^{-1}
R	reflectivity	ξ	electric polarizibility
$\mathbf{S}$	Poynting vector, $W\,m^{-2}$	ρ	net charge density, $C\,m^{-3}$
$\overline{\overline{S}}$	strain tensor	ρ_s	surface charge density, $C\,m^{-2}$
t	time, s; or transmission coefficient	σ_e	electrical conductivity, $\Omega^{-1}\,m^{-1}$
T	temperature, K	τ	transmissivity
T_e	temperature of phonons coming toward interface, i.e., temperature of emitted phonons, K	ϕ	azimuthal angle, rad
		φ	phase factor
		χ	electric susceptibility
$\mathbf{u}$	displacement, m	Ψ	wavefunction
U_0	potential barrier height, J	ω	angular frequency, rad.Hz
v_F	Fermi velocity, $m\,s^{-1}$	$\mathfrak{R}''$	specific thermal boundary resistance, $K\,m^2\,W^{-1}$
$\mathbf{v}$	velocity, $m\,s^{-1}$		

Subscripts

0	vacuum
1, 2	medium 1 or medium 2
12	from medium 1 into medium 2
21	from medium 2 into medium 1
a	amplitude
c	complex
e	based on emitted phonon temperature
i	incident wave
L	longitudinal, or Lamb constant
r	reflected wave
s	surface
t	transmitted wave
T	transverse wave
x, y, z	Cartesian components

Superscripts

*	complex conjugate
=	second-order tensor
-	average

5.9 References

Anderson, P.W., 1958, "The Absence of Diffusion in Certain Random Lattices," *Physical Review*, vol. 109, pp. 1492–1505.

Angelescu, D.E., Cross, M.C., and Roukes, M.L., 1998, "Heat Transport in Mesoscopic Systems," *Superlattices and Microstructures*, vol. 23, pp. 673–689.

Ashcroft, N.W., and Mermin, N.D., 1976, *Solid State Physics*, Saunders College Publishing, Fort Worth, TX.

Auld, B.A., 1990, *Acoustic Fields and Waves in Solids, I, II*, 2nd ed., Krieger, FL.

Binnig, G., and Rohrer, H., 1982, "Scanning Tunneling Microscopy," *Helvetica Physica Acta*, vol. 55, pp. 726–735.

Binnig, G., Quate, C.F., and Gerber, C., 1986, "Atomic Force Microscope," *Physical Review Letters*, vol. 56, pp. 930–933.

Bohren, C.F., and Huffman, D.R., 1983, *Absorption and Scattering of Light by Small Particles*, Wiley, New York, pp. 235–239.

Born, M., and Wolf, E., 1980, *Principles of Optics*, 6th ed., Pergamon Press, Oxford, chapters 1 and 10.

Brillouin, L., 1960, *Wave Propagation and Group Velocity*, Academic Press, New York.

Capinski, W.S., Maris, H.J., Ruf, T., Cardona, M., Ploog, K., and Katzer, D.S., 1999, "Thermal-Conductivity Measurements of GaAs/AlAs Superlattices Using a Picosecond Optical Pump-and-Probe Technique," *Physical Review B*, vol. 59, pp. 8105–8113.

Chen, G., 1996, "Wave Effects on Radiative Transfer in Absorbing and Emitting Thin-Film Media," *Microscale Thermophysical Engineering*, vol. 1, pp. 215–224.

Chen, G., 1997, "Size and Interface Effects on Thermal Conductivity of Superlattices and Periodic Thin-Film Structures," *Journal of Heat Transfer*, vol. 119, pp. 220–229.

Chen, G., 1998, "Thermal Conductivity and Ballistic Phonon Transport in Cross-Plane Direction of Superlattices," *Physical Review B*, vol. 57, pp. 14958–14973.

Chen, G., 1999, "Phonon Wave Effects on Heat Conduction in Thin Films and Superlattices," *Journal of Heat Transfer*, vol. 121, pp. 945–953.

Chen, G., 2001, "Phonon Transport in Low-Dimensional Structures," *Semiconductors and Semimetals*, vol. 71, pp. 203–259.

Chen, G., 2003, "Diffusion–Transmission Interface Conditions," *Applied Physics Letters*, vol. 82, pp. 991–993.

Chen, G., and Tien, C.L., 1992, "Partial Coherence Theory of Thin Film Radiative Properties," *Journal of Heat Transfer*, vol. 114, pp. 636–643.

Chen, G., and Zeng, T., 2001, "Nonequilibrium Phonon and Electron Transport in Heterostructures and Superlattices," *Microscale Thermophysical Engineering*, vol. 5, pp. 71–88.

Chen, G., Tien, C.L., Wu, X., and Smith, J.S., 1994, "Measurement of Thermal Diffusivity of GaAs/AlGaAs Thin-Film Structures," *Journal of Heat Transfer*, vol. 116, pp. 325–331.

Cohen-Tannoudji, C., Diu, B., and Laloe, F., 1977, *Quantum Mechanics*, vol. I, Wiley, New York.

Costescu, R.M., Wall, M.A., and Cahill, D.G., 2003, "Thermal Conductance of Epitaxial Interfaces," *Physical Review B*, vol. 67, pp. 054302/1–5.

Cravalho, E.G., Tien, C.L., and Caren, R.P., 1967, "Effect of Small Spacing on Radiation Transfer between Two Dielectrics," *Journal of Heat Transfer*, vol. 89, pp. 351–358.

Dames, C., and Chen, G., 2004, "Theoretical Phonon Thermal Conductivity of Si/Ge Superlattice Nanowires," *Journal of Applied Physics*, vol. 95, pp. 682–693.

DiMatteo, R.S., Greiff, P., Finberg, S.L., Young-Waithe, K.A., Choy, H.K.H., and Masaki, M.M., 2001, "Enhanced Photogeneration of Carriers in a Semiconductor Via Coupling Across a Nonisothermal Nanoscale Vacuum Gap," *Applied Physics Letters*, vol. 79, pp. 1894–1896.

Domoto, G.A., Boehm, R.F., and Tien, C.L., 1970, "Experimental Investigation of Radiative Transfer between Metallic Surfaces at Cryogenic Temperatures," *Journal of Heat Transfer*, vol. 92, pp. 412–417.

Esaki, L., 1958, "New Phenomenon in Narrow Germanium p–n Junctions," *Physical Review*, vol. 109, pp. 603–604.

Esaki, L., and Tsu, R., 1970, "Superlattice and Negative Differential Conductivity in Semiconductors," *IBM Journal of Research and Development*, vol. 14, pp. 61–65.

Ferry, D.K., and Goodnick, S.M., 1997, *Transport in Nanostructures*, Cambridge University Press, Cambridge, UK.

Garcia-Martin, A., and Niero-Vesperinas, M., Sáenz, J.J., 2000, "Spatial Field Distributions in the Transition from Ballistic to Diffusive Transport in Randomly Corrugated Waveguides," *Physical Review Letters*, vol. 84, pp. 3578–3581.

Hargreaves, C.M., 1969, "Anomalous Radiative Transfer between Closely-Spaced Bodies," *Physics Letters*, vol. 30A, pp. 491–492.

Hu, L., Schmidt, A., Narayanaswamy, A., and Chen, G., 2005, "Effects of Periodic Structure on Coherence Properties of Blackbody Radiation," *Journal of Heat Transfer*, in press.

Huang, K.-M., and Wu, G.Y., 1992, "Transfer-Matrix Theory of the Energy Levels and Electron Tunneling in Heterostructures under an In-Plane Magnetic Field," *Physical Review B*, vol. 45, pp. 3461–3464.

Imry, Y., and Landauer, R., 1999, "Conductance Viewed as Transmission," *Review of Modern Physics*, vol. 71, pp. S306–S312.

James, D.F.V., and Wolf, E., 1991, "Spectral Changes Produced in Young's Interference Experiment," *Optics Communications*, vol. 81, pp. 150–154.

Ju, Y.S., and Goodson, K.E., 1999, "Phonon Scattering in Silicon Films of Thickness Below 100 nm," *Applied Physics Letters*, vol. 74, pp. 3005–3007.

Kapitza, P.L., 1941, "The Study of Heat Transfer in Helium II," *J. Phys.* (USSR) vol. 4, pp. 181–210; in *Collected Papers of P.L. Kapitza*, vol. 2, p. 581, ed. der Haar, D., 1965, Pergamon, Oxford.

Katerberg, J.A., Reynolds, C.L., Jr., and Anderson, A.C., 1977, "Calculations of the Thermal Boundary Resistance," *Physical Review B*, vol. 16, pp. 672–679.

Knittl, Z., 1976, *Optics of Thin Films*, Wiley, New York, chapter 4.

Koyama, F., Kinoshita, S., and Iga, K., 1989, "Room-Temperature Continuous Wave Lasing Characteristics of a GaAs Vertical-Cavity Surface-Emitting Laser," *Applied Physics Letters*, vol. 55, pp. 221–222.

Lee, S.M., Cahill, D.G., and Venkatasubramanian, R., 1997, "Thermal Conductivity of Si–Ge Superlattices," *Applied Physics Letters*, vol. 70, pp. 2957–2959.

Little, W.A., 1959, "The Transport of Heat between Dissimilar Solids at Low Temperatures," *Canadian Journal of Physics*, vol. 37, pp. 334–349.

Majumdar, A., Lai, J., Chandrachood, M., Nakabeppu, O., and others, 1995, "Thermal Imaging by Atomic Force Microscopy Using Thermocouple Cantilever Probes," *Review of Scientific Instruments*, vol. 66, pp. 3584–3592.

Mandel, L., and Wolf, E., 1995, *Optical Coherence and Quantum Optics*, Cambridge University Press, Cambridge, UK, pp. 307–318.

Mehta, C.L., 1963, "Coherence-Time and Effective Bandwidth of Blackbody Radiation," *Nuovo Cimento*, vol. 21, pp. 401–408.

Mulet, J.-P, Joulain, K.L., Carminati, R., and Greffet, J.-J., 2002, "Enhanced Radiative Heat Transfer at Nanometric Distances," *Microscale Thermophysical Engineering*, vol. 6, p. 209–222.

Narayanamurti, V., Stormer, J.L., Chin, M.A., Gossard, A.C., and Wiegmann, W., 1979, "Selective Transmission of High-Frequency Phonons by a Superlattice: the 'Dielectric' Phonon Filter," *Physical Review Letters*, vol. 43, pp. 2012–2016.

Narayanaswamy, A., and Chen, G., 2003, "Surface Modes for Near-Field Thermophotovoltaics," *Applied Physics Letters*, vol. 82, pp. 3544–3546.

Narayanaswamy, A., and Chen, G., 2004, "Direct Computation of Thermal Emission from Nanostructures," *Annual Review of Heat Transfer*, vol. 14, in press.

Nayfeh, A.H., 1995, *Wave Propagation in Layered Anisotropic Media*, Elsevier, Amsterdam.

Nulman, J., 1989, "Emissivity Issues in Pyrometric Temperature Monitoring for RTP Systems," *SPIE*, vol. 1189, *Rapid Thermal Processing*, pp. 72–82.

Odom, T.W., Huang, J., Kim, P., and Lieber, C.M., 1998, "Atomic Structure and Electronic Properties of Single-Walled Carbon Nanotubes," *Nature*, vol. 391, pp. 62–64.

Palik, E., 1985, *Handbook of Optical Constants of Solids*, Academic Press, San Diego, CA.

Pendry, J.B., 1996, "Calculating Photonic Band Structure," *Journal of Physics–Condensed Matter*, vol. 8, pp. 1085–1108.

Pendry, J.B., 1999, "Radiative Exchange of Heat between Nanostructures," *Journal of Physics—Condensed Matter*, vol. 11, pp. 6621–6633.

Pendry, J.B., 2000, "Negative Refraction Makes a Perfect Lens," *Physical Review Letters*, vol. 85, pp. 3966–3969.

Polder, D., and Van Hove, M., 1971, "Theory of Radiative Heat Transfer between Closely Spaced Bodies," *Physical Review B*, vol. 4, pp. 3303–3314.

Raether, H., 1987, *Surface Plasmons on Smooth and Rough Surfaces and on Gratings*, Springer-Verlag, Berlin.

Rayleigh, J.W.S., 1945, *The Theory of Sound*, Dover, New York.

Reddick, R.C., Warmack, R.J., and Ferrell, T.L., 1989, "New Form of Scanning Optical Microscopy," *Physical Review B*, vol. 39, pp. 767–770.

Rytov, S.M., Kravtsov, Y. A., and Tatarski, V.I., 1987, *Principles of Statistical Radiophysics*, vol. 3, Springer-Verlag, Berlin.

Santarsiero, M., and Gori, F., 1992, "Spectral Change in Young Interference Pattern," *Physics Letters A*, vol. 167, p. 123.

Schwab, K., Henriksen, E.A., Worlock, J.M., and Roukes, M.L., 2000, "Measurement of the Quantum of Thermal Conductance," *Nature*, vol. 404, pp. 974–977.

Shelby, R.A., Smith, D.R., and Schultz, S., 2001, "Experimental Verification of a Negative Index of Refraction," *Science*, vol. 292, pp. 77–79.

Sheng, P., ed., 1990, *Scattering and Localization of Classical Waves in Random Media*, World Scientific, Singapore.

Siegel, R., and Howell, J., 1972, *Thermal Radiation Heat Transfer*, 3rd ed., Hemisphere, Washington, DC, p. 928.

Simkin, M.V., and Mahan, G.D., 2000, "Minimum of Thermal Conductivity of Superlattices," *Physical Review Letters*, vol. 84, pp. 927–930.

Slater, J.C., 1967, *Quantum Theory of Molecules and Solids*, vol. 3, McGraw-Hill, New York, chapter 2 and appendix 1.

Stoner, R.J., and Maris, H.J., 1993, "Kapitza Conductance and Heat Flow between Solids at Temperatures from 50 to 300 K," *Physical Review B*, vol. 48, p. 16373.

Swartz, E.T., and Pohl, R.O., 1989, "Thermal Boundary Resistance," *Reviews of Modern Physics*, vol. 61, pp. 605–668.

Tamura, S., Tanaka, Y., and Maris, H.J., 1999, "Phonon Group Velocity and Thermal Conduction in Superlattices," *Physical Review B*, vol. 60, pp. 2627–2630.

Tien, C.L., and Cunnington, G.R., 1973, "Cryogenic Insulation Heat Transfer," *Advances in Heat Transfer*, vol. 9, pp. 349–417.

Tsu, R., and Esaki, L., 1973, "Tunneling in a Finite Superlattice," *Applied Physics Letters*, vol. 22, pp. 562–564.

Valanju, P.M., Walser, R.M., and Valanju, A.P., 2002, "Wave Refraction in Negative-Index Media: Always Positive and Very Inhomogeneous," *Physical Review Letters*, vol. 88, pp. 187401-1 to 187401 (1–4).

Venkatasubramanian, R., 2000, "Lattice Thermal Conductivity Reduction and Phonon Localization Phenomena in Superlattices," *Physical Review B*, vol. 61, pp. 3091–3097.

Veselago, V.G., 1968, "Electrodynamics of Substances with Simultaneously Negative Electrical and Magnetic Permeabilities," *Soviet Physics Usp.*, vol. 10, p. 509.

Walker, J.D., 1993, Vertical-Cavity Laser Diodes Fabricated by Phase-Locked Epitaxy, Ph.D. thesis, Berkeley, CA.

Whale, M.D., and Cravalho, E.G., 2000, "Modeling and Performance of Microscale Thermophotovoltaic Energy Conversion Devices," *IEEE Transactions on Energy Conversion*, vol. 17, pp. 130–142.

Wong, P.Y., Hess, C.K., and Miaoulis, I.N., 1995, "Coherence Thermal Radiation Effects on Temperature Dependent Emissivity of Thin-Film Structures on Optically Thick Substrates," *Optical Engineering*, vol. 34, pp. 1776–1781.

Xu, J.B., Lauger, K., Moller, R., Dransfeld, K., and Wilson, I.H., 1994, "Heat Transfer between Two Metallic Surfaces at Small Distances," *Journal of Applied Physics*, vol. 76, pp. 7209–7216.

Yablonovitch, E., 1986, "Inihibited Spontaneous Emission in Solid-State Physics and Electronics," *Physical Review Letters*, vol. 58, pp. 2059–2062.

Yang, B., and Chen, G., 2001, "Anisotropy of Heat Conduction in Superlattices," *Microscale Thermophysical Engineering*, vol. 5, pp. 107–116.

Yang, B., and Chen, G., 2003, "Partially Coherent Heat Conduction in Superlattices," *Physical Review B*, vol. 67, pp. 195311 (1–4).

Yao, T., 1987, "Thermal Properties of AlAs/GaAs Superlattices," *Applied Physics Letters*, vol. 51, pp. 1798–1780.

Zhang, Z.M., Fu, C.J., and Zhu, Q.Z., 2003, "Optical and Thermal Radiative Properties of Semiconductors Related to Micro/Nanotechnology." *Advances in Heat Transfer*, vol. 37, pp. 179–296.

5.10 Exercises

5.1 *Surface emissivity.* The refractive index of silicon at 0.63 μm is (3.882, 0.019). Calculate the surface reflectivity, transmissivity, and emissivity of a semi-infinite silicon wafer (a) at normal incidence, (b) at 30° angle of inidence, and (c) at 60° angle of incidence, for both TE and TM waves. Also, estimate the penetration depth for normal incidence.

5.2 *Inhomogeneous wave in an absorbing medium.* A plane wave in vacuum is reflected by a medium with a complex refractive index $N = n + i\kappa$ at an angle of incidence θ. Derive an expression for the electric and magnetic fields inside the medium. Show that the constant amplitude and constant phase surfaces of the wave do not coincide with each other. Such waves are called inhomogeneous waves. Derive an expression for the Poynting vector inside the medium.

5.3 *Heat generation distribution due to absorption.* A plane wave with an intensity of 10^4 W m^{-2} at 0.517 μm meets a gold surface at 30° of incidence. Determine the heat generation distribution inside the gold specimen. The refractive index of gold at 0.517 μm is $N = 0.608 + 2.12i$.

5.4 *Fresnel formula for TE wave.* Derive the Fresnel formula for a transverse electric wave incident onto a plane surface, that is, eqs. (5.73) and (5.74).

5.5 *Transmissivity into an absorbing medium.* If the medium is absorbing, one must be careful in writing down the Poynting vector. Examining eq. (5.76) and assuming that only n_2 is complex, $n_2 + i\kappa_2$, derive an expression for the transmissivity, using n_2 and κ_2 explicitly.

5.6 *Interference effects in thin films—Color of thin film.* Experienced workers in thin-film deposition can tell the film thickness from its color. At 0.5 μm, the refractive index of SiO_2 is $N = (1.46, 0)$ and that of silicon is $N = (4.14, 0.045)$. Calculate the reflectivity of a thin film of SiO_2 deposited on the silicon wafer for a film thickness between 500 Å and 2000 Å at normal incidence. Mark down a few colors you expect to see at normal incidence for a few film thickness values in the given range.

5.7 *Optical interference effects in thin films—Angle effects.* A substrate coated with a film may have different colors when looked at from different directions. At 0.5 μm, the refractive index of SiO_2 is $N = (1.46, 0)$ and that of silicon is $N = (4.14, 0.045)$. Calculate the reflectivity of a 500 Å SiO_2 film deposited on silicon wafer for the angles of incidence 0°, 30°, and 45°.

5.8 *Critical angle of incidence for optical waves.* For radiation going from a high refractive index medium into air, calculate the critical angle if the refractive index of the medium is (a) 1.4 and (b) 3.5.

5.9 *Acoustic wave reflection and transmission—SH wave.* For a transverse acoustic wave polarized in the direction perpendicular to the plane of incidence (an SH wave), calculate the reflectivity and transmissivity of the wave at an interface between two isotropic materials at the following angles of incidence: (a) normal, (b) 15°, and (c) 60°. The materials' properties are: material 1: $\rho_1 = 5.33 \times 10^3$ kg m^{-3}, $v_{T1} = 3900$ m s^{-1}; material 2: $\rho_2 = 2.33 \times 10^3$ kg m^{-3}, $v_{T2} = 6400$ m s^{-1}.

5.10 *Reflection of electron wave.* Calculate the reflectivity of a free electron with an energy of 1 eV propagating toward a potential barrier with the following barrier heights: (a) 0.2 eV, (b) 0.8 eV, (c) 1.5 eV.

5.11 *Thermal boundary resistance.* Estimate the thermal boundary resistance between two materials with the following properties on the basis of the diffuse interface scattering model: material 1: $v_1 = 3900$ m s^{-1}, $C_1 = 1.67 \times 10^6$ J m^{-3} K^{-1}; material 2: $v_2 = 6400$ m s^{-1}, $C_2 = 1.66 \times 10^6$ J m^{-3} K^{-1}. For a heat flux of 10^8 W m^{-2}, estimate the temperature drop occurring at the interface?

5.12 *Reflection of longitudinal acoustic wave.* A longitudinal acoustic wave is incident from medium 1 into medium 2. Derive an expression for the reflection and transmission coefficients of the excited longitudinal and transverse waves as a function of the angle of incidence. Both media are assumed to be isotropic and their properties are: $v_{T1} = 6400$ m s^{-1}, $v_{L1} = 8000$ m s^{-1}, $\rho_1 = 2.3 \times 10^3$ kg m^{-3}; $v_{T2} = 3900$ m s^{-1}, $v_{L2} = 5000$ m s^{-1}, $\rho_2 = 5.3 \times 10^3$ kg m^{-3}. Use Auld's book (1990) as a reference for solving this problem.

5.13 *Thermal boundary resistance at low temperature.* Thermal boundary resistance is a phenomenon that is important at low temperatures even for bulk materials and becomes important even at room temperature in nanostructures. Treating the transmissivity in eq. (5.92) as independent of angle and frequency, derive an expression for the proportionality coefficient in eq. (5.93) at low temperatures.

5.14 *Analogy of thermal boundary resistance for photons.* Reflection of carriers can be regarded as an additional resistance, as in the case of thermal boundary resistance. Photons can be reflected at an interface too, as we discussed in this chapter. Now we want to develop an analogy of thermal boundary resistance for photons by considering a partially reflecting and partially transmitting interface located between two parallel black walls maintained at temperatures T_1 and T_2. The transmissivity of the interface is τ_{12}. Derive an expression for the net radiation heat transfer exchange between the two walls, and a corresponding expression for the photon thermal boundary resistance at the interface. In radiation, however, we do not call such a phenomenon thermal boundary resistance.

5.15 *Interference in multilayer structures.* Two layers of thin films are grown on a silicon substrate. At the optical wavelength of 1 μm, the refractive index of silicon is (3.6,0). The refractive index of the layer grown directly on silicon is (2.4,0) and its thickness is 2000 Å. The refractive index of the subsequent layer is (1.3,0) and its thickness varies in the range of 0.1–1 μm. Calculate the reflectivity of the structure at the given wavelength, using the transfer matrix method, for normal incidence.

5.16 *Tunneling of electrons.* For a potential barrier of height 1.0 eV, plot the transmissivity of a free electron with an energy of 0.5 eV through the barrier for a barrier width ranging from 1 Å to 50 Å.

5.17 *Tunneling of photons.* A vacuum gap of 0.2 μm is formed between two glass substrates. Plot the transmissivity of light from one glass substrate into another as a function of angle of incidence for an incident TM wave with a wavelength at 0.5 μm. The refractive index of the glass is taken as 1.46. Compare the results with the situation if a thin film of glass, of 0.2 μm thick, is sandwiched within a vacuum.

5.18 *Landauer formula for phonon heat conduction.* A freestanding thin film of thickness d is suspended between two thermal reservoirs at temperatures T_1 and T_2. The dispersion can be approximated as

$$\omega = v\left[k_x^2 + k_y^2 + \left(\frac{n\pi}{d}\right)^2\right]^{1/2}$$

Assuming that the phonon transmissivity is one and neglecting scattering, derive an expression for the thermal conductance for heat conduction along the thin film plane (x-direction).

5.19 *Landauer formulation for electron thermal conduction.* A metallic square nanowire is placed between two thermal reservoirs at temperatures T_1 and T_2. Assume that electron transmissivity is equal to one. Derive an expression for the thermal conductivity of the nanowire contributed by the electrons.

5.20 *Coherence length of blackbody radiation.* Estimate the coherence length of a blackbody radiation source at 10 K and 300 K.

5.21 *Coherence length of laser radiation.* Estimate the coherence of a laser radiation with a central wavelength of 1.06 μm and a spectral width of 10 Å.

5.22 *Coherence properties of electrons.* At low temperatures, the Fermi velocity in a material is 2.76×10^5 m s^{-1}, the electron relaxation time is 3.8 ps (1 ps = 10^{-12} s), and the phase-breaking time is 18 ps. Calculate the mean free path and the phase coherence length of an electron.

5.23 *Phonon group velocity.* The phonon dispersion for a monatomic lattice chain is

$$\omega = 2\sqrt{\frac{K}{m}} \left| \sin \frac{ka}{2} \right|$$

Derive an expression of its group velocity. Prove that the group velocity at the zone boundary is zero.

5.24 *Difference between wave and particle approaches (project type).* In section 5.6 we stated that wave optics and geometrical optics do not lead to the same results for the radiative properties of periodic multilayer structures for blackbody radiation. Consider a periodic structure made of two alternating layers with refractive indices of (4,0) and (2,0), that is, nonabsorbing films. Blackbody radiation at 1000 K comes toward the periodic multilayer structure at normal incidence. Assuming both sides of the multilayer structure are vacuum, calculate the reflectivity and transmissivity averaged over the blackbody spectrum for the following cases, using wave optics and ray tracing:

(a) For each layer thickness of 1 μm, 10 μm, and 100 μm calculate the variation of reflectivity and transmissivity as a function of the number of periods in the structure. Compare the results for wave and ray tracing.

(b) For 10, 100, 1000 periods, calculate the average reflectivity and transmissivity as a function of the thickness of each layer, assuming all layers are of equal thickness, for the layer thickness range of 1 μm to 100 μm.

Geometrical optics can be obtained using the following recursive formula for the addition of every interface (Siegel and Howell, 1992, p. 928)

$$R_{n+m} = R_m + \frac{R_n T_m^2}{1 - R_m R_n} \qquad \tau_{m+n} = \frac{\tau_m \tau_n}{1 - R_m R_n}$$

where the subscript m refers to the total reflectivity and transmissivity of the first m interfaces (counted from the incident side) and n represents those of the subsequent n additional interfaces. For example, for one layer with two interfaces (the reflectivity and transmissivity at the first interface are R_1 and τ_1 and those at the second interface are R_2 and τ_2), the above formula becomes

$$R_{1+1} = R_1 + \frac{R_2 \tau_1^2}{1 - R_1 R_2} \qquad \tau_{1+1} = \frac{\tau_1 \tau_2}{1 - R_1 R_2}$$

Hint: one numerical problem with the transfer matrix method for thick film is that the exponential function may blow up. One must find ways to solve this problem for calculating thick films using the transfer matrix method.

6

Particle Description of Transport Processes: Classical Laws

We discussed in the previous chapter when we can ignore the coherence effects and treat heat carriers as individual particles without considering their phase information. In the next few chapters, we will describe how to deal with energy transfer under the particle picture. Most constitutive equations for macroscale transport processes, such as the Fourier law and the Newton shear stress laws, are obtained under such particle pictures. These equations are often formulated as laws summarized from experiments. In this chapter, we will see that most of the classical laws governing transport processes can be derived from a few fundamental principles.

In chapter 4, we studied systems at equilibrium and developed the equilibrium distribution functions (Fermi–Dirac, Bose–Einstein, and Boltzmann distributions). The distribution function for a quantum state at equilibrium is a function of the energy of the quantum state, the system temperature, and the chemical potential. When the system is not at equilibrium, these distribution functions are no longer applicable. Ideally, we would like to trace the trajectory of all the particles in the system, as in the molecular dynamics approach that we will discuss in chapter 10. This approach, however, is not realistic for most systems, because they have a large number of atoms or molecules. Thus, we resort to a statistical description of the particle trajectory.

In the statistical description we use nonequilibrium distribution functions, which depend not only on the energy and temperature of the system but also on positions and other variables. We will develop in this chapter the governing equations for the nonequilibrium distribution functions. In particular, we will rely on the Boltzmann equation, also called the Boltzmann transport equation. From the Boltzmann equation we will derive familiar constitutive equations such as the Fourier law, the Newton shear stress law, and the Ohm law. We will also demonstrate that conservation equations, such as the Navier–Stokes equations for fluids and electrohydrodynamic equations for charged particles, can be obtained from the Boltzmann equation. Special attention will be paid to the approximations made in these derivations, which will be relaxed in the next chapter when we consider various classical size effects. A discussion is also presented in this chapter on thermal waves and their appropriate descriptions.

6.1 The Liouville Equation and the Boltzmann Equation

We discussed, in chapter 4, the probability distribution of an equilibrium system occupying a specific accessible quantum state. Because the system is at equilibrium, the probability distribution take a simple form. For example, the Boltzmann distribution depends only on the energy of the quantum state and on the system temperature. Transport occurs, however, only when the system is in a nonequilibrium state and consequently the equilibrium distribution can no longer describe the state of the system. Conceivably, to describe the state of such a nonequilibrium system, more information is needed. In this section, we will introduce nonequilibrium distribution functions that describe the states of systems and the governing equations for the evolution of the nonequilibrium distribution functions. We will start from the general Liouville equation, which is valid for all classical systems but is difficult to solve, and move on to the simpler Boltzmann equation that serves as the basis for our future analysis. We will also discuss the assumptions made in the Boltzmann equation and see, consequently, its limitations.

6.1.1 The Phase Space and Liouville's Equation

Consider a system with N particles, where each particle can be described by the generalized coordinate $\mathbf{r}$ and momentum $\mathbf{p}$. For example, the generalized coordinates of a diatomic molecule, $\mathbf{r}_1$, include the position (x_1, y_1, z_1), the vibrational coordinate (the separation between the two atoms, $\triangle x_1$), the rotational coordinates (polar and azimuthal angles, θ_1 and φ_1); likewise, the generalized momentum, $\mathbf{p}_1$, includes the translational momenta $(mv_{x1}, mv_{y1}, mv_{z1})$, the vibrational momentum proportional to the relative velocity of the two atoms $(md\triangle x_1/dt)$, and the rotational momenta (angular momenta of rotation corresponding to θ and φ directions). We assume here that there are m degrees of freedom in space, that is, m generalized spatial coordinates, and m degrees of freedom in momentum for each particle. The number of the degree of freedom of the whole system is $2n = 2m \times N$. These $2n$ variables form a $2n$-dimensional space that is called a *phase space*. The system at any instant can be described as one point in such a space. The time evolution of the system, that is, the time history of all the particles in the system, traces one line in such a $2n$-dimensional *phase space*, which we will call the flow line as in fluid mechanics.

Now we consider an ensemble of systems—a collection of many systems satisfying the same macroscopic constraints—as we did in chapter 4. At time $t = 0$, each system in the ensemble is represented by a different point in the phase space, as shown in figure 6.1. From classical mechanics, we know that with a given initial condition the trajectory of the system is uniquely determined. Since the initial condition for each system differs from that of other systems in the ensemble, the traces of systems in

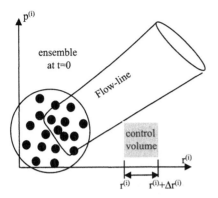

Figure 6.1 Phase space, and an ensemble in the phase space.

such an ensemble never intersect, so that the flow lines in phase space do not intersect each other.

The number of systems in an ensemble is usually very large, much larger than the number of the particles in one system. Because of the large number of systems in one ensemble, we can treat the points of the ensemble, each representing one microstate of the original macroscopic system, as forming a continuum in the phase space, just as we treat atoms or molecules in a macroscopic system as a continuous medium in real space. We define a particle density $f^{(N)}$ such that, surrounding any point $(\mathbf{r}^{(n)}, \mathbf{p}^{(n)})$ in the phase space, where $\mathbf{r}^{(n)} = (\mathbf{r}_1, \mathbf{r}_2, \ldots, \mathbf{r}_N) = (r^{(1)}, r^{(2)}, r^{(3)}, \ldots, r^{(n)})$ includes all the space coordinates of N particles and similarly $\mathbf{p}^{(n)}$ represents all the momentum coordinates, the number of systems is

$$\text{No. of systems} = f^{(N)}(t, \mathbf{r}^{(n)}, \mathbf{p}^{(n)}) \Delta \mathbf{r}^{(n)} \Delta \mathbf{p}^{(n)} \tag{6.1}$$

in a small volume of the phase space, $\Delta \mathbf{r}^{(n)} \Delta \mathbf{p}^{(n)}$, where $\Delta \mathbf{r}^{(n)} = \Delta \mathbf{r}_1 \Delta \mathbf{r}_2 \ldots \Delta r_N = \Delta r^{(1)} \Delta r^{(2)} \ldots \Delta r^{(n)}$ and $\Delta \mathbf{p}^{(n)} = \Delta \mathbf{p}_1 \Delta \mathbf{p}_2 \ldots \Delta \mathbf{p}_N = \Delta p^{(1)} \Delta p^{(2)} \ldots \Delta p^{(n)}$. We use superscript (n) to denote the generalized space and momentum coordinates, and superscript (N) to represent the N particles. The particles density in the phase space $f^{(N)}(t, \mathbf{p}^{(n)}, \mathbf{p}^{(n)})$ is called the N-particle distribution function, which represents the probability density of finding a particular system at a specific state defined by $\mathbf{r}^{(n)}$ and $\mathbf{p}^{(n)}$. If we assume that the ensemble is ergodic for all time, this distribution function also represents the probability of observing one system at a particular state $\mathbf{r}^{(n)}$ and $\mathbf{p}^{(n)}$ over a period of time (such a time period should be smaller than the characteristic time we use in tracing the trajectory, or the relaxation time that we will discuss later).

The time evolution of $f^{(N)}(t, \mathbf{r}^{(n)}, \mathbf{p}^{(n)})$ in the phase space is governed by the *Liouville equation*, which can be derived on the basis that the flow lines of systems in the ensemble do not intersect. Consider a tube formed by the traces of a set of points (a subset of systems in the ensemble) as shown in figure 6.1. Since the flow lines do not intersect, the points in the phase space are conserved. We want to derive an equation for the distribution function $f^{(N)}$ based on this conservation requirement. Recall that in fluid mechanics or heat transfer, we often use the control volume method rather than tracing the trajectory of individual fluid particles. We could do the same for the points in phase space and examine a small control volume in phase space, as shown in figure 6.1.

The net rate of points flowing into the control volume should equal the rate of change of points inside the control volume. This leads to

$$-\sum_{i=1}^{n} \frac{\partial}{\partial r^{(i)}}\left(f^{(N)} \frac{\partial r^{(i)}}{\partial t}\right) - \sum_{i=1}^{n} \frac{\partial}{\partial p^{(i)}}\left(f^{(N)} \frac{\partial p^{(i)}}{\partial t}\right) = \frac{\partial f^{(N)}}{\partial t} \quad (6.2)$$

where $\partial p^{(i)}/\partial t = \dot{p}^{(i)}$ and $\partial r^{(i)}/\partial t = \dot{r}^{(i)}$ represents the flow rate of the points. The left-hand side of eq. (6.2) is the net rate of points flowing into the control volume and the right-hand side is the rate of change of the points in the control volume. The above relation can be further written as

$$\frac{\partial f^{(N)}}{\partial t} + \sum_{i=1}^{n} \dot{r}^{(i)} \times \frac{\partial f^{(N)}}{\partial r^{(i)}} + \sum_{i=1}^{n} \dot{p}^{(i)} \times \frac{\partial f^{(N)}}{\partial p^{(i)}} = 0 \quad (6.3)$$

where we have used $\partial \dot{r}^{(i)}/\partial r^{(i)} + \partial \dot{p}^{(i)}/\partial p^{(i)} = 0$, which is a result that we will obtain in chapter 10 on the basis of the Hamilton equations of motion.

Equation (6.3) is the Liouville equation that governs the time evolution of the N-particle distribution function $f^{(N)}$. The equation is valid for all classical systems and has quantum mechanical counterparts for quantum systems (Liboff, 1998). It has a large number of variables since n is of the order of 10^{23}, that is, the Avogadro constant, in macroscale systems. Direct solution of the Liouville equation for nonequilibrium systems is impossible, not only because of the large number of variables, but also because we are hampered by the necessity to determine the exact initial states of the ensemble. However, the Liouville equation provides a good starting point for further simplification. The Boltzmann equation, to be discussed below, is one example. In chapter 10, we will develop another approach for transport problems, the linear response theory, based on the perturbation analysis of the Liouville equation.

6.1.2 The Boltzmann Equation

The Liouville equation involves $2n$ variables in the phase space, plus time. This large number of variables makes it impractical in terms of the boundary and initial conditions, as well as for analytical and numerical solutions. One way to simplify the Liouville equation is to consider one particle in a system. This is a representative particle having coordinate $\mathbf{r}_1$ and momentum $\mathbf{p}_1$; each of the vectors has m components, that is, m degrees of freedom. We introduce a one-*particle distribution function* by averaging the N-particle distribution function over the rest of the $(N-1)$ particles in the system,

$$f^{(1)}(t, \mathbf{r}_1, \mathbf{p}_1) = \frac{N!}{(N-1)!} \int \cdots \int f^{(N)}(t, \mathbf{r}^{(n)}, \mathbf{p}^{(n)}) d\mathbf{r}_2 \ldots d\mathbf{r}_N d\mathbf{p}_2 \ldots d\mathbf{p}_N \quad (6.4)$$

where, again, each vector $\mathbf{r}_i$ and $\mathbf{p}_i$ has m degrees of generalized freedom, so that $n = m \times N$, as we discussed before. For a monatomic atom, $m = 3$, and for a diatomic atom, $m = 6$ (neglecting the electronic states). The factorials are normalization factors. For simplicity in notation, we will drop the subscript 1 and use $(\mathbf{r}, \mathbf{p})$ as the coordinates and momenta of the particle. Since $f^{(N)}(t, \mathbf{r}^{(n)}, \mathbf{p}^{(n)})$ represents the number density of systems having generalized coordinates $(\mathbf{r}^{(n)}, \mathbf{p}^{(n)})$ in the ensemble, the one-particle distribution function represents the number density of systems having $(\mathbf{r}, \mathbf{p})$,

$$f(t, \mathbf{r}, \mathbf{p}) d^3\mathbf{r} d^3\mathbf{p} = \text{number of systems in } d^3\mathbf{r} d^3\mathbf{p}$$

This one-particle distribution function features a significant reduction of variables. For one mole of monatomic gas with 6×10^{23} atoms, the number of variables in the generalized phase space is $6 \times 6 \times 10^{23}$, because of the three space and three momentum coordinates. The one-particle phase space for monatomic atoms, however, has only three space coordinates and three momentum coordinates. The one-particle phase space can be thought of as the projection of the N-particle phase space, similarly to the projection of a volume in three-dimensional space into the area of a two-dimensional space.

With the introduction of the averaging method to obtain the one-particle distribution function, one can start from the Liouville equation, eq. (6.3), and carry out the averaging over the space and momentum coordinates of the other $(N-1)$ particles. This procedure leads to (Liboff, 1998)

$$\frac{\partial f}{\partial t} + \frac{d\mathbf{r}}{dt} \bullet \nabla_\mathbf{r} f + \frac{d\mathbf{p}}{dt} \bullet \nabla_\mathbf{p} f = \left(\frac{\partial f}{\partial t}\right)_c \quad (6.5)$$

where the subscripts ($\mathbf{r}$ and $\mathbf{p}$) in the gradient operators represent the variables of the gradients:

$$\nabla_\mathbf{r} f = \frac{\partial f}{\partial x}\hat{\mathbf{x}} + \frac{\partial f}{\partial y}\hat{\mathbf{y}} + \frac{\partial f}{\partial z}\hat{\mathbf{z}}$$

$$\nabla_\mathbf{p} f = \frac{\partial f}{\partial p_x}\hat{\mathbf{p}}_x + \frac{\partial f}{\partial p_y}\hat{\mathbf{p}}_y + \frac{\partial f}{\partial p_z}\hat{\mathbf{p}}_z$$

Unlike the $2n$-phase space for the derivation of the Liouville equation, in which one point represents a system and the flow lines of the points do not intersect, the particle as represented by the one-particle distribution function interacts with other particles in the system, and thus the number of particles along a flow line in the one-particle phase space is no longer conserved. The right-hand side of eq. (6.5) lumps the interaction of this one particle with the rest of the particles in the system and represents the nonconserving nature of the one-particle distribution function. This scattering term should not be considered as a derivative, but rather as a symbol representing the net rate of gaining particles at point $(\mathbf{r}, \mathbf{p})$. We will give more detailed expressions for the scattering term in section 6.2.

Equation (6.5), together with the expressions to be given in section 6.2 for $\left(\frac{\partial f}{\partial t}\right)_c$, is called the *Boltzmann equation* or Boltzmann transport equation. Rather than using momentum $\mathbf{p}$, we can also use velocity $\mathbf{v}(\mathbf{p} = m\mathbf{v})$ or wavevector $\mathbf{k}(\mathbf{p} = \hbar \mathbf{k})$ to rewrite the Boltzmann equation as

$$\frac{\partial f}{\partial t} + \mathbf{v} \bullet \nabla_\mathbf{r} f + \frac{\mathbf{F}}{m} \bullet \nabla_\mathbf{v} f = \left(\frac{\partial f}{\partial t}\right)_c \quad (6.6)$$

$$\frac{\partial f}{\partial t} + \mathbf{v} \bullet \nabla_\mathbf{r} f + \frac{\mathbf{F}}{\hbar} \bullet \nabla_\mathbf{k} f = \left(\frac{\partial f}{\partial t}\right)_c \quad (6.7)$$

where $\mathbf{F} = d\mathbf{p}/dt$ is the external force acting on the particle. The use of $\hbar$ in eq. (6.7) implies that the Boltzmann equation can also be applied to quantum particles as long as we do not consider the phase coherence of the particles. In connection with the discussion in the previous chapter on the group velocity and the crystal momentum, it should be understood that $\mathbf{v}$ in these forms of the Boltzmann equation is the group velocity, while $\hbar \mathbf{k}$ in eq. (6.7) is the crystal momentum. In the following treatment, we will use $\mathbf{p}$, $\mathbf{v}$, and $\mathbf{k}$ as momentum variables interchangeably. Variables $\mathbf{v}$ or $\mathbf{p}$ are often used as continuum variables when treating gases, and variables $\mathbf{k}$ are often used for treating electron and phonon transport in crystals. Please be reminded that $\mathbf{v} = \nabla_{\mathbf{k}}\omega$ is the group velocity.

The Boltzmann equation gives the distribution function at $\mathbf{r}$ and $\mathbf{p}$ (or $\mathbf{k}$ or $\mathbf{v}$) in the phase space. Typically, we are interested in the quantities averaged over $\mathbf{p}$, for example, the average velocity and energy. If the solution of the Boltzmann equation is known for a problem, we can calculate the volume-average of any microscopic property X of the particle from

$$\langle X(\mathbf{r}) \rangle = \frac{1}{V} \sum_{\mathbf{k},s} X(\mathbf{r},\mathbf{k}) f = \frac{1}{(2\pi)^3} \int X(\mathbf{r},\mathbf{k}) f d^3\mathbf{k} \quad (6.8)$$

where s is the polarization, if appropriate. When going from the summation over discrete wave vector $\mathbf{k}$ to the integration, the factor $(2\pi)^3$ comes from the fact that the volume of one quantum state in k-space is $V/(2\pi)^3$.

There are two directions in exploring the solution of the Boltzmann equation. One is to solve for f and to calculate the average quantities of interest according to eq. (6.8). This approach will be used in section 6.3 to derive constitutive equations such as the Fourier law and the Ohm law, and the Newton shear stress law. It will also be used in the next chapter when considering the classical size effects. The other approach is to take the moments of the Boltzmann equation, from which conservation equations such as the Navier–Stokes equations can be derived.

The above arguments leading to the Boltzmann equation are by no means a rigorous derivation. The derivation of the Boltzmann equation from the Liouville equation is a fundamental topic in statistical physics (Kubo et al., 1991; Liboff, 1998). Here it is appropriate to comment on the range of validity of the Boltzmann equation. We use the one-particle distribution instead of the N-particle distribution function and assume that this one-particle distribution function is an appropriate representation of all the particles in the system. This will only be true if the N-particle distribution function can be factorized as the product of the distribution function for each particle, that is,

$$f^{(N)}(t, \mathbf{r}^{(n)}, \mathbf{p}^{(n)}) = f^{(1)}(t, \mathbf{r}_1, \mathbf{p}_1) f^{(1)}(t, \mathbf{r}_2, \mathbf{p}_2) \ldots f^{(1)}(t, \mathbf{r}_N, \mathbf{p}_N) \quad (6.9)$$

Such a factorization means that the particles in the system are quite independent of each other, even though collisions between particles can affect the one-particle distribution function. Consider the collision of two particles. Before and after the collision, the distribution functions of one particle are independent of the coordinates and momentum of the other particle. This is the so-called *molecular chaos* assumption. Such factorization is only valid when the interactions of the particles are infrequent. Thus, the Boltzmann equation is appropriate only for dilute systems such as molecular gases, electron gases, phonon gases, and photon gases. It is not valid for dense fluids such as liquids. The

Boltzmann equation does not include explicitly wave effects such as interference and tunneling. Extension of the classical particle picture to the quantum wave picture involves the so-called Wigner function (Liboff, 1998), which we will not discuss here. Despite these restrictions, the Boltzmann equation is powerful and can be applied to a wide range of problems from nanoscale to macroscale.

6.1.3 Intensity for Energy Flow

The single-particle distribution function f is a scalar in the one-particle phase space. Sometimes we try to map this and related quantities into the real space $\mathbf{r}$. At each point in real space, the possible wavevectors lie in all directions. Along each wavevector direction, the particle moves at the group velocity $\mathbf{v}_g(\mathbf{k})$ and the energy flows at the rate of $E \times \mathbf{v}_g(\mathbf{k}) \times f$ for the specific quantum state. In section 3.4.4, we introduced the differential density-of-states $dD(E, \mathbf{k})$ as

$$dD(E, \mathbf{k}) = \frac{\text{No. of States within } (E, E + dE) \text{ and } d\Omega}{V \, dE \, d\Omega} = \frac{D(E)}{4\pi} \quad (6.10)$$

where the last equality is valid only for isotropic media. The rate of energy propagating along this direction per unit solid angle is then

$$I(t, E, \mathbf{k}) = E \times v_g(\mathbf{k}) f(t, \mathbf{r}, \mathbf{k}) dD(E, \mathbf{k})$$
$$= \frac{1}{4\pi} E \times v_g(\mathbf{k}) f(t, \mathbf{r}, \mathbf{k}) D(E) \quad (6.11)$$

In thermal radiation, $E = h\nu$ for photons, and I is called the intensity. Majumdar (1993) extended the intensity concept to phonons. Equation (6.11) shows that intensity is a simple transformation of the distribution function. It is usually defined, without referring to the phase space, as the power flowing along direction $\hat{\Omega}$ per unit solid angle, per unit frequency interval, and per unit area normal to the direction of propagation:

$$I_\nu = \frac{\text{Power}}{dA_\perp \, d\Omega \, d\nu} \quad (6.12)$$

where $dA_\perp$ is a differential area perpendicular to the direction of propagation.

Comparing eq. (6.12) with (6.11), we see that the solid angle, which is usually considered as an angle in real space, is actually sustained by the wavevectors in the phase space. In phase space, intensity is a scalar. Without considering the phase space, it is difficult to tell whether intensity is a scalar or a vector. Although the concept of intensity is widely used in thermal radiation, it is not very common in the treatment of charge transport since the major concern is not energy but the flux of charges. In the following treatment, we will use both f and the intensity, in accordance with the customs in each field, while attempting to present different carriers in a parallel fashion.

6.2 Carrier Scattering

The key to the Boltzmann equation lies in the description of the scattering term. This term, in its most general form, is a complex multi-variable integral that contains the

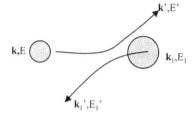

Figure 6.2 Collision of two particles with initial wavevector and energy (**k**, E) and (**k**$_1$, E_1). After the collision, the two particles are at states (**k**$'$, E') and (**k**$'_1$, E'_1).

distribution function, making the Boltzmann equation an integro-differential equation that is extremely difficult to solve. The relaxation time approximation is often made to simplify the scattering term. In this section, we will first give the general expression for the scattering integral and then introduce the relaxation time approximation, followed by a more detailed discussion of the scattering mechanisms of various carriers.

6.2.1 Scattering Integral and Relaxation Time Approximation

We consider the collision process between two particles as shown in figure 6.2. After the collision, the energy and the velocity of each particle may change. Clearly, the collision is a time-dependent process. The rigorous way of dealing with the collision process is to solve the corresponding time-dependent Schrödinger equation for the combined system of both particles. This approach is, however, usually very complicated and not practical. A simpler way to treat the collision is to use the perturbation method (Landau and Lifshitz, 1977). This method considers the time-dependent interaction between the two particles as a small perturbation in energy, $H'(\mathbf{r}, t)$, from the original steady-state, non-interacting energy H_0 of the two particles, such that the total system energy is

$$H = H_0(\mathbf{r}) + H'(\mathbf{r}, t)$$

For example, when we use the harmonic oscillator approximation for the actual interatomic potentials, the higher order term $O[(x' - x_0)^3]$ in eq. (2.51) can be considered as the perturbation from the harmonic potential. In quantum mechanics, we must treat H as an operator and solve the Schrödinger equation for the two-particle system with the new H, the Hamiltonian of the system, as in eq. (2.22). By treating H' as a small perturbation to the unperturbed Hamiltonian H_0, the solution of the Schrödinger equation for the new H can be obtained through the perturbation method and expressed in terms of the wave functions Ψ of the unperturbed two-particle system with Hamiltonian H_0. Using the perturbation solution, one can calculate the probability for the system jumping from one quantum state Ψ_i to another quantum state Ψ_f, both being accessible quantum states of the original two-particle system. The rate of this transition probability is the transition rate and is given by

$$\begin{aligned}
W_i^f &= \frac{2\pi}{\hbar} \left| \int \Psi_f^* H' \Psi_i d^3\mathbf{r} \right|^2 \delta(E_f - E_i) \\
&= \frac{2\pi}{\hbar} |\langle i | \mathrm{H'} | f \rangle|^2 \delta(E_f - E_i) \\
&= \frac{2\pi}{\hbar} M_{if}^2 \delta(E_f - E_i)
\end{aligned} \qquad (6.13)$$

where $d^3\mathbf{r} = dx\,dy\,dz$ means integration over the whole volume of the system and

$$M_{if} \equiv \langle i|H'|f\rangle \equiv \int \Psi_f^* H' \Psi_i d^3\mathbf{r} \tag{6.14}$$

is called the scattering matrix. The delta function, $\delta(E_f - E_i)$, defined as

$$\delta(E_f - E_i) = \begin{cases} 1 & E_i = E_f \\ 0 & E_i \neq E_f \end{cases} \quad \text{and} \quad \int_{-\infty}^{\infty} \delta(x)dx = 1 \tag{6.15}$$

is a manifestation of the requirements of the conservation of energy. According to eq. (6.15), $\delta(E_f - E_i)$ has a unit of J^{-1}. Equation (6.13) is often-referred to the *Fermi golden rule*. It should be kept in mind that E_i and E_f are the initial and final total energies of the two-particle system.

The Fermi golden rule gives the transition rate from one set of quantum states of the two particles into another set due to the scattering. The scattering term in the Boltzmann equation is the net gain of particles in one quantum state. This net gain consists of two components: one is the increase in the number of particles due to scattering from other quantum states into the quantum state under consideration ("in-scattering"); the other is the decrease of the number of particles due to scattering from the current quantum state to other quantum states ("out-scattering"). We again take the two-particle scattering process as an example. The initial wavevector of one particle is $\mathbf{k}$ and it collides with another particle with a wavevector $\mathbf{k}_1$. The corresponding distribution functions for the two particles are $f(t, \mathbf{r}, \mathbf{k})$ and $f(t, \mathbf{r}_1, \mathbf{k}_1)$. After scattering, the momenta of the two particles are $\mathbf{k}'$ and $\mathbf{k}_1'$ and their distribution functions are $f(t, \mathbf{r}', \mathbf{k}')$ and $f(t, \mathbf{r}_1', \mathbf{k}_1')$, respectively. The scattering term for the particle at state $\mathbf{k}$ can be expressed as

$$\begin{aligned}\left(\frac{\partial f}{\partial t}\right)_c &= -\sum_{\mathbf{k}_1,\mathbf{k}',\mathbf{k}_1'} f(t,\mathbf{r},\mathbf{k})f(t,\mathbf{r},\mathbf{k}_1)W(\mathbf{k},\mathbf{k}_1 \to \mathbf{k}',\mathbf{k}_1') \\ &\quad + \sum_{\mathbf{k}_1,\mathbf{k}',\mathbf{k}_1'} f(t,\mathbf{r},\mathbf{k}')f(t,\mathbf{r},\mathbf{k}_1')W(\mathbf{k}',\mathbf{k}_1' \to \mathbf{k},\mathbf{k}_1) \\ &= -K\int f(t,\mathbf{r},\mathbf{k})f(t,\mathbf{r},\mathbf{k}_1)W(\mathbf{k},\mathbf{k}_1 \to \mathbf{k}',\mathbf{k}_1')d^3\mathbf{k}_1 d^3\mathbf{k}' d^3\mathbf{k}_1' \\ &\quad + K\int f(t,\mathbf{r},\mathbf{k}')f(t,\mathbf{r},\mathbf{k}_1')W(\mathbf{k}',\mathbf{k}_1' \to \mathbf{k},\mathbf{k}_1)d^3\mathbf{k}_1 d^3\mathbf{k}' d^3\mathbf{k}_1' \end{aligned} \tag{6.16}$$

where $K = V^3/(2\pi)^9$ is a factor that converts the summation over wavevector into integration over the phase space. The first term represents the rate of particles being scattered out of quantum states determined by $\mathbf{k}$ and $\mathbf{k}_1$, and the second term represents the rate of particles scattered into the quantum state. We have used the same $\mathbf{r}$, assuming that at the point of scattering all the particles are at the same location. This means that the particles do not have a finite volume. The integration must be done over all other possible particles in the initial states $\mathbf{k}_1$ and counts all possibilities of the final states $\mathbf{k}'$ and $\mathbf{k}_1'$. For a particle with only translational motion, as we will assume from here on, $d^3\mathbf{k} = dk_x dk_y dk_z$. Equation (6.16) thus contains a nesting of nine integrals.

However, the conservation of energy and momentum, which is included in the transition probability, eq. (6.13), and the scattering matrix

$$E(\mathbf{k}) + E(\mathbf{k}_1) = E(\mathbf{k}') + E(\mathbf{k}'_1) \tag{6.17}$$

$$\mathbf{k} + \mathbf{k}_1 = \mathbf{k}' + \mathbf{k}'_1 \tag{6.18}$$

lead to a reduction in the number of integrals. Also, the following reciprocity relation

$$W(\mathbf{k}, \mathbf{k}_1 \rightarrow \mathbf{k}', \mathbf{k}'_1) = W(\mathbf{k}', \mathbf{k}'_1 \rightarrow \mathbf{k}, \mathbf{k}_1) \tag{6.19}$$

arising from the principle of detailed balance, is valid and can be used to write eq. (6.16) as

$$\left(\frac{\partial f}{\partial t}\right)_c = -K \int W \times [f(t, \mathbf{r}, \mathbf{k}) f(t, \mathbf{r}, \mathbf{k}_1) \\ - f(t, \mathbf{r}, \mathbf{k}') f(t, \mathbf{r}, \mathbf{k}'_1)] d^3 \mathbf{k}_1 \, d^3 \mathbf{k}' d^3 \mathbf{k}'_1 \tag{6.20}$$

Combining eqs. (6.20) and (6.7), we see that the *Boltzmann equation*

$$\frac{\partial f}{\partial t} + \mathbf{v} \bullet \nabla_\mathbf{r} f + \frac{\mathbf{F}}{\hbar} \bullet \nabla_\mathbf{k} f \\ = -K \int W \times [f(t, \mathbf{r}, \mathbf{k}) f(t, \mathbf{r}, \mathbf{k}_1) \\ - f(t, \mathbf{r}, \mathbf{k}') f(t, \mathbf{r}, \mathbf{k}'_1)] d^3 \mathbf{k}_1 d^3 \mathbf{k}' d^3 \mathbf{k}'_1 \tag{6.21}$$

is a complicated integral-differential equation with seven variables $(t, \mathbf{r}, \mathbf{k})$, due to our assumption of translational motion only.

The integral-differential Boltzman equation, eq. (6.21), is very difficult to solve in general. Most solutions rely on a drastic simplification of the scattering integral by the relaxation time approximation

$$\left(\frac{\partial f}{\partial t}\right)_c = -\frac{f - f_0(T, E, \mu)}{\tau(\mathbf{r}, \mathbf{k})} \tag{6.22}$$

where $\tau(\mathbf{r}, \mathbf{k})$ is the *relaxation time*, and f_0 represents the equilibrium distribution of the carriers, such as the Boltzmann, the Fermi–Dirac, and the Bose–Einstein distributions given in chapter 4. The relaxation time approxi-mation is also called the BGK approximation in gas dynamics in honor of the joint work of Bhatnagar, Gross, and Krook (1954). In chapter 8, we will go through the scattering integrals more carefully for the case of electron–phonon scattering and show that the approximation is valid only for elastic scattering. Despite this limitation, the relaxation time approximation is actually used widely, even for processes including inelastic scattering, with correct end results for most situations.

We can understand the meaning of τ easily by neglecting the spatial nonuniformity of the distribution function. Equation (6.7) becomes

$$\frac{\partial f}{\partial t} = -\frac{f - f_0}{\tau} \tag{6.23}$$

and thus

$$f - f_0 = Ce^{-t/\tau} \tag{6.24}$$

So the relaxation time is a measure of how long it takes for a nonequilibrium system to relax back to an equilibrium distribution. Often, the relaxation time is expressed in terms of the energy rather than the velocity, $\tau = \tau(E)$, which implies isotropic scattering.

The scattering may be caused by the coexistence of different processes and a relaxation time can be defined for each process. The total relaxation time, τ_t, can be calculated from individual relaxation times, τ_j, according to the *Matthiessen rule*,

$$\frac{1}{\tau_t} = \sum_j \frac{1}{\tau_j} \tag{6.25}$$

The Matthiessen rule assumes that the scattering mechanisms are independent of each other (Ashcroft and Mermin, 1976).

Under the relaxation time approximation, the Boltzmann equation becomes

$$\frac{\partial f}{\partial t} + \mathbf{v} \bullet \nabla_\mathbf{r} f + \frac{\mathbf{F}}{m} \bullet \nabla_\mathbf{v} f = -\frac{f - f_0}{\tau} \tag{6.26}$$

where we have used $\mathbf{v}$ rather than $\mathbf{k}$ as the variable. The corresponding equation using $\mathbf{k}$ as the variable is, from eq. (6.7),

$$\frac{\partial f}{\partial t} + \mathbf{v} \bullet \nabla_\mathbf{r} f + \frac{\mathbf{F}}{\hbar} \bullet \nabla_\mathbf{k} f = -\frac{f - f_0}{\tau} \tag{6.27}$$

Equations (6.26) and (6.27) are also called the *Krook equation* in gas dynamics (Chapman and Cowling, 1970). In the rest of this section, we will discuss in greater detail the scattering mechanisms and the relaxation time of various energy carriers.

6.2.2 Scattering of Phonons

The derivation of the phonon modes in chapter 3 is based on the assumption of harmonic interatomic potential. Under this assumption, the lattice waves are decomposed into normal modes which do not interact with each other. For such an ideal case, there is no resistance to heat flow and the thermal conductivity is infinite. In contrast, real crystals clearly have a finite thermal conductivity, which is caused by the scattering of phonons.

In a pure dielectric crystal, the phonon scattering is primarily due to the scattering of phonons among themselves. Anharmonic force interaction is the source of scattering among phonons. The second-order term in the Taylor expansion of the interatomic potential around the equilibrium point, as in eq. (2.51), gives the harmonic oscillator model that we used to represent phonons. By considering the third-order term in the potential as a perturbation to the original Hamiltonian, $H' \sim x^3$, and through the use of the Fermi golden rule, it is found that this anharmonic force term acts as a mechanism for two phonons to merge into a third phonon or for one phonon to split into two phonons, as shown in figure 6.3 (Ziman, 1960). Such scattering processes are called three-phonon scattering. The two-particle collision picture shown in figure 6.2 must now be modified

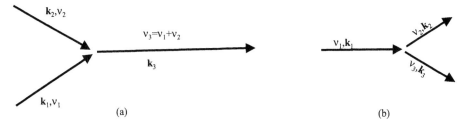

Figure 6.3 Three-phonon scattering processes: (a) two phonons merge into one (annihilation process); (b) one phonon splits into two (creation process).

in accordance with figure 6.3, with corresponding changes to the general scattering integral expressed by eq. (6.20) and the energy and momentum conservation rules expressed by eqs. (6.17) and (6.18). For the merging of two phonons into one, energy conservation gives

$$h\nu_1 + h\nu_2 = h\nu_3 \tag{6.28}$$

and a similar equation can be written for the process in which one phonon splits into two.

Momentum conservation during the three-phonon interaction processes takes a special form. For the phonon merging process, the momentum conservation can be written as

$$\mathbf{k}_1 + \mathbf{k}_2 - \mathbf{k}_3 = \mathbf{G} \tag{6.29}$$

where the reciprocal lattice vector $\mathbf{G}$ can be zero or a linear combination of the reciprocal lattice vectors. If $(\mathbf{k}_1 + \mathbf{k}_2)$ falls within the first Brillouin zone wavevector, then $\mathbf{G} = \mathbf{0}$; otherwise, $\mathbf{G} \neq \mathbf{0}$ (figure 6.4). The latter result comes from the requirement that the phonon wavelength cannot be smaller than the lattice constant, as discussed in chapter 3. The $\mathbf{G} = \mathbf{0}$ phonon scattering process is called the *normal process* and the $\mathbf{G} \neq \mathbf{0}$ is the *umklapp process*. Without the umklapp process, the thermal conductivity of a crystal would still be infinite because in a normal scattering process, the generated third phonon preserves both the energy and the direction of the two original phonons. The extra reciprocal lattice wavevector in the umklapp process changes the net direction of phonon propagation and thus creates resistance to the heat flow (Peierls, 1929).

The evaluation of the scattering integral for phonons is very difficult (Ziman, 1960). Instead, approximate expressions for the relaxation time, based on eq. (6.20), have been developed (Klemens, 1958). For example, an often-used expression for the three-phonon umklapp process is

$$\tau_u^{-1} = Be^{-\theta_D/bT}T^3\omega^2 \tag{6.30}$$

where B and b are constants and θ_D is the Debye temperature. The values of B and b for different materials can be obtained by matching the model predictions for thermal conductivity with experimental results, as we will see later.

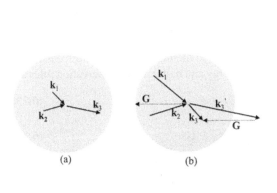

Figure 6.4 Three-phonon (a) normal and (b) umklapp scattering processes, using the merging of two phonons into one as an example. The gray region represents the first Brillouin zone. In the umklapp process, $\mathbf{k}'_3$, which is the sum of $\mathbf{k}_1$ and $\mathbf{k}_2$, is outside the first Brillouin zone. It is brought back into the first Brillouin zone by the reciprocal lattice wavevector $\mathbf{G}$. Energy must be conserved during both scattering processes. The normal processes in (a) do not create thermal resistance because the merged phonon carries the same amount of energy and momentum as the original two phonons. Umklapp scattering causes thermal resistance because the momenta of the two original phonons are changed after scattering.

In real crystals, there are defects such as impurities, dislocations, and grain boundaries. These can all scatter phonons; details can be found in the given references (Ziman, 1960; Klemens, 1958). For impurities, the scattering obeys the familiar Rayleigh law

$$\tau_I^{-1} = A\omega^4 \tag{6.31}$$

Boundary scattering is also sometimes included in the total relaxation time, using the Matthiessen rule. The relaxation time due to boundary scattering is of the order of

$$\tau_b^{-1} = b_s v/L \tag{6.32}$$

where L is a characteristic length, such as the diameter of a circular rod for heat conduction along the axial direction, and b_s is a shape factor that can be modeled similarly to the radiation shape factor (Berman et al., 1955).

The total relaxation time is obtained by combining the expressions for individual relaxation processes according to the Matthiessen rule. We should point out, however, that incorporating the boundary scattering with the Matthiessen rule is questionable because the boundary scattering is a surface process whereas the phonon–phonon and phonon–impurity scattering occur inside the volume. In chapter 7, we will consider many size effects by imposing interfaces and boundaries as boundary conditions of the Boltzmann equation rather than being based on the Matthiessen rule. Although τ_t can include several potential mechanisms, typically, in a certain temperature range, there is a dominant scattering mechanism. We will come back to this point when discussing thermal conductivity.

Phonons cannot be acted upon by external force. Under the relaxation time approximation, the force term in eq. (6.27) drops out in the phonon Boltzmann equation.

6.2.3 Scattering of Electrons

Electrons are predominantly scattered by phonons. Electron–electron scattering is typically much weaker. An electron can create or annihilate a phonon in the scattering process, and must obey the energy conservation and momentum conservation during the process. For a phonon creation process, the energy and momentum conservation rules are

$$E_i = E_f + h\nu_p \tag{6.33}$$

$$\mathbf{k}_i = \mathbf{k}_f + \mathbf{k}_p + \mathbf{G} \tag{6.34}$$

where $h\nu_p$ and $\mathbf{k}_p$ are the energy and wavevector of the created phonon, respectively. Again, the process can be a normal or an umklapp one, depending on whether $\mathbf{G} = \mathbf{0}$ or not. In most cases, the dominant scattering process has $\mathbf{G} = \mathbf{0}$.

In metals, at temperatures higher than the Debye temperature, the number of phonons is proportional to temperature T, as the temperature independent specific heat suggests. The more phonons, the more chance that the electron will experience scattering by phonons; consequently, the electron–phonon relaxation time is (Ashcroft and Mermin, 1976)

$$\frac{1}{\tau} \propto T \tag{6.35}$$

Scattering in semiconductors is more complicated and one must determine whether the scattering is caused by acoustic or optical phonons. The optical phonons can be further divided into nonpolar, such as in silicon and germanium, or polar, such as in gallium arsenide (GaAs). In chapter 8, we will discuss in more detail the electron–phonon scattering in relation to energy exchange mechanisms. Impurity scattering in semiconductors is also a very important mechanism. Refer to Lundstrom (2000) and Hess (2000) for a more detailed discussion of various electron scattering mechanisms.

6.2.4 Scattering of Photons

Photon scattering is often divided into two parts: the inelastic and the elastic processes. In the inelastic process, photons are absorbed or emitted. The absorption coefficient is often used to represent the process. Under the relaxation time approximation, inelastic scattering that includes absorption and emission of photons can be expressed as

$$\left(\frac{\partial f}{\partial t}\right)_{c,\text{inelastic}} = -\frac{f - f_0}{\tau} \tag{6.36}$$

and the steady-state Boltzmann equation becomes

$$v\frac{\partial f}{\partial s} = -\frac{f - f_0}{\tau} + \left(\frac{\partial f}{\partial t}\right)_{c,\text{elastic}} \tag{6.37}$$

where we have used

$$\mathbf{v} \bullet \nabla_{\mathbf{r}} f = v \frac{\partial f}{\partial s}$$

s is the distance along the direction of propagation and v is the speed of light in the medium. The term $\nabla_{\mathbf{k}} f$ in the Boltzmann equation for photons drops out because photons do not interact with external force. The elastic scattering term can be obtained from solving the Maxwell equations. A familiar example is the Mie scattering theory, which represents the full solution of the Maxwell equations for a plane wave interacting with a spherical particle with given optical constants (Bohren and Huffman, 1983).

As mentioned before, for thermal radiation it is customary to use intensity rather than the distribution function. Using the intensity notation, eq. (6.37) can be written as

$$\frac{\partial I_\nu}{\partial s} = -\frac{I_\nu - I_{\nu 0}}{\Lambda} + \frac{1}{v}\left(\frac{\partial I_\nu}{\partial t}\right)_{c,\text{elastic}} \tag{6.38}$$

where $\Lambda = v\tau$ is the photon inelastic scattering mean free path, and I_0 is the blackbody radiation intensity, as we have proven in chapter 4. Here we have added the subscript ν to denote that the quantities are frequency dependent. Using the terminology that is more frequently used in thermal radiation, the absorption coefficient is the inverse of inelastic scattering mean free path,

$$\alpha_\nu = \frac{1}{\Lambda} = \frac{1}{v\tau} = \frac{4\pi\kappa}{\lambda_0} \tag{6.39}$$

where the last equality is the expression we introduced in eq. (5.40) and is valid only for a homogeneous medium with κ as the imaginary part of the complex refractive index. For other systems such as a system with particulates, the absorption coefficient can be obtained from solving the Maxwell equations (Bohren and Huffman, 1983; Siegel and Howell, 1992). The elastic scattering term is also divided into two parts: the outgoing scattering, which is proportional to the scattering coefficient, and the incoming scattering,

$$\frac{1}{v}\left(\frac{\partial I_\nu}{\partial t}\right)_{c,\text{elastic}} = -\sigma_{s\nu} I_\nu + \frac{1}{v}\left(\frac{\partial I_\nu}{\partial t}\right)_{c,\text{elastic, in}} \tag{6.40}$$

where $\sigma_{s\nu}$ is the scattering coefficient. The incoming scattering is often expressed as

$$\frac{1}{v}\left(\frac{\partial I_\nu}{\partial t}\right)_{c,\text{elastic,in}} = \frac{\sigma_{s\nu}}{4\pi} \int I'_\nu(\hat{\Omega}')\phi(\hat{\Omega}' \to \Omega) d\Omega' \tag{6.41}$$

where ϕ is called the scattering phase function, representing the fraction of photons scattered from direction Ω' to Ω per unit solid angle of the incident direction. The integration in eq. (6.41) is thus the total radiation scattered into the Ω direction. The final equation, which is called the equation of radiative transfer, becomes (Siegel and Howell, 1992)

$$\frac{\partial I_\nu}{\partial s} = -K_{e\nu} I_\nu + \alpha_\nu I_{\nu 0} + \frac{\sigma_{s\nu}}{4\pi} \int I'_\nu(\hat{\Omega}')\phi(\hat{\Omega}' \to \Omega) d\Omega' \tag{6.42}$$

where the extinction coefficient, $K_{e\nu} = \alpha_\nu + \sigma_{s\nu}$, combines absorption and outgoing scattering into an extinction term.

Although the equation of radiative heat transfer, eq. (6.42), looks quite different from the Boltzmann equation, eq. (6.21), the above explanation illustrates that it does come from the Boltzmann equation. This point was exploited by Majumdar (1993), who transformed the phonon Boltzmann equation into a form that is similar to the equation of radiative heat transfer by introducing phonon intensity. We see that the analogy is natural because all of these equations originate from the Boltzmann equation. For electron transport, the intensity concept is not customarily used, although it can be similarly introduced, not in terms of the energy flux but in terms of the particle flux.

6.2.5 Scattering of Molecules

In eq. (1.37), we gave the mean free path between successive collisions of two molecules as

$$\Lambda = \frac{m}{\sqrt{2}\pi\rho d^2} \tag{6.43}$$

where d is the molecule diameter, m is the molecular weight, and ρ is the density. From the mean free path, the relaxation time can be obtained,

$$\frac{1}{\tau} = \frac{\bar{v}}{\Lambda} = \frac{\sqrt{2}\pi\rho\bar{v}d^2}{m} \tag{6.44}$$

where $\bar{v}$ is the average speed of the molecules,

$$\bar{v} = \int_0^\infty \int_0^\infty \int_0^\infty v f_0(v)\, dv_x\, dv_y\, dv_z = \sqrt{\frac{8\kappa_B T}{\pi m}} \tag{6.45}$$

and f_0 is the Maxwell velocity distribution given by eq. (1.26).

6.3 Classical Constitutive Laws

Using the Boltzmann equation under the relaxation time approximation, we can investigate the transport of energy carriers. We will show in this section that classical laws, such as the Fourier law, the Newton shear stress law, the Ohm law, and so on, are special solutions of the Boltzmann equation under the assumption of local thermal equilibrium. The limitations of this assumption can be appreciated through the derivations.

Consider the Boltzmann equation under the relaxation time approximation, that is, eqs. (6.26) and (6.27). Let us introduce a deviation function g,

$$g = f - f_0 \tag{6.46}$$

and write eq. (6.26) as

$$\frac{\partial g}{\partial t} + \frac{\partial f_0}{\partial t} + \mathbf{v} \bullet \nabla_\mathbf{r} f_0 + \mathbf{v} \bullet \nabla_\mathbf{r} g + \frac{\mathbf{F}}{m} \bullet \nabla_\mathbf{v} f_0 + \frac{\mathbf{F}}{m} \bullet \nabla_\mathbf{v} g = -\frac{g}{\tau} \tag{6.47}$$

PARTICLE DESCRIPTION OF TRANSPORT PROCESSES: CLASSICAL LAWS

All the diffusion laws can be obtained under the following assumptions: (1) the transient terms are negligible; (2) the gradient of g is much smaller than the gradient of f_0; and similarly, (3) g is much smaller than f_0. Under these assumptions, eq. (6.47) becomes

$$g = -\tau\left(\mathbf{v}\bullet\nabla_r f_0 + \frac{\mathbf{F}}{m}\bullet\nabla_v f_0\right) \quad (6.48)$$

or

$$f = f_0 - \tau\left(\mathbf{v}\bullet\nabla_r f_0 + \frac{\mathbf{F}}{m}\bullet\nabla_v f_0\right) \quad (6.49)$$

This solution for the distribution function can also be obtained by treating g as the first-order expansion of f (f_0 is the 0th-order expansion) and neglecting higher order terms. The Boltzmann equation thus obtained is said to be the linearized Boltzmann equation. From the distribution function, we can calculate the flux of various quantities of interest (charge, momentum, energy). We will narrow our focus next to examine some of the fluxes associated with various carriers.

6.3.1 Fourier Law and Phonon Thermal Conductivity

We first consider the heat conduction by phonons. In this case, there is no external force. Since the Bose–Einstein distribution

$$f_0 = \frac{1}{\exp(\hbar\omega/k_B T) - 1} \quad (6.50)$$

depends only on temperature, we can write eq. (6.49) as

$$f(\mathbf{r},\mathbf{k}) = f_0 - \tau\frac{df_0}{dT}\mathbf{v}\bullet\nabla T \quad (6.51)$$

where T is a function of coordinate $\mathbf{r}$, i.e., $T(t,\mathbf{r})$. We have dropped the subscript $\mathbf{r}$ in the gradient operator. The nonequilibrium carrier distribution depends on both $\mathbf{r}$ and $\mathbf{v}$.

For simplicity, we consider a temperature gradient along the x-direction without loss of generality. We can calculate the heat flux from

$$J_{qx}(x) = \sum_s\left[\frac{1}{V}\sum_{k_{x1}=-\infty}^{\infty}\sum_{k_{y1}=-\infty}^{\infty}\sum_{k_{z1}=-\infty}^{\infty}v_x\hbar\omega f\right] \quad (6.52)$$

where s represents the summation over all polarizations. It is interesting to compare this expression with eq. (5.152) which we used in deriving the Landauer formalism. In eq. (5.152), we are considering only the heat flux going from point 1 to point 2 but there also exists a reverse heat flux from point 2 to point 1. In eq. (6.52), we are considering the net heat flux at any constant x-plane inside the domain. There are carriers going across the plane in both directions, as determined by eq. (6.51) and sketched in figure 6.5. Following a similar procedure to that used before, we can transform eq. (6.52) first into an integration over all wavevectors and then into an integration over energy and solid angle, using a spherical coordinate system for the wavevectors as shown in figure 6.6.

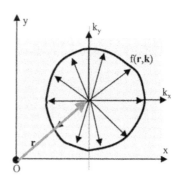

Figure 6.5 In the calculation of heat flux at a constant x plane, carriers moving in all directions may contribute to the net heat flux in the x-direction, depending on the velocity component v_x of the carriers. Note that we align k_x parallel to x, but the k-space origin is located at $\mathbf{r}$.

$$J_{qx}(x) = (1/V) \sum_s \int_{-\infty}^{\infty}\int_{-\infty}^{\infty}\int_{-\infty}^{\infty} v_x \hbar \omega f \, dk_x \, dk_y \, dk_z/(2\pi/L)^3$$

$$= \int_0^{\omega_{\max}} d\omega \left[\int_0^{2\pi} \left\{ \int_0^{\pi} v\cos\theta \hbar\omega f \frac{D(\omega)}{4\pi} \sin\theta \, d\theta \right\} d\varphi \right] \quad (6.53)$$

where $\omega_{\max}$ represents the highest phonon frequency, such as the Debye frequency in the Debye model. Substituting eq. (6.51) into the above equation, we obtain

$$J_{qx}(x)$$
$$= \int_0^{\omega_{\max}} d\omega \left[\int_0^{2\pi} \left\{ \int_0^{\pi} v\cos\theta \hbar\omega \left[f_0 - \tau \frac{df_0}{dT}\frac{dT}{dx} v\cos\theta \right] \frac{D(\omega)}{4\pi} \sin\theta \, d\theta \right\} d\varphi \right]$$
$$= -\frac{1}{2}\frac{dT}{dx} \int_0^{\omega_{\max}} d\omega \left\{ \int_0^{\pi} \tau v^2 \sin\theta \cos^2\theta \times \hbar\omega D(\omega) \frac{df_0}{dT} d\theta \right\} \quad (6.54)$$

We see that the first term f_0 in eq. (6.51) naturally drops out of the integration. This is because f_0 represents the equilibrium distribution and it contributes an equal amount of energy going from left to right as in the reverse direction. Equation (6.54) can be written as the *Fourier law*

$$J_{qx} = -k\frac{dT}{dx} \quad (6.55)$$

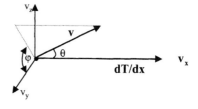

Figure 6.6 Polar coordinates for the momentum components. Note that we align the momentum and space Cartesian components in the same direction.

with the following expression for thermal conductivity

$$k = \frac{1}{2} \int_0^{\omega_{max}} \left\{ \int_0^\pi \tau v^2 C_\omega \sin\theta \times \cos^2\theta \, d\theta \right\} d\omega \quad (6.56)$$

where $C_\omega = \hbar\omega D(\omega)\, df_0/dT$ is the specific heat per unit frequency at frequency ω and temperature T. When v and τ are isotropic, eq. (6.56) can be integrated to

$$k = \frac{1}{3} \int \tau v^2 C_\omega \, d\omega \quad (6.57)$$

In the case that both τ and v are independent of frequency, the above expression reverts to the kinetic relation, eq. (1.35), that we obtained in chapter 1,

$$k = \frac{1}{3} C v \Lambda \quad (6.58)$$

where $\Lambda = \tau v$ is the phonon mean free path.

Our previous discussion suggests that the relaxation time is highly frequency dependent. On the basis of the Mathiessen rule, and considering the following phonon scattering mechanisms: (1) phonon–phonon umklapp scattering, (2) phonon–impurity scattering, and (3) phonon–boundary scattering, we have

$$\frac{1}{\tau} = \frac{b_s v}{L} + A\omega^4 + B e^{-\theta_D/bT} T^3 \omega^2 \quad (6.59)$$

We cautioned before that our treatment of boundary scattering is very crude and should be taken as a rough approximation. Substituting eq. (6.59) into (6.57), we obtain an expression to calculate the thermal conductivity. There are three unknown parameters, particularly A, B and b, since b_s can be modeled (Berman et al., 1955). These parameters can be determined from fitting eq. (6.57) with experimental temperature-dependent thermal conductivity data. In figure 6.7, we show a fit of the thermal conductivity of GaAs (Chen and Tien, 1993). The thermal conductivity of a crystalline solid typically shows a dome shape with a peak around 20 K, depending on the size of the crystals. At high temperature, the dominant scattering mechanism is due to phonon–phonon scattering and thermal conductivity is approximately inversely proportional to temperature,

$$k \propto \frac{1}{T} \quad (6.60)$$

In practice, the high temperature dependence is often T^{-n} with n $= 1-1.5$. At low temperature, phonon–boundary scattering dominates heat conduction. Thermal conductivity is proportional to specific heat and also to the size of the crystal (Casimir, 1938),

$$k \propto T^3 \quad (6.61)$$

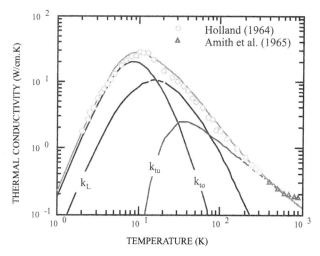

Figure 6.7 Thermal conductivity of GaAs, based on a model that considers phonon dispersion and contributions from different phonon branches, k_{to} from low-frequency transverse phonons and k_{tu} from high-frequency ones, and k_L from longitudinal phonons (Chen and Tien, 1993).

The reader should be reminded that the fitting of the thermal conductivity model based on eq. (6.57) and (6.59) with experimental data has some arbitrariness since the model also depends on the treatment of the density of states, the phonon group velocity v, and the dispersion of phonons. Often, the Debye model is used with a constant phonon group velocity. In addition, expressions for phonon–phonon scattering such as eq. (6.30) are obtained with a number of assumptions that also involve phonon dispersion. The fitting of the GaAs thermal conductivity data in figure 6.7 was based on Holland's model (Holland, 1963; Chen and Tien, 1993), which considered, approximately, the changing group velocity of phonons as a function of frequency, and used different relaxation time expressions for different branches of phonons. It was found that, depending on how the dispersion is approximated, the experimental data can be fitted equally well with different sets of parameters. Thus, extracting the exact relaxation time is, to a large extent, a still unsolved question even for bulk materials. Another point that should be mentioned is that although the normal three-phonon scattering process does not create resistance, it can redistribute phonons and thus indirectly affect the umklapp scattering process. A phenomenological model, based on displaced equilibrium distribution for the normal process, was established by Callaway (1959) to take the normal scattering process into consideration and this model has become a standard in thermal conductivity modeling.

Rather than being a convenient way of calculating thermal conductivity, eq. (6.58) is often used to estimate the mean free path on the basis of the experimental values of thermal conductivity, the specific heat, and the speed of sound in a material. This way of estimating the phonon mean free path, however, usually leads to an underestimation of the mean free path for those phonons that are actually carrying the heat because of the following reasons (Chen, 2001a):

1. Phonons are dispersive and their group velocity varies from the speed of sound at the Brillouin zone center to zero at the zone edge. The average phonon group velocity is much smaller than the speed of sound.
2. Optical phonons contribute to the specific heat but typically contribute little to heat flux, due to their low group velocity and their high scattering rates.
3. Phonon scattering is highly frequency dependent. High-frequency phonons are usually scattered more strongly than low-frequency phonons.

For example, the simple kinetic theory based on eq. (6.58) leads to a mean free path for silicon of ~410 Å. More careful consideration of the phonon dispersion and optical phonons (Chen, 1998) and experimental results (Ju and Goodson, 1999) indicate that the mean free path of those phonons actually carrying the heat is ~2500–3000 Å, much longer than what simple kinetic theory would give.

6.3.2 Newton's Shear Stress Law

To derive the Newton shear stress law for gas, we again consider a one-dimensional flow along the x-direction with the average velocity variation along the y-direction, as shown in figure 6.8. Because the molecules have an average velocity superimposed on their random velocity, we can no longer use the Maxwell velocity distribution as given in eq. (1.26), which would lead to a zero average velocity. As an approximation, the following *displaced Maxwell velocity distribution* is often used for the probability of finding one particle having velocity **v**

$$P(v_x, v_y, v_z) = \left(\frac{m}{2\pi \kappa_B T}\right)^{3/2} e^{-m[(v_x-u)^2+v_y^2+v_z^2]/2\kappa_B T} \quad (6.62)$$

where u is the average velocity along the x-direction. On the basis of this distribution, it can be shown that the average velocity is indeed u (this is left as an exercise). Assuming that the number density of particles is n, the number density of particles having velocity **v** is

$$f_0(v_x, v_y, v_z) = n P(v_z, v_y, v_z) = n \left(\frac{m}{2\pi \kappa_B T}\right)^{3/2} e^{-m[(v_x-u)^2+v_y^2+v_z^2]/2\kappa_B T} \quad (6.63)$$

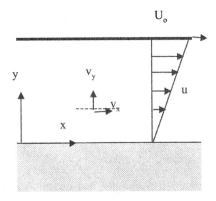

Figure 6.8 One-dimensional laminar viscous flow for deriving the Newton shear stress law.

From eq. (6.49), the distribution function is

$$f = f_0 - \tau v_y \frac{\partial f_0}{\partial y} = f_0 - \tau v_y \frac{\partial f_0}{\partial u}\frac{\partial u}{\partial y} \quad (6.64)$$

The shear stress along the x-direction on a plane perpendicular to the y-axis can be calculated by considering the momentum exchange across the plane,

$$\begin{aligned}\tau_{xy} &= \int_{-\infty}^{\infty}\int_{-\infty}^{\infty}\int_{-\infty}^{\infty} v_y[mv_x] f \, dv_x \, dv_y \, dv_z \\ &= -\frac{\partial u}{\partial y}\int_{-\infty}^{\infty}\int_{-\infty}^{\infty}\int_{-\infty}^{\infty} \tau v_y^2 m v_x \frac{\partial f_0}{\partial u} dv_x \, dv_y \, dv_z = \mu \frac{\partial u}{\partial y}\end{aligned} \quad (6.65)$$

where mv_x is the momentum of one particle along the x-direction and the $v_y(mv_x)f$ represents the rate of momentum change along the x-direction due to the flow of molecules across the constant y-plane. Such a rate of change of momentum equals the shear force acting on the constant y-plane, according to Newton's second law. The dynamic viscosity is

$$\begin{aligned}\mu &= -\int\int\int \tau v_y^2 (mv_x)\frac{\partial f_0}{\partial u} dv_x \, dv_y \, dv_z \\ &= m^2 \tau n \left(\frac{m}{2\pi \kappa_B T}\right)^{3/2} \int_{-\infty}^{\infty} e^{-mv_z^2/(2\kappa_B T)} dv_z \\ &\times \int_{-\infty}^{\infty} v_y^2 e^{-mv_y^2/(2\kappa_B T)} dv_y \int_{-\infty}^{\infty} \frac{v_x'^2}{\kappa_B T} e^{-mv_x'^2/(2\kappa_B T)} dv_x'\end{aligned} \quad (6.66)$$

In the last equation, we assume that the relaxation time is a constant and use $v_x' = v_x - u$. Carrying out the above integration, we obtain the following expression for the dynamic viscosity:

$$\mu = n\tau\kappa_B T = \frac{1}{4d^2}\sqrt{\frac{m\kappa_B T}{\pi}} \quad (6.67)$$

In writing the last step, we have used the mean free path expression [eq. (1.36)] and $\tau = \Lambda/\langle v \rangle$, where the average speed of molecules is $\langle v \rangle = [8\kappa_B T/(\pi m)]^{1/2}$, as given in example 1.1.

Following a similar procedure, we can also calculate the energy flux due to molecular heat conduction and obtain the thermal conductivity for a gas as

$$k = \frac{5}{2}\left(\frac{k_B}{m}\right)n\tau\kappa_B T = \frac{5}{2}\left(\frac{\kappa_B}{m}\right)\mu \quad (6.68)$$

Thus the thermal conductivity and viscosity are related to each other, because all these quantities arise from the same microscopic carrier motion. The difference is that for the shear stress we examine the momentum of the carriers, whereas for heat conduction we examine the energy of the carriers. Relationships as such are often found between kinetic coefficients. We will see similar relations in the next few sections between the electrical conductivity and mobility, the electrical and thermal conductivity of electrons, and the Seebeck and Peltier coefficients.

Equations (6.67) and (6.68) for viscosity and thermal conductivity are obtained on the basis of the relaxation time approximation. Enskog and Chapman, independently, solved the Boltzmann equation in its integral form (Chapman and Cowling, 1970) by series expansion of the distribution function. The final results for the viscosity and thermal conductivity for gas molecules, approximated as elastic spheres, are (Vincenti and Kruger, 1986)

$$\mu = \frac{5}{16d^2}\sqrt{\frac{m\kappa_B T}{\pi}} \quad \text{and} \quad k = \frac{15}{4}\left(\frac{\kappa_B}{m}\right)\mu \qquad (6.68a)$$

6.3.3 Ohm's Law and the Wiedemann–Franz Law

Having considered phonon and molecule transport, let us turn our attention now to electron transport. We first limit our consideration to electron flow in an isothermal conductor driven by an external electric field. The force acting on the electron from the external field is

$$\mathbf{F} = -e\mathscr{E} = e\nabla\varphi_e \qquad (6.69)$$

where e is the unit charge, the charge of an electron is $(-e)$, $\mathscr{E}$ is the electric field, and φ_e is the electrostatic potential that is related to the field by $\mathscr{E} = -\nabla\varphi_e$. Consider the one-dimensional case with charge flow in the x-direction due to a field of magnitude $\mathscr{E}$. Substituting eq. (6.69) into (6.49), we have

$$f = f_0 - \tau\left(v_x \frac{\partial f_0}{\partial x} - \frac{e\mathscr{E}}{m}\frac{\partial f_0}{\partial v_x}\right) \qquad (6.70)$$

where f_0 obeys the Fermi–Dirac distribution

$$f_0(E, E_f, T) = \frac{1}{\exp\left(\frac{E-E_f}{\kappa_B T}\right) + 1} \qquad (6.71)$$

Here we are using E_f to represent the chemical potential. In chapter 4, we used E_f for the Fermi level and μ for the chemical potential, but in this chapter μ is used for the dynamic viscosity. Sometimes the distinction between the Fermi level and the chemical potential is not rigorously made. In electrical engineering, chemical potential is usually called the Fermi level. To calculate $\partial f_0/\partial x$ in eq. (6.70), we should be careful where we place the reference for E and E_f. In a semiconductor, the conduction band energy is [Eq. (4.60)]

$$E = E_c + \frac{\hbar^2}{2m^*}(k_x^2 + k_y^2 + k_z^2)$$

where E_c is the location of the band edge. If we choose a common flat reference point for E_c, E_f, and E, such as shown in figure 6.9(a), it is clear that all three quantities

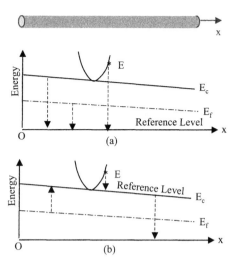

Figure 6.9 Choice of reference level for E_f and E. In (a), E_f, E_c, and E are relative to an absolute reference level. In this case all three quantities are x-dependent. In (b), E_f and E are relative to E_c and E_c is relative to an absolute reference level. The arrows in the dashed lines mark the reference point of the quantities.

are dependent on x. It is easier, however, to measure E_f relative to E_c. For a given carrier concentration, for example, the chemical potential should be measured relative to the bottom of the conduction band for electrons, as is clearly seen from eq. (4.64). In this reference system, E_f is actually the sum of the chemical potential $(E_f - E_c)$ and E_c, that is, $E_f = (E_f - E_c) + E_c$. The bottom of the conduction band E_c is related to the electrostatic potential φ_e from the relation between force and potential (for electrons with a negative charge)

$$F = -\frac{dE}{dx} = -\frac{dE_c}{dx} = e\frac{d\varphi_e}{dx} \qquad (6.72)$$

Thus, $-E_c/e$ is parallel to the *electrostatic potential*. We can usually take $\varphi_e = -E_c/e$.

Note the difference of the "potential" used in electricity from what we normally call the potential energy. The sum of the electrostatic potential φ_e and chemical potential (divided by charge) $(E_f - E_c)/(-e)$ is often called the *electrochemical potential*, Φ,

$$\Phi = \varphi_e - (E_f - E_c)/e = -E_f/e \qquad (6.73)$$

Thus, in a reference system as shown in figure 6.9(a), E_f itself includes both electrostatic and chemical potential contributions.

A different reference system, as shown in figure 6.9(b), is to have E_f and E always relative to E_c while E_c refers to an absolute potential level. This reference system has the advantage that E_f always represents the chemical potential and E is independent of x. This can also be seen from eq. (6.71), if we write $(E - E_f)$ in the Fermi–Dirac distribution as

$$E - E_f = (E - E_c) - (E_f - E_c) = \frac{\hbar^2}{2m^*}(k_x^2 + k_y^2 + k_z^2) - (E_f - E_c)$$

Under this reference, the electrochemical potential is then

$$\Phi = -(E_c + E_f)/e = \varphi_e - E_f/e \qquad (6.73a)$$

where we have taken $\varphi_e = -E_c/e$.

PARTICLE DESCRIPTION OF TRANSPORT PROCESSES: CLASSICAL LAWS

Because of its simplicity, we will choose figure 6.9(b) as the reference system in our subsequent derivations. The end results are independent of the reference system. In the chosen reference system

$$\frac{\partial f_0}{\partial x} = \frac{\partial f_0}{\partial E_f} \frac{dE_f}{dx} = -\frac{\partial f_0}{\partial E} \frac{dE_f}{dx} \tag{6.74}$$

and eq. (6.70) becomes

$$f = f_0 + \tau \left(v_x \frac{dE_f}{dx} \frac{\partial f_0}{\partial E} + \frac{e\mathscr{E}}{m} \frac{\partial f_0}{\partial E} \frac{\partial E}{\partial v_x} \right)$$

$$= f_0 + \tau v_x \left(\frac{dE_f}{dx} + e\mathscr{E} \right) \frac{\partial f_0}{\partial E} \tag{6.75}$$

In eq. (6.74), we used the relation $\partial f/\partial E_f = -\partial f/\partial E$. The current density ($\mathrm{A\,m^{-2}}$) is then

$$J_e = \frac{1}{V} \sum_{k_z=-\infty}^{\infty} \sum_{k_y=-\infty}^{\infty} \sum_{k_z=-\infty}^{\infty} (-e) v_x f$$

$$= -\frac{2}{(2\pi)^3} \int_{-\infty}^{\infty} \int_{-\infty}^{\infty} \int_{-\infty}^{\infty} e v_x f \, dk_x \, dk_y \, dk_z$$

$$= -\frac{e}{4\pi} \int_{4\pi} \left\{ \int_0^{\infty} v_x^2 \tau \left(\frac{dE_f}{dx} + e\mathscr{E} \right) \frac{\partial f_0}{\partial E} D(E) dE \right\} d\Omega$$

$$= -\frac{e}{3} \left(\frac{dE_f}{dx} + e\mathscr{E} \right) \int_0^{\infty} \tau v^2 D(E) \frac{\partial f_0}{\partial E} dE = \frac{e^2}{3} \frac{d\Phi}{dx} \int_0^{\infty} \tau v^2 D(E) \frac{\partial f_0}{\partial E} dE \tag{6.76}$$

where we have used the same angle notation as in figure 6.6 and the integration over the entire solid angle (4π) is carried out over θ and φ similar to eq. (6.54). We see from eq. (6.76) that the driving force for current flow is the electrochemical potential, not just the electrostatic potential alone. Because the chemical potential is related to the carrier concentration, the chemical potential gradient is representative of the carrier concentration gradient and the current due to the chemical potential gradient is thus the diffusion current. In metals and semiconductors, the relative importance of the two terms in the electrochemical potential is different, so we will discuss metals and semiconductors separately.

6.3.3.1 Metals

The electron density in metals is very large, such that the transport does not affect the chemical potential since the chemical potential is a measure of the electron number density. Equation (6.76) becomes

$$J_e = -\frac{e^2}{3} \mathscr{E} \int v^2 \tau \frac{\partial f_0}{\partial E} D(E) dE = \sigma \mathscr{E} \tag{6.77}$$

where σ is the electrical conductivity. Equation (6.77) is the Ohm law. The electrical conductivity is

$$\sigma = -\frac{e^2}{3}\int v^2\tau\frac{\partial f_0}{\partial E}D(E)dE = L_{11} \quad (6.78)$$

where L_{11} is a notation to be used later. Because $\partial f_0/\partial E$ is zero everywhere except close to the chemical potential or Fermi level, we can use a delta function to approximate it, $\partial f_0/\partial E = -\delta(E - E_f)$. Equation (6.78) becomes

$$\sigma = \frac{\tau_F D_F e^2 v_F^2}{3} \quad (6.79)$$

where subscript F represents the values at the Fermi level. Equation (6.79) means that not all the electrons in a metal actually participate in carrying the current. Only those close to the Fermi level are actively contributing to current flow. From eq. (3.53), we have

$$n = \frac{2}{3}E_F D_F = \frac{1}{3}mv_F^2 D_F$$

so that

$$\sigma = \frac{ne^2}{m}\tau_F \quad (6.80)$$

In the above derivation, we have used the relation $E_f = mv_F^2/2$.

6.3.3.2 Semiconductors

For transport in semiconductors, the carrier concentration changes with position and thus the chemical potential is not constant. We can start from eq. (6.70) to rewrite eq. (6.76) as

$$J_e = -\frac{e}{3}\int v^2\tau\left(e\mathscr{E}\frac{\partial f_0}{\partial E}\right)D(E)dE + \frac{e}{3}\int v^2\tau\frac{\partial f_0}{\partial x}D(E)dE \quad (6.81)$$

The first term represents current flow caused by the electrostatic field and the second term arises from the concentration gradient. Equation (6.81) is often written as

$$J_e = en\mu_e\mathscr{E} + e\frac{\partial(an)}{\partial x} \approx en\mu_e\mathscr{E} + ea\frac{\partial n}{\partial x} \quad (6.82)$$

where μ_e is called the *mobility* [m^2 V^{-1} s^{-1}] and a is the *diffusivity* [m^2 s^{-1}]. The approximation is valid only when the diffusivity is independent of location (Hess, 2000). The latter expression in eq. (6.82) is called the drift–diffusion equation. The electrostatic field causes the drifting of electrons while the concentration gradient drives the diffusion of electrons. The product of the mobility and the electric field

$$v_d = \mu_e\mathscr{E} \quad (6.83)$$

defines a velocity which is called the drift velocity. It is the average velocity that the electrons gain under the external field. Equation (6.83) is often used as the definition of mobility (Sze, 1981). Because the drift current density is equal to env_d, the electrical conductivity is related to mobility through

$$\sigma = en\mu_e \quad (6.84)$$

For semiconductors, because the carrier concentration is not fixed, the mobility is a measure of the mobile properties of individual electrons and thus is more useful. From eqs. (6.81) and (6.82), we can write the following expressions for the mobility and the diffusivity

$$\mu_e = -\frac{\frac{e}{3}\int_0^\infty v^2\tau(\partial f_0/\partial E)D(E)dE}{\int_0^\infty f_0 D(E)dE} = \frac{e\tau_m}{m} \quad (6.85)$$

$$a = \frac{\frac{1}{3}\int_0^\infty v^2\tau f_0 D(E)dE}{\int_0^\infty f_0 D(E)dE} \approx \frac{\kappa_B T}{e}\mu_e \quad (6.86)$$

where τ_m is called the momentum relaxation time and the approximate relationship between the diffusivity and the mobility, eq. (6.86), which is valid only when the distribution function obeys the Boltzmann distribution, is called the *Einstein relation*.

6.3.3.3 Wiedemann–Franz Law

The thermal conductivity of electrons can be derived in a similar manner to the derivation for phonons. Assuming no current flow, the thermal conductivity due to electrons can be expressed as

$$k_e = \frac{1}{3}C_e v_F^2 \tau_E \quad (6.87)$$

where τ_E is the energy relaxation time of electrons and is an average of τ weighed against the energy of the electrons, and C_e is the volumetric specific heat of electrons. τ_E represents the average time for an electron to lose its excess energy. In general, the energy relaxation time can be different from the momentum relaxation time. Typically, however, the two relaxation times are very close. From eqs. (6.80) and (6.87), and neglecting the difference between the relaxation times, we get

$$L = \frac{k_e}{\sigma T} = \frac{mC_e v_F^2}{3nTe^2} = \frac{\pi^2}{3}\left(\frac{\kappa_B}{e}\right)^2 = 2.45 \times 10^{-8} (W\,\Omega\,K^{-2}) \quad (6.88)$$

where we have used the specific heat of metal obtained in chapter 4, eq. (4.71). This is called the *Wiedemann–Franz law* and L is called the Lorentz number. Many metals obey this law, with slight changes in the value of the Lorentz number. For semiconductors, the Lorentz number should be calculated since the relation between n and the Fermi level depends on doping, but the magnitude of the Lorentz number remains close to the value. The Wiedemann–Franz law is often used to estimate the electron contribution to the thermal conductivity. For metals, it is sometimes used to calculate the thermal conductivity directly from the electrical conductivity because electrons are the dominant heat carriers in most metals.

6.3.4 Thermoelectric Effects and Onsager Relations

We have considered electron transport under the condition of either uniform temperature or no current flow (heat conduction by electrons). Now, let's examine the coupling of the temperature gradient with the electric field. In this case, both the Fermi level and temperature are functions of location, and $\partial f_0/\partial x$ can be expressed as

$$\frac{\partial f_0}{\partial x} = -\frac{\partial f_0}{\partial E} \cdot \frac{dE_f}{dx} - \frac{E - E_f}{T} \frac{\partial f_0}{\partial E} \frac{dT}{dx} \quad (6.89)$$

Substituting the above equation into eq. (6.70) and further into eq. (6.76), we obtain

$$J_e = -\frac{e}{3} \int v^2 \tau \left(\frac{dE_f}{dx} + \frac{E - E_f}{T} \frac{dT}{dx} + e\mathscr{E} \right) \frac{\partial f_0}{\partial E} D(E) dE \quad (6.90)$$

or

$$J_e = L_{11} \left(\mathscr{E} + \frac{1}{e} \frac{dE_f}{dx} \right) + L_{12} \left(-\frac{dT}{dx} \right)$$

$$= L_{11} \left(-\frac{d\Phi}{dx} \right) + L_{12} \left(-\frac{dT}{dx} \right) \quad (6.91)$$

where L_{11} is the electrical conductivity as given by eq. (6.78), and L_{12} is the coupling coefficient between current and the temperature gradient

$$L_{12} = \frac{e}{3T} \int v^2 \tau (E - E_f) \frac{\partial f_0}{\partial E} D(E) dE \quad (6.92)$$

The first term in eq. (6.91) is the normal electrical conduction due to the electrochemical potential gradient. The second term arises from the thermal diffusion of electrons under a temperature gradient. Under an open circuit, equation (6.91) leads to

$$\frac{d\Phi}{dx} = -\frac{L_{12}}{L_{11}} \frac{dT}{dx} = -S \frac{dT}{dx} \quad (6.93)$$

where S [V K^{-1}] is called the *Seebeck coefficient*, defined as

$$S = \frac{-d\Phi/dx}{dT/dx} = \frac{L_{12}}{L_{11}}$$

$$= -\frac{1}{eT} \frac{\int v^2 \tau (E - E_f) \frac{\partial f_0}{\partial E} D(E) dE}{\int v^2 \tau \frac{\partial f_0}{\partial E} D(E) dE} \quad (6.94)$$

where the negative sign arises because we are dealing with electrons. Similar treatment for holes would lead to a positive sign. This expression shows that the Seebeck coefficient is a measure of the average energy of an electron above the Fermi level under the open circuit condition, weighted against the differential electrical conductivity at each energy level. As we will show later, $(E - E_f)$ is related to the heat carried by an electron and $(E - E_f)/T$ is related to the entropy. Thus the Seebeck coefficient is a measure of the average heat current carried per electron.

Equation (6.93) can also be written as

$$-(\Phi_2 - \Phi_1) = S(T_2 - T_1) = -V \tag{6.95}$$

where V is the voltage drop measured from point 2 (hot point) to point 1 (cold point). Thus, in a conductor or semiconductor, a temperature difference generates a voltage difference. Physically, when one side of the conductor (or semiconductor) is hot, electrons have higher thermal energy and will diffuse to the cold side. The higher charge concentration in the cold side builds an internal electric field that resists the diffusion. The Seebeck voltage is the steady-state voltage accumulated under the open-circuit condition. If the conductor is a uniform material such that S is constant, the voltage difference does not depend on the temperature profile. This is the principle behind the thermocouple for temperature measurement. A thermocouple employs two conductors for ease of measuring the voltage difference. The same effect can also be used for power generation. We will present more discussion in chapter 8 on thermoelectric effects for energy conversion applications (Goldsmid, 1986).

We can also examine the heat flow when temperature and voltage gradients coexist in the conductor. When calculating the heat flow, we must carefully distinguish the energy flux from the heat flux because we are treating an open system with particles flowing across the boundaries. Consider a small control volume of fixed volume. The first law of thermodynamics should be written as

$$dU = dQ + E_f dN \tag{6.96}$$

In terms of energy flux, the above equation can be expressed as

$$d\mathbf{J}_q = d\mathbf{J}_E - E_f d\mathbf{J}_n \tag{6.97}$$

where $\mathbf{J}_q$ is the heat flux, $\mathbf{J}_E$ the energy flux, and $\mathbf{J}_n$ the particle flux. Considering again one-dimensional flow along the x-direction, these fluxes can be expressed as

$$J_E = \int E v_x f dv_x dv_y dv_z \text{ and } J_n = \int v_x f dv_x dv_y dv_z \tag{6.98}$$

so that the heat flux along the x-direction can be calculated from

$$J_q = \int (E - E_f) v_x f dv_x dv_y dv_z \tag{6.99}$$

Substituting eqs. (6.70) and (6.89) into the above expression and following the same procedure as we used for the electrical current, we obtain the following expression for the heat flux

$$J_q = L_{21}\left(\mathscr{E} + \frac{1}{e}\frac{dE_f}{dx}\right) + L_{22}\left(-\frac{dT}{dx}\right) \tag{6.100}$$

The first term is the energy carried due to the convection of electrons under an electrochemical potential gradient, and the second term is due to the diffusion of electrons under a temperature gradient. The expressions for the coefficients are

$$L_{21} = \frac{e}{3}\int (E - E_f) v^2 \tau \frac{\partial f_0}{\partial E} D(E) dE = T L_{12} \tag{6.101}$$

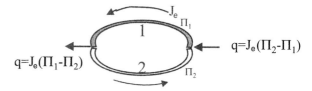

Figure 6.10 Cooling or heating at the junction of two materials occurs because of the difference between the Peltier coefficients of the two materials.

and

$$L_{22} = -\frac{1}{3T}\int (E-E_f)^2 v^2 \tau \frac{\partial f_0}{\partial E} D(E)dE \qquad (6.102)$$

Eliminating the electrochemical potential in eqs. (6.91) and (6.100) leads to

$$J_q = \frac{L_{21}}{L_{11}} J_e + \left(L_{22} - \frac{L_{12}L_{21}}{L_{11}}\right)\left(-\frac{dT}{dx}\right) = \Pi J_e - k\frac{dT}{dx} \qquad (6.103)$$

where

$$\Pi = \frac{L_{21}}{L_{11}} = TS \text{ and } k = L_{22} - \frac{L_{12}L_{21}}{L_{11}} \qquad (6.104)$$

Π [J K^{-1}] is called the *Peltier coefficient* and k is the electronic thermal conductivity. The relationship between the Peltier coefficient and the Seebeck coefficient is one of the *Kelvin relations*.

Equation (6.103) shows that in addition to the normal heat conduction by electrons, the charge flow also carries another heat that is proportional to the current. When two materials are joined together and a current passes through the junction, heat must be supplied or rejected at the interface because of the difference between the Peltier coefficients of the two materials, as shown in figure 6.10. The energy absorbed ($q > 0$) or rejected ($q < 0$) is

$$q = (\Pi_2 - \Pi_1)J_e$$

depending on the sign of q. The rejection or absorption of heat depends on the current direction and therefore, unlike heat conduction, the Peltier heat is reversible. This effect has been used to make thermoelectric refrigerators and heat pumps (Goldsmid, 1986).

A third thermoelectric effect, the Thomson effect, refers to reversible heating or cooling along a conductor when both a current and a temperature gradient are applied to the conductor. The energy deposited inside a differential volume along the conductor includes contributions from the heat flux variation and the electrochemical potential drop,

$$\dot{q} = -\frac{dJ_q}{dx} + J_e\left(-\frac{d\Phi}{dx}\right)$$

$$= -\frac{d\Pi}{dx} \bullet J_e + \frac{d}{dx}\left(k\frac{dT}{dx}\right) + J_e\left(-\frac{d\Phi}{dx}\right)$$

$$= -J_e \frac{d\Pi}{dT}\frac{dT}{dx} + \frac{d}{dx}\left(k\frac{dT}{dx}\right) - SJ_e\left(\frac{dT}{dx}\right) - J_e\frac{d\Phi}{dx}$$

$$= -\left(T\frac{dS}{dT}\right)J_e\frac{dT}{dx} + \frac{d}{dx}\left(k\frac{dT}{dx}\right) + \frac{J_e^2}{\sigma} \quad (6.105)$$

In the above derivation we have used eqs. (6.91), (6.103), and (6.104). In the last equation, the second term is due to heat conduction and the third term is due to Joule heating. These two terms are quite familiar in a heat conduction equation. The first term, however, is not familiar. It shows that heat can be absorbed or released, depending on the current direction. This reversible heat absorption or rejection is called the Thomson effect. The *Thomson coefficient* [V K^{-1}] is defined as the rate of cooling

$$\beta = \dot{q}_c \Big/ \left(J_e \frac{dT}{dx}\right) = T\frac{dS}{dT} \quad (6.106)$$

where the negative sign in the first term of eq. (6.105) does not appear because a positive Thomson effect is based on cooling whereas $\dot{q}$ is the heat generation. Equations (6.104) and (6.106), relating the three thermoelectric coefficients, S, Π, and β, are called the Kelvin relations.

Throughout this section, we have seen that the transport coefficients are often related, as for example in the Kelvin relations between the thermoelectric coefficients and the Einstein relation for the electrical diffusivity and the mobility. The fact that many of these coefficients are related has a more profound origin than a result from the Boltzmann equation. It is a requirement of the "time reversal invariance" of the mechanical equations of motion, that is, the particles retrace their former paths if all velocities are reversed. On the basis of this principle, Onsager (1931) derived the famous *Onsager reciprocity relations*. Here we will give a brief explanation of the reciprocity relations without proof (Callen, 1985). The flux of any extensive variable of a system (such as energy flux, particle flux) or at a local point of a system can be expressed as a linear combination of all the generalized driving forces F_j,

$$J_k = \sum_j L_{jk} F_j \quad (6.107)$$

where L_{jk} are called the kinetic coefficients. The generalized forces are the driving forces for entropy production. The Onsager reciprocal relations are

$$L_{jk} = L_{kj} \quad (6.108)$$

For local thermoelectric transport, the generalized forces are $\nabla(1/T)$ for heat flow and $(-\nabla\Phi)/T$ for electrical current, which leads to a relation between the two coefficients L_{12} and L_{21} as given by eq. (6.101).

Example 6.1

The relaxation time usually depends on the electron energy as $\tau \sim E^\gamma$, where r differs among scattering mechanisms for electron transport ($\gamma = -1/2$ for acoustic phonon scattering, $\gamma = 1/2$ for optical phonon scattering, and $\gamma = 3/2$ for impurity scattering). Derive an expression for the Seebeck coefficient of a nondegenerate semiconductor.

258   NANOSCALE ENERGY TRANSPORT AND CONVERSION

Solution: A nondegenerate semiconductor is one with the Fermi level inside the bandgap. In this case, the Fermi–Dirac distribution function can be approximated by the Boltzmann distribution

$$f = \frac{1}{\exp\left(\frac{E-E_f}{\kappa_B T}\right) + 1} \approx \exp\left(-\frac{E-E_f}{\kappa_B T}\right) \quad \text{(E6.1.1)}$$

The Seebeck coefficient can be calculated from eq. (6.94). Assuming a parabolic band, the density of states is

$$D(E) = \frac{1}{2\pi^2}\left(\frac{2m^*}{\hbar^2}\right)^{3/2} E^{1/2} \quad \text{(E6.1.2)}$$

Substituting (E6.1.2) and the relaxation time into eq. (6.94), we obtain the Seebeck coefficient as

$$S = -\frac{1}{eT} \frac{\int_0^\infty (E-E_f) E^{\gamma+3/2} \exp\left(-\frac{E-E_f}{\kappa_B T}\right) dE}{\int_0^\infty E^{\gamma+3/2} \exp\left(-\frac{E-E_f}{\kappa_B T}\right) dE}$$

$$= -\frac{\kappa_B}{e}\left[-\frac{E_f}{\kappa_B T} - \left(\gamma + \frac{5}{2}\right)\right] \quad \text{(E6.1.3)}$$

where E_f is the chemical potential, which can be controlled by doping. Using eq. (4.64) ($E_c = 0$ for the reference system here), we can write the above equation as

$$S = -\frac{\kappa_B}{e}\left[\ln\left(\frac{n}{N_c}\right) - \left(\gamma + \frac{5}{2}\right)\right] \quad \text{(E6.1.4)}$$

Comment. The value of κ_B/e is 86 μV K^{-1}, which gives an idea of the order of the magnitude of the Seebeck coefficient in many materials.

6.3.5 Hyperbolic Heat Conduction Equation and Its Validity

One assumption we made in the derivation of the classical constitutive equations, such as the Fourier law, is that the transient effect on the distribution function is negligible,

$$\tau \frac{\partial f}{\partial t} \ll \tau \mathbf{v} \bullet \nabla f \quad (6.109)$$

This will be valid if the variation of the distribution function in the time scale is much smaller than the variation of the distribution function in the length scale. Now, let's relax this approximation but still make the assumption that deviation from spatial equilibrium is small. Equation (6.49) becomes

$$\tau \frac{\partial f}{\partial t} + f = f_0 - \tau\left(\mathbf{v} \bullet \nabla_r f_0 + \frac{\mathbf{F}}{m} \bullet \nabla_\mathbf{v} f_0\right) \quad (6.110)$$

Let's consider one-dimensional phonon transport as the case for study and neglect the force term. We can transform the above expression by considering the energy flux of the left- and right-hand sides:

$$\int_{4\pi} \left[\int v_x \hbar \omega \left(\tau \frac{\partial f}{\partial t} + f \right) \frac{D(\omega)}{4\pi} d\omega \right] d\Omega$$

$$= \int_{4\pi} \left[\int v_x \hbar \omega \left(f_0 - \tau v_x \frac{df_0}{dx} \right) \frac{D(\omega)}{4\pi} d\omega \right] d\Omega \quad (6.111)$$

where we have used 4π under the integral to denote that the solid angle integration is over all directions. Eq. (6.111) can be written as

$$\bar{\tau} \frac{\partial J_q}{\partial t} + J_q = -k \frac{\partial T}{\partial x} \quad (6.112)$$

where $\bar{\tau}$ is a weighted average of the relaxation time relative to the heat flux expression. Equation (6.112) is the *Cattaneo equation* (Cattaneo, 1958; Joseph and Preziosi, 1989; Tamma and Zhou, 1997). Combining this equation with the energy conservation equation (no heat generation considered)

$$\frac{\partial J_q}{\partial x} = \rho c \frac{\partial T}{\partial t} \quad (6.113)$$

and eliminating J_q, we arrive at the following governing equation for the temperature distribution

$$\bar{\tau} \frac{\partial^2 T}{\partial t^2} + \frac{\partial T}{\partial t} = \frac{k}{\rho c} \frac{\partial^2 T}{\partial x^2} \quad (6.114)$$

This is a *hyperbolic* type of equation, or telegraph equation. It differs from the parabolic heat conduction equation obtained under the Fourier law, eq. (1.19), by adding the first term on the left-hand side. The parabolic heat conduction equation implies that if a temperature perturbation is applied at the boundary, it will be immediately felt through the whole region (the temperature rise at infinity may be infinitely small but it is still not absolutely zero). The hyperbolic heat conduction equation overcomes this dilemma since the heat propagation is in the form of a wave and the temperature rise is zero on the other side of the wave front. The solution is typically a damped wave due to the existence of the second term on the left-hand side. In addition to the Cattaneo equation, there are also other accepted modifications such as the Jeffreys type of equation (Joseph and Preziosi, 1989),

$$\bar{\tau} \frac{\partial q}{\partial t} + q = -k \frac{\partial T}{\partial x} - \bar{\tau} k_1 \frac{\partial}{\partial t} \left(\frac{\partial T}{\partial x} \right) \quad (6.115)$$

where k_1 is another physical property similar to thermal conductivity.

We comment here that although these equations can overcome the dilemma of infinite speed of Fourier's heat conduction equation, neither should be taken as generally applicable. There are many mathematical studies on the solution of the hyperbolic type of heat

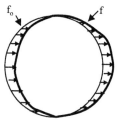

Figure 6.11 Local distribution f under the local-equilibrium distribution deviates slightly from the equilibrium distribution f_0. The difference between f and f_0, that is, the arrowed area, is the driving force for the heat and current flux.

conduction equation with various boundary conditions, but few experimental reports on the observation of hyperbolic heat conduction. Most of the theoretical studies are for conditions that are difficult to realize in practice. This is because heat conduction in a fast heat transfer process is typically confined in a very small region such that the assumption of small deviation from equilibrium in space is no longer valid. For such cases, one should resort to the Boltzmann equation (Majumdar, 1993) or to other approximations such as the ballistic–diffusive equations that take the spatial deviation from local equilibrium into consideration (Chen, 2001b). Although there is also experimental evidence on the wave type of response in different media, these are mostly caused by energy exchange between two fluids such as electrons and phonons (Qiu and Tien, 1993) or between solids and liquids in porous media.

Thermal waves in dielectrics have been observed experimentally (Landau, 1941; Ackerman et al., 1966; Narayanamurti and Dynes, 1972) and are often used as support for the hyperbolic heat conduction equations. These thermal waves can only be observed under special conditions when the mean free path of the umklapp scattering is long compared to the specimen size and that of the normal scattering process is relatively short (Guyer and Krumhansl, 1966). In this case, heat waves will propagate at a speed of $v/\sqrt{3}$; this speed is called the second sound. In section 6.4.3, we will discuss in more detail the origin of the second sound. It will be seen there that the second sound cannot be attributed to the approximation made in eq. (6.110). Rather, it is due to the fact that in a transport process dominated by normal scattering, the equilibrium distribution must be replaced by the displaced function, as in the displaced Maxwell distribution we used in section 6.3.2. Thus it is justifiable to say that the hyperbolic heat conduction equation, as derived in this section, cannot be used in most situations.

6.3.6 Meaning of Local Equilibrium and Validity of Diffusion Theories

From the derivation of the classical constitutive laws, the meaning of local equilibrium that underlies all these relations becomes clear. From eq. (6.47) to eq. (6.49) we assumed that the deviation of the distribution function from equilibrium is small, such that the local distribution function can be represented by its equilibrium value modified by a small deviation term that is proportional to the local gradient, as implied by eq. (6.49). In this sense, the local equilibrium is not equilibrium at all. In figure 6.11, we illustrate the equilibrium distribution function f_0 and the actual distribution function represented by eq. (6.49). The distortion of f from f_0 is small under the local equilibrium assumption. Because f_0 is isotropically distributed, it does not contribute to any net flux. Only the difference between f and f_0, that is, the arrowed area, contributes to the next current or heat fluxes.

Thus, for the (drift) diffusion theories to be valid, the deviation from the equilibrium must be small:

$$f_0 \gg \tau \left(\mathbf{v} \bullet \nabla_{\mathbf{r}} f_0 + \frac{\mathbf{F}}{m} \bullet \nabla_{\mathbf{v}} f_0 \right) \tag{6.116}$$

Taking phonon heat conduction as an example, multiplying the above inequality by $\hbar\omega$ and summing over all the phonon states, we have

$$\sum_{\mathbf{k}} \hbar\omega f_0 \gg \sum_{\mathbf{k}} \hbar\omega \tau \mathbf{v} \bullet \nabla_{\mathbf{r}} f_0 \tag{6.117}$$

The left-hand side is the phonon internal energy and the right-hand side is of the order of $q/\bar{v}$, where $\bar{v}$ is the average phonon velocity; thus,

$$U \gg -\frac{k}{\bar{v}} \frac{dT}{dx} \sim -C\Lambda \frac{dT}{dx} \tag{6.118}$$

where we used $k = C\bar{v}\Lambda/3$.

If we make the approximation that the specific heat is independent of temperature such that $U = CT$, the above condition becomes

$$\frac{\Lambda}{T} \frac{dT}{dx} \ll 1 \tag{6.119}$$

Equation (6.119) means that the temperature variation within one mean free path must be small compared to the absolute temperature for the diffusion theory to be valid.

Another assumption that we made in our derivations, in going from eq. (6.47) to (6.48), is that the deviation term g is much larger than the gradient of g:

$$\frac{g}{\tau} \gg \mathbf{v} \bullet \nabla g \tag{6.120}$$

If we approximate $\nabla g \approx g/L$, where L is a characteristic length that can be associated with, for example, the film thickness, the above inequality becomes

$$\frac{\tau v}{L} = \frac{\Lambda}{L} = Kn \ll 1 \tag{6.121}$$

where Kn is called the *Knudsen number*. The above relation means the characteristic length must be much larger than the mean free path for the diffusion theory to be valid.

Temporal wise, for the term $\partial f/\partial t$ to be negligible, we must have

$$\frac{\partial f}{\partial t} \ll \frac{f - f_0}{\tau} \tag{6.122}$$

For a transient phenomenon occurring with a characteristic time scale τ_c (such as a laser pulse width), the above inequality requires

$$\tau_c \gg \tau \tag{6.123}$$

which means that the transient process must be slow compared to the relaxation time.

If any of the conditions (6.119), (6.121), or (6.123) are not satisfied, one must be careful about whether the drift–diffusion relations discussed in this section are still valid. Sometimes, eq. (6.121) can be violated but (6.119) is still valid, as in the case of heat conduction along a thin film. In this case, the Fourier law, for example, is still valid for heat conduction along the film, but the thermal conductivity must be modified. In chapter 7, we will discuss various size effects for which one or several of these conditions are no longer valid.

6.4 Conservation Equations

In the previous section we saw how constitutive equations, for example, relations between heat flux and temperature gradient, or between electric current density and the electrochemical potential gradient, can be derived from the Boltzmann equation. In this section we will derive conservation equations, such as the particle continuity equation, Navier–Stokes equations, and so on, from the Boltzmann equation. For simplicity in notation, we will change the Cartesian coordinate notation from (x, y, z) to (x_1, x_2, x_3) and from (v_x, v_y, v_z) to (v_1, v_2, v_3). This will permit us to write long summations using the so-called *Einstein summation convention*, for example

$$\frac{\partial v_x}{\partial x} + \frac{\partial v_y}{\partial y} + \frac{\partial v_z}{\partial z} = \frac{\partial v_1}{\partial x_1} + \frac{\partial v_2}{\partial x_2} + \frac{\partial v_3}{\partial x_3} = \frac{\partial v_i}{\partial x_i} \qquad (6.124)$$

where i is a dummy index and the repeating of i means summation over i. As another example, $v_i v_i = v_1^2 + v_2^2 + v_3^2$.

From the probability distribution function f we can calculate the average quantities of every microscopic variable X,

$$\langle X \rangle = \frac{\int X f \, d^3\mathbf{v}}{\int f \, d^3\mathbf{v}} = \frac{\int X f d^3\mathbf{v}}{n} = \frac{m}{\rho} \int X f \, d^3\mathbf{v} \qquad (6.125)$$

where we have used the short notation $d^3\mathbf{v} = dv_1 dv_2 dv_3$, ρ is the density, m is the mass per particle, and the integration, which is a triple integration, is over all the possible values of v_1, v_2, v_3, that is, $(-\infty, \infty)$. In the previous section we solved the distribution function first and then proceeded to find the average quantities such as the heat and current fluxes, or the shear stress as a function of the generalized driven force. Here, we do not seek a solution for the distribution function. Rather, we will seek equations governing the average value of X. We multiply both sides of the Boltzmann equation by X and integrate over the momentum space (Reif, 1965; Vincenti and Kruger, 1986)

$$\int X \left(\frac{\partial f}{\partial t} + \mathbf{v} \bullet \nabla_\mathbf{r} f + \frac{\mathbf{F}}{m} \bullet \nabla_\mathbf{v} f \right) d^3\mathbf{v} = \int X \left(\frac{\partial f}{\partial t} \right)_c d^3\mathbf{v} \qquad (6.126)$$

From the above equation, one can derive the Navier–Stokes equations for gas flow, and similar "convective" type of equation for electron and phonon transport, as will be demonstrated below.

PARTICLE DESCRIPTION OF TRANSPORT PROCESSES: CLASSICAL LAWS 263

6.4.1 Navier–Stokes Equations

If we take the quantity X as meaning conserved quantities (such as mass, momentum, and energy), the net scattering term, that is, the right-hand side of eq. (6.126), should vanish. In these cases, the averaged Boltzmann equation can be significantly simplified. The microscopic expressions for these conserved quantities per particle of a dilute gas are

$$\text{Mass } X = m \tag{6.127}$$

$$\text{Momentum } \mathbf{X} = m\mathbf{v} \tag{6.128}$$

$$\text{Energy } X = mv^2/2 + m\psi_{\text{int}} \tag{6.129}$$

Where ψ_{int} is the potential energy per unit mass of a particle.

Substituting eq. (6.127) into (6.126), the three terms on the left-hand side of eq. (6.126) become, respectively,

$$\int_{-\infty}^{\infty}\int_{-\infty}^{\infty}\int_{-\infty}^{\infty} m\frac{\partial f}{\partial t} dv_1 dv_2 dv_3 = \frac{\partial(mn)}{\partial t} = \frac{\partial\rho}{\partial t} \tag{6.130}$$

$$\int_{-\infty}^{\infty}\int_{-\infty}^{\infty}\int_{-\infty}^{\infty} m\mathbf{v}\bullet\nabla_r f\, dv_1 dv_2 dv_3$$

$$= \int_{-\infty}^{\infty}\int_{-\infty}^{\infty}\int_{-\infty}^{\infty} \left(\frac{\partial(mv_1 f)}{\partial x_1} + \frac{\partial(mv_2 f)}{\partial x_2} + \frac{\partial(mv_3 f)}{\partial x_3}\right) dv_1 dv_2 dv_3$$

$$= \frac{\partial(mnu_i)}{\partial x_i} = \frac{\partial(\rho u_i)}{\partial x_i} \tag{6.131}$$

$$\int_{-\infty}^{\infty}\int_{-\infty}^{\infty}\int_{-\infty}^{\infty} m\frac{\mathbf{F}}{m}\bullet\nabla_v f\, dv_1 dv_2 dv_3$$

$$= \int_{-\infty}^{\infty}\int_{-\infty}^{\infty}\int_{-\infty}^{\infty} \left(\frac{\partial(F_i f)}{\partial v_i}\right) dv_1 dv_2 dv_3 = 0 \tag{6.132}$$

where $\mathbf{u}$ is the average velocity. Equation (6.132) ends in zero because f approaches zero as v approaches infinity. We also used the fact that t, $\mathbf{r}$, $\mathbf{v}$ are independent variables. Although the average quantity, such as $\mathbf{u}$, depends on $\mathbf{r}$, this dependence is due to the dependence of f on $\mathbf{r}$. Using eqs. (6.126, 6.130–6.132), the mass conservation equation can be written as

$$\frac{\partial\rho}{\partial t} + \frac{\partial}{\partial x_i}(\rho u_i) = 0 \tag{6.133}$$

Next, we examine the momentum equation. In this case, $m\mathbf{v}$ is a vector and as an example, we choose one component, $X = mv_j$. The three terms on the left-hand side of eq. (6.126) become

$$\int_{-\infty}^{\infty}\int_{-\infty}^{\infty}\int_{-\infty}^{\infty} mv_j \frac{\partial f}{\partial t} dv_1 dv_2 dv_3 = \frac{\partial}{\partial t}\int_{-\infty}^{\infty}\int_{-\infty}^{\infty}\int_{-\infty}^{\infty} \frac{\partial (mv_j f)}{\partial t} dv_1 dv_2 dv_3$$

$$= \frac{\partial (m \times nu_j)}{\partial t} = \frac{\partial (\rho u_j)}{\partial t} \qquad (6.134)$$

$$\int_{-\infty}^{\infty}\int_{-\infty}^{\infty}\int_{-\infty}^{\infty} mv_j \mathbf{v} \bullet \nabla_\mathbf{r} f \, dv_1 dv_2 dv_3$$

$$= \int_{-\infty}^{\infty}\int_{-\infty}^{\infty}\int_{-\infty}^{\infty} \left(\frac{\partial (mv_j v_i f)}{\partial x_i} \right) dv_1 dv_2 dv_3 \qquad (6.135)$$

$$\int_{-\infty}^{\infty}\int_{-\infty}^{\infty}\int_{-\infty}^{\infty} mv_j \frac{\mathbf{F}}{m} \bullet \nabla_\mathbf{v} f \, dv_1 dv_2 dv_3$$

$$= \int_{-\infty}^{\infty}\int_{-\infty}^{\infty}\int_{-\infty}^{\infty} \left(v_j \frac{\partial (F_i f)}{\partial v_i} \right) dv_1 dv_2 dv_3$$

$$= -\int_{-\infty}^{\infty}\int_{-\infty}^{\infty}\int_{-\infty}^{\infty} F_j f \, dv_1 dv_2 dv_3 = -nF_j = -\Psi_j \qquad (6.136)$$

where Ψ is the force per unit volume or the body force. To further simplify eq. (6.135), we now decompose the velocity into a random component and an average component,

$$\mathbf{v} = \mathbf{u} + \mathbf{v}' \qquad (6.137)$$

such that

$$\mathbf{u} = \frac{\int \mathbf{v} f d^3 \mathbf{v}}{\int f d^3 \mathbf{v}} = \mathbf{u}(t, \mathbf{r}) \qquad (6.138)$$

$$\langle \mathbf{v}' \rangle = \frac{\int (\mathbf{v} - \mathbf{u}) f d^3 \mathbf{v}}{\int f d^3 \mathbf{v}} = 0 \qquad (6.139)$$

Equation (6.135) can be written as

$$\int m \frac{\partial (f v_i v_j)}{\partial x_i} d^3 v = \frac{\partial (\rho u_i u_j)}{\partial x_i} + \frac{\partial (m \int v'_i v'_j f d^3 v)}{\partial x_i}$$

$$= \frac{\partial (\rho u_i u_j)}{\partial x_i} + \frac{\partial (\rho \langle v'_i v'_j \rangle)}{\partial x_i} \tag{6.140}$$

With eqs. (6.134), (6.136), and (6.140), the averaged Boltzmann equation, (6.126), becomes

$$\frac{\partial (\rho u_j)}{\partial t} + \frac{\partial (\rho u_i u_j)}{\partial x_i} + \frac{\partial (\rho \langle v'_i v'_j \rangle)}{\partial x_i} = \Psi_i \tag{6.141}$$

We can write the cross term as

$$\rho \langle v'_i v'_j \rangle = P - \tau_{ij} \tag{6.142}$$

where P is the pressure, which comprises the normal components of the random thermal velocity

$$P = \frac{\rho}{3} (\langle v'^2_1 \rangle + \langle v'^2_2 \rangle + \langle v'^2_3 \rangle) = \frac{\rho}{3} \langle v'_i v'_i \rangle \tag{6.143}$$

and τ_{ij} is the shear stress

$$\tau_{ij} = -[\rho \langle v'_i v'_j \rangle - P \delta_{ij}] \tag{6.144}$$

where δ_{ij} is again the Kronecker delta function, which equals 1 when $i = j$ and zero when $i \neq j$. Equation (6.141) then becomes

$$\rho \frac{\partial (\rho u_j)}{\partial t} + \rho \frac{\partial (\rho u_i u_j)}{\partial x_i} = -\frac{\partial P}{\partial x_j} + \frac{\partial \tau_{ij}}{\partial x_i} + \Psi_j \tag{6.145}$$

This is the momentum conservation equation for the component in the j-direction. Similarly, we can derive the energy conservation equation by setting $X = mv^2/2 + m\psi_{\text{int}}$ (Vincenti and Kruger, 1986).

$$\frac{\partial}{\partial t} \left(\rho \psi + \frac{1}{2} \rho u_i u_i \right) + \frac{\partial}{\partial x_i} \left[\rho u_i \left(H + \frac{1}{2} u_j u_j \right) \right]$$

$$= \frac{\partial}{\partial x_i} (\tau_{ik} u_k - J_{qi}) + \rho \langle F_i v_i \rangle \tag{6.146}$$

where

$$\psi = \frac{1}{2} \langle v'_j v'_j \rangle + \psi_{\text{int}} \tag{6.147}$$

is the total internal energy (translational plus other forms of internal energy) and H is the enthalpy per unit mass [J kg^{-1}]

$$H = \psi + \frac{P}{\rho} \tag{6.148}$$

The heat flux is defined as

$$J_{q_i} = \frac{1}{2}\rho\langle v_i' v'^2\rangle + \rho\langle v_i' \psi_{\text{int}}\rangle \qquad (6.149)$$

Equations (6.133), (6.145), and (6.146) are the conservation equations for mass, momentum, and energy. To further simply the equation, we need the relation between the shear stress and the average velocity, and between heat flux and temperature. The previous section has shown how to derive these constitutive equations. Following eqs. (6.63–6.65) and maintaining a more general three-dimensional velocity profile, one can show that the shear stress can be expressed as

$$\tau_{ij} = n\kappa_B T \tau \left[\frac{\partial u_i}{\partial x_j} + \frac{\partial u_j}{\partial x_i} - \frac{2}{3}\frac{\partial u_k}{\partial x_k}\delta_{ij}\right] = \mu\left[\frac{\partial u_i}{\partial x_j} + \frac{\partial u_j}{\partial x_i} - \frac{2}{3}\frac{\partial u_k}{\partial x_k}\delta_{ij}\right] \qquad (6.150)$$

Substituting the above-generalized Newton shear stress law into eq. (6.145), we obtain the familiar *Navier–Stokes* equations.

6.4.2 Electrohydrodynamic Equation

When electrons flow in an electric field, they acquire a nonzero average velocity. We can perform similar operations for electron transport as we did above for gas molecules to derive the governing equations for electron transport. There is, however, one complication. In a semiconductor, the numbers of electrons and holes can vary. An electron in the conduction band can fall back into the valence band, a process called recombination which we will discuss in more detail in chapter 8. Due to electron and hole recombination, the number of electrons in the conduction band and the number of holes in the valence band are reduced. During this process, the excess energy of the electron is lost, either by emitting a photon or by generating phonons. The latter becomes heat dissipation. The reverse process, that an electron is excited from the valence band to the conduction band by absorbing photons and phonons, or kicked by other electrons, thus creating an electron in the conduction band and a hole in the valence band, is called generation. The existence of the generation and recombination processes means that the scattering term is no longer zero. Other than this major difference, the derivation of the electron mass, momentum, and energy conservation equation is very similar to the derivation of the Navier–Stokes equations given in the previous section (Blotekjaer, 1970; Lundstrom, 2000). Without the detailed derivation, we write down the continuity equation as

$$\frac{\partial n}{\partial t} + \frac{\partial(nu_i)}{\partial x_i} = G - R \qquad (6.151)$$

where n is the number density of electrons or holes, **u** is the average velocity of the electrons defined according to eq. (6.138) and is called the drift velocity, G is the rate of generation of electrons, and R is the rate of recombination. We can also write the above equation in terms of the current density for electrons

$$\frac{\partial n}{\partial t} + \frac{1}{(-e)}\frac{\partial J_{ei}}{\partial x_i} = G - R \qquad (6.152)$$

The momentum equation can be written as (Lundstrom, 2000)

$$\frac{\partial(nu_j)}{\partial t} + \frac{\partial(nu_i u_j)}{\partial x_i} + \frac{ne\mathscr{E}_j}{m} + \frac{1}{m}\frac{\partial(n\kappa_B \overline{\overline{T}}_{ij})}{\partial x_i} = \left[\frac{\partial p_j}{\partial t}\right]_c \qquad (6.153)$$

where the *temperature tensor* is defined as

$$\kappa_B \overline{\overline{T}}_{ij} = \frac{1}{2}m\int (v_i - u_i)(v_j - u_j) f d^3\mathbf{v} \qquad (6.154)$$

Equation (6.154) actually corresponds to eq. (6.142). In fluid mechanics, this term is split into the normal stress (pressure) and shear stress. In electrohydrodynamics, this term is often directly related to the electron temperature T by treating the tensor as diagonal and isotropic so that

$$\frac{\partial \overline{\overline{T}}_{ij}}{\partial x_i} = \frac{\partial T}{\partial x_j} \qquad (6.155)$$

The right-hand side of eq. (6.153) is the rate of momentum scattering, which is often expressed using the relaxation time approximation

$$\left[\frac{\partial p_j}{\partial t}\right]_c = -\frac{nu_j}{\tau_m} \qquad (6.156)$$

Using eqs. (6.155) and (6.156), we can write the momentum conservation equation as

$$\frac{\partial(nu_j)}{\partial t} + \frac{\partial(nu_i u_j)}{\partial x_i} + \frac{ne\mathscr{E}_j}{m} + \frac{1}{m}\frac{\partial(n\kappa_B T)}{\partial x_i} = -\frac{nu_j}{\tau_m} \qquad (6.157)$$

The second term on the left-hand side represents the inertia effects of the electrons. When these effects are negligible, the steady-state solution for the electron velocity is

$$nu_j = -\frac{\tau_m ne\mathscr{E}_j}{m} - \frac{\tau_m}{m}\frac{\partial(n\kappa_B T)}{\partial x_j} \qquad (6.158)$$

The current is $\mathbf{J}_e = -en\mathbf{u}$, and thus

$$J_{ej} = \frac{\tau_m ne^2}{m}\mathscr{E}_j + \frac{e\tau_m}{m}\frac{\partial(nk_B T)}{\partial x_j}$$

$$\approx \sigma\mathscr{E}_j + ea\frac{\partial n}{\partial x_j} \qquad (6.159)$$

where we have neglected the variation of T with x. If this variation is included, we obtain an additional term corresponding to the thermoelectric effect. Equation (6.159) is identical to eq. (6.82), the drift–diffusion equation.

Taking the moment from the Boltzmann equation for the kinetic energy of electrons leads to the following energy conservation equation

$$\frac{\partial \psi_e}{\partial t} + \frac{\partial(u_i \psi_e)}{\partial x_i} = J_{ei}\mathscr{E}_i - \frac{\partial J_{qi}}{\partial x_i} + \left(\frac{\partial \psi_e}{\partial t}\right)_c - \dot{q} \qquad (6.160)$$

where the energy Ψ_e comprises the thermal energy plus the average kinetic energy:

$$\psi_e = \frac{3}{2}n\kappa_B T + \frac{m}{2}nu_i u_i = \frac{3}{2}n\kappa_B T + \frac{m}{2}n\mathbf{u}^2 \tag{6.161}$$

The heat flux is expressed by the Fourier law

$$J_{qi} = -k\frac{\partial T}{\partial x_i} \tag{6.162}$$

and the energy scattering rate must be determined for different scattering processes (Blotekjaer, 1970). One example, which considers the electron and phonon at nonequilibrium temperatures, is

$$\left(\frac{\partial \psi_c}{\partial t}\right)_c = -\frac{n\kappa_B (T - T_p)}{\tau_e} + \frac{nm\mathbf{u}^2}{3}\left(\frac{2}{\tau_m} - \frac{1}{\tau_e}\right) \tag{6.163}$$

where τ_e is the energy relaxation time, T_p is the phonon temperature, and τ_m is the momentum relaxation time. This equation is valid only for electrons in the same conduction band. Similar equations can be written for holes and for electrons in different bands. The equation can be coupled to the phonon heat conduction equation to form a closed set of equations. Further consideration of different phonon groups, such as optical and acoustic phonons, has also been undertaken (Fushinobu et al., 1995; Lai and Majumdar, 1996).

Equations (6.152), (6.157), and (6.160) form a set of closed equations, called electrohydrodynamic equations, that can be used to solve for the electron density, drift velocity, and temperature distributions. These equations were studied quite intensively in investigations of "hot electron" effects, that is, when the electron temperature is significantly higher than the phonon temperature. Such hot electrons can be generated under a high electric field. Small electronic devices, such as the field-effect transistors used in integrated circuits, are often operated with a high electric field and often have an electron temperature much higher than that of phonons. The electrohydrodynamic equations are sometimes used to study submicron devices. The applicability of the electrohydrodynamic equations to very small devices, however, is highly questionable because these equations are derived under the assumption of local equilibrium, which may not be valid when the electron mean free path is much larger than the characteristic length of the device.

The above summary shows although the electrohydrodynamic equations share many similarities with the Navier–Stokes equation, there are clearly places where different concepts are used, such as the temperature tensor rather than the stress tensors. The consequences of these subtle differences have not been examined in detail.

6.4.3 Phonon Hydrodynamic Equations

In analogy to the hydrodynamic equations for molecules and electrons, theories have been developed for the hydrodynamics of phonons (Gurevich, 1986). The motivation behind phonon hydrodynamics is the relative importance of the normal scattering process (N-process) that conserves crystal momentum versus the umklapp scattering that does not conserve the momentum. Denoting the relaxation times for these two processes

PARTICLE DESCRIPTION OF TRANSPORT PROCESSES: CLASSICAL LAWS   269

as τ_N and τ_u, the phonon hydrodynamic equations are established for the regime when $\tau_N \ll \tau_u$, and the time constant of heat transfer τ_c is comparable with or smaller than τ_u. When $\tau_c \ll \tau_u$, that is, there exist no momentum-destroying scattering processes, the phonon energy is redistributed in the normal scattering processes but the total momentum along the direction of heat flow does not change. In this case, phonons have a nonzero average velocity, $\mathbf{u}$, as do molecules or electrons. The equilibrium distribution function for the normal process is the drifted Bose–Einstein distribution

$$f_d = \frac{1}{\exp\left(\frac{\hbar\omega - \mathbf{k} \cdot \mathbf{u}}{\kappa_B T}\right) - 1} \tag{6.164}$$

where $\mathbf{u}$ is the average phonon drift velocity and $\mathbf{k}$ is the phonon wavevector. The phonon Boltzmann equation can be written as

$$\frac{\partial f}{\partial t} + \mathbf{v} \bullet \nabla_r f = -\frac{f - f_0}{\tau_u} - \frac{f - f_d}{\tau_N} \tag{6.165}$$

We define the local energy and momentum density variables as

$$U = \frac{3}{(2\pi)^3} \int \hbar\omega f d^3\mathbf{k} \tag{6.166}$$

$$\mathbf{P} = \frac{3}{(2\pi)^3} \int \hbar\mathbf{k} f d^3\mathbf{k} \tag{6.167}$$

Here, for simplicity in notation, we have assumed that the three phonon polarizations are identical. Because phonons can be created and annihilated, the number density is not conserved and no continuity equation is needed. Multiplying eq. (6.165) by $\hbar k_i$ and integrating over all wavevectors leads to the following momentum equation,

$$\frac{\partial P_i}{\partial t} + \frac{\partial \Sigma_{ij}}{\partial x_j} = -\frac{3}{(2\pi)^3} \int \frac{f - f_0}{\tau_u} \hbar k_i d^3\mathbf{k} \tag{6.168}$$

where the term related to τ_N is zero because momentum is conserved for the N-process, and

$$\Sigma_{ij} = \frac{3}{(2\pi)^3} \int \hbar k_i v_j f d^3\mathbf{k} \tag{6.169}$$

is similar to the inertia terms in the Navier–Stokes equations.

Multiplying eq. (6.165) by $\hbar\omega$ and integrating over all wavevectors, we obtain the energy equation as

$$\frac{\partial U}{\partial t} + \frac{\partial J_{qi}}{\partial x_i} = 0 \tag{6.170}$$

where

$$J_{qi} = \frac{3}{(2\pi)^3} \int \hbar\omega v_i f d^3\mathbf{k} \tag{6.171}$$

The above derivations have not made any assumption and are thus generally applicable. Now we will follow strategies similar to those used in deriving the thermal conductivity in order to further evaluate various terms in the momentum and energy equations. From eq. (6.165), we see that the distribution function approaches

$$f \to \frac{f_0/\tau_u + f_d/\tau_N}{1/\tau_u + 1/\tau_N} \tag{6.172}$$

When $\tau_N \ll \tau_u$, the dominant term is f_d. As a 0th-order approximation, we first examine how a nonzero drift velocity affects the momentum and energy transport. In this case, we can approximate

$$f \approx f_d \approx f_0 - \frac{\partial f_0}{\partial \omega}\mathbf{k} \bullet \mathbf{u} = f_0 + f_0(1+f_0)\frac{\hbar \mathbf{k} \bullet \mathbf{u}}{\kappa_B T} \tag{6.173}$$

Using this distribution function, eq. (6.167) becomes

$$P_{i0} = \frac{3}{(2\pi)^3}\int \hbar k_i f_d d^3\mathbf{k}$$

$$\approx \frac{3}{(2\pi)^3}\int \hbar k_i \left(f_0 + f_0(1+f_0)\frac{\hbar \mathbf{k} \bullet \mathbf{u}}{\kappa_B T}\right) d^3\mathbf{k}$$

$$= \frac{3}{(2\pi)^3}\int \hbar k_i \left(f_0(1+f_0)\frac{\hbar k_j u_j}{\kappa_B T}\right) d^3\mathbf{k}$$

$$= u_j \frac{3}{(2\pi)^3}\int \hbar k_i \left(f_0(1+f_0)\frac{\hbar k_j}{\kappa_B T}\right) d^3\mathbf{k} = \eta_{ij}u_j \tag{6.174}$$

where η_{ij} is a second-order tensor,

$$\eta_{ij} = \frac{3}{(2\pi)^3}\int \hbar k_i \left(f_0(1+f_0)\frac{\hbar k_j}{\kappa_B T}\right) d^3\mathbf{k} \tag{6.175}$$

and has units of kg m^{-3}, which makes the **P** term in eq. (6.168) similar to the $\rho\mathbf{u}$ term in the Navier–Stokes equations. Substituting eq. (6.173) into (6.169) and maintaining only the leading terms, we obtain the following expression for the second derivative term in eq. (6.168):

$$\frac{\partial \Sigma_{ij0}}{\partial x_j} = \frac{3}{(2\pi)^3}\int \hbar k_i v_j \frac{\partial f_d}{\partial x_j} d^3\mathbf{k}$$

$$= \frac{3}{(2\pi)^3}\int \hbar k_i v_j \left(\frac{\partial f_d}{\partial T}\frac{\partial T}{\partial x_j} + \frac{\partial f_d}{\partial u_k}\frac{\partial u_k}{\partial x_j}\right) d^3\mathbf{k}$$

$$\approx \frac{3}{(2\pi)^3}\int \hbar k_i \frac{\partial \omega}{\partial k_j}\left(f_0(1+f_0)\frac{\hbar \omega}{\kappa_B T^2}\frac{\partial T}{\partial x_j} + f_0(1+f_0)\frac{\hbar k_k}{\kappa_B T}\frac{\partial u_k}{\partial x_j}\right) d^3\mathbf{k}$$

$$= \frac{3}{(2\pi)^3}\int \hbar k_i \frac{\partial \omega}{\partial k_j}\left(-\frac{1}{\hbar}\frac{\partial s}{\partial \omega}\frac{\partial T}{\partial x_j}\right) d^3\mathbf{k} = -\frac{\partial T}{\partial x_j}\frac{3}{(2\pi)^3}\int k_i \frac{\partial s}{\partial k_j} d^3\mathbf{k}$$

$$= \delta_{ij}\frac{3}{(2\pi)^3}\int s\, d^3\mathbf{k}\left(\frac{\partial T}{\partial x_j}\right) = S_p \frac{\partial T}{\partial x_j} \tag{6.176}$$

In the above derivation, s is the entropy of one phonon quantum state and S_p is the phonon entropy density. We have used integration by parts in carrying out the second-to-last step, and the following relationship for the entropy of one phonon state,*

$$\frac{\hbar\omega}{T} f_0(1+f_0) = -\frac{\kappa_B T}{\hbar} \frac{\partial s}{\partial \omega} \qquad (6.177)$$

Substituting eq. (6.173) into (6.171), we obtain a 0th-order expression for the second term in eq. (6.170):

$$\frac{\partial J_{qi}}{\partial x_i} = \frac{3}{(2\pi)^3} \int \hbar\omega v_i \frac{\partial f}{\partial x_i} d^3\mathbf{k}$$

$$= \frac{3}{(2\pi)^3} \int \hbar\omega \frac{\partial \omega}{\partial k_i} \left(f_0(1+f_0) \frac{\hbar\omega}{\kappa_B T^2} \frac{\partial T}{\partial x_i} + f_0(1+f_0) \frac{\hbar k_k}{\kappa_B T} \frac{\partial u_k}{\partial x_i} \right) d^3\mathbf{k}$$

$$= \frac{3}{(2\pi)^3} \int \hbar\omega \frac{\partial \omega}{\partial k_i} f_0(1+f_0) \frac{\hbar k_k}{\kappa_B T} \frac{\partial u_k}{\partial x_i} d^3\mathbf{k}$$

$$= \frac{\partial u_k}{\partial x_i} \frac{3}{(2\pi)^3} \int \frac{\partial \omega}{\partial k_i} \left(-k_k T \frac{\partial s}{\partial \omega} \right) d^3\mathbf{k}$$

$$= \frac{\partial u_k}{\partial x_i} \frac{3}{(2\pi)^3} \int \left(-k_k T \frac{\partial s}{\partial k_i} \right) d^3\mathbf{k}$$

$$= T \frac{\partial u_k}{\partial x_i} \frac{3}{(2\pi)^3} \int (\delta_{ki} s) d^3\mathbf{k} = T S_p \frac{\partial u_i}{\partial x_i} \qquad (6.178)$$

Substituting eqs. (6.174), (6.176), and (6.178) into (6.168) and (6.170), we obtain

$$\frac{\partial(\eta_{ij} u_j)}{\partial t} + S_p \frac{\partial T}{\partial x_i} = 0 \qquad (6.179)$$

$$C \frac{\partial T}{\partial t} + T S_p \frac{\partial u_i}{\partial x_i} = 0 \qquad (6.180)$$

where we have used

$$\frac{\partial U}{\partial t} = C \frac{\partial T}{\partial t} \qquad (6.181)$$

Equations (6.179) and (6.180) constitute the 0th-order phonon hydrodynamic equation, that is, the inviscid phonon flow since we have completely neglected the umklapp scattering. Elminating u_i from eqs. (6.179) and (6.180), we obtain

$$\frac{\partial^2 T}{\partial t^2} = \frac{T S_p^2}{C \eta_{ij}} \frac{\partial^2 T}{\partial x_i \partial x_j} \qquad (6.182)$$

---

*Equation (6.177) can be proven based on eqs. (4.14) and (4.40). See exercise 4.20.

This equation is a hyperbolic one that implies that the temperature field propagates as a wave. In an isotropic medium, η_{ij} degenerates into a scalar η and the corresponding speed of wave propagation is called the *second sound* (Ward and Wilks, 1952):

$$v_s = \sqrt{\frac{TS_p^2}{C\eta}} \qquad (6.183)$$

In helium-II (Gurevich, 1986),

$$\eta = \frac{2\pi^2 \kappa_B^4 T^4}{45\hbar^3 v^5}; \quad C = \frac{2\pi^2 \kappa_B^4 T^3}{15\hbar^3 v^3} \text{ and } C = 3S_p \qquad (6.184)$$

where v is the speed of sound. Substituting the above expressions into eq. (6.183) leads to $v_s = v/\sqrt{3}$. This thermal wave has a very different origin to that of the hyperbolic heat conduction equation discussed in section 6.3.5. The latter is derived under the relaxation time approximation while keeping the time derivative of the distribution function. The former is derived on the basis of the displaced Bose–Einstein distribution for processes dominated by normal scattering. Although the hyperbolic heat conduction equation can be applied to umklapp-dominated processes, its derivation implies that the spatial deviation from equilibrium is small. This condition is unlikely to be realized in fast transport processes, which are usually accompanied by a steep temperature gradient or masked by other carrier transport, as in fast laser experiments (Qiu and Tien, 1993). The derivation of the phonon hydrodynamic equation does not consider the steep temperature gradient either. However, its premise is built on the assumption that the normal scattering is much faster than the umklapp scattering, such that phonons have a large drift velocity. This condition can be satisfied only at low temperatures. Thus, at room temperature, the wave types of equation, either the hyperbolic heat conduction equation discussed in section 6.3.5 or the hydrodynamic equation discussed here, are unlikely to be applicable.

If the umklapp scattering is included, the phonon hydrodynamic equations can be expressed as (Gurevich, 1986)

$$\frac{\partial P_i}{\partial t} + S_p \frac{\partial T}{\partial x_i} - \xi_{ijmn} \frac{\partial^2 u_j}{\partial x_m \partial x_n} + \frac{TS_p^2}{k_{ij}} u_j = 0 \qquad (6.185)$$

$$C \frac{\partial T}{\partial t} + \frac{\partial J_{qi}}{\partial x_i} = 0 \qquad (6.186)$$

$$J_{qi} = TS_p u_i - \chi_{ij} \frac{\partial T}{\partial x_j} \qquad (6.187)$$

$$P_i = \eta_{ij} u_j \qquad (6.188)$$

where ζ_{ijmn} is analogous to viscosity for normal scattering processes, and χ_{ij} is the thermal conductivity tensor due to normal scattering processes, whereas κ_{ij} is the thermal conductivity tensor due to non-momentum-conserving scattering processes such as

Umklapp scattering. These coefficients can be derived similarly to what we have done in section 6.3 for other transport coefficients, that is, by substituting eq. (6.164) in the left-hand side of the Boltzmann equation and solving for from eq. (6.165), followed by substituting f into eqs. (6.169) and (6.171). Details can be found in Gurevich (1986).*
Here we will give only the final results:

$$\xi_{\ell mnp} = \frac{\hbar^2}{\kappa_B T} \frac{3}{(2\pi)^3} \int \tau_n f_0(f_0 + 1)$$
$$\times \left(k_\ell v_m - \delta_{\ell m}\omega\frac{S_p}{C}\right)\left(k_n v_p - \delta_{np}\omega\frac{S_p}{C}\right) d^3\mathbf{k} \quad (6.189)$$

$$\chi_{ij} = \frac{\hbar^2}{\kappa_B T^2} \frac{3}{(2\pi)^3} \int \tau_n f_0(f_0+1) \left(\omega v_i - \frac{TS_p}{\eta_{im}}k_m\right)\left(\omega v_j - \frac{TS_p}{\eta_{jm}}k_m\right) d^3\mathbf{k} \quad (6.190)$$

$$k_{ij}^{-1} = \frac{\hbar^2}{\kappa_B T^2 S_p^2} \frac{3}{(2\pi)^3} \int \frac{1}{\tau_u} f_0(f_0+1) k_i k_j d^3\mathbf{k} \quad (6.191)$$

where v_i is the component of the group velocity. Their orders of magnitude are

$$\xi \sim \kappa_B T \left(\frac{T}{\hbar v}\right)^3 \tau_n, \quad \chi \sim \kappa_B^4 v^2 \tau_n \left(\frac{T}{\hbar v}\right)^3,$$

$$\eta \sim \frac{\kappa_B^4 T^4}{\hbar^3 v^5}, \quad k \sim \kappa_B^4 \left(\frac{T}{\hbar v}\right)^3 v^2 \tau_u \quad (6.192)$$

Equations (6.185) and (6.186) are very similar to the Navier–Stokes equations. Similar flow regimes can also be expected. For example, the phonon Poiseuille flow has been observed and discussed in literature (Guyer and Krumhansl, 1966; Gurevich, 1986).

6.5 Summary of Chapter 6

This chapter has two major aims: one is to introduce the Boltzmann equation, and the other is to illustrate that classical laws can be derived from the Boltzmann equation under appropriate approximations.

The Boltzmann equation, or the Boltzmann transport equation, can be derived from the general Liouville equation. These equations are all established in phase space, which is multidimensional. The state of a system, described by the generalized coordinates $\mathbf{r}^{(n)}$ and momentum $\mathbf{p}^{(n)}$ of all the particles in the system, is one point in the phase space at any specific time. The Liouville equation of motion describes the evolution of the distribution function for an ensemble of systems in phase space. The Boltzmann equation is simplified from the Liouville equation through the use of the one-particle distribution function. The interaction of this particle with the rest of the particles in the system is represented by the

---

*Gurevich's book does not carry κ_B in the Boltzmann factor. Expressions given here included κ_B for consistency.

scattering term. The Boltzmann equation is valid only for a dilute system of particles, such as gases, electrons, photons, and phonons in the particle regime. Because the Boltzmann equation is based on the one-particle distribution function, the phase-space coordinates comprise six coordinates: position **r** and translational momentum **p** (we did not consider other modes of motion, such as rotation).

The key to the Boltzmann equation is the scattering term. Quantum mechanical principles are often used to deal with scattering. The time-dependent perturbation treatment in quantum mechanics leads to the Fermi golden rule for calculating the scattering probability from one quantum state to another. A general expression of the scattering integral can be formally written down on the basis of the scattering probability and the distribution function. This leads to an integral-differential form of the Boltzmann equation, which is difficult to solve but has often been treated in thermal radiation transport in the form of the equation of radiative transfer. For phonon and electron transport, as well as gas transport, we often use the relaxation time approximation. In chapter 8, we will further examine the cases in which the relaxation time approximation is invalid. The relaxation times for different carriers, including electrons, phonons, photons, and molecules, are discussed. These typically involve unknown constants that are determined by fitting experimental data on transport properties. When multiple scattering coexists, the Matthiessen rule is often used to obtain the total relaxation time.

Starting from the Boltzmann equation under the relaxation time approximation, we proceed to derive the classical constitutive equations including the Fourier law, the Newton shear stress law, the Ohm law, the drift–diffusion equations, and the thermoelectric relations. The common assumptions made in all these relations are that (1) the transport process occurs in a time scale much longer than the relaxation time, and (2) deviation from equilibrium at every point is small—that is, the local equilibrium assumption. We showed that the kinetic coefficients of a particular type of carrier are often related, because of their common origin, such as the relationship between viscosity and thermal conductivity, the Einstein relation between diffusivity and mobility, and the Wiedmann–Franz law linking electrical and thermal conductivity. The relationship of kinetic coefficients culminates in the Onsager reciprocal relations. We also commented on the appropriateness of the hyperbolic heat conduction equation. The key message is that all the constitutive relations are derived under certain approximations, which may no longer be valid for transport at micro- and nanoscales, as we will discuss in more detail in the next chapter.

From the Boltzmann equation, we can also derive the familiar conservation equations. We explained the derivation of the Navier–Stokes equations. Along a similar line of derivation, one can obtain the electrohydrodynamic equations for charged carriers and the phonon hydrodynamic equations. For phonons, we showed the phonon hydrodynamic equations that originate from the difference between the normal scattering and momentum-destroying scattering processes. The phonon hydrodynamic equations lead to second sound and temperature waves, which occur only when the time scale of the transport is longer than the relaxation time of normal scattering but much shorter than the umklapp or other momentum-destroying processes. From our discussion, it should become clear that the hyperbolic heat conduction equation, which often invokes the existence of the second sound at low temperatures as a proof of its validity, has a limited validity range that is difficult to realize through experiments, at least for single-carrier systems. In multiple-carrier systems, such as electron–phonon interactions, a subject we will discuss in more detail in chapter 8, the coupling of electrons and phonons, with their different heat capacities, can lead to hyperbolic types of heat conduction equation, even though the governing equation for each carrier is still of the diffusion type. Such wave-like equations should not be confused with the wave behavior of single carrier systems.

We again followed the tradition of parallel development for electrons, photons, phonons, and molecules. Through this effort, we hope that the reader can see that divisions between different disciplines are quite arbitrary. Although the languages are very different, due to the historical developments within each field, common grounds exist among them.

6.7 Nomenclature for Chapter 6

a	diffusivity, m² s⁻¹	H	enthalpy, J kg⁻³
A	coefficient in eq. (6.31), s³	I	intensity, W m⁻² srad⁻¹
b	coefficient in eq. (6.30)	J	flux of heat transfer rate, W m⁻²; energy transfer rate, W m⁻²; current, A m⁻², and particles, s⁻¹ m⁻²
B	coefficient in eq. (6.30), K⁻³s		
c	specific heat, J kg⁻¹ K⁻¹		
C	volumetric specific heat, J m⁻³ K⁻¹		
d	molecule diameter, m	k	thermal conductivity, W m⁻¹ K⁻¹
D	density of states per unit volume, m⁻³	k_{ij}	thermal conductivity tensor due to umklapp process, W m⁻¹ K⁻¹
e	unit charge, C		
E	energy of one particle, J	$\mathbf{k}$	wavevector, m⁻¹
E_c	conduction band edge, J	$\hat{\mathbf{k}}$	unit vector along the wavevector direction
E_f	chemical potential, J		
$\mathscr{E}$	magnitude of electric field, V m⁻¹	K_e	extinction coefficient, m⁻¹
		L	characteristic length or crystal length, m; Lorentz number, W Ω K⁻²; coefficients
$\boldsymbol{\mathscr{E}}$	electric field vector, V m⁻¹		
f	one-particle distribution function		
f_0	equilibrium distribution function	m	space or momentum degrees of freedom of one particle; mass, kg
$f^{(N)}$	N-particle distribution function		
		M_{if}	scattering matrix element, J
$\mathbf{F}$	external force on the particle, N	n	total degree of freedom in space or momentum for N particles; particle number density, m⁻³
g	deviation from equilibrium distribution		
G	rate of generation per unit volume, s⁻¹ m⁻³	N	number of particles in the system
h	Planck constant, J s	$p^{(i)}$	ith component of the momentum in phase space, kg m s⁻¹
$\hbar$	Planck constant divided by 2π, J s		

p	momentum coordinate vector of one particle, kg m s^{-1}	θ	polar angle
p$_i$	momentum coordinate vector of particle i, kg m s^{-1}	θ_D	Debye temperature, K
		Λ	mean free path, m
P	pressure, N m^{-2}	μ	dynamic viscosity, N s m^{-2}
P	average momentum per unit volume, kg m^{-2} s^{-1}	μ_e	electron mobility, m^2 V^{-1} s^{-1}
q	heat absorbed or rejected, W m^{-2}	ξ	viscosity defined by eq. (6.189), kg m^{-1} s^{-1}
$\dot{q}$	heat generation, W m^{-3}	Π	Peltier coefficient, V
$r^{(i)}$	ith component of the space coordinates, m	ρ	density, kg m^{-3}
		σ	electrical conductivity, Ω^{-1} m^{-1}
r	space coordinate vector of one particle, m	$\sigma_{s\nu}$	frequency-dependent scattering coefficient, m^{-1}
r$_i$	space coordinate vector of particle i, m	Σ_{ij}	defined by eq. (6.169), J m^{-4}
		τ	relaxation time, s
R	rate of recombination, s^{-1} m^{-3}	τ_c	characteristic time of a process, s
		ϕ	scattering phase function
s	polarization index; entropy J K^{-1}	φ	azimuthal angle
S	Seebeck coefficient, V K^{-1}	φ_e	electrostatic potential, V
S_p	entropy density, J K^{-1} m^{-3}	Φ	electrochemical potential, V
t	time, s	χ_{ij}	thermal conductivity tensor due to normal process, W m^{-1} K^{-1}
T	temperature, K		
$\overline{\overline{T}}$	temperature tensor, K		
u	average velocity, m s^{-1}		
U	total energy density, J	Ψ	internal energy of molecules, J kg^{-3}
v	velocity, m s^{-1}	Ψ_e	electron energy density, J m^{-3}
v_d	drift velocity, m s^{-1}		
v_F	Fermi velocity, m s^{-1}	Ψ_{int}	potential energy, J kg^{-3}
v$_g$	group velocity, m s^{-1}		
v_s	second sound, m s^{-1}	$\boldsymbol{\psi}$	body force, N m^{-3}
V	volume, m^3	ω	angular frequency, Hz
W	transition rate from initial state i to final state f, s^{-1}	Ω	solid angle, srad
		$\langle \rangle$	ensemble average
x, y, z	Cartesian coordinates		
X	microscopic quantity		**Subscripts**
α	absorption coefficient, m^{-1}		
β	Thomson coefficient, V K^{-1}	0	zeroth order
γ	parameter in the energy dependence of electron scattering	1, 2, 3	components of Cartesian coordinates, corresponding to x, y, z
δ	delta function	b	boundary
η_{ij}	second-order tensor defined by eq. (6.175), kg m^{-3}	d	drift
		e	electron

E	energy	t	total
f	final state	u	umklapp scattering
F	Fermi level	x, y, z	Cartesian component
g	group velocity	v	frequency dependent, spectral quantity
i	initial state; coordinate index		
I	impurity		Superscripts
m	momentum	(n)	n component in the phase space
n	total degree		
N	normal process	(N)	N-particle
p	phonon	·	time derivative
q	heat	‾	average

6.8 References

Ackerman, C.C., Bertman, B., Fairbank, H.A., and Guyer, R.A., 1966, "Second Sound in Solid Helium," *Physical Review Letters*, vol. 16, pp. 789–791.

Amith, A., Kudman, I., and Steigmeier, E.F., 1965, "Electron and Phonon Scattering in GaAs at High Temperatures," *Physical Review*, vol. 138, pp. A1270–A1276.

Ashcroft, N.W., and Mermin, N.D., 1976, *Solid State Physics*, Saunders College Publishing, Fort Worth, TX, p. 525.

Berman, R., Foster, E.L., and Ziman, J.M., 1955, "Thermal Conduction in Artificial Sapphire Crystals at Low Temperatures," *Proceedings of the Royal Society A* (London), vol. 231, pp. 130–144.

Bhatnagar, P.L., Gross, E.P., and Krook, M., 1954, "A Model for Collision Process in Gases, I. Small Amplitude Processes in Charged and Neutral One-Component Systems," *Physical Review*, vol. 94, pp. 511–525.

Blotekjaer, K., 1970, *Transport Equations for Electrons in Two-Valley Semiconductors*, IEEE Transactions on Electron Devices, vol. ED-17, pp. 38–47.

Bohren, C.F., and Huffman, D.R., 1983, *Absorption and Scattering of Light by Small Particles*, Wiley, New York.

Callaway, J., 1959, "Model for Lattice Thermal Conductivity at Low Temperatures," *Physical Review*, vol. 113, pp. 1046–1051.

Callen, H.B., 1985, *Thermodynamics and an Introduction to Thermostatistics*, 2nd ed., chapter 14, Wiley, New York.

Casimir, H.B.G., 1938, "Note on the Conduction of Heat in Crystals," *Physica*, vol. 5, pp. 495–500.

Cattaneo, C., 1958, "A Form of Heat Conduction Equation which Eliminates the Paradox of Instantaneous Propagation," *Comptes Rendus*, Paris Academy of Sciences, vol. 247, pp. 431–433.

Chapman, S., and T.G. Cowling, 1970, *The Mathematical Theory of Non-Uniform Gases*, 3rd ed., Cambridge University Press, Cambridge, UK.

Chen, G., 1996, "Heat Transfer in Micro- and Nanoscale Photonic Devices," *Annual Review of Heat Transfer*, vol. 7, pp. 1–58.

Chen, G., 1998, "Thermal Conductivity and Ballistic Phonon Transport in Cross-Plane Direction of Superlattices," *Physical Review B*, vol. 57, pp. 14958–14973.

Chen, G., 2001a, "Phonon Heat Conduction in Low-Dimensional Structures," *Semiconductors and Semimetals*, vol. 71, pp. 203–259.

Chen, G., 2001b, "Ballistic-Diffusive Heat-Conduction Equations," *Physical Review Letters*, vol. 86, pp. 2297–2300.

Chen, G., and Tien, C.L., 1993, "Thermal Conductivity of Quantum Well Structures," *Journal of Thermophysics and Heat Transfer*, vol. 7, pp. 311–318.

Fushinobu, K., Majumdar, A., and Hijikata, K., 1995, "Heat Generation and Transport in Submicron Semiconductor Devices," *Journal of Heat Transfer*, vol. 117, pp. 25–31.

Goldsmid, H.J., 1986, *Electronic Refrigeration*, Pion Limited, London.

Gurevich, V.L., 1986, *Transport in Phonon Systems*, chapter 2, North-Holland, Amsterdam.

Guyer, R.A., and Krumhansl, J.A., 1966, "Thermal Conductivity, Second Sound, and Phonon Hydrodynamic Phenomena in Nonmetallic Crystals," *Physical Review*, vol. 148, pp. 778–788.

Hess, K., 2000, *Advanced Theory of Semiconductor Devices*, chapters 7 and 11, IEEE Press, Piscataway.

Holland, M.G., 1963, "Analysis of Lattice Thermal Conductivity," *Physical Review*, vol. 132, pp. 2461–2471.

Holland, M.G., 1964, "Phonon Scattering in Semiconductors from Thermal Conductivity Studies," *Physical Review*, vol. 134, pp. A471–A480.

Joseph, D.D., and Preziosi, L., 1989, "Heat Waves," *Review of Modern Physics*, vol. 61, pp. 41–73.

Ju, Y.S., and Goodson, K.E., 1999, "Phonon Scattering in Silicon Films with Thickness of Order 100 nm," *Applied Physics Letters*, vol. 74, pp. 3005–3007.

Klemens, P.G., 1958, "Thermal Conductivity of Lattice Vibrational Modes," *Solid State Physics*, vol. 7, pp. 1–98.

Kubo, R., Toda, M., and Hashitsume, N., 1991, *Statistical Physics II, Nonequilibrium Statistical Mechanics*, 2nd ed., Springer, Berlin.

Lai, J., and Majumdar, A., 1996, "Concurrent Thermal and Electrical Modeling of Sub-Micrometer Silicon Devices," *Journal of Applied Physics*, vol. 79, pp. 7353–7361.

Landau, L.D., 1941, "The Theory of Superfluidity of Helium II," *Journal of Physics* (Moscow), vol. 5, pp. 71–90.

Landau, L.D., and Lifshitz, E.M., 1977, *Quantum Mechanics*, chapter 6, Pergamon, Oxford.

Liboff, R.L., 1998, *Kinetic Theory*, 2nd ed., chapter 3, pp. 338–341, Wiley, New York.

Lundstrom, M., 2000, *Fundamentals of Carrier Transport*, 2nd ed., chapters 2 and 7, Cambridge University Press, Cambridge, UK.

Majumdar, A., 1993, "Microscale Heat Conduction in Dielectric Thin Films," *Journal of Heat Transfer*, vol. 115, pp. 7–16.

Narayanamurti, V., and Dynes, R.C., 1972, "Observation of Second Sound in Bismuth," *Physical Review Letters*, vol. 28, pp. 1461–1465.

Onsager, L., 1931, "Reciprocal Relations in Irreversible Processes I," *Physical Review*, vol. 37, pp. 405–426.

Peierls, R.E., 1929, "Zur Kinetischen Theorie der Wärmeleitung in Kristallen," *Annalen der Physik*, vol. 3, pp. 1055–1101.

Qiu, T.Q., and Tien, C.L., 1993, "Heat Transfer Mechanism during Short-Pulse Laser Heating of Metals," *Journal of Heat Transfer*, vol. 115, pp. 835–841.

Reif, F., 1965, *Fundamentals of Statistical and Thermal Physics*, p. 527, McGraw-Hill, New York.

Siegel, R., and Howell, J.R., 1992, *Thermal Radiation Heat Transfer*, Hemisphere, Washington, DC.

Sze, S.M., 1981, *Physics of Semiconductor Devices*, Wiley, New York.

Tamma, K.K., and Zhou, X., 1997, "Macroscale and Microscale Thermal Transport and Thermo-Mechanical Interactions: Some Noteworthy Perspectives," *Journal of Thermal Stress*, vol. 21, pp. 405–449.

Vincenti, W.G., and Kruger, C.H., 1986, *Introduction to Physical Gas Dynamics*, pp. 325–348, Krieger, Malabar.

Ward, J.C., and Wilks, J., 1952, "Second Sound and the Thermo-Mechanical Effect at Very Low Temperature," *Philosophical Magazine*, vol. 43, pp. 48–50.

Ziman, J.M., 1960, *Electrons and Phonons*, chapter 3, Clarendon Press, Oxford.

6.9 Exercises

6.1 *Phonon thermal conductivity at intermediate temperature.* The phonon–phonon scattering relaxation time in the intermediate range of temperature (when $T < \theta_D$) can be approximated as

$$\frac{1}{\tau} = A \exp\left[-\frac{\Theta}{aT}\right] T^3 \omega^2$$

On the basis of the Debye model (linear dispersion), derive an expression for the thermal conductivity and discuss its dependence on temperature.

6.2 *High-temperature thermal conductivity.* At high temperature, the phonon relaxation time in a crystal is

$$\frac{1}{\tau} = \frac{\kappa_B T}{mva}$$

where a is of the order of distance between atoms and m is the atomic weight.

(a) Prove that the high-temperature thermal conductivity is proportional to $1/T$.

(b) The thermal conductivity of silicon at 300 K is 145 W m^{-1}K^{-1}. Estimate its thermal conductivity at 400 K.

6.3 *Rosseland diffusion approximation for phonon transport.* Consider an absorbing and emitting medium for thermal radiation transport. When the photon mean free path is much smaller than the characteristic length in the transport direction, the local equilibrium approximation is valid. Prove that under this condition (called optically thick) the radiative heat flux can be expressed as

$$q = -\frac{4\pi}{3\alpha}\frac{dI_b}{dx}$$

where α is the absorption coefficient. This is called the Rosseland diffusion approximation.

6.4 *Wiedemann–Franz law.* The electrical resistivity of gold at 300 K is 3.107×10^{-8} Ω m. Estimate its thermal conductivity.

6.5 *Wiedemann–Franz law.* The thermal conductivity of copper is 401 W m^{-1} K^{-1} at 300 K. Estimate its electrical conductivity at the same temperature.

6.6 *Energy and momentum relaxation time.* The electrical resistivity and thermal conductivity of gold at 300 K are 3.107×10^{-8} Ω m and 315 W m^{-1} K^{-1}. Estimate the momentum and energy relaxation time, and the momentum and energy relaxation length, of electrons in gold.

6.7 *Thermal conductivity and viscosity.* The thermal conductivity of air is 0.025 W m^{-1} K^{-1}. Estimate its dynamic viscosity.

6.8 *Electrons in semiconductors.* An n-type semiconductor has a carrier concentration of 10^{18} cm^{-3} and a mobility of 200 cm^2 V^{-1} s^{-1} at 300 K. Estimate the following: (a) electrical conductivity; (b) electron diffusivity; (c) momentum relaxation time; and (d) electron mean free path. Take the electron effective mass as that of a free electron.

6.9 *Thermal conductivity of gases.* Prove that the thermal conductivity of a dilute monatomic gas is

$$k = \frac{5}{2}\left(\frac{\kappa_B}{m}\right) n\tau\kappa_B T$$

6.10 *Thermoelectric cooler.* A thermoelectric device is typically made of p–n junctions as shown in figure P6.10. When a current flows through the p–n junction, both electrons and holes carry energy from the cold side to the hot side. The Peltier coefficients of both p and n materials are equal in magnitude, Π, but of opposite sign. The cooling rate due to current flow is $2\Pi \times I$. In addition to this cooling, there is also Joule heating and reverse heat conduction. Assuming that the electrical and thermal conductivities of both legs are the same, derive an expression for the net cooling power at the cold side in terms of the temperatures at the cold and the hot side, the current, and the cross-sectional area and length of the leg. Show that the cooling power reaches a maximum at a certain optimum current value.

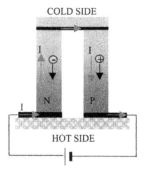

Figure P6.10 Figure for problem 6.10.

6.11 *Seebeck coefficients of a quantum well.* Derive an expression for the Seebeck coefficient of a quantum well of well-width d and with an infinite potential barrier height, as a function of the doping concentration.

6.12 *Power factor of a quantum well.* Because of Joule heating and reverse heat conduction, the efficiency of a thermoelectric device is determined by the figure of merit, defined as $Z = S^2\sigma/k$, where S is the Seebeck coefficient, σ the electrical conductivity, and k the thermal conductivity. The numerator $S^2\sigma$ is also called the power factor.

(a) Derive an expression for the power factor $S^2\sigma$ for a quantum well of width d and with an infinite barrier height, in terms of electron effective mass, relaxation time, chemical potential, and quantum well width.

(b) Assuming constant relaxation time and Boltzmann distribution, simplify the results obtained.

6.13 *Seebeck coefficient of nondegenerate silicon.* For silicon with doping concentration between 10^{16} and 10^{18} cm^{-3}, the Boltzmann distribution can be used instead of the Fermi–Dirac distribution. Silicon has six identical conduction bands with an effective mass of $0.33\, m_e$ for each conduction band, where m_e is the mass of a free electron. Assume a constant relaxation time.

(a) Calculate the Fermi level as a function of the carrier concentration from both the Fermi–Dirac and the Boltzmann distribution, and show that the levels do not differ much in the given doping range.

(b) Calculate the Seebeck coefficient as a function of the dopant concentration.

6.14 *Seebeck coefficient of a metal.*

(a) Assuming a constant relaxation time, prove that the Seebeck coefficient of a metal is given by

$$|S| = \frac{\pi^2 \kappa_B^2 T}{2 e E_f}$$

(b) Prove that ZT for a metal satisfies the following inequality

$$ZT \leq \frac{3\pi^2 \kappa_B^2 T^2}{4 E_f^2}$$

(c) Estimate the Seebeck coefficient of copper.

6.15 *Einstein relation.* When the Boltzmann approximation is valid, prove the Einstein relation between mobility and diffusivity for electrons

$$a = \frac{k_B T}{e} \mu$$

7

Classical Size Effects

In section 6.3.6, we discussed the premises of the diffusion laws. Generally, for the diffusion laws to be applicable, the length scale of the structures in which the transport occurs must be much larger than the mean free path and the time scale of the process must be much longer than the relaxation time. When one or both of these conditions is violated, the transport can no longer be described by diffusion theory. Chapter 5 examined interface and size effects from the wave propagation point of view. In this chapter, we will study size effects on transport processes by treating energy carriers as particles to which the Boltzmann equation is applicable. In this sense, we are dealing with classical size effects. However, even under the particle picture, quantum effects can be included to a certain extent, based on consideration of factors such as the density of states and statistical distributions. For example, when we take the energy of one phonon as $h\nu$, we are implicitly using the quantum concept.

Classical size effects on transport processes and transport properties of all the energy carriers have been studied in the past. Classical size effects on electron transport were summarized in the book of Tellier and Tosser (1982). The transport of photons is inherently related to the conditions at the boundaries, due to the long photon mean free path in weakly absorbing and scattering media, and thus the vast literature on photon transport is always a good source to consult for solutions and solution methods of the Boltzmann equation. Another useful source of reference is the past literature on neutron transport (Davison, 1957), built on the neutron Boltzmann equation. Flow and heat transfer of rarefied gas has also been studied extensively in the past (Cercignani, 2000). Studies of size effects on phonon transport began in the 1930s through Casimir's pioneering work (Casimir, 1938). The last two decades have seen a surge of investigations on phonon size effects in thin films (Tien et al., 1998).

Although these topics are diverse and often studied in different fields under different names, the starting point in most cases is the Boltzmann equation and thus treatments are often similar. One of our objectives in this chapter is to bring these seemingly diverse subjects under one roof, as we have been doing in previous chapters for electrons, phonons, photons, and gas molecules. Size effect problems typically involve integral equations because field quantities such as temperature and chemical potential in the domain of interest are affected directly by other points in the domain or points at the boundaries. Problems of this type are also called nonlocal. An example is thermal radiation in an evacuated enclosure. Photon density inside the enclosure at every point depends on emissions from the boundaries, rather than the local temperature gradient of the photons as in the Fourier law. We will see further that, for nonlocal transport, the local equilibrium conditions are no longer satisfied and thus the meanings of field quantities, usually thermodynamic variables definable only under equilibrium conditions, are different from our conventional understanding and their appropriateness can be rightfully questioned.

We divide this chapter along the following lines. Section 7.1 treats electron and phonon transport parallel to thin film planes. Section 7.2 deals with transport perpendicular to boundaries and interfaces in layered media. Rarefied gas flow in planar geometries is discussed in section 7.3. These first three sections can be categorized as transport in planar geometries. Ideally, sections 7.1 and 7.3 should be treated together as transport parallel to interfaces and boundaries. The mathematical solution of rarefied gas flow, however, is more complicated and is actually closer to the solution method used for transport perpendicular to the interfaces and boundaries. Thus, rarefied gas flow is treated in a separate section. Transport in nonplanar geometries is introduced in section 7.4. Except for simple cases such as heat and electrical conduction along a thin–film plane, these transport problems typically involve integral equations with no analytical solutions. Integral equations are usually more difficult to solve numerically than differential equations, because while the derivative of a quantity in a differential equation depends only on the immediate neighboring points, the integral of a quantity in the integral equation depends on the distribution of the same quantity over the whole domain. We will introduce two approximations that convert integral equations into differential equations. One is the diffusion approximation (section 7.5), coupled with discontinuous boundary conditions. The other is a ballistic–diffusive treatment (section 7.6).

7.1 Size Effects on Electron and Phonon Conduction Parallel to Boundaries

When the electron and/or phonon mean free paths are comparable to or larger than the thickness of a thin film, the carriers will collide more with the boundaries. The random motion of the carriers, and thus their possible collision with the boundaries, means that we need to consider multidimensional transport. In this section, we consider transport parallel to boundaries, such as thermal and electrical conduction along a thin film plane

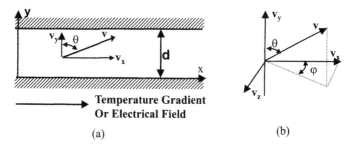

Figure 7.1 (a) One-dimensional electrical or thermal conduction along a thin film; and (b) momentum (velocity) coordinates.

or a wire axis. We start from the steady-state Boltzmann equation under the relaxation time approximation

$$\mathbf{v} \bullet \nabla_r f + \frac{\mathbf{F}}{m} \bullet \nabla_v f = -\frac{f - f_0}{\tau} \qquad (7.1)$$

where f_0 is the equilibrium distribution that depends on the spatial coordinates through the spatial variations of temperature T and/or chemical potential μ for electrons and phonons, and on the average velocity for fluid flow. To simplify the analysis, we further limit the consideration to two-dimensional geometries, as shown in figure 7.1, such as thin films for electrical and thermal conduction. We also assume that along the x-direction the length is long, so that the end effects can be neglected. The Boltzmann equation for such a two-dimensional problem can be written as

$$v_x \frac{\partial f}{\partial x} + v_y \frac{\partial f}{\partial y} + \frac{F_x}{m} \frac{\partial f}{\partial v_x} = -\frac{f - f_0}{\tau} \qquad (7.2)$$

We introduce a deviation function g such that

$$g = f - f_0 \qquad (7.3)$$

Equation (7.2) can be written as

$$v_x \frac{\partial f_0}{\partial x} + v_x \frac{\partial g}{\partial x} + v_y \frac{\partial f_0}{\partial y} + v_y \frac{\partial g}{\partial y} + \frac{F_x}{m} \frac{\partial f_0}{\partial v_x} + \frac{F_x}{m} \frac{\partial g}{\partial v_x} = -\frac{g}{\tau} \qquad (7.4)$$

In the above equation, $\partial f_0/\partial y = 0$ because f_0 is a function of x only for transport along the film. We can also set $\partial g/\partial v_x = 0$ on the basis that the spatial size effect does not affect f in the moment space. Along the x-direction, we will use the same approximation as we did in $\partial f/\partial x$ term deriving the diffusion equations and keep only the $\partial f_0/\partial x$ term while neglecting $\partial g/\partial x$ term, which is justified since we assumed that the length in this direction is long. This approximation leads to

$$v_x \frac{\partial f_0}{\partial x} + v_y \frac{\partial g}{\partial y} + \frac{F_x}{m} \frac{\partial f_0}{\partial v_x} = -\frac{g}{\tau} \qquad (7.5)$$

or

$$\tau v \cos\theta \frac{\partial g}{\partial y} + g = -\frac{\tau F_x}{m}\frac{\partial f_0}{\partial v_x} - \tau v_x \frac{\partial f_0}{\partial x} \equiv -S_0 \tag{7.6}$$

in the coordinate system shown in figure 7.1. Note that we have chosen the polar angle θ oriented from the v_y direction instead of the x-direction as we did in chapter 6. We call S_0 the source function. The major difference of eq. (7.6) from the derivation of the diffusion equation is that, due to the existence of boundaries in the y-direction, the distribution function f or g becomes dependent on y.

Equation (7.6) is a first-order differential equation and only one boundary condition is needed. If this is the case, how will the two surfaces, such as the two surfaces of a film in figure 7.1, affect the transport? The answer lies in that we need one boundary condition for all the velocity components, that is, for all directions θ in eq. (7.6). Taking a boundary point at $y = 0$, we need the boundary condition for carriers both coming toward the boundary and leaving the boundary. Because the carriers coming toward the boundary at $y = 0$ can originate from the other boundary at $y = d$, the distribution of these carriers is again unknown. Thus, it is usually difficult to specify the carriers coming toward the boundary as a known boundary condition. Instead, the boundary conditions are often specified for only the carriers leaving the boundary and at both $y = 0$ and $y = d$, each covering half the space, and this is equivalent to specifying a boundary condition at one boundary for the entire space. This point will become clearer in the subsequent sections.

7.1.1 Electrical Conduction along Thin Films

For electron flow along the x-direction of an isothermal conductor, the chemical potential E_f is a function of x only, such that $\partial f_0/\partial y$ is zero and the source function S_0 depends only on x. Using the results in chapter 6, such as eq. (6.75), we can write the source function as

$$S_0 = -e\tau v_x \frac{\partial f_0}{\partial E}\left(\mathscr{E} + \frac{1}{e}\frac{dE_f}{dx}\right) = S_0(x) \tag{7.7}$$

which does not depend on y. Introducing a nondimensional coordinate $\eta = y/(\tau v)$, the general solution of eq. (7.6) is

$$g^+(\eta, \mu) = g(\eta, \mu) = g_1 \exp\left(-\frac{\eta}{\mu}\right) - S_0 \quad (0 < \mu < 1) \tag{7.8}$$

$$g^-(\eta, \mu) = g(\eta, \mu) = g_2 \exp\left(\frac{\xi - \eta}{\mu}\right) - S_0 \quad (-1 < \mu < 0) \tag{7.9}$$

where $\mu (= \cos\theta)$ is the directional cosine, $\xi = d/(v\tau)$ is the nondimensional thickness, and $\eta = y/(v\tau)$ is the nondimensional y-coordinate. We split the solution into two parts because, as discussed before, at the two boundaries $y = 0$ and $y = d$, it is easier to specify the boundary conditions for the particles leaving the surfaces. To evaluate the electrical current, we must specify these boundary conditions such that g_1 and g_2 can be determined. Carrier reflection at an interface or surface was dealt with in chapter 5. If the interface is very smooth in comparison with the wavelength, the carrier reflection is *specular* and the reflectivity and transmissivity can be calculated using relations such as the Fresnel coefficients. Specular interfaces, however, are usually difficult to realize because of material imperfections. In many cases, the interfaces can be considered as *diffuse*, meaning that an incident carrier has equal probability to be reflected in all directions, or somewhere between specular and diffuse. We will consider two special cases: in the first limit, the boundaries scatter electrons diffusely; in the other limit, the boundaries scatter electrons specularly. The solution for partially specular and partially diffuse surfaces will be given without mathematical details.

When the boundary scatters electrons diffusely, and when there is no current flowing out of the boundary, electrons leaving the boundaries are distributed isotropically and follow the local Fermi–Dirac distribution; that is, $f = f_0$. We have

$$g^+(y=0) = 0 \text{ and } g^-(y=d) = 0 \tag{7.10}$$

Substituting the above equations into eqs. (7.8) and (7.9), we obtain

$$g_1 = g_2 = S_0 \tag{7.11}$$

and

$$g^+(\eta, \mu) = g(\eta, \mu) = S_0 \left[\exp\left(-\frac{\eta}{\mu}\right) - 1 \right] \quad (0 \le \mu \le 1) \tag{7.12}$$

$$g^-(\eta, \mu) = g(\eta, \mu) = S_0 \left[\exp\left(\frac{\xi - \eta}{\mu}\right) - 1 \right] \quad (-1 \le \mu \le 0) \tag{7.13}$$

The total electrical current (per unit depth perpendicular to the xy-plane) in the x-direction is

$$\begin{aligned} j_e &= \int_0^d \left\{ \int \left[\int (-e) v_x f D(E) d\Omega / 4\pi \right] dE \right\} dy \\ &= -\int_0^d \left\{ \int \left[e D(E) dE \int_0^{2\pi} d\varphi \left(\int_0^\pi v_x g(\eta, \theta, \varphi) \sin\theta \, d\theta / 4\pi \right) \right] dE \right\} dy \\ &= -\frac{1}{4\pi} \int_0^d \left\{ \int \left[e D(E) dE \int_0^{2\pi} d\varphi \left(\int_0^1 v_x [g^+(y, \mu) + g^-(y, -\mu)] d\mu \right) \right] dE \right\} dy \end{aligned}$$
$$\tag{7.14}$$

where we have again used the fact that f_0 is isotropic and cancels after carrying out the angular integrations, and have used the variable transformation $\mu = \cos\theta$.

Substituting eqs. (7.12) and (7.13) (with $v_x = v \sin\theta \cos\varphi$) into eq. (7.14) and carrying out the integration over φ, we obtain the following expression

$$\frac{j_e/d}{\left(\mathscr{E} + \frac{1}{e}\frac{dE_f}{dx}\right)} = \frac{1}{4d}\int_0^d\left\{\int\left[e^2v^2\tau\frac{\partial f_0}{\partial E}D(E)\right.\right.$$

$$\left.\left.\times \int_0^1 (1-\mu^2)[e^{-\eta/\mu} + e^{-(\xi-\eta)/\mu} - 2]d\mu\right]dE\right\}dy \quad (7.15)$$

The left-hand side is the current density divided by the electrochemical potential gradient, which gives the effective electrical conductivity of the film [eq. (6.76)]. Carrying out the integration over y in eq. (7.15) yields the following expression for the electrical conductivity

$$\sigma = -e^2\int \tau v^2 \frac{\partial f_0}{\partial E}D(E)\,dE\left\{\frac{1}{3} - \frac{\Lambda}{8d}\left[1 - 4E_3\left(\frac{d}{\Lambda}\right) + 4E_5\left(\frac{d}{\Lambda}\right)\right]\right\} \quad (7.16)$$

where the electron mean free path $\Lambda = \tau v$ and the *integral exponential function* of nth order is defined as

$$E_n(x) = \int_0^1 \mu^{n-2}\exp\left(-\frac{x}{\mu}\right)d\mu \quad (7.17)$$

The first term, $\frac{1}{3}$, inside the curly bracket of eq. (7.16) is the bulk electrical conductivity and the rest represents the size effects. Generally, $\Lambda = \tau v$ depends on frequency (energy), so that we cannot pull the exponential functions out of the integration over energy. Under a simplified picture, we can use the average mean free path to approximate τv. In this case, the ratio of the electrical conductivity of the film to the bulk value is

$$\frac{\sigma}{\sigma_b} = 1 - \frac{3}{8\xi}[1 - 4\langle E_3(\xi) - E_5(\xi)\rangle] \quad (7.18)$$

where the nondimensional film thickness $\xi = d/\Lambda$ will be called the electronic thickness. Its inverse $Kn = \Lambda/d$ can be called the *electron Knudsen number*, in accordance with the convention of rarefied gas dynamics (Kennard, 1938).

For specular interfaces, the boundary conditions are

$$f^+(y=0,\mu) = f^-(y=0,-\mu) \text{ and } f^+(y=d,\mu) = f^-(y=d,-\mu) \quad (7.19)$$

For this case, one can prove that the electrical conductivity of the film equals the bulk value because the electrons change direction only in the y-direction upon reflection at the boundary. The flux along the x-direction remains the same.

For interfaces that are partially diffuse and partially specular, solutions have been given for the electrical conductivity,

$$\frac{\sigma}{\sigma_b} = 1 - \frac{3(1-p)}{2\xi}\int_0^1 (\mu - \mu^3)\frac{1 - \exp(-\xi/\mu)}{1 - p\exp(-\xi/\mu)}d\mu \quad (7.20)$$

where p represents the fraction of specularly reflected electrons, which we will call the specularity parameter. The rest of the electrons, $(1-p)$, experience diffuse reflections at the interfaces. Such solutions are generally called Fuchs–Sondheimer solutions (Fuchs, 1938; Sondheimer, 1952; Tellier and Tosser, 1982). Following the same procedure, one can derive an identical expression for the electronic contribution to the thermal conductivity (Kumar and Vradis, 1994).

We should emphasize that the above derivations consider the size effects due to reflection at the boundaries. In reality, the solid structure may change for films of different thicknesses and deposition conditions. For example, the grain size typically increases with increasing thickness, and the change of internal structure will affect the relaxation time. An example is the thermal conductivity of metallic films (Nath and Chopra, 1973; Qiu and Tien, 1993). Experimental data on the thickness dependence of the thermal conductivity of metallic films (Nath and Chopra, 1973) can be explained only after considering the grain size dependence on the film thickness (Qiu and Tien, 1993).

7.1.2 Phonon Heat Conduction along Thin Films

For a freestanding single layer thin film (that is, no heat goes across the boundary), the heat flux solution is essentially the same as the solution for electrical flux. Following the same procedure, we can arrive at the following solution for phonon thermal conductivity with diffusely scattering boundaries,

$$\frac{k}{k_b} = 1 - \frac{3}{8\xi}[1 - 4(E_3(\xi) - E_5(\xi))] \qquad (7.21)$$

where $\xi = d/\Lambda$, Λ being the phonon mean free path, can be called the acoustic thickness (Majumdar, 1993) and its inverse can be called the *phonon Knudsen number* $Kn = \Lambda/d$. For partially specular and partially diffuse boundaries, the thermal conductivity can be calculated from

$$\frac{k}{k_b} = 1 - \frac{3(1-p)}{2\xi} \int_0^1 (\mu - \mu^3) \frac{1 - \exp(-\xi/\mu)}{1 - p\exp(-\xi/\mu)} d\mu \qquad (7.22)$$

Phonon heat conduction in thin silicon films has been measured and modeled by Volklein and co-workers on polycrystalline thin films (Volklein and Kessler, 1986; Dilliner and Volklein, 1990) and by Goodson's group for single crystalline silicon thin films (Ju and Goodson, 1999; Goodson and Ju, 1999). A model based on geometrical considerations rather than on solving the Boltzmann equation was developed by Flik and Tien (1990).

In multilayer structures, phonons experience not only reflection but also transmission at the interface. There is no general analytical solution for multilayer structures for transport along the thin film plane. Chen and Tien (1993) analyzed quantum well structures with diffuse interface scattering and showed that the result is identical to that of a freestanding film with diffuse interfaces. Chen (1997) considered phonon transport in superlattices for partially diffuse and partially specular interfaces. In this case, the reflection and transmission of phonons at the interface must be considered. Using the phonon mean free path and the gray medium approximation, that is, the mean free path is independent of frequency, the effective thermal conductivity of a superlattice can be expressed in terms of the bulk properties of each layer as (Chen, 1997)

$$k = \sum_{i=1}^{2} \chi_i k_i [1 - 1.5p(1 - a_j/a_i)A_{si}/\xi_i - 1.5(1 - p)A_{di}/\xi_i] \quad (7.23)$$

where k_i and a_i are the thermal conductivity and diffusivity of the ith layer ($i = 1, 2$) of one period of a superlattice, respectively, x_i is the fractional thickness of the ith layer, $\chi_i = d_i/(d_1 + d_2)$. The nondimensional integral functions A_{s1} and A_{d1} are given below:

$$A_{s1} = \int_0^1 [\tau_{12}\mu_1(1-\mu_1^2)(1 - e^{-\xi_1/\mu_1})(1 - pe^{-\xi_2/\mu_2})/N]d\mu_1 \quad (7.24)$$

$$A_{d1} = \int_0^1 \{\mu_1(1-\mu_1^2)(1 - e^{-\xi_1/\mu_1})$$
$$\times [1 - p(R_{21} - \tau_{12}a_2/\alpha_1)e^{-\xi_2/\mu_2}]/N\}d\mu_1 \quad (\tau_{12} \neq 0) \quad (7.25)$$

$$A_{d1} = \int_0^1 \{\mu_1(1-\mu_1^2)(1 - e^{-\xi_1/\mu_1})$$
$$\times [1 + pe^{-\xi_1/\mu_1}]/[1 - p^2 e^{-2\xi_1/\mu_1}]\}d\mu_1 \quad (\tau_{12} = 0) \quad (7.26)$$

$$N = 1 - p(R_{12}e^{-\xi_1/\mu_1} + R_{21}e^{-\xi_2/\mu_2})$$
$$+ p^2 (R_{12}R_{21} - \tau_{12}\tau_{21}) e^{-\xi_1/\mu_1 - \xi_2/\mu_2} \quad (7.27)$$

where τ_{12} and R_{12} represent the phonon transmissivity and reflectivity at the interface for phonons going from medium 1 to 2 and vice versa. Models for these coefficients have been discussed in chapter 5. Similar expressions can be written for layer 2 by replacing and permuting the subscript index.

In the above expressions, we have used the phonon mean free path to express the results. For phonons, the estimation of the mean free path must be done very carefully. Often, an estimation of the mean free path is obtained from the simple kinetic theory derived in chapter 1,

$$k = \frac{1}{3}Cv\Lambda$$

using the measured volumetric specific heat, C, and the speed of sound. This method of estimating the phonon mean free path, however, usually leads to an underestimation of the mean free path for those phonons that are actually carrying the heat, for the following reasons (Hyldgaard and Mahan, 1996; Chen, 1997):

1. Phonons are dispersive and their group velocities vary from the speed of sound at the Brillouin zone center to zero at the zone edge. The average phonon group velocity is much smaller than the speed of sound.

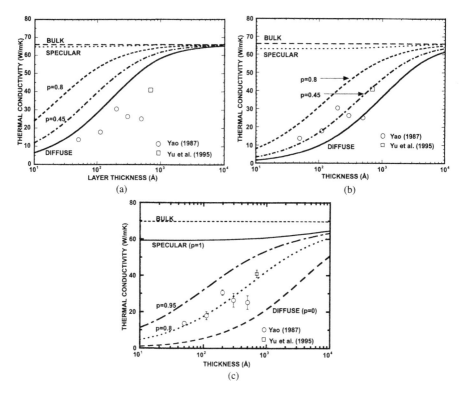

Figure 7.2 Calculated thermal conductivity of GaAs/AlAs superlattices based on different ways of estimating the phonon mean free path in bulk materials: (a) kinetic theory, (b) averaging the phonon mean free path over the acoustic branches and neglecting contributions from optical phonons, and (c) based on frequency-dependent phonon relaxation time. Points are experimental data (Chen, 1997).

2. Optical phonons contribute to the specific heat but typically contribute little to heat flux, due to their small group velocity and their high scattering rates.
3. Phonon scattering is highly frequency dependent. High–frequency phonons are usually scattered more strongly than low–frequency phonons.

We use figures 7.2(a)–(c) to strengthen these points (Chen, 1997). These figures compare the modeling results with experimental data of the in-plane thermal conductivity for GaAs/AlAs superlattices, based on different ways of estimating the phonon mean free path. In figure 7.2(a), the mean free path is estimated on the basis of simple kinetic theory, the bulk specific heat, and the speed of sound, which leads to a phonon mean free path of about 209 Å in GaAs and 369 Å in AlAs. The underestimation of phonon mean free path is clear since results for even the worst scenario, totally diffuse interfaces, lead to theoretical values much larger than the experimental results. In figure 7.2(b), we estimate the average phonon group velocity and the mean free path on the basis of a sine-function approximation to the acoustic phonon dispersion relation, totally neglecting optical phonon contributions to thermal transport. The phonon mean free paths thus obtained are 1058 Å for GaAs and 2248 Å for AlAs. The agreement with experimental data is much better. In figure 7.2(c), we show the calculated thermal conductivity on the basis of the frequency-dependent phonon relaxation time, and again good agreement with experimental data is obtained, albeit with a very different interface specularity parameter from figure 7.2 (b). For silicon, simple kinetic theory will give a phonon mean free path of ~400 Å, but more careful estimation leads to a value between 2500 and 3000 Å (Chen, 1998; Ju and Goodson, 1999).

Naturally, one would think that the frequency-dependent treatment is more accurate. It should be noted, however, that the frequency-dependent relaxation time is obtained from fitting the bulk thermal conductivity data of those materials. These bulk thermal conductivity models are subject to uncertainty (as we explained in the previous chapter) that will propagate into the thin-film models.

The solutions given above are for planar geometries. Phonon heat conduction in wire geometries, based on the Boltzmann equation, has also been discussed in literature (Ziman, 1960; Volz and Chen, 1999; Walkauskas et al., 1999). All these solutions are based on the assumption that the length of the wire is much longer than the mean free path. In very low temperatures, the phonon mean free path is long and the finite-length of the sample should also be considered (Berman et al., 1955).

An unsolved issue is that of the specularity parameter p. This is the case not only for phonon interface transport but also for all types of carriers. Most of the chapter 5 treatments on interface reflection and transmission processes deal with specular interfaces that obey the Snell law. In section 5.2.3, for thermal boundary resistance, we pointed out that there exists no rigorous model for interface reflectivity and transmissivity when interface scattering is not specular. The phonon diffuse scattering model (Swartz and Pohl, 1989) was introduced as a limit for diffuse scattering, but we also cautioned that the model is crude and not generally applicable. It is accepted that if the boundary roughness is much larger than the dominant carrier wavelength, the scattering can be approximated as diffuse and, in the opposite limit, specular. It should be realized, however, that interface roughness is not the only reason for diffuse scattering and different carriers have different interface scattering mechanisms. For example, it is the contrast of the refractive index that scatters photons but it is the electrostatic potential that scatters electrons. An interface may be very flat structurally but has "bumpy" electrostatic potentials. The scattering of gas molecules at an interface will depend strongly on the interaction potential of the molecules with the solid wall. For phonons, while the acoustic mismatch can be a major interface scattering mechanism, and thus the interface roughness reflects the fluctuations of the acoustic mismatch, it is also possible that inelastic force interaction provides another channel for diffuse interface scattering (Stoner and Maris, 1993). There has been some modeling in the past for interface roughness scattering of electrons (Weisbuch and Vinter, 1991) and phonons (Majumdar, 1991). A relatively large amount of literature exists for the effects of roughness on radiative properties (Modest, 2003; Siegel and Howell, 1992). However, as is often the case for complex problems, there are no general answers. An approximate formula was derived by Ziman (1960) for the fraction of specularly scattered phonons,

$$p(\lambda) = \exp\left(-\frac{16\pi^3 \Delta^2}{\lambda^2}\right) \tag{7.28}$$

where Δ is the root mean square deviation of the height of the surface from the reference plane, also called the asperity parameter, and λ is the phonon wavelength. Because the dominant phonon wavelength is very short at room temperature, typically ~10–20 Å, the above expression leads to a small fraction of the specularly scattered phonons even for an interface with only ~ 1 Å asperity. Although Ziman's expression is often used (Chen and Tien, 1993; Asheghi et al., 1998), it should be viewed with caution and more work needs to be done on the effects of interface structures on phonon transport.

7.2 Transport Perpendicular to the Boundaries

In this section, we will examine the energy transport perpendicular to boundaries as shown in figure 7.3. This situation is most often seen in thermal radiation where photons experience reflection and refraction at the interfaces. In chapter 5 we discussed the thermal boundary resistance for phonons, which is a process that occurs at a single interface. In thin film structures, electron and phonon transport is affected by processes occurring both inside the film and at the interfaces. Heat conduction in a rarefied gas between two parallel plates also has similar solutions. This section will exclude the flow of molecules since boundaries are typically impermeable to molecules. The external flow geometry may be classified into our current situation but, since we will be limiting our consideration to planar geometry only, we opt to spend another section to discuss the external transport.

For transport along the boundary, as treated in section 7.1, T and E_f are independent of y. This simplification makes the solution of the Boltzmann equation relatively straight-foward. In the direction perpendicular to the boundary, however, temperature and/or chemical potential may depend on y. We will need to treat these quantities as unknowns and solve for their distributions, which significantly complicates the mathematics of the problem.

7.2.1 Thermal Radiation between Two Parallel Plates

Let's consider radiation heat transfer between two parallel plates separated by an absorbing, emitting, and scattering medium. For simplicity, we limit our discussion to isotropic scattering; that is, the phase function in eq. (6.42) equals one. In this case, the equation of radiative transfer, eq. (6.42), becomes

$$\mu \frac{\partial I_\nu}{\partial y} = -K_{e\nu}(I_\nu - I_{0\nu}) \tag{7.29}$$

where $K_{e\nu}$ is the extinction coefficient, that is, the inverse photon mean free path. We have opted to use the intensity rather than the distribution function and the extinction coefficient rather than the photon mean free path for consistency with the literature in

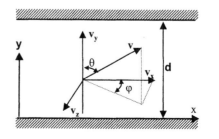

Figure 7.3 Coordinate system for photon, phonon, and molecular heat transfer perpendicular to the plane of a thin film.

radiation heat transfer (Modest, 2003, Siegel and Howell, 1992). Intensity is related to the distribution function through a simple transformation, as discussed in section 6.1.3. Correspondingly, $I_{0\nu}$ is the blackbody radiation intensity. We further assume that the boundaries are diffuse and that the medium is gray, which means that the extinction coefficient is independent of wavelength, and thus we will drop the spectral symbol in the subscript and treat the intensity as spectrally integrated. The solution of eq. (7.29) is

$$I^+(\eta, \mu) = \frac{J_{q1}^+}{\pi} \exp\left[-\frac{\eta}{\mu}\right] + \int_0^\eta I_0(\eta') \exp\left[-\frac{\eta - \eta'}{\mu}\right] \frac{d\eta'}{\mu} \quad (\mu > 0) \quad (7.30)$$

$$I^-(\eta, \mu) = \frac{J_{q2}^-}{\pi} \exp\left[\frac{\xi - \eta}{\mu}\right] - \int_\eta^\xi I_0(\eta') \exp\left[\frac{\eta' - \eta}{\mu}\right] \frac{d\eta'}{\mu} \quad (\mu < 0) \quad (7.31)$$

where $\eta = K_e y$ and $\xi = K_e d$ (also called the optical thickness), J_{q1}^+, and J_{q2}^- are the heat fluxes leaving surface 1 and surface 2 respectively. The inverse of ξ is the photon Knudsen number $Kn = 1/(K_e d)$. In thermal radiation, J_{q1}^+ and J_{q2}^- are called the radiosity of the surfaces (Siegel and Howell, 1992; Modest, 1993). The first term on the right-hand side represents the contribution to the intensity at location η in the direction $\mu (= \cos \theta)$ from the wall. The second term represents the contribution from the emission along the ray propagation path, or the optical path for photons. The heat flux at any point is given by

$$J_q(\eta) = \int_{4\pi} I(\eta, \mu) \cos\theta \, d\Omega = \int_0^{2\pi} \int_0^\pi I \sin\theta \cos\theta \, d\theta \, d\varphi$$

$$= 2\pi \int_0^{\pi/2} I^+(\eta, \mu) \sin\theta \cos\theta \, d\theta + 2\pi \int_{\pi/2}^\pi I^-(\eta, \mu) \sin\theta \cos\theta \, d\theta$$

$$= 2\pi \int_0^1 I^+(\eta, \mu) \mu \, d\mu - 2\pi \int_0^1 I^-(\eta, -\mu) \mu \, d\mu$$

$$= J_q^+(\eta) - J_q^-(\eta) \quad (7.32)$$

where

$$J_q^+(\eta) = 2\pi \int_0^1 \mu I^+(\eta, \mu) d\mu \quad \text{and} \quad J_q^-(\eta) = 2\pi \int_0^1 \mu I^-(\eta, -\mu) d\mu \quad (7.33)$$

represent the photon heat flux propagating in the positive and negative y-directions, respectively. Substituting the intensity solutions, eqs. (7.30) and (7.31), into eqs. (7.33) and carrying out the integration, we get

$$J_q^+ = 2 \int_0^1 \left\{ J_{q1}^+ \exp\left[-\frac{\eta}{\mu}\right] + \int_0^\eta e_0(\eta') \exp\left[-\frac{\eta - \eta'}{\mu}\right] \frac{d\eta'}{\mu} \right\} \mu \, d\mu \quad (7.34)$$

and

$$J_q^- = 2\int_0^1 \left\{ J_{q_2}^- \exp\left[-\frac{\xi-\eta}{\mu}\right] + \int_\eta^\xi e_0(\eta') \exp\left[-\frac{\eta'-\eta}{\mu}\right] \frac{d\eta'}{\mu} \right\} \mu\, d\mu \quad (7.35)$$

where e_0 is the blackbody emissive power, $e_0 = \pi I_0 = \sigma T^4$. The net heat flux in the y direction at location η is

$$\begin{aligned} J_q &= J_q^+ - J_q^- \\ &= 2\left\{ J_{q_1}^+ E_3(\eta) + \int_0^\eta e_0(\eta') E_2(\eta - \eta') d\eta' \right. \\ &\quad \left. - J_{q_2}^- E_3(\xi - \eta) - \int_\eta^\xi e_0(\eta') E_2(\eta' - \eta) d\eta' \right\} \end{aligned} \quad (7.36)$$

When no other forms of heat transfer are present, that is, no conduction or convection, and no other forms of internal heat generation, the radiation heat flux must be constant, or

$$dJ_q/d\eta = 0 \quad (7.37)$$

Substituting eq. (7.36) into (7.37) leads to

$$J_{q_1}^+ E_2(\eta) + \int_0^\eta e_0(\eta') E_1(\eta - \eta') d\eta' - 2e_0(\eta) + J_{q_2}^- E_2(\xi - \eta)$$

$$+ \int_\eta^\xi e_0(\eta') E_1(\eta' - \eta) d\eta' = 0 \quad (7.38)$$

The above equation determines the distribution of e_0 or temperature. We can write the equation in a more amenable form by introducing

$$e_0^*(\eta) = \frac{e_0(\eta) - J_{q_2}^-}{J_{q_1}^+ - J_{q_2}^-} \text{ and } J_q^* = \frac{J_q}{J_{q_1}^+ - J_{q_2}^-} \quad (7.39)$$

Equation (7.38) becomes

$$2e_0^*(\eta) = E_2(\eta) + \int_0^\xi e_0^*(\eta') E_1(|\eta - \eta'|) d\eta' \quad (7.40)$$

This is a linear, nonhomogeneous, *Fredholm integral equation* of the second kind. The function inside the integral beside the unknown e_0^*, that is, $E_1(|\eta - \eta'|)$, is called the kernel. Such a Fredholm integral equation does not have an analytical solution, although

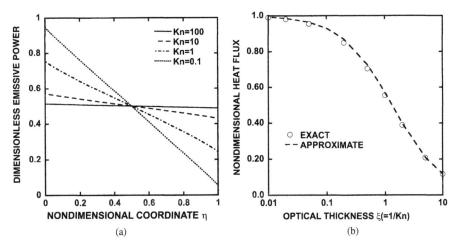

Figure 7.4 (a) Nondimensional emissive power; (b) heat flux for radiation transfer between two parallel plates. Approximate solution will be discussed in section 7.5.1.

approximate solution methods have been developed. In example 7.1, we introduce a direct numerical solution method for such integral equations. With e_0^* known, the heat flux can be calculated from eq. (7.36) as

$$J_q^* = 1 - 2 \int_0^{\xi} e_0^*(\eta) E_2(\eta') d\eta' \tag{7.41}$$

Figure 7.4(a) shows the distribution of nondimensional blackbody emissive power e_0^* for different optical thicknesses ξ, and figure 7.4(b) shows the nondimensional heat flux distribution as a function of the optical thickness ξ. The approximate solution will be explained in section 7.5.1.

We notice from figure 7.4(a) that, for a small optical thickness, there exist discontinuities in the temperature distribution at the two boundaries. Let's take an extreme case where both walls are black and are at temperatures T_1 and T_2, respectively. In this case, $J_{q1}^+ = e_{01}(T_1)$ and $J_{q2}^- = e_{02}(T_2)$. We then have

$$e_0^*(\eta) = \frac{e_0(\eta) - e_{02}(T_2)}{e_{01}(T_1) - e_{02}(T_2)} \tag{7.42}$$

At $\eta = 0$, e_0^* is not equal to 1. This means that $e_0(y=0) \neq e_{01}(T_1)$; in other words, the photon gas temperature at $y = 0$ is not equal to the wall temperature. A similar situation exists at the other wall. This phenomenon is called *temperature slip*. Such discontinuous phenomena are common in the interface regions. It is very important that we understand why the discontinuity occurs, as will be discussed below.

In chapter 5, we carefully explained the origins of the temperature discontinuity at an interface in the thermal boundary resistance phenomenon, which include (1) the interface reflection of phonons and (2) the use of equilibrium quantities to represent the highly nonequilibrium process at the boundaries. We also emphasized the importance

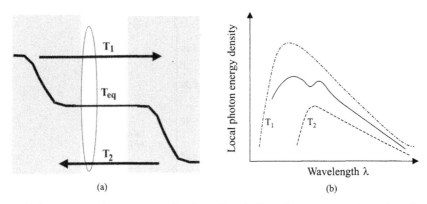

Figure 7.5 Meaning of temperature for thermal radiation when no scattering exists. In (a), T_{eq} is a measure of the photon energy denisty in between the two objects. The photon distribution deviates significantly from the Planck law and the photons are at highly nonequilibrium states as illustrated in (b).

of using a consistent temperature in calculating the thermal boundary resistance. The photon temperature slips observed in figure 7.4 are of similar origins, but with a subtle difference regarding which temperature should be used. We consider the limit $\xi = K_e d \to 0$, when there is no scattering in the medium between the plates. In this case, photons propagating from one wall to the other do not interact with photons emitted from the other wall, nor do they interact with the medium, as shown in figure 7.5(a). There exists no equilibrium among photons inside the medium. The local energy density distribution at any point between the space is the summation of the two groups of photons originating from each wall, which gives $e_0^* = 0.5$, agreeing with the results shown in figure 7.4(a). However, the local photon spectral distribution, as sketched in figure 7.5(b), deviates far from the Planck distribution. The photon temperature, which can be shown to be $T_{\text{ave}} = (T_1^4 + T_2^4)^{1/4}$, cannot be understood as a near-equilibrium quantity but can be regarded as a measure of the average local photon energy. Such a photon temperature can be understood as the equivalent equilibrium photon temperature, similar to what we defined for phonons in chapter 5. It is also clear that T_1 and T_2 are the temperatures of photons emitted into the medium from the two boundaries and are not the local photon temperatures at the other sides of the two interfaces; that is, they correspond to the emitted photon temperature, as we discussed for phonons in section 5.2.3. Clearly, we are using two inconsistent temperature definitions. In the case of black walls, the reflectivity is zero and thus photons incident toward an interface are all transmitted into the wall. If we could measure the local equivalent equilibrium photon temperature immediately inside the wall, this temperature should equal T_{ave} with no temperature jump, as shown in figure 7.5(a), similarly to our discussion on thermal boundary resistance in the limit that no interface exists. For thermal radiation, however, we cannot measure the photon temperature immediately inside the wall. The emitted photon temperatures T_1 and T_2 actually represent temperatures of the solids. When the temperature of the solid is uniform due to heat conduction, the photons coming toward the interfaces are at temperatures T_1 and T_2, although these are not the equivalent equilibrium photon temperatures at $y = 0$ and $y = d$, respectively. In this case, the inconsistent use of temperature is actually warranted.

For nonblack but diffuse–gray walls, we can use additional energy conservation relations at the walls. For example, at surface 1 we have

$$J_q = J_{q1}^+ - J_{q1}^- \text{ and } J_{q1}^+ = \varepsilon_1 e_{01} + (1 - \varepsilon_1) J_{q1}^- \tag{7.43}$$

where J_{q1}^- is the radiative flux coming toward the wall (irradiation) and ε_1 is the emissivity. Similar relations can be written for the other surface. Eliminating J_{q1}^- and similarly J_{q2}^- and using eq. (7.39), we arrive at

$$\frac{e_0(\eta) - e_{02}}{e_{01} - e_{02}} = \frac{e_0^*(\eta) + \frac{1-\varepsilon_2}{\varepsilon_2} J_q^*}{1 + q^*(\frac{1}{\varepsilon_1} + \frac{1}{\varepsilon_2} - 2)} \tag{7.44}$$

$$J_q = \frac{J_q^*(e_{01} - e_{02})}{1 + J_q^*(\frac{1}{\varepsilon_1} + \frac{1}{\varepsilon_2} - 2)} \tag{7.45}$$

In the general case, the temperature discontinuities are a combination of the reflection effect and the inconsistent use of temperature definitions. A physical discontinuity always exists unless the reflectivity is zero. The inconsistent use of temperature definition leads to an artificial discontinuity. However, in the case of thermal radiation, as discussed above, the inconsistent use of temperature is necessary since T_1 and T_2, which are the temperatures of the solids and also represent the temperatures of the photons away from the interface, are the temperatures that can actually be measured. Similar situations exist for other cases such as the electrochemical potential for electron transport. Readers must develop their own judgments on when to use which definition, depending on the problem they are dealing with.

If the properties are spectrally dependent, eq. (7.36) is then the spectral heat flux. The total heat flux is obtained by integrating eq. (7.36) over all frequencies. Equation (7.37) applies only to the integrated total heat flux. Strategies for the solution of spectral-dependent problems are similar but the mathematics is much more complicated.

Example 7.1 *Develop a numerical method to solve integral equation* (7.40)

Solution: The key steps to solve an integral equation such as (7.40) are: (1) choosing a numerical integration scheme; (2) dealing with the singularity point at $\eta' = \eta$; (3) formulating the matrix equation; and (4) inverting the matrix, as explained below.

1. *Choosing a numerical integration scheme.* There are many ways to carry out the numerical integration of a function, with different levels of simplicity and efficiency. Consider the integration of a function $f(x)$ from x_1 to x_n; the trapezoidal method is to approximate the function $f(x)$ as small, equally spaced trapezoids with spacing Δx, and to approximate the integral as

$$\int_{x_1}^{x_n} f(x) dx = f(x_1) \frac{\Delta x}{2} + \Delta x \sum_{i=2}^{n-1} f(x_i) + f(x_n) \frac{\Delta x}{2} \tag{E7.1.1}$$

The trapezoidal integration method, however, is not very efficient. A more efficient integration method is the Gauss–Legendre method,

$$\int_{x_0}^{x_n} f(x)dx = \sum_{i=1}^{n} f(x_i)w_i \tag{E7.1.2}$$

where x_i are the roots of the Gauss quadrature and w_i are the weights, both depending on the choice of the number of integration points n. There are standard numerical routines to find these roots and weights (Press et al., 1992).

2. *Dealing with singularities.* If we apply Gauss–Legendre points directly to eq. (7.40), we have

$$2e_0^*(\eta) = E_2(\eta) + \sum_{i=1}^{n} e_0^*(\eta_i) E_1(|\eta - \eta_i|)w_i \tag{E7.1.3}$$

The above equation is valid for every point η. To solve for the n unknowns of $e_0^*(\eta_i)$, we should set η equal to every point η_i and thus obtain n algebraic equations. There exists, however, one glitch. The kernel $E_1(|\eta - \eta_i|)$ approaches infinity, $E_1(0) \to \infty$, at $\eta = \eta_i$. Such singularities are often encountered in integral equations. To deal with this singularity point inside the kernel, one common method is to rewrite eq. (7.40) as

$$2e_0^*(\eta) = E_2(\eta) + \int_0^\xi [e_0^*(\eta') - e_0^*(\eta)] E_1(|\eta - \eta'|) d\eta'$$

$$+ e_0^*(\eta) \int_0^\xi E_1(|\eta - \eta'|) d\eta' \tag{E7.1.4}$$

The advantage of this treatment is that $[e_0^*(\eta') - e_0^*(\eta)]$ will be zero at the singularity point, and the third term on the right-hand side can be integrated directly,

$$\int_0^\xi E_1(|\eta - \eta'|) d\eta' = \int_0^\xi d\eta' \int_0^1 \frac{1}{\mu} e^{-|\eta-\eta'|/\mu} d\mu$$

$$= \int_0^1 \left[\int_0^\eta \frac{1}{\mu} e^{-(\eta-\eta')/\mu} d\eta' + \int_\eta^\xi \frac{1}{\mu} e^{-(\eta'-\eta)/\mu} d\eta' \right] d\mu$$

$$= \int_0^1 (2 - e^{-\eta/\mu} - e^{-(\xi-\eta)/\mu}) d\mu$$

$$= 2 - E_2(\eta) - E_2(\xi - \eta) = c(\eta) \tag{E7.1.5}$$

3. On the basis of the above discussion, we can now discretize eq. (E7.1.4) as

$$2e_0^*(\eta_j) = E_2(\eta_j) + \sum_{i=1}^{n} w_i E_1(|\eta_i - \eta_j|)[e_0^*(\eta_i) - e_0^*(\eta_j)] + e_0^*(\eta_j)c(\eta_j)$$

(E7.1.6)

The above equation can be written in a matrix form

$$A_{ij} e_0^*(\eta_j) = B_i \qquad \text{(E7.1.7)}$$

and solved by any standard matrix inversion method for the unknown distribution function. Figure 7.4(a) shows such distribution functions for different ξ values.

Comments.

1. When using the finite difference method to solve differential equations, most of the matrix elements are zero and the matrix is typically tridiagonal; that is, nonzero elements exist only in the diagonal and its immediately adjacent elements. In contrast, the matrix elements A_{ij} for integral equations are not zero in most cases, and the matrix is no longer a sparse matrix.
2. Because the Gauss–Legendre integration points x_i and the weights change with the integration limits, it is often more convenient to transform the integration limits to fixed values, for example, [0, 1] rather than [0, ξ].

7.2.2 Heat Conduction across Thin Films and Superlattices

Heat conduction perpendicular to a thin film was dealt with in several papers (Majumdar, 1993; Chen and Tien; 1993; Jen and Chieng, 1998; Chen and Zeng, 2001; Zeng and Chen, 2001). Majumdar (1993) solved the phonon Boltzmann equation by assuming that the two surfaces of the film are black phonon emitters. Chen and Tien (1993) employed solutions from radiative heat transfer to express the thermal conductivity of a thin film sandwiched between two thermal reservoirs. The boundary conditions used in these two initial studies (Majumdar, 1993; Chen and Tien, 1993) were analogous to those used in thermal radiation, which also implies the use of inconsistent temperature definitions as discussed in the previous section. For radiation, this inconsistency is not only tolerable but justified. For heat conduction, however, this treatment is prone to errors because the temperature that one can measure is usually the equivalent equilibrium temperature, not the emitted phonon temperature. In thermal boundary resistance measurements at very low temperatures, measurements of emitted phonon temperature are possible with care in placing the thermometer (Swartz and Pohl, 1989) because the phonon mean free path is very long at low temperatures. At room temperature, the short phonon mean free path makes the direct measurement of the emitted phonon temperatures difficult. Also, all the simulations based on either the Fourier law or the Boltzmann equation are made on the basis of equivalent equilibrium temperature. Chen (1998) employed a consistent temperature definition when solving the Boltzmann equation for superlattice structures with multiple interfaces. Heat conduction perpendicular to a thin film plane, using a consistent temperature definition based on equivalent equilibrium temperature, was reconsidered by Zeng and Chen (2001).

Heat conduction in superlattices is an interesting example of solving the phonon Boltzmann equation. We discussed the superlattice phonon spectrum in chapter 5. The superlattice modes are formed if the phases of phonon waves are preserved over several periods. If the interface scattering is strongly phase destroying, which is likely to be the case if diffuse scattering is predominant, the superlattice modes are not formed (Yang et al., 2003). In this case, we can use the bulk phonon properties for each layer and consider the impacts of interfaces on the phonon transport based on the Boltzmann equation. For each layer, we write down the Boltzmann equation, again in intensity form, as

$$\cos\theta_i \frac{\partial i_i}{\partial y_i} + \frac{i_i}{\Lambda_i} = -\cos\theta_i \frac{dI_{0i}}{dy_i} \tag{7.46}$$

where $i_i(y_i, \theta_i) = I_i(y_i, \theta_i) - I_{0i}(T_i)$ is the deviation of intensity for the ith layer ($i = 1, 2$), corresponding to the deviation function g we used in section 7.1. The solutions of eq. (7.46) are similar to eqs. (7.30) and (7.31),

$$i_i^+(\eta_i, \mu_i) = i_i^+(0, \mu_i)e^{-\eta_i/\mu_i} - \int_0^{\eta_i} \frac{dI_{0i}}{d\eta_i'} e^{-(\eta_i-\eta_i')/\mu_i} d\eta_i' \quad \text{(for } 0 < \mu_i < 1\text{)} \tag{7.47}$$

$$i_i^-(\eta_i, \mu_i) = i_i^-(\xi, \mu_i)e^{(\xi_i-\eta_i)/\mu_i}$$
$$+ \int_{\eta_i}^{\xi_i} \frac{dI_{0i}}{d\eta_i'} e^{-(\eta_i-\eta_i')/\mu_i} d\eta_i' \quad \text{(for } -1 < \mu_i < 0\text{)} \tag{7.48}$$

where $\mu_i (= \cos\theta_i)$ is the directional cosine, $\eta_i (= y_i/\Lambda_i)$ the nondimensional y-coordinate, and $\xi_i (= d_i/\Lambda_i)$ the nondimensional layer thickness of the ith layer. The local heat flux in the y-direction can be obtained, according to eq. (7.32), from

$$J_{q_i}(\eta_i) = 2\pi \int_0^1 [i_i^+(\eta_i, \mu_i) - i_i^-(\eta_i, -\mu_i)] \mu_i d\mu_i \tag{7.49}$$

To establish boundary conditions, we use the energy balance at the interface between two adjacent layers. We consider the case of diffuse interfaces only. For the interface between layer $2p$ and layer 1, as shown in figure 7.6, the energy balance gives

$$\int_{2\pi} I_1^+(0, \mu_1) \mu_1 d\Omega_1 = R_{d12} \int_{2\pi} I_1^-(0, -\mu_1) \mu_1 d\Omega_1$$
$$+ \tau_{d21} \int_{2\pi} I_{2p}^+(\xi_2, \mu_2) \mu_2 d\Omega_2 \tag{7.50}$$

where the integration with respect to the solid angle is over the half-space, and R_{dij} and τ_{dij} ($i, j = 1, 2$) are the energy reflectivity and transmissivity, respectively, at an interface for phonons incident from the ith layer toward the jth layer, which are independent of

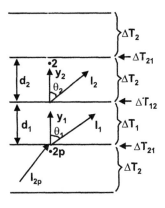

Figure 7.6 Model for phonon heat conduction in periodic thin film structures such as superlattices.

direction for diffuse interface scattering. The symbol $I_{2p}^+(\xi_2, \mu_2)$ in eq. (7.50) represents the phonon intensity at point $2p$, which corresponds to $\eta_2 = \xi_2$ in the layer preceding layer 1. Because phonons are scattered diffusely at the interfaces, phonons leaving an interface are isotropically distributed. Thus, the left-hand side of eq. (7.50) can be integrated analytically, leading to

$$I_1^+(0, \mu_1) = 2R_{d12} \int_{2\pi} I_1^-(0, -\mu_1)\mu_1 d\mu_1 + 2\tau_{d21} \int_{2\pi} I_{2p}^+(\xi_2, \mu_2)\mu_2 d\mu_2 \quad (7.51)$$

We can establish similar equations for the other three quantities by applying the energy balance to each side of the two interfaces in one period:

$$I_1^-(\xi_1, -\mu_1) = 2R_{d12} \int_{2\pi} I_1^+(\xi_1, \mu_1)\mu_1 d\mu_1 + 2\tau_{d21} \int_{2\pi} I_2^-(0, -\mu_2)\mu_2 d\mu_2 \quad (7.52)$$

$$I_2^+(0, \mu_1) = 2R_{d21} \int_{2\pi} I_2^-(0, -\mu_2)\mu_2 d\mu_2 + 2\tau_{d12} \int_{2\pi} I_1^+(\xi_1, \mu_1)\mu_1 d\mu_1 \quad (7.53)$$

$$I_2^-(\xi_2, -\mu_2) = 2R_{d21} \int_{2\pi} I_2^+(\xi_2, \mu_2)\mu_2 d\mu_2 + 2\tau_{d12} \int_{2\pi} I_{1n}^-(0, -\mu_1)\mu_1 d\mu_1 \quad (7.54)$$

where $I_{1n}^-(0, -\mu_1)$ represents the phonon intensity corresponding to $\eta_1 = 0$ in the layer next to layer 2. The above set of interface conditions is applicable to multilayer structures and is not subject to the periodicity requirement of superlattices. For a periodic structure with heat flow in the y-direction, temperatures at the corresponding locations of two identical layers and thus the equilibrium intensities, $I_{02p}^+(\xi_2)$ and $I_{02}^+(\xi_2)$, are not equal to each other. Chen (1998) reasoned that the following conditions must be satisfied

$$i_{2p}(\eta_2, \mu_2) = i_2(\eta_2, \mu_2) \text{ and } i_{1n}(\eta_1, \mu_1) = i_1(\eta_1, \mu_1) \quad (7.55)$$

due to the periodicity of the superlattice.

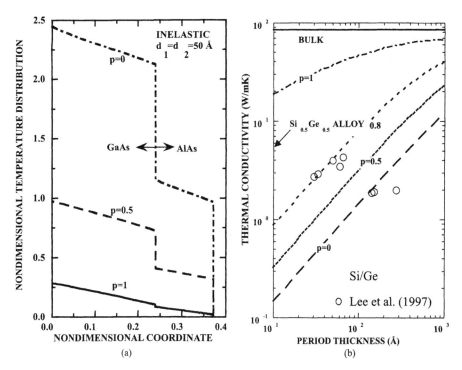

Figure 7.7 (a) Temperature distribution in one period of a GaAs/AlAs superlattice. (b) Effective thermal conductivity of Si/Ge superlattices (Chen, 1998).

On the basis of the four boundary conditions and eq. (7.75), the unknown coefficients $i_i^+(0, \mu_i)$ and $i_i^-(\xi_i, \mu_i)$ in eqs. (7.47) and (7.48) can, in principle, be expressed in terms of the unknown equilibrium phonon distributions $I_{01}(\eta_1)$ and $I_{02}(\eta_2)$ and their derivatives. Following the steps used in determining the temperature distribution and nondimensional heat flux for radiative heat transfer between two parallel plates, the phonon temperature distributions and the effective thermal conductivity of the superlattices can be obtained. Detailed solutions are complicated and will not be given here. Interested readers can refer to the original work for diffuse interfaces (Chen, 1996b), specular interfaces (Chen and Neagu, 1997), and partially diffuse and partially specular interfaces (Chen, 1998). Figure 7.7(a) shows the temperature distribution inside one period of a GaAs/AlAs superlattice and figure 7.7(b) shows the effective thermal conductivity of Si/Ge superlattice structures. These models indicate that the largest temperature drop actually occurs at the interfaces rather than inside the films [figure 7.7(a)]. The large interface resistance causes orders-of-magnitude reduction in the effective thermal conductivity of these structures in comparison with the Fourier heat conduction theory, as shown in figure 7.7(b).

7.2.3 Rarefied Gas Heat Conduction between Two Parallel Plates

The study of rarefied gas flow and heat transfer has a long history (Knudsen, 1909; Maxwell, 1953). Because of the long molecular mean free path at low pressures, the subject gained significant attention with its applications in vacuum and space technologies. The Boltzmann-equation-based solution for rarefied gas heat conduction between two parallel plates was formulated by Willis (1962). Here, we will recast it into our current formulation, based on the description of Patterson (1971).

We consider heat conduction in the y-direction as shown in figure 7.3. The Boltzmann equation under the relaxation time approximation can be written as

$$v_y \frac{\partial f}{\partial y} = -\frac{f - f_0}{\tau} \qquad (7.56)$$

The equilibrium distribution f_0 is given by the Maxwell distribution,

$$f_0 = n \left(\frac{m}{2\pi \kappa_B T}\right)^{3/2} \exp\left[-\frac{v_x^2 + v_y^2 + v_z^2}{2\kappa_B T/m}\right]$$

$$= \frac{n}{(2\pi RT)^{3/2}} \exp\left(-\frac{v^2}{2RT}\right) \qquad (7.57)$$

where $v^2 = v_x^2 + v_y^2 + v_z^2$, $R = \kappa_B N_A/(m N_A) = R_u/M_A$ [J K^{-1} kg^{-1}] is the ideal gas constant, M_A being the molar weight and R_u the universal gas constant. For the heat conduction problem, both $T(y)$ and the local particle density $n(y)$ are functions of y. The dependences of both T and n on y significantly complicate the mathematics, as we will see. In addition, the relaxation time may also depend on the local particle number density. In chapter 1, we derived the mean free path as $\Lambda = 1/(n\pi d_a^2)$, where d_a is the molecular diameter, which means that the relaxation time $\tau = \Lambda/v_{th}$ is also dependent on y through its dependency on n and T.

Similarly to electrons and phonons, the reflection of molecules can also be approximated as specular or diffuse. In a specular reflection, the momentum and energy of the reflected molecules equal that of the incident ones. In a diffuse reflection, it is normally assumed that the reflected molecules obey the local Boltzmann distribution as determined by the local temperature and velocity of the wall. Thus, for a partially specular and partially diffuse interface, the distribution function of the molecules leaving a wall ($v_y > 0$) is (Maxwell, 1953)

$$f(y=0, v_y > 0) = pf(y=0, v_y < 0) + (1-p)f_0(T_w, \mathbf{v}_w, n_w) \qquad (7.58)$$

where f_0 is the Boltzmann distribution as determined by the wall temperature, velocity and gas molecule density. From this relation, one can also derive relationships between the momentum (M) and energy (or temperature) of the reflected molecules and those of the incoming ones,

$$T_i - T_r = \alpha_T (T_i - T_w), \quad M_i - M_r = \alpha_v (M_i - M_w) \qquad (7.59)$$

where i, r, and w represent corresponding quantities for incident, reflected, and wall, respectively. The prefactor α_T is called the energy (temperature) accommodation coefficient and α_v the momentum accommodation coefficient. Their relationship with p can be derived by taking the moments of eq. (7.58) with respect to energy and tangential momentum, respectively, as we will show in section 7.5.4.

Next, we will assume here that when a molecule collides with the surface, it is completely thermalized with the surface and bounces back diffusely with the equilibrium distribution as determined by the wall temperature. The boundary conditions are then

$$y = 0, \quad f_1 = \frac{n_1}{(2\pi RT_1)^{3/2}} \exp\left[-\frac{v^2}{2RT_1}\right] \quad \text{(for } v_y > 0\text{)} \tag{7.60}$$

$$y = d, \quad f_2 = \frac{n_2}{(2\pi RT_2)^{3/2}} \exp\left[-\frac{v^2}{2RT_2}\right] \quad \text{(for } v_y < 0\text{)} \tag{7.61}$$

The solutions of eq. (7.56) for f are

$$f^+(y) = f_1 \exp\left(-\frac{\eta}{c_y}\right) + \int_0^\eta \frac{f_0}{c_y} \exp\left(-\gamma(y)\frac{\eta - \eta'}{c_y}\right) d\eta' \quad \text{(for } c_y > 0\text{)} \tag{7.62}$$

$$f^-(y) = f_2 \exp\left(\gamma_2 \frac{\xi - \eta}{c_y}\right) - \int_\eta^\xi \frac{f_0}{c_y} \exp\left(-\gamma(y)\frac{\eta - \eta'}{c_y}\right) d\eta' \quad \text{(for } c_y < 0\text{)}$$

$$\tag{7.63}$$

where we have defined the following nondimensional variables

$$\eta = \int_0^y \frac{dy}{\tau(2RT_1)^{1/2}} \quad c_y = \frac{v_y}{(2RT_1)^{1/2}} \quad \gamma(y) = \left(\frac{T_1}{T(y)}\right)^{1/2} \tag{7.64}$$

and $\xi = \eta(y = d)$, $\gamma_2 = (T_1/T_2)^{1/2}$. The governing equations determining the density and temperature distributions can be obtained from

$$n(y) = \int_{-\infty}^{\infty}\int_{-\infty}^{\infty}\int_{-\infty}^{\infty} f\,dv_x dv_y dv_z \tag{7.65}$$

$$\frac{3n(y)\kappa_B T(y)}{2} = \int_{-\infty}^{\infty}\int_{-\infty}^{\infty}\int_{-\infty}^{\infty} \left(\frac{m(\mathbf{v} - \mathbf{u})^2}{2}\right) f\,dv_x dv_y dv_z \tag{7.66}$$

For heat conduction, the average velocity of the gas molecules is zero, $\mathbf{u} = \mathbf{0}$.

Equation (7.66) needs more explanation. In previous sections, we have used the requirement that heat flux J_q is a constant to determine the temperature distribution. These two methods are equivalent as long as the relaxation time does not depend on random velocity. We can show this by multiplying both sides of equation (7.56) by $mv^2/2$ and integrating over all velocities, which leads to

$$\frac{\partial}{\partial y}\left(\int_{-\infty}^{\infty}\int_{-\infty}^{\infty}\int_{-\infty}^{\infty} \left(\frac{mv^2}{2}\right) v_y f\,dv_x dv_y dv_z\right)$$

$$= -\int_{-\infty}^{\infty}\int_{-\infty}^{\infty}\int_{-\infty}^{\infty} \left(\frac{mv^2}{2}\right)\left(\frac{f - f_0}{\tau}\right) dv_x dv_y dv_z \tag{7.67}$$

The left-hand side of eq. (7.67) is dJ_q/dy. According to eq. (7.66), if τ is independent of v, the right-hand side equals zero, and thus J_q is constant. On the other hand, if τ depends on the random velocity, it cannot be pulled out of the integral and the right-hand side of eq. (7.67) is not zero. In that case, we must redefine the instantaneous local temperature on the basis of

$$\int_{-\infty}^{\infty}\int_{-\infty}^{\infty}\int_{-\infty}^{\infty} \frac{f_0}{\tau}\left(\frac{mv^2}{2}\right)dv_x dv_y dv_z = \int_{-\infty}^{\infty}\int_{-\infty}^{\infty}\int_{-\infty}^{\infty} \frac{f}{\tau}\left(\frac{mv^2}{2}\right)dv_x dv_y dv_z \quad (7.68)$$

to reflect the different rate of the relaxation at different velocities. Such a definition should lead to a consistent result with the requirement that the heat flux is continuous. Equation (7.66) defines temperature directly from the solution for f.

Substituting eqs. (7.62) and (7.63) into (7.65), and performing integration over v_x and v_z, we obtain the following equation

$$\sqrt{\pi}\, n(\eta) = n_1 H_0(\eta) + n_2 H_0[\gamma_2(\xi - \eta)] + \int_0^{\xi} n(\eta) H_{-1}[\gamma(y)|\eta - \eta'|] d\eta' \quad (7.69)$$

where the integral function $H_n(x)$ is defined as

$$H_n(x) = \int_0^{\infty} t^n \exp\left(-t^2 - \frac{x}{t}\right) dt \quad (7.70)$$

Similarly, substituting eqs. (7.62) and (7.63) into (7.66) and performing the integration over v_x and v_z yields

$$\frac{3}{2}\frac{\sqrt{\pi}\, n(\eta)}{\gamma(\eta)^2} = K + n_1 H_0(\eta) + \frac{n_2}{\gamma_2^2} H_0[\gamma_2(\xi - \eta)]$$

$$+ \int_0^{\xi} \frac{n(\eta')}{\gamma(\eta')} H_{-1}[\gamma(\eta)|\eta - \eta'|] d\eta' \quad (7.71)$$

where

$$K = n_1 H_2(\eta) + \frac{n_2}{\gamma_2^2} H_2[\gamma_2(\xi - \eta)] + \int_0^{\xi} \frac{n(\eta')}{\gamma(\eta')} H_1[\gamma(\eta')|\eta - \eta'|] d\eta' \quad (7.72)$$

Although K seems to be dependent on η, it actually should be constant since this term is proportional to $\iiint v_y^2 f dv$. We can prove this by multiplying eq. (7.56) by v_y and integrating over all velocities,

$$\frac{\partial}{\partial y}\left(\int_{-\infty}^{\infty}\int_{-\infty}^{\infty}\int_{-\infty}^{\infty} v_y^2 f dv_x dv_y dv_z\right) = \int_{-\infty}^{\infty}\int_{-\infty}^{\infty}\int_{-\infty}^{\infty} v_y f dv_x dv_y dv_z \quad (7.73)$$

The right-hand side of eq. (7.73) gives the average velocity in the y-direction, which should be zero since there is no convection. Substituting f into the left-hand side of eq. (7.73) leads to

$$n_1 H_1(\eta) - \frac{n_2}{\gamma_2} H_1[\gamma_2(\xi - \eta)] + \int_0^\xi \text{sign}(\eta - \eta') n(\eta') H_0[\gamma(\eta')|\eta - \eta'|] d\eta' = 0$$
(7.74)

where $\text{sign}(x)$ takes a positive sign when $x > 0$ and a negative sign when $x < 0$. One can further show, by differentiating eq. (7.72), that if K is constant eq. (7.74) is automatically satisfied.

Now we have three equations, eqs. (7.69), (7.71), and (7.72), that can be used to determine the distributions $n(\eta)$ and $\gamma(\eta)$. In addition, we also need to determine n_1 and n_2, the densities of gas molecules on the wall. Two concerns arise. One is whether we have too many equations. The other is whether we have enough to determine n_1 and n_2. These questions can be answered by noticing that eq. (7.74) provides only one additional condition for the two unknowns n_1 and n_2 which can be taken by setting η equal to any allowable value. This reduction can be seen by integrating eq. (7.56) over all velocities to show that if eq. (7.69) is satisfied, du/dy must be zero and thus u must be constant. Since there is no convection, u must be zero. Thus, when solving the distributions of n and γ, eqs. (7.69) and (7.71) should be used. Equation (7.74) provides only one condition for the two unknown densities n_1 and n_2. The other condition can be found by specifying the average density $\bar{n}$ in the region,

$$\bar{n} = \frac{1}{d} \int_0^d n(y) dy = \frac{\Lambda}{d} \int_0^\xi n(\eta) d\eta$$
(7.75)

Substituting eq. (7.69) into (7.75) leads to the additional required condition. With a given $\bar{n}$, the values of ξ can also be evaluated,

$$\xi = \int_0^d \frac{dy}{\tau (2RT_1)^{1/2}} \approx \frac{d}{\Lambda}$$
(7.76)

where $\Lambda = 1/(\bar{n}\pi d_a^2)$. We used an approximation sign in eq. (7.76) because we assumed that τ does not depend on temperature, and $v_{th} = \sqrt{2RT_1}$.* In conjunction with this assumption, it is interesting to note that in the previous discussions for photons, phonons, and electrons, we assumed that the mean free path is independent of the random velocity, which is reasonable in view of the derivation of mean free path in chapter 1. Based on the relation $\Lambda = v\tau$, the relaxation time is inversely proportional to the random velocity. In rarefied gas literature, however, it is typically assumed that τ, $\propto 1/(nA)$, is independent of the random velocity, where A is a constant. Whether one assumes Λ or τ independent

---

*$v_{th} = \sqrt{2RT}$ is the most probable speed of molecules. The average speed of molecules is $v = \sqrt{8RT/\pi}$. Rigorously, the Knudsen number should be defined on the basis of average speed. However, v_{th} is more convenient to use because of its natural appearance in the Maxwell distribution, and is thus frequently used.

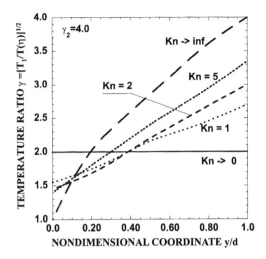

Figure 7.8 Nondimensional temperature distribution at different values of the Knudsen number for heat conduction of rarefied gas between two parallel plates (data from Patterson, 1971).

of random velocity has significant consequences on the form of exponential integrals. When Λ is assumed to be independent of random velocity, we encounter exponential integral functions $E_n(x)$. When τ is assumed to be independent of random velocity, the integral function takes the form of $H_n(x)$. It is suspected that the two approaches lead to similar results because of the averaging of both types of integral functions. Interested readers may reformulate the problem and check whether this is indeed the case or not.

Figure 7.8 gives an example of the temperature distribution obtained by Willis (Patterson, 1971) for a given set of Knudsen number $Kn(= 1/\xi)$ values at $T_2/T_1 = \gamma_2 = 4$. The figure shows the familiar temperature discontinuities at the two surfaces, due to similar reasons to those we discussed for photons and phonons.

7.2.4 Current Flow across Heterojunctions

The equivalent electron transport problem to the above-discussed photon, phonon, and molecule energy transfer perpendicular to a thin film is electrical current flow through double heterojunction structures (Hess, 2000), which are used frequently in electronics and photonics. Examples of transport perpendicular to interfaces are applications in semiconductor lasers, photodiodes, and heterojunction bipolar transistors. Surprisingly, no treatment based on the Boltzmann equation for such structures can be found. Most treatments are based on the drift-diffusion equation coupled to thermionic emission interface conditions, which we will discuss in more detail later when introducing approximations.

Treatment of electron transport is complicated by the fact that the Boltzmann distribution contains the force term, as shown in eqs (7.1) and (7.2). For electron transport, the local force is proportional to the local electrical potential, which is described by the Poisson equation, which can be obtained by substituting eqs. (5.8) and (6.69) into eq. (5.13). Thus the Boltzmann equation is coupled to the Poisson equation and is highly nonlinear. The closest study we have identified is the solution of the Boltzmann equation for a heterojunction structure with one interface (Stettler and Lundstrom, 1994). At the interface region, because of the reflection of electrons, a discontinuity in the Fermi level and electron temperature appears. The behavior is similar in its physical origins to what we have shown consistently for photons, phonons, and molecules in this section, all reflecting the highly nonequilibrium nature of transport near the interfaces.

7.3 Rarefied Poiseuille Flow and Knudsen Minimum

We originally planned to include rarefied gas flow between two parallel plates (Poiseuille flow) in section 7.1, analogously to the treatment of electron and phonon heat conduction inside a thin film. It soon became clear that such a reconciliation is not feasible on the basis of existing solutions for these problems. The key difference is that the molecules have a nonzero average velocity and thus they obey the displaced Maxwellian velocity distribution. For fully developed Poiseuille flow between two parallel plates along the x-direction, as in figure 7.9, and with a constant temperature throughout the channel, the displaced Maxwell distribution given by eq. (6.63) is repeated here,

$$f_d(v_x, v_y, v_z) = n(x)\left(\frac{m}{2\pi\kappa_B T}\right)^{3/2} e^{-m[(v_x-u)^2+v_y^2+v_z^2]/2\kappa_B T} \qquad (7.77)$$

where $u(y)$ is the average velocity to be determined and $n(x)$ is the local density along the x-direction. Because of this y dependency of u, the solution for the Poiseuille flow of rarefield gas will bear a stronger resemblance to transport perpendicular to a thin film plane, although the transport is along the film plane.

A peculiar phenomenon associated with rarefied gas flow inside a duct is the Knudsen minimum in the flow rate through the duct, which occurs when the Knudsen number is around $Kn = 1$. This minimum was experimentally observed by Knudsen in the early 1900s (Knudsen, 1909; 1950). The existence of the minimum was a topic of debate for a long time until Cercignani and Deneri (1963) solved the Boltzmann equation for Poiseuille flow between two parallel plates and indeed found such a minimum in the solution. Rarefied gas flow and heat conduction become important for some micro- and nanodevices, even at atmospheric pressure (Karniadakis and Beskok, 2002). An example is the air-bearing problem in hard-drive data storage devices (Fukui and Kaneko, 1988). To increase the data storage density, the distance between the magnetic head and the magnetic disk is pushed down to tens of nanometers, much smaller than the molecular mean free path. Describing the fluid flow under such stringent conditions is a topic of

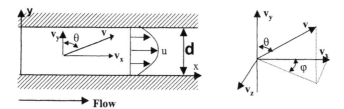

Figure 7.9 Poiseuille flow between two parallel plates.

current interest. We will recap here the solution of Cercignani and Deneri (1963) for Poiseuille flow between two parallel plates (see also Patterson, 1971).

The Boltzmann equation under the relaxation time approximation is

$$v_x \frac{\partial f}{\partial x} + v_y \frac{\partial f}{\partial y} = -\frac{f - f_d}{\tau} \tag{7.78}$$

where the z-direction has been eliminated since f does not depend on z. Cercignani and Deneri's approach is based on a Taylor expansion of the displaced Maxwell distribution as used by Bhatnagar et al. (1954)—the BGK model. Considering the general case that temperature is not uniform at each cross-section and assuming that the deviations of n and T from reference values at each constant x cross-section, $n_0(x)$ and $T_0(x)$, are small, that is,

$$n = n_0 + n', \ T = T_0 + T' \tag{7.79}$$

the Taylor expansion of f_d is

$$f_d(v_x, v_y, v_z) = (n_0 + n') \left(\frac{m}{2\pi \kappa_B (T_0 + T')}\right)^{3/2} e^{-m(\mathbf{v}-\mathbf{u})^2 / 2\kappa_B (T_0+T')}$$

$$\approx f_0 \left\{ 1 + \frac{n'}{n_0} + \left[\frac{\mathbf{v}^2}{2RT_0} - \frac{3}{2}\right] \frac{T'}{T_0} + \frac{u v_x}{RT_0} \right\} \tag{7.80}$$

where we have kept only the first-order terms from the expansion, $(\mathbf{v} - \mathbf{u})^2 = (v_x - u)^2 + v_y^2 + v_z^2$, and

$$f_0(x) = n_0(x) \left(\frac{1}{2\pi RT_0(x)}\right)^{3/2} \exp\left(-\frac{v_x^2 + v_y^2 + v_z^2}{2RT_0(x)}\right) \tag{7.81}$$

is the local Maxwell distribution that depends on the local values of $n_0(x)$ and $T_0(x)$. We further expand the distribution function f as

$$f = f_0(1 + \Phi) \tag{7.82}$$

Then, from eqs. (7.65) and (7.66),

$$n' = \iiint f_0 \Phi d^3 \mathbf{v} \tag{7.83}$$

$$3R(T_0 n' + n_0 T') = \iiint (\mathbf{v} - \mathbf{u})^2 f_0 \Phi d^3 \mathbf{v} \approx \iiint \mathbf{v}^2 f_0 \Phi d^3 \mathbf{v} \tag{7.84}$$

which gives

$$R n_0 T' = \iiint \left[\frac{1}{3}\mathbf{v}^2 - RT_0\right] f_0 \Phi d^3 \mathbf{v} \tag{7.85}$$

Because the relaxation time also depends on density, we can write it as

$$\frac{1}{\tau} = \left(1 + \frac{n'}{n_0}\right)\frac{1}{\tau_0} \tag{7.86}$$

Substituting eqs. (7.80) and (7.86) into Eq. (7.78) and retaining only the first-order term, we have

$$v_x \frac{\partial(f_0 + f_0\Phi)}{\partial x} + v_y \frac{\partial(f_0 + f_0\Phi)}{\partial y}$$

$$= -\frac{f_0\Phi}{\tau_0} + \frac{f_0}{\tau_0}\left\{\frac{n'}{n_0} + \left[\frac{\mathbf{v}^2}{2RT_0} - \frac{3}{2}\right]\frac{T'}{T_0} + \frac{v_x u}{RT_0}\right\} \tag{7.87}$$

which is one form of BGK approximation of the Boltzmann equation.

For the Poiseuille flow under consideration, we assume a linear pressure gradient. For a channel of length L,

$$P(x) = P_1 - \left(\frac{P_1 - P_2}{L}\right)x = P_1(1 + \Pi x) \tag{7.88}$$

where P_1 and P_2 are the inlet and exit pressure, respectively, and $\Pi = -(P_1 - P_2)/(P_1 L)$ is a constant. For ideal gas, $P = mnRT$, we also have

$$n_0 = n_1(1 + \Pi x) \tag{7.89}$$

In reality, the local number density also depends on y in addition to x, as represented by the deviation n'. The left-hand side of eq. (7.87) becomes

$$v_x \frac{\Pi}{1 + \Pi x} f_0 (1 + \Phi) + v_x f_0 \frac{\partial \Phi}{\partial x} + v_y f_0 \frac{\partial \Phi}{\partial y}$$

Retaining only the first-order term in the above expression and substituting back into eq. (7.87), we have

$$\Pi v_x + v_y \frac{\partial \Phi}{\partial y} = -\frac{\Phi}{\tau_0} + \frac{1}{\tau_0}\left\{\frac{n'}{n_0} + \left[\frac{\mathbf{v}^2}{2RT_0} - \frac{3}{2}\right]\frac{T'}{T_0} + \frac{v_x u}{RT_0}\right\} \tag{7.90}$$

where we have dropped the term $\partial \Phi/\partial x$ as we did for thermal and electrical conduction along x direction. The diffuse boundary condition at $y = 0$ for f is

$$f^+(y=0) = n_1(1 + \Pi x)\left(\frac{1}{2\pi RT_0}\right)^{3/2} \exp\left(-\frac{v_x^2 + v_y^2 + v_z^2}{2RT_0}\right) (v_y > 0) \tag{7.91}$$

The boundary condition at $y = d$ is identical with $v_y < 0$. The corresponding conditions for Φ are

$$\Phi(y=0 \text{ or } y=d) = 0 \text{ (for } v_y > 0 \text{ or } v_y < 0) \tag{7.92}$$

Now our task is to solve eq. (7.90) with the boundary condition (7.92) for Φ. Note that n' and T' can be expressed in terms of Φ as given by eqs. (7.83) and (7.85). Similarly, the average velocity u can be expressed in terms of Φ,

$$u = \frac{1}{n} \iiint v_x f d^3\mathbf{v} = \frac{1}{n} \iiint v_x f_0 \Phi d^3\mathbf{v} \qquad (7.93)$$

Rather than solving Φ directly, which is the strategy we used in treating phonon and photon transport problems, we will introduce a new function $Z(y, v_y)$,

$$Z(y, v_y) = \frac{1}{2\pi R T_0} \int_{-\infty}^{\infty}\int_{-\infty}^{\infty} v_x \exp\left(-\frac{v_x^2 + v_z^2}{2RT_0}\right) \Phi \, dv_x \, dv_z \qquad (7.94)$$

and transform eq. (7.90) into an equation for Z by multiplying both sides of the equation by $v_x \exp[-(v_x^2 + v_z^2)/(2RT_0)]/(2\pi RT_0)$ and integrating over allowable values of v_x and v_z, which leads to

$$\frac{1}{2}A + v_y \frac{\partial Z}{\partial y} = \frac{u - Z}{\tau_0} \qquad (7.95)$$

where $A = 2RT_0\Pi$. Note that in such an operation the terms associated with n' and T' drop out because the integrands are odd functions of v_x. The corresponding boundary conditions can be derived from eq. (7.92) as

$$Z(y = 0, v_y \geq 0 \text{ or } y = d, v_y < 0) = 0 \qquad (7.96)$$

The solutions for eq. (7.95) with boundary conditions (7.96) are

$$Z(y, v_y > 0) = \int_0^y \frac{(u - A\tau_0/2)}{v_y \tau_0} \exp\left(-\frac{y - y'}{v_y \tau_0}\right) dy' \qquad (7.97)$$

$$Z(y, v_y < 0) = \int_d^y \frac{(u - A\tau_0/2)}{v_y \tau_0} \exp\left(-\frac{y - y'}{v_y \tau_0}\right) dy' \qquad (7.98)$$

The average velocity u can be calculated from Z according to the definition

$$u = \frac{1}{n} \iiint v_x f d^3\mathbf{v} = \frac{1}{n} \iiint v_x f_0 \Phi d^3\mathbf{v}$$

$$= \frac{1}{\sqrt{2\pi RT_0}} \int_{-\infty}^{\infty} Z \exp\left(-\frac{v_y^2}{2RT_0}\right) dv_y \qquad (7.99)$$

Substituting eqs. (7.97) and (7.98) into the above equation yields

$$u = \frac{1}{\Lambda_0 \sqrt{\pi}} \int_0^d (u - A\tau_0/2) H_{-1}\left(\frac{|y - y'|}{\Lambda_0}\right) dy' \qquad (7.100)$$

where $\Lambda_0 = \tau_0\sqrt{2RT_0}$ is the product of the relaxation time and the most probable speed, v_{th}, of the molecules. Using the following nondimensional variables,

$$u = \frac{A\tau_0}{2}(1-\Psi) \quad \eta = \frac{y}{\Lambda_0} \tag{7.101}$$

eq. (7.100) can be written as

$$\Psi(\eta) = 1 + \frac{1}{\sqrt{\pi}} \int_0^{\xi} H_{-1}(|\eta - \eta'|)\Psi(\eta')\,d\eta' \tag{7.102}$$

where $\xi = d/\Lambda_0 = 1/Kn$ is the inverse Knudsen number. Equation (7.102) is again the standard Fredholm integral equation and can be solved similarly to example 7.1. The mass flow rate per unit depth can be calculated from

$$j_m = \rho \int_0^d u\,dy = \frac{\rho A\tau_0}{2}\int_0^d (1-\Psi)\,dy = \frac{dP}{dx}\tau_0\left(d - \Lambda_0\int_0^{\xi}\Psi\,d\eta\right) \tag{7.103}$$

or, in a nondimensional form,

$$j_m^* = \frac{\sqrt{2RT_0}\,j_m}{-d^2 dP/dx} = -\frac{1}{\xi} + \frac{1}{\xi^2}\int_0^{\xi}\Psi\,d\eta \tag{7.104}$$

Figure 7.10(a) shows the nondimensional velocity profile at different Knudsen numbers (Fukui and Kaneko, 1988) and figure 7.10(b) shows the nondimensional mass

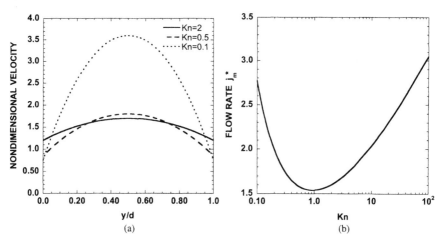

Figure 7.10 (a) Velocity profile (data from Fukui and Kaneko, 1988) and (b) nondimensional mass flow rate as a function of Knudsen number for rarefied gas flow between parallel plates (data from Cercignani and Daneri, 1963).

flow rate (Cercignani and Deneri, 1963) as a function of the Knudsen number. The velocity profile shows that, at the wall, the average velocity is not zero; in other words, there is a velocity slip. The mass flow rate demonstrates the existence of the Knudsen minimum around $Kn = 1$ and is in reasonably good agreement with experimental results of Dong (1956).

The above derivation follows the original work of Cercignani and Deneri (1963). Identical integral equations can also be obtained following a similar treatment as for electron and phonon conduction in thin films. This alternative will be left as an exercise for interested readers.

The key difference between the rarefied gas flow and our treatment of electron and phonon conduction along a thin film is the use of the displaced Maxwell distribution. It is legitimate to ask why we did not use displaced distributions for electrons and phonons and whether there are similar Knudsen minima for electrical and thermal conductivity. In theory, this treatment should be possible, because electrons obey displaced distributions, and phonons also can have similar displaced distributions when the normal scattering processes are faster than umklapp scattering, as we have seen in chapter 6. It is not known whether Knudsen minima can appear for electron and phonon transport. We will make the following comments and encourage interested readers to pursue further.

The electrons flowing inside a conductor or semiconductor have a drift velocity similar to that of molecules. They should obey the displaced Boltzmann distribution or the drifted Fermi–Dirac distribution. The original BGK model (Bhatnagar et al., 1954) actually dealt with charged particles. If we assume a displaced distribution function and follow a similar derivation, it can be shown that the final equations are similar to those of molecular flow. Thus, a Knudsen minimum likely exists for the electrical current.

The equilibrium phonon distribution depends on the scattering processes. For umklapp scattering processes, the equilibrium distribution is the normal Bose–Einstein distribution f_0. For normal scattering processes, phonons tend to obey the displaced Bose–Einstein distribution f_d as given in eq. (6.164) and the corresponding Boltzmann equation is given by eq. (6.165). We mentioned in section 6.4.3 that phonon Poiseuille flow has been considered (Guyer and Krumhansl, 1966; Gurevich, 1986) without size effects. It is possible that at very low temperatures, when τ_n is dominant and size effects play a role, a phonon Knudsen transport minimum may exist.

7.4 Transport in Nonplanar Structures

Sections 7.1–7.3 deal with solving the Boltzmann equation in planar geometries. Some of the problems have already become very involved mathematically. One can rightfully expect that transport problems in nonplanar geometries and multidimensions are more complicated. We will mainly discuss in this section two nonplanar structures: spheres and cylinders, with little mathematical detail. Some comments on multidimensional problems will be made whenever appropriate. Our discussions are limited to photons, phonons, and rarefied gases, but one could easily anticipate a similar discussion applied to electron transport. There are not many treatments on electron transport in such configurations, presumably because they are of little practical interest.

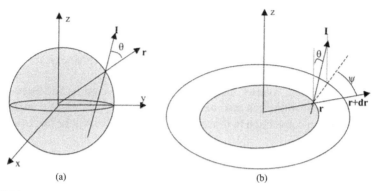

Figure 7.11 (a) Spherical and (b) cylindrical coordinates for the Boltzmann equation.

7.4.1 Thermal Radiation between Concentric Cylinders and Spheres

Photon thermal radiation in nonplanar structures has been extensively studied (Siegel and Howell, 1992; Modest, 1993). These solutions and the solution methods can often be consulted for studying the transport of other types of carriers.

In a spherical coordinate system, the photon Boltzmann equation or the equation of photon radiative transfer is (Modest, 1993)

$$\mu \frac{\partial I}{\partial r^*} + \frac{1-\mu^2}{r^*} \frac{\partial I}{\partial \mu} = S(r^*, \mu) - I \qquad (7.105)$$

where S is the source term, $r^* = r/\Lambda = K_e r$ is the nondimensional radial coordinate, and $\mu (= \cos \theta)$ is the directional cosine formed between the direction of intensity and the radial vector as shown in figure 7.11(a). For an isotropically scattering medium without an internal heat source, $S = I_b(T)$ is given by the blackbody intensity and is ultimately related to the local temperature. The numerical solution for radiation heat transfer between concentric spheres was considered by Sparrow et al. (1961), Ryhming (1966), Viskanta and Crosbie (1967), and well summarized by Modest (1993).

Radiative transfer in cylindrical coordinates has also been studied extensively. In these coordinates, the equation of radiative transfer can be expressed as

$$\sin \theta \left[\cos \psi \frac{\partial I}{\partial r^*} - \frac{\sin \psi}{r^*} \frac{\partial I}{\partial \psi} \right] = S - I \qquad (7.106)$$

with the angular notations shown in figure 7.11(b) (Modest, 1993). Solution of radiative transfer in cylindrical coordinates is considerably more difficult than in spherical coordinates. The Monte Carlo simulation results obtained by Perlmutter and Howell (1964) are considered as exact and often used as a standard.

7.4.2 Rarefied Gas Flow and Convection

Rarefied gas flow and heat transfer have drawn considerable attention in the past (Eckert and Drake, 1972; Cercignani, 2000). For nonplanar geometries, exact solutions based on the Boltzmann equation are again difficult to find because of the mathematical difficulties. Rarefied gas flow is characterized by the Knudsen number, defined as the ratio of the gas molecule mean free path Λ to a characteristic length L,

$$Kn = \frac{\Lambda}{L} \qquad (7.107)$$

In eq. (7.100), we defined Λ_0 as the relaxation time multiplied by the most probable speed of the gas molecules. But more often, the average gas molecule velocity is used in the definition of the mean free path,

$$\Lambda = \bar{v}\tau \qquad (7.108)$$

where the average velocity

$$\bar{v} = \frac{\int_{-\infty}^{\infty} \int_{-\infty}^{\infty} \int_{-\infty}^{\infty} v f_0 dv_x dv_y dv_z}{\int_{-\infty}^{\infty} \int_{-\infty}^{\infty} \int_{-\infty}^{\infty} f_0 dv_x dv_y dv_z} = \sqrt{\frac{8\kappa_B T}{\pi m}} = \sqrt{\frac{8}{\pi} RT} \qquad (7.109)$$

Depending on the magnitude of the Knudsen number, the flow is often divided into the following regimes (Schaaf and Chambre, 1961)

Continuum regime:	$Kn \leq 10^{-2}$
Slip flow regime:	$10^{-2} \leq Kn \leq 0.1$
Transition flow regime:	$0.1 \leq Kn \leq 10$
Free molecular flow regime:	$Kn \geq 10$

In the slip flow regime, slip boundary condition, which we will discuss in more detail later, is often used in combination with the Navier–Stokes equations. In the free-molecule flow regime, scattering among molecules is totally neglected and only the gas molecule–surface interaction is counted for. In the transition regime, the full Boltzmann equation should be used but its solutions are difficult to obtain for complicated geometries.

7.4.3 Phonon Heat Conduction

Chen (1996a) considered the issue of rarefied phonon gas heat conduction surrounding a nanoscale heat generation region by approximating the region as a spherical nanoparticle. The solution follows the formulation for radiative heat transfer between two concentric spheres (Viskanta and Crosbie, 1967) and showed the following characteristics: (1) a temperature discontinuity at the interface between the nanoparticle and its surrounding media; and (2) a reduced effective thermal conductivity that the nanoparticle "feels" about its surrounding. Although these conclusions are in general valid, it was also pointed out that one must be careful about the temperature definition since the emitted phonon temperature was assumed to be the temperature at the boundary whereas the equivalent equilibrium temperature is used inside the surrounding (Yang et al., 2005). Later, we will show that, with consistent use of the temperature definition, the temperature of a nanoscale heat generation region is still higher than predicted by the Fourier law. Goodson's group recently provided some experimental evidence on the existence of such a nonlocal, rarefied heat conduction phenomenon (Sverdrup et al., 2001b), and simulated the effect on the basis of a numerical solution of the Boltzmann equation (Sverdrup et al., 2001a).

7.4.4 Multidimensional Transport Problems

Multidimensional problems are much harder to deal with. Typical solution methods involve starting directly from the Boltzmann equation, rather than solving the integral equations as we did in sections 7.2 and 7.3. Here we will briefly introduce the ideas behind three numerical methods: (1) discrete ordinates method; (2) spherical harmonics method; and (3) Monte Carlo simulations. Detailed descriptions of these methods can be found in thermal radiation textbooks (Modest, 1993).

The discrete ordinates method (Chandrasekhar, 1960) solves the Boltzmann equation for specific wavevector directions, that is, for specific directional cosine values μ_i as in eq. (7.29). The values of these discrete wavevector directions are determined by the Gauss quadrature integration scheme. If $f(\mu)$ is the function to be integrated over the interval $-1 < \mu < 1$, such as the integrand in eq. (7.47), the Gaussian quadrature formula is

$$\int_{-1}^{1} f(\mu) d\mu = \sum_{i=1}^{N} w_i f(\mu_i) \quad (7.110)$$

where w_i are the weight factors and μ_i are the zeroes of the Legendre polynomial of order N, and thus are different for different choices of N. There are standard programs to calculate the values of w_i and μ_i (Press et al., 1992). The numerical accuracy usually increases with increasing N. The spatial gradients in the Boltzmann equation for each discrete ordinate direction can be solved by finite difference or other numerical schemes for differential equations. For lower values of N (such as $N = 2$), approximate analytical solutions are also possible.

Similar to the discrete ordinates method, the strategy behind the spherical harmonics method (Jeans, 1917) is to decouple the spatial coordinate from the wavevector coordinates. The spherical harmonics method arrives at such a decoupling by expanding the distribution function $f(\mathbf{r},\mathbf{v})$ or intensity $I(\mathbf{r}, \hat{\mathbf{s}})$ (where $\hat{\mathbf{s}}$ is a unit vector in the direction of propagation) into a series with the spherical harmonic functions $Y_l^m(\hat{\mathbf{s}})$ as the basis, similar to the sine or cosine functions used in Fourier expansion. We have encountered the spherical harmonics function in chapter 2. Using intensity as an example, the spherical harmonics expansion is

$$I(\mathbf{r}, \hat{\mathbf{s}}) = \sum_{\ell=0}^{\infty} \sum_{m=-\ell}^{m=\ell} I_\ell^m(\mathbf{r}) Y_\ell^m(\hat{\mathbf{s}}) \quad (7.111)$$

Substituting such an expansion into the Boltzmann equation and exploiting the orthogonality of the spherical harmonics functions, one can obtain a set of differential equations for I_ℓ^m. The expansion can be of arbitrary accuracy, depending on the order to which ℓ is truncated. In reality, either one (called P_1 approximation) or three (called P_3 approximation) terms in ℓ are retained.

Monte Carlo simulations are a stochastic approach that is totally different from the previous deterministic approaches. In a typical Monte Carlo scheme, the particles (electrons, photons, phonons, or molecules) are divided into bundles. The trajectories of these bundles are traced. Their interactions with each other or with boundaries are described with probabilities that are determined by the relaxation time (or more detailed scattering rules) and the boundary conditions. The Monte Carlo method is most suited for complex geometries for which it is difficult to solve the Boltzmann equation. Monte Carlo methods have been widely used for transport of many different carriers, such as photons (Siegel and Howell, 1992), electrons (Hess, 2000), molecules (Bird, 1994), and phonons (Klitsner et al., 1988; Peterson, 1994; Mazumder and Majumdar, 2001; Song and Chen, 2003). Because Monte Carlo simulations are stochastic methods, a large number of samples is needed to reduce the statistical variation of the computation, which means long computational time.

7.5 Diffusion Approximation with Diffusion–Transmission Boundary Conditions

Because the Boltzmann equation is difficult to solve, various approximate treatments have been developed. Many of these techniques are discussed in detail in radiation heat transfer textbooks (Modest, 1993; Siegel and Howell, 1992). This section and the one that follows will introduce two levels of approximations that are believed to be particularly convenient (Chen, 2001, 2002, 2003).

The approximation discussed in this section attributes all the nonlocal transport effects to the interfacial and boundary regions by modifying the boundary condition. Away from the boundary, normal diffusion laws are used. Similar boundary conditions have been introduced in different fields for the transport of photons, molecules, phonons, and electrons. We will first give a generalized treatment and introduce the transmission–diffusion boundary condition (Chen, 2003) and then discuss different carriers and, along the way, show that previously established boundary conditions are included in such a generalized boundary condition, and how the boundary condition can be used to simplify transport problems.

We have shown in the previous chapter that we can derive all the diffusion laws, including the drift-diffusion equation for electrons, from the Boltzmann equation under the local equilibrium assumption, eq. (6.49), reproduced here:

$$f = f_0 - \tau \left(\mathbf{v} \bullet \nabla_\mathbf{r} f_0 + \frac{\mathbf{F}}{m} \bullet \nabla_\mathbf{v} f_0 \right) \quad (7.112)$$

The first term on the right-hand side cancels in all the transport fluxes (heat flux, current, shear stress, etc.) because the equilibrium distribution function f_0 is isotropic and does not contribute to the net flux. At the boundaries or interfaces, however, the equilibrium distribution functions at the two sides are no longer the same. Examples are the thermal boundary resistance that we considered in chapter 5. The boundary conditions for phonons at an interface, eqs. (7.51–7.54), and for molecules, (7.60) and (7.61), are other examples. In general, we do not expect that f_0 will cancel at the interfaces or boundaries, and this leads to the discontinuities in temperature, velocity, and Fermi level that we are now familiar with.

The classical way to deal with interfacial transport is to consider the transmission of particles across the interface, as shown in figure 5.7. On each side of the interface, incoming carriers are assumed to be at an equilibrium state as determined by the local temperature, the local Fermi level, and so on. The net flux for the microscopic quantity of interest, X (X can be the charge, energy, momentum, etc. of a particle), can be obtained from

$$J_X = \frac{1}{(2\pi)^3} \sum_p \left[\int_{k_x>0} \tau_{12} X f_e(E, T_{e1}, E_{ef1}) v_{x1} d^3\mathbf{k}_1 \right.$$
$$\left. + \int_{k_x<0} \tau_{21} X f_e(E, T_{e2}, E_{ef2}) v_{x2} d^3\mathbf{k}_2 \right] \quad (7.113)$$

where we use f_e to denote that it presents the equilibrium distribution function for carriers coming toward the interface with characteristic temperature and chemical potentials at T_e and E_{ef}; τ_{12} and τ_{21} are the transmissivities of carriers from one side to the other as denoted by the order of the subscript. Using the principle of detailed balance we introduced following eq. (5.89), eq. (7.113) can be written as

$$J_X = \sum_p \int_{k_x>0} \tau_{12} X [f_e(E, T_{e1}, E_{ef1}) - f_e(E, T_{e2}, E_{ef2})] v_{x1} d^3\mathbf{k}_1/(2\pi)^3 \quad (7.114)$$

which is the familiar Landauer formalism discussed in section 5.5. Equation (7.114) implies that a nonzero flux must be sustained by a difference in either the temperature or the Fermi level. However, eq. (7.114) presents a dilemma in the limit of no interface because it leads to an artificial drop in the Fermi level or temperature at such a fictitious interface. We discussed this dilemma in section 5.2.3 when using this method to treat thermal boundary resistance. The discussion in section 7.2.1 on temperature jump at the interface is also related to this issue.

From eqs. (5.99) to (5.101) we introduced a way to remedy this deficiency by distinguishing the local equivalent equilibrium temperature from the temperature of incoming phonons for a consistent formulation of the thermal boundary resistance. The equivalent equilibrium distributions at the two sides of the carriers, f_0, are typically different from f_e. There is another delicate difference between the treatments of interfacial transport in eq. (7.114) and the diffusion laws derived from eq. (7.112) that we did not consider in eq. (5.103). In eq. (5.103), the carriers coming toward the interface are assumed to have an isotropic distribution. The diffusion approximation, eq. (7.112), however, clearly has an anisotropic distribution. To combine the diffusion approximation with the interface treatment, it becomes apparent that we should use eq. (7.112) to replace f_e in eq. (7.113). We call the boundary conditions thus established the diffusion-transmission conditions (Chen, 2003). For simplicity, we consider one-dimensional transport along the x-direction. Replacing f_e in eq. (7.113) with f in eq. (7.112), we obtain

$$J_X = \frac{1}{(2\pi)^3} \sum_p \int_{k_x>0} \tau_{12} X [f_{01}(E, T_1, E_{f1}) - f_{01}(E, T_2, E_{f2})] v_{x1} d^3\mathbf{k}_1$$
$$+ \frac{\tau_{12}'' J_{X1}}{2} + \frac{\tau_{21}'' J_{X2}}{2} \quad (7.115)$$

where J_{X1} and J_{X2} are diffusion fluxes of quantity X in regions 1 and 2, respectively. The average transmissivity $\tau_{12''}$ is defined as

$$\tau''_{12} = -(2\pi)^{-3} \sum_p \int_{k_x>0} \tau_{12}\tau_1 X v_{x1}^2 (\partial f_{01}/\partial x + F_x \partial f_{01}/\partial E) d^3\mathbf{k}_1/(J_{X1}/2) \quad (7.116)$$

with a similar definition for $\tau_{21''}$. With the assumption of flux continuity (no source or sink at the interface), that is, $J_X = J_{X1} = J_{X2}$, eq. (7.115) becomes

$$[1 - (\tau''_{12} + \tau''_{21})/2] J_X$$
$$= \frac{\tau'_{12}}{(2\pi)^3} \sum_p \int_{k_x>0} X[f_{01}(E, T_1, E_{f1}) - f_{01}(E, T_2, E_{f2})] v_{x1} d^3\mathbf{k}_1 \quad (7.117)$$

where

$$\tau'_{12} = \sum_p \int_{k_x>0} \tau_{12} X v_{x1} f_0(E, T_1, E_{f1}) d^3\mathbf{k}_1 / \sum_p \int_{k_x>0} X v_{x1} f_0(E, T, E_{f1}) d^3\mathbf{k}_1 \quad (7.118)$$

Equation (7.117) resolves the previously mentioned dilemmas inherent in eq. (7.114). One can appreciate this statement by examining the case where no interface is in existence. In this case,

$$\tau_{12} = \tau_{21} = \tau'_{12} = \tau'_{21} = \tau''_{12} = \tau''_{21} = 1 \quad (7.119)$$

and we have the following solutions

$$T_1 = T_2, E_{f1} = E_{f2} \quad (7.120)$$

for any arbitrary flux J_X. If one uses eq. (7.114), however, an artificial temperature or chemical potential drop develops over such a completely transmitting—that is, nonexisting—interface. In addition, equilibrium properties T and E_f are consistent with those used by the diffusion approximation, and thus eq. (7.117) can be used as the boundary condition in combination with the diffusion laws inside the medium away from the boundaries. We should point out that eq. (7.117) is not the only way to express the diffusion-transmission boundary conditions, as the following applications will show.

7.5.1 Thermal Radiation between Two Parallel Plates

Now let's consider radiation between two parallel plates, which we treated in section 7.2.1. We can apply the diffusion approximation for radiative transfer and obtain the following expression for heat flux (exercise 6.3)

$$J_q = -\frac{4}{3K_e} \frac{de_0}{dy} \quad (7.121)$$

This equation is essentially the Fourier law for thermal radiation, but in radiative parlance it is called the Rosseland diffusion approximation. Like the Fourier law, this approximation should be valid between two parallel plates if the spacing is large compared to the optical thickness. Within a distance of a few photon mean free paths from the boundary, however, this approximation is clearly not applicable.

The application of the diffusion–transmission boundary condition to thermal radiation between two parallel plates is complicated by the question of what happens within the plates. As discussed in section 7.2.1, the temperatures of the plates are not the same as the temperatures of the photons inside the plates due to heat conduction. Since photons coming toward the surface within the plates are emitted by the solid, their distribution follows the Bose–Einstein distribution as determined by the solid temperature, which will be assumed to be uniform, although the photon temperature, as sketched in figure 7.5, varies inside the plate. We can address this situation by removing the second term in the right-hand side of eq. (7.115), because this term represents the diffusion photon transport inside the plate. The corresponding boundary condition is, by setting $X = h\nu$ and assuming diffuse gray surfaces,

$$\left(\frac{1}{\varepsilon_1} - \frac{1}{2}\right) J_q = e_{01}(T_1) - e_0[T(y=0)] \tag{7.122}$$

$$\left(\frac{1}{\varepsilon_2} - \frac{1}{2}\right) J_q = e_0[T(y=d)] - e_{02}(T_2) \tag{7.123}$$

where ε_1 and ε_2 are the emissivities and are equivalent to the transmissivities of photons through surface 1 and surface 2, respectively, and e_{01} and e_{02} are the emissive powers of the emitted photons while $e_0(y)$ represents the local emissive power at point y in between the plates. We have assumed that the two averaged transmissivities τ''_{21} and τ'_{12} are the same and equal to ε. In thermal radiation literature (Siegel and Howell, 1992; Modest, 1993), eqs. (7.122) and (7.123) are called the Deissler boundary conditions (Probstein, 1963; Deissler, 1964). If we solve eq. (7.121) for a constant radiation heat flux, assuming heat conduction is negligible in the medium of interest, subject to eqs. (7.122) and (7.123) as the boundary conditions, we obtain the following solutions for the nondimensional temperature distributions and the nondimensional heat flux:

$$e_0^*(\eta, \xi) = \frac{e_0(\eta) - e_{02}}{e_{01} - e_{02}} = \frac{1}{1 + 4/(3\xi)} \left[1 + \frac{2}{3\xi} - \frac{\eta}{\xi}\right] \tag{7.124}$$

$$J_q^* = \frac{J_q}{e_{01} - e_{02}} = \frac{1}{\frac{3dK_e}{4} + \frac{1}{\varepsilon_1} + \frac{1}{\varepsilon_2} - 1} \tag{7.125}$$

This result compares very well with the exact solution from eqs. (7.40) and (7.41), as shown in figure 7.4(b).

In passing, we point out that sometimes second-order terms, that is, $\partial^2 f_0/\partial^2 x$, are used in deriving the Deissler boundary condition, as exemplified in the treatment of radiation heat transfer between two concentric cylinders and two concentric spheres (Siegel and Howell, 1992; Modest, 1993). For these geometries, however, the inclusion of higher order terms actually leads to results worse than including the first-order term only because the diffusion approximation itself used for regions away from the interfaces is only the first-order approximation, which is incompatible with the use of second-order approximation at the boundaries. If eqs. (7.122) and (7.123) only are used, one can show that the solutions obtained for radiation heat transfer between two concentric cylinders or two concentric spheres are actually close to the first-order spherical harmonic solution, which are at least correct in trend with the exact solutions, although the error is large when the spacing is much smaller than the photon mean free path.

7.5.2 Heat Conduction in Thin Films

7.5.2.1 Phonon Heat Conduction

We can combine the Fourier law for heat conduction with the diffusion–transmission boundary condition to deal with heat conduction in thin films and superlattices. For phonon heat conduction, this leads to the effective thermal conductivity of a thin film as (Chen and Zeng, 2001)

$$\frac{k}{k_b} = \frac{3d/(4\Lambda)}{\frac{[1-(\tau'_{12}+\tau'_{21})/2]}{\tau''_{21}} + \frac{[1-(\tau'_{23}+\tau'_{32})/2]}{\tau''_{23}} + \frac{3d}{4\Lambda}} \quad (7.126)$$

In figure 7.12 we give an example of the comparison of eq. (7.126) with the exact solution of the Boltzmann equation (Chen and Zeng, 2001). Similar treatment leads to a simple expression for the cross-plane thermal conductivity of superlattices that is again in reasonably good agreement with the solution of the Boltzmann equation.

7.5.2.2 Rarefied Gas Heat Conduction

In eq. (7.126), the effective thermal conductivity of the thin film is based on measuring the equivalent equilibrium phonon temperature on the other sides of the two interfaces

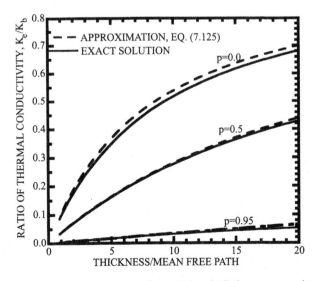

Figure 7.12 Thermal conductivity obtained from solving the Boltzmann equation and eq. (7.126) (Chen and Zeng, 2001).

sandwiching the film. If we consider rarefied heat conduction of a gas sandwiched between two solid surfaces, we run into a similar problem to that discussed for photon thermal radiation. For diffuse surfaces, that is, fully accommodating surfaces, the distribution function of gas molecules leaving the surface should obey the local equilibrium distribution, and the corresponding effective thermal conductivity of the gas slab is

$$\frac{k}{k_b} = \frac{1}{1 + \frac{4\Lambda}{3d}} \tag{7.127}$$

One can compare this result with those existing in literature. When there is no scattering, the heat flux between two plates can be expressed as (Luikov, 1964, Corrucini, 1958)

$$\frac{J_q}{T_2 - T_1} = \frac{\varsigma + 1}{\varsigma - 1}\left(\frac{R}{8\pi MT}\right)^{1/2} P \frac{1}{1/\alpha_{T1} + 1/\alpha_{T2} - 1} \tag{7.128}$$

where α_T is the accommodation coefficient defined in eq. (7.59), ς is the ratio of the constant pressure specific heat to the constant volume specific heat, R is the ideal gas constant, M the molar weight, and P the pressure. Typical values of the accommodation coefficients for air are between 0.87 and 0.97 on bronze, cast iron and aluminum surfaces intermediately finished (Springer, 1971). In the intermediate range of Knudsen number, the ratio of the thermal conductivity k of the gas in the gap to its bulk thermal conductivity k_b can be expressed as (Springer, 1971)

$$\frac{k}{k_b} = \frac{1}{1 + K\frac{\Lambda}{d}} \tag{7.129}$$

where K is a constant. This equation is identical to eq. (7.127) with $K = 4/3$.

7.5.3 Electron Transport across an Interface: Thermionic Emission

Electrical current flow across an interface is a basic problem in electronics, encompassing a large number of applications. A metal–semiconductor interface has a typical potential profile as shown in figure 7.13(a) and is called the Schottky barrier. A similar situation exists at a semiconductor–semiconductor interface, figure 7.13(b), such as one of the interfaces of a quantum well shown in figure 3.28. Before the two materials are joined together, their chemical potentials are independent of each other [figure 7.13(c)]. The

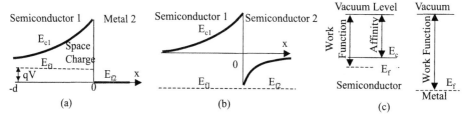

Figure 7.13 Electron transport at a heterojunction interface: (a) a Schottky barrier under an external bias; (b) a semiconductor–semiconductor interface at equilibrium; (c) energy levels before the formation of an interface.

energy difference between the chemical potential and the vacuum, that is, the amount of energy needed to eject one electron inside the metal into vacuum, is called the work function, W. The electron affinity χ is the energy needed to eject one electron from the conduction band edge of a semiconductor to vacuum. When the two materials are joined together and form an interface, the chemical potentials in both materials must be equal at equilibrium [figure 7.13(b)]. Electrons in the conduction band of the semiconductor have a higher energy than the chemical potential inside the metal and thus tend to migrate into the metal when the interface is formed (a similar argument can be made for one of the two semiconductors in the case of a semiconductor–semiconductor interface). Thus there are fewer electrons in the semiconductor sides near the interface, which translates into a larger separation between E_f and E_c, creating a bent conduction band as shown in figure 7.13(a) [or figure 7.13(b) for a semiconductor–semiconductor interface]. This bent band creates an electrostatic potential which balances the migration of the electrons from the semiconductor into the metal.

The equation that determines the profile of E_c is one of the Maxwell equations, eq. (5.13). The right-hand side of eq. (5.13) is the net charge density. Let us consider an n-type semiconductor with dopant concentration N_d, each dopant contributing one free electron. The dopant ions are then positively charged and the net charge density is then $e(N_d - n)$, where n is the electron density. Using eqs. (5.8) and (6.72) for the electrostatic potential, eq. (5.13) becomes

$$\frac{d}{dx}\left(\varepsilon_e \frac{1}{e}\frac{dE_c}{dx}\right) = e(N_d - n) \tag{7.130}$$

where ε_e is the electrical permittivity of the semiconductor. The electron density n is given by eq. (4.64), that is,

$$n = \int_0^\infty fD(E)dE \approx N_c \exp\left(-\frac{E_c - E_f}{\kappa_B T}\right) \tag{7.131}$$

Here Boltzmann statistics have been used to approximate the Fermi–Dirac distribution. This approximation is valid when the doping concentration is relatively low so that $E_c - E_f \gg \kappa_B T$. This approximation will be used throughout this section, even though some of the situations to be discussed should really use the Fermi–Dirac distribution. Substituting eq. (7.131) into (7.130), we obtain

$$\frac{d^2 E_c}{dx^2} = \frac{e^2}{\varepsilon_e}\left(N_d - N_c \exp\left(-\frac{E_c - E_f}{\kappa_B T}\right)\right) \tag{7.132}$$

with the following boundary conditions:

$$x = 0, \; E_c - E_f = e(W - \chi) = e\phi \tag{7.133}$$

$$x \to \infty, \; E_c - E_f = E_{c\infty} - E_{f\infty} \tag{7.134}$$

where ϕ is the electrostatic potential discontinuity at the interface, called the barrier height, and $(E_{c\infty} - E_{f\infty})$ are undisturbed values determined by setting n equal to N_d in eq. (7.131). Equation (7.132) is also called the Poisson–Boltzmann equation and is applicable to many electrostatic phenomena at interfaces such as the semiconductor–semiconductor interface and even the solid–liquid interface, as we will see in chapter 9. The equation does not, however, have an explicit solution. Often, it is assumed that the nonuniform distribution of E_c occurs only in a small region $[-d, 0]$ in the sketch of figure 7.13(a). In this region, called the space charge region, n is much smaller than N_d and can be neglected; in other words, very few mobile electrons exist in this region. With this approximation, eq. (7.132) can be easily solved to obtain the profile of E_c as a function of x. For the following presentation, however, an explicit solution for E_c is not necessary.

There are three standard theories to describe the current flow through a Schottky barrier as described in textbooks (for example, Sze, 1981). The Bethe theory assumes no scattering across the space charge region and thus transport is completely ballistic. In this case, the external voltage drop occurs only at the interface. The Schottky theory assumes that diffusion inside the space charge region dominates and thus no voltage drop occurs at the interface. The Crowell–Sze theory combines these two limits by dividing the space charge region into two parts. One region is approximately one mean free path within the interface where thermionic emission dominates, and the other is the rest of the space charge region where diffusion dominates. An effective recombination velocity is introduced to treat the thermionic region. All these theories assume that the barrier height ϕ is much larger than $\kappa_B T$. We will use the transmission-diffusion boundary condition in combination with the drift-diffusion equation to derive an expression for the themionic emission current that includes both thermionic emission at the interface and diffusion in the space charge region, and is therefore similar to the Crowell–Sze theory, but without requiring that the barrier height be much larger than $\kappa_B T$.

Assuming a uniform electron temperature on both sides of the interface, the corresponding interface condition, eq. (7.117), becomes

$$[1 - (\tau''_{12} + \tau''_{21})/2] J_e$$
$$= \frac{-2e\tau'_{21}}{(2\pi)^3} \int_{k_x > 0} [f_{01}(E, T, E_{f1}(0)) - f_{01}(E, T, E_{f2}(0))] v_{x1} d^3 \mathbf{k}_1 \qquad (7.135)$$

Our choice of coordinate in figure 7.13(a) leads to $\tau'_{12} = \tau''_{12} = 1$. We now need to evaluate the transmission coefficient τ''_{21}, defined similarly to eq. (7.116). We assume a constant relaxation time and that only carriers with velocity component $v_x > (2e\phi/m)^{1/2}$ can go across the barrier, that is,

$$\tau_{21} = \begin{cases} 1 & \text{for} \quad v_{x2} > (2e\phi/m)^{1/2} \\ 0 & \text{for} \quad v_{x2} < (2e\phi/m)^{1/2} \end{cases} \qquad (7.136)$$

Such a transmissivity expression neglects quantum reflection and also the possibility of tunneling. Equation (7.116) becomes

$$\tau''_{21} = \frac{\int_{k_{x2}<0} \tau_{21} \tau_2 e v_{x2}^2 (\partial f_{02}/\partial x + F_x \partial f_{02}/\partial E) d^3 \mathbf{k}}{\int_{k_{x2}<0} \tau_2 e v_{x2}^2 (\partial f_{02}/\partial x + F_x \partial f_{02}/\partial E) d^3 \mathbf{k}}$$
$$= \frac{\int_{-\infty}^{\infty} dv_y \int_{-\infty}^{\infty} dv_z \int_{-\infty}^{-(2\phi/m)^{1/2}} v_{x2}^2 \partial f_0/\partial E \, dv_x}{\int_{-\infty}^{\infty} dv_y \int_{-\infty}^{\infty} dv_z \int_{-\infty}^{0} v_{x2}^2 \partial f_0/\partial E \, dv_x}$$

$$= \frac{2}{\sqrt{\pi}} \int_{x_0}^{\infty} e^{-x^2} dx + \frac{2}{\sqrt{\pi}} x_0 e^{-x_0^2}$$

$$= \operatorname{erfc}(x_0) + (2/\sqrt{\pi}) x_0 e^{-x_0^2} \tag{7.137}$$

where

$$x_0 = (e\phi/\kappa_B T)^{1/2} \quad \text{and} \quad \operatorname{erfc}(x_0) = \frac{2}{\sqrt{\pi}} \int_{x_0}^{\infty} e^{-x^2} dx \tag{7.138}$$

and $\operatorname{erfc}(x)$ is the complementary error function. In deriving eq. (7.137), the Boltzmann distribution was used instead of the Fermi–Dirac distribution. Equation (7.135) can be easily integrated to give

$$[1 - (\tau_{12}'' + \tau_{21}'')/2] J_e$$
$$= -A_R T^2 \left[\exp\left(-\frac{E_{c1}(0) - E_{f1}(0)}{\kappa_B T}\right) - \exp\left(-\frac{E_{c1}(0) - E_{f2}(0)}{\kappa_B T}\right) \right] \tag{7.139}$$

where

$$A_R = \frac{4\pi e m \kappa_B^2}{h^3} \tag{7.140}$$

is called the Richardson constant (Sze, 1981), E_c and E_f are the conduction band edge and the Fermi level, respectively, with the subscripts denoting which side of the interface is considered.

As a result of the application of the general diffusion–transmission interface condition, the interface condition, eq. (7.139), is applicable to both metal–semiconductor and semiconductor–semiconductor interfaces, with the requirement that $E_{f1} - E_{c1}(0) \gg \kappa_B T$ for the Boltzmann distribution to be valid. Because we used the principle of detailed balance to express the right-hand side of eq. (7.135) in terms of the properties of side one only, the Boltzmann statistics are only imposed on this side. However, the transmissivity expression, eq. (7.137), also used the Boltzmann statistics and thus it is subject to the same requirement. More general cases can be treated with the Fermi–Dirac distribution correspondingly.

If we apply expression (7.139) to a metal–semiconductor Schottky barrier and assume that there is no additional voltage drop in the space charge region, eq. (7.139) becomes

$$[1 - (\tau_{12}'' + \tau_{21}'')/2] J_e = A_R T^2 \exp\left(-\frac{e\phi}{\kappa_B T}\right) \left[\exp\left(\frac{E_{f1}(0) - E_{f2}(0)}{\kappa_B T}\right) - 1 \right]$$

$$= A_R T^2 \exp\left(-\frac{e\phi}{\kappa_B T}\right) \left[\exp\left(\frac{eV_i}{\kappa_B T}\right) - 1 \right] \tag{7.141}$$

where

$$V_i = [E_{f1}(0) - E_{f2}(0)]/e$$

is the voltage drop at the interface. The right-hand side of the eq. (7.141) identical to the Richardson formula for thermionic emission current over a potential barrier. The prefactor in front of J_e on the left-hand side, however, is new. This prefactor should affect the magnitude of the Richardson constant A_R, which has a standard value of 120 A cm^{-2} K^{-2}. The initial experiments on electron emission were done on metals, for which one cannot easily measure the equivalent chemical potential at the interface. Instead, the measured voltage can be associated directly with the emitted electrons. Under such experimental conditions the prefactor should be one, analogously to our discussion on photon radiation between two plates in section 7.5.1. For semiconductor–metal junctions, the Richardson constant and the barrier heights are often used as adjustable parameters to fit the experiments such that one cannot check the validity of the extra factor.

The advantage of the interface condition, eq. (7.141), lies in that it can be used in conjunction with diffusion theory for transport in the space charge region so that the voltage drop in the space charge region is consistently included. On the semiconductor side, we apply the drift-diffusion equation and the Einstein relation to write eq. (6.82) as

$$J_e = -e\left[n(x)\mu\mathscr{E} + a\frac{\partial n}{\partial x}\right] = -\frac{ea}{\kappa_B T} n \frac{dE_f}{dx} \quad (7.142)$$

Substituting eq. (7.131) into eq. (7.142), rearranging the resultant equation, and carrying out integration, we get

$$J_e F_1 \equiv J_e \int_0^{-d} \exp\left(\frac{E_{c1}}{\kappa_B T}\right) dx = -eaN_c\left[\exp\left(\frac{E_{f1}(-d)}{\kappa_B T}\right) - \exp\left(\frac{E_{f1}(0)}{\kappa_B T}\right)\right]$$

(7.143)

In the above expressions, F_1 is an integral function as defined by the first equality in eq. (7.143). F_1 is known once the electric potential distribution is obtained from the solution of the Poisson–Boltzmann equation, (7.132). On the metal side, the Fermi level is approximately constant, $E_{f2}(0) = E_{f2}$. Combining eqs. (7.139) and (7.143), we get (Chen, 2003)

$$\left[\frac{1 - (\tau_{12}'' + \tau_{21}'')}{A_R T^2 \exp(-E_{c1}(0)/\kappa_B T)} + \frac{F_1}{eaN_c}\right] J_e = -\exp\left(\frac{E_{f2}}{\kappa_B T}\right)\left[\exp\left(\frac{eV}{\kappa_B T}\right) - 1\right]$$

(7.144)

where $V = [E_{f2}(-d) - E_{f1}(0)]/e$ is the external voltage drop and the negative sign arises from the negative charge of electrons. The above expression is very similar to the Crowell–Sze theory, as rederived by Hess (2000), except that the first term on the left-hand side includes the transmissivity, which makes the formula applicable to any barrier height. It should be cautioned, however, that we have used the Boltzmann distribution whereas the Fermi–Dirac distribution should be used in a more vigorous treatment.

The issue of interface conditions for electron transport has been discussed in various literatures related to electron transport in conventional semiconductor–semiconductor interfaces (Schroeder, 1992; Ferry and Goodnick, 1997). Our discussion on electrons, phonons, photons, and molecules in this section again shows the similarities among these carriers.

7.5.4 Velocity Slip for Rarefied Gas Flow

The above application of the diffusion–transmission interface condition suggests that a similar strategy should be applicable to fluid flow. This, however, is only partially true, because the flow is parallel to the surfaces rather than perpendicular to them. In this section, we will start from the classical slip boundary condition and move on to discuss the problems with the diffusion–transmission treatments. From such discussion, we hope to show consistencies between all existing treatments, starting with the work of Maxwell (Kennard, 1938).

In the slip-flow region, a velocity and a temperature jump are often used in combination with the continuum description for flow and heat transfer inside the fluids (Kennard, 1938),

$$u_o - u_w = \frac{2-\alpha_v}{\alpha_v} \tau \sqrt{\frac{\pi \kappa_B T}{2m}} \left(\frac{\partial u}{\partial y}\right)_o \tag{7.145}$$

$$T_o - T_w = \frac{2-\alpha_T}{\alpha_T} \frac{\sqrt{2\pi \kappa_B T_w/m}}{(\varsigma+1) c_v P} \left(k\frac{\partial T}{\partial x}\right)_o \tag{7.146}$$

where ς is the ratio of constant pressure to constant volume specific heat, α_T and α_v are the energy and momentum accommodation coefficients, and the subscript o represents the values of the fluid adjacent to the wall while w is for the corresponding values on the wall.

The slip boundary conditions given by eqs. (7.145) and (7.146) are similar to the discontinuities on the temperatures and Fermi levels that we discussed above for transport perpendicular to interfaces. We can indeed derive these slip conditions from the distribution function, under certain approximations. Our derivation here will neglect the effects of heat transfer, which can be dealt with following a similar line. We can start from two distribution functions. One is eq. (7.82), with Φ obtained from eq. (7.90) by dropping the $\partial \Phi/\partial y$ term. The other approach is to directly approximate ∇f in the Boltzmann eq. (7.78) by ∇f_d, as we did before for phonon and photon transport. Both approaches lead to similar results and conclusions. We choose the latter approach for its consistency with our treatments of other carriers and also for its consistency with historical development (Kennard, 1938).

The distribution function coming toward the wall can be approximated as

$$f = f_d + \Pi v_x \tau f_d + v_y \tau \frac{\partial f_d}{\partial y} \tag{7.147}$$

where τ is the relaxation time. The shear stress across any surface $y = $ constant can be calculated from Newton's second law [eq. (6.65)]

$$\tau_{xy} = -\int_{v_y<0} fmv_x v_y d^3\mathbf{v} - \int_{v_y>0} fmv_x v_y d^3\mathbf{v} \tag{7.148}$$

At an internal point inside the channel, substituting eq. (7.147) into (7.148) leads to the Newton shear stress law. At the boundary, eq. (7.147) represents only those molecules coming toward the boundary ($v_y < 0$ at $y = 0$). The scattered carriers leaving the boundary are assumed to consist of a specular part and a diffuse part,

$$f_0(0, v_y > 0) = (1 - p)(1 + \Pi x)n_0 \left(\frac{m}{2\pi \kappa_B T}\right)^{3/2}$$
$$\times \exp\left[-\frac{v_x^2 + v_y^2 + v_z^2}{2\kappa_B T/m}\right] + pf(0, v_y < 0) \quad (7.149)$$

where p is again the specularity parameter we used before for electrons and phonons. Substituting eq. (7.147) for $v_y < 0$ and eq. (7.149) for $v_y > 0$ into eq. (7.148) and performing the integration, we obtain

$$\tau_{xy}(y=0) = (1-p)\left\{n_0 m \left(\frac{\kappa_B T}{2\pi m}\right)^{1/2}\left[u(y=0) - \frac{\tau}{mn_0}\frac{dP}{dx}\right] + \frac{1}{2}\mu\left(\frac{\partial u}{\partial y}\right)_{y=0}\right\} \quad (7.150)$$

where $\mu = n_0 \tau \kappa_B T$ is the dynamic viscosity as given by eq. (6.67). Again, our strategy is to extend the diffusion approximation (the Newton shear stress law) inside the medium at the wall such that $\tau_{xy}(y=0) = \mu(\partial u/\partial y)_{y=0}$. This leads to the following boundary condition

$$u(y=0) = \frac{1+p}{2(1-p)}\tau\sqrt{\frac{2\pi\kappa_B T}{m}}\left(\frac{\partial u}{\partial y}\right)_{y=0} + \frac{\tau}{mn_0}\frac{dP}{dx} \quad (7.151)$$

Without the second term, the above expression reduces to the standard slip boundary condition

$$u(y=0) \approx \frac{1+p}{(1-p)}\tau\sqrt{\frac{\pi\kappa_B T}{2m}}\left(\frac{\partial u}{\partial y}\right)_{y=0} \quad (7.152)$$

By setting $1 - p = \alpha_v$, we see that eq. (7.152) is exactly eq. (7.145) (here $u_w = 0$ since the wall is stationary). Thus the commonly used slip condition (Maxwell, 1952; Millikan, 1923) can be derived by neglecting the pressure term in eq. (7.151) and the relationship between p and α_v becomes very clear. The velocity slip boundary condition, eq. (7.145), introduced by Maxwell is widely used to treat the flow in the slip regime. One cannot, however, predict the Knudsen minimum on the basis of this expression for the velocity slip. Can the extra term in eq. (7.151), obtained from the diffusion–transmission interface condition, explain the Knudsen minimum? We have solved the Navier–Stokes equations for Poiseuille flow subject to the boundary condition given by eq. (7.151) on both surfaces for flow between two parallel plates. Unfortunately, the additional term does not lead to the prediction of the Knudsen minimum. A Knudsen minimum would be predicted only if the sign in front of the pressure gradient changed to negative. In fact, without such a sign change, the velocity slip can become negative because dP/dx is negative.

The reason for such a dilemma is that in this case the flow is parallel to the interface. The diffusion–transmission interface condition we introduced earlier assumes that the distribution within approximately one mean free path of the interface obeys a profile determined by the diffusion approximation. When the flow is parallel to the interface

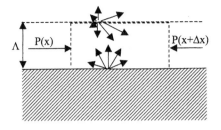

Figure 7.14 Pressure gradient effects on the slip boundary condition near a solid surface.

there is an additional pressure force acting in this region, as shown in figure 7.14. Dufresne (1996) and Wu and Bogy (2001) realized this point and derived a modified boundary condition by considering that the thickness of this layer is one mean free path (called the Knudsen layer). Their derivations, however, did not include the distortion of the distribution due to pressure, that is, the second term in eq. (7.151). If we include the pressure force within the Knudsen layer, eq. (7.150) should be written as

$$\tau_{xy}(y=0) = (1-p)\left\{ n_0 m \left(\frac{\kappa_B T}{2\pi m}\right)^{1/2} \left[u(y=0) - \frac{\tau}{mn_0}\frac{dP}{dx} \right] \right.$$
$$\left. + \frac{1}{2}\mu\left(\frac{\partial u}{\partial y}\right)_{y=0} \right\} + \Lambda\frac{dP}{dx} \qquad (7.153)$$

where Λ is the mean free path

$$\Lambda = \bar{v}\tau \qquad (7.154)$$

and $\bar{v}$ being the average velocity of the molecules. With eq. (7.153), we can show that

$$u(y=0) = \frac{1+p}{(1-p)}\tau\sqrt{\frac{\pi\kappa_B T}{2m}}\left(\frac{\partial u}{\partial y}\right)_{y=0} - \left(\frac{4}{1-p} - 1\right)\frac{\tau}{mn_0}\frac{dP}{dx} \qquad (7.155)$$

The above expression is similar to the results of Dufresne (1996) and Wu and Bogy (2001), albeit different in the coefficient for the pressure term. With the above equation as the boundary condition, one can demonstrate the existence of the Knudsen minimum. The proof for the diffuse interface case will be given in example (7.2). From such a derivation, we see that the recovery of the flow rate in figure 7.10(b) is due to the pressure gradient.

We should point out that there are also efforts to expand the velocity slip at the boundaries to second order directly. These models, however, do not start with the basic distribution profiles (Hsia and Domoto, 1983; Mitsuya, 1993). Although these higher order models can also lead to a prediction of the Knudsen minimum phenomenon, the physical origin of the Knudsen minimum becomes less obvious.

Example 7.2

Using the slip boundary condition, eq. (7.155), derive an analytical solution for the Poiseuille flow of gas between two parallel plates for a diffuse interface ($p = 0$).

Solution: For Poiseuille flow, we can neglect the acceleration, that is, the left-hand side of eq. (6.145), of the fluids and consider only the balance of the pressure against

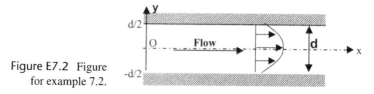

Figure E7.2 Figure for example 7.2.

the viscous force. In the coordinates given as shown in figure E7.2 and based on the Newton shear stress law [eq. (6.65)], eq. (6.145) becomes

$$\frac{dp}{dx} = \mu \frac{d^2 u}{dy^2} \quad (E7.2.1)$$

with the following boundary conditions

$$y = 0, \frac{du}{dy} = 0 \quad (E7.2.2)$$

$$y = -d/2, u = \tau\sqrt{\frac{\pi \kappa_B T}{2m}}\left(\frac{\partial u}{\partial y}\right) - \frac{3\tau}{mn_0}\frac{dP}{dx} \quad (E7.2.3)$$

The solution of eq. (E 7.2.1) under the above boundary conditions is

$$u(y) = -\frac{1}{2\mu}\frac{dP}{dx}\left[\left(\frac{d}{2}\right)^2 - y^2\right] - \frac{d\tau}{2\mu}\sqrt{\frac{\pi \kappa_B T}{2m}}\frac{dP}{dx} - \frac{3\tau}{mn_0}\frac{dP}{dx} \quad (E7.2.4)$$

where the first term on the right-hand side is the continuum solution for Poiseuille flow and the other two terms are due to slip at the boundaries. The mass flow rate per unit depth is

$$\dot{m} = \int_{-d/2}^{d/2} \rho u \, dy \quad (E7.2.5)$$

Following eq. (7.104), we can write the mass flow rate in a nondimensional form as

$$\frac{\sqrt{2RT}\dot{m}}{-d^2 dP/dx} = \frac{d}{6\tau\sqrt{2RT}} + \frac{\sqrt{\pi}}{2} + \frac{3\tau\sqrt{2RT}}{d}$$

$$= \frac{1}{6Kn} + \frac{\sqrt{\pi}}{2} + 3Kn \quad (E7.2.6)$$

This flow rate has a minimum when

$$Kn = \frac{\tau\sqrt{2RT}}{d} = 0.23 \quad (E7.2.7)$$

Comment. To compare the Kn number obtained from this simple treatment with that obtained by solving the Boltzmann equation, figure 7.10(b), we can relate ξ and Kn through

$$\xi = \frac{1}{Kn} = 4.3 \quad (E7.2.8)$$

Although the boundary condition can predict the phenomenon of the Knudsen minimum, the location of this minimum is quite far from the solution of the Boltzmann equation that we discussed in section 7.3 (Cercignani and Daneri, 1963).

7.6 Ballistic–Diffusive Treatments

Although the diffusion approximation in conjunction with the diffusion–transmission boundary conditions leads to excellent results for planar geometries, as discussed above, they are not as accurate for nonplanar geometries and multidimensional problems. When the mean free path is much larger than the characteristic length, energy carriers at any points come mostly from the boundaries and their distributions are highly nonuniform and unsymmetric, deviating far from the diffusion picture as shown in figure 6.11. One solution method is to treat the carriers from the boundaries separately from those emitted or scattered inside the medium. For radiation heat transfer, this approach is called modified differential approximation and was originally proposed by Olfe (1967a, 1967b, 1970) and further developed by several other groups (Wu et al., 1987; Modest, 1989, 1993). Chen (2001, 2002) used a similar treatment for phonons and named the derived heat conduction equations the ballistic–diffusive equations. Although such a strategy has not been applied to rarefied gas, nor to electron transport, it is reasonable to expect that extension to these carriers is feasible and useful. Interested readers are encouraged to explore such possibilities.

7.6.1 Modified Differential Approximation for Thermal Radiation

For thermal radiation, we again use the intensity notation. Assuming isotropic scattering and a gray medium, the equation of radiative transfer can be written as

$$\hat{\mathbf{\Omega}} \cdot \nabla I = -K_e(I - I_0) \tag{7.156}$$

where $\hat{\mathbf{\Omega}}$ is the unit vector in the direction of carrier propagation. Treatment for more generally anisotropic scattering and non-gray medium has been given in other references (Modest, 1993).

In Olfe's approach (1967a, 1967b), named the modified differential approximation, the intensity at any point is written as the sum of two components:

$$I(\mathbf{r}, \hat{\mathbf{\Omega}}) = I_w(\mathbf{r}, \hat{\mathbf{\Omega}}) + I_m(\mathbf{r}, \hat{\mathbf{\Omega}}) \tag{7.157}$$

where I_m is due to the emission and scattering within the medium and I_w originates from the wall, which can be the emission from the wall or the reflection of the incoming radiation toward the wall. I_w includes outgoing scattering and absorption along the photon path from the boundary to the point of interest,

$$\hat{\mathbf{\Omega}} \cdot \nabla I_w = -K_e I_w \tag{7.158}$$

A general solution for eq. (7.158) is

$$I_w(\mathbf{r}, \hat{\mathbf{\Omega}}) = I_{w0}(\mathbf{r}_w) e^{-K_e|\mathbf{r}-\mathbf{r}_w|} \tag{7.159}$$

where $I_{w0}(\mathbf{r}_w)$ is intensity at the boundary point that intersects with the $\hat{\mathbf{\Omega}}$ direction at $\mathbf{r}$, as shown in figure 7.15. The heat flux carried by I_w is

$$\mathbf{J}_{q_w} = \int I_w(\mathbf{r}, \hat{\mathbf{\Omega}}) \hat{\mathbf{\Omega}} d\Omega \tag{7.160}$$

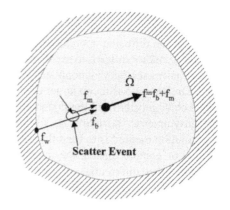

Figure 7.15 In the ballistic–diffusive approximation, local carrier distribution is divided into two parts. One originates from the boundary (or internal sources) and experiences outscattering only. The rest is classified as the diffusive part and is treated with the diffusion approximation (Chen, 2001).

The other part of eq. (7.157), I_m, represents those photons originating from inside the medium due to emission or scattering,

$$\hat{\Omega} \bullet \nabla I_m = -K_e (I_m - I_0) \tag{7.161}$$

For this part of the radiation, we can use the diffusion approximation that is familiar in thermal radiation. This can be derived from the spherical harmonic expansion of the intensity (Modest, 1993),

$$I_m(\hat{\Omega}) = I_{m0}(\mathbf{r}) + \mathbf{I}_{m1}(\mathbf{r}) \bullet \hat{\Omega} \tag{7.162}$$

Substituting eq. (7.162) into (7.161) yields

$$\hat{\Omega} \bullet \nabla I_{m0} + \hat{\Omega} \bullet \nabla (\mathbf{I}_{m1} \bullet \hat{\Omega}) = K_e(I_0 - I_{m0}) - K_e \mathbf{I}_{m1} \bullet \hat{\Omega} \tag{7.163}$$

Integrating the above equation over the solid angle of entire space leads to

$$\frac{1}{3} \nabla \bullet \mathbf{I}_{m1} = -K_e (I_{m0} - I_0) \tag{7.164}$$

Multiplying eq. (7.163) by unit vector $\hat{\Omega}$ and integrating the obtained equation over the solid angle of entire space gives

$$\nabla I_{m0} = -K_e \mathbf{I}_{m1} \tag{7.165}$$

The heat flux carried due to I_m is

$$\mathbf{J}_{q_m}(\mathbf{r}) = \int I_m(\mathbf{r}, \hat{\Omega}) \hat{\Omega} \, d\Omega = \frac{4\pi}{3} \mathbf{I}_{m1} = -\frac{4\pi}{3K_e} \nabla I_{m0} \tag{7.166}$$

This equation is nothing but the Rosseland diffusion approximation. However, in the current treatment, the diffusion heat flux $\mathbf{J}_{qm}$ is only part of the total heat flux. The other part comprises contributions direct from the boundaries, given by eq. (7.160).

Appropriate boundary conditions for the modified diffusion approximation have been the subject of few studies (Wu et al., 1987; Modest, 1989). Olfe considered only blackbody radiation from the wall. Modest (1989) considered a diffuse gray wall and argued that the diffusive part of the intensity, I_m, upon reflection at the interface, should remain as the diffusive part. This reasoning leads to the diffusive heat flux at the wall

$$\mathbf{J}_{q_m} \bullet \hat{\mathbf{n}} = \varepsilon \int I_m(\mathbf{r}, \hat{\mathbf{\Omega}}) \hat{\mathbf{\Omega}} \bullet \hat{\mathbf{n}} \, d\Omega \qquad (7.167)$$

Substituting the expression for I_m into the above equation yields

$$2\left(\frac{2}{\varepsilon} - 1\right) \mathbf{J}_{q_m} \bullet \hat{\mathbf{n}} + \frac{4\pi}{3} I_{m1} = 0 \qquad (7.168)$$

which is the Deissler boundary condition [eq. (7.122)] for the diffusive photons. The boundary condition for the ballistic part includes the radiation emitted by the wall and reflection of the incident radiation that originates from other boundaries:

$$I_w(r_w) = \varepsilon I_{bw}(r_w) + (1 - \varepsilon) \int_A I(r_w') \frac{\cos\theta \cos\theta'}{\pi s^2} e^{-K_e s} dA \qquad (7.169)$$

or in finite difference form, for a point on surface i,

$$I_w(r_{wi}) = \varepsilon I_{bw}(r_{wi}) + (1 - \varepsilon) \sum_{j=1}^{N} I_j e^{-K_e s_{ij}} F_{ij} \qquad (7.170)$$

where I_{bw} is the blackbody radiation at the wall, F_{ij} is the radiation view factor (Modest, 1993), and s_{ij} is the distance between surfaces i and j.

The above treatment of the boundary conditions implies that ballistic and diffusive parts remain ballistic and diffusive upon reflection at the boundaries. The appropriateness of this treatment can be questioned. If the wall is diffuse, it will be impossible to distinguish which part of the radiation leaving the wall is from the ballistic photons that originate from other parts of the wall and which is from the diffusive photons inside the medium that reach the wall.

7.6.2 Ballistic–Diffusive Equations for Phonon Transport

Since the strategy of separating radiation intensity into two parts works well for photon transport, it should be applicable to phonon transport. One major difference is that, for phonon transport, we may encounter transient problems. Chen (2001, 2002) started from the transient Boltzmann equation under the relaxation time approximation and followed a similar procedure to treat transient phonon heat conduction. An extension to include a nanoscale internal heat generation region was given by Yang et al. (2002).

The transient Boltzmann equation under the relaxation time approximation is

$$\frac{\partial f}{\partial t} + \mathbf{v} \bullet \nabla_\mathbf{r} f = -\frac{f - f_0}{\tau_\omega} + f_g \qquad (7.171)$$

where f_g is the phonon generation source, which we will discuss in more detail in the next chapter. We assume here that the distribution of f_g is known. As in the previous section, we divide the carrier distribution function at any internal point along a specific direction into two parts

$$f(t, \mathbf{r}, \hat{\mathbf{\Omega}}) = f_w(t, \mathbf{r}, \hat{\mathbf{\Omega}}) + f_m(t, \mathbf{r}, \hat{\mathbf{\Omega}}) \tag{7.172}$$

and, consequently, the heat flux and local internal energy density can be expressed as the sum of two parts,

$$\mathbf{J}_q(t, \mathbf{r}) = \mathbf{J}_{q_w}(t, \mathbf{r}) + \mathbf{J}_{q_m}(t, \mathbf{r}) \tag{7.173}$$

$$U(t, \mathbf{r}) = U_w(t, \mathbf{r}) + U_m(t, \mathbf{r}) \tag{7.174}$$

where the ballistic terms are

$$\mathbf{J}_{q_w}(t, \mathbf{r}) = 3\left(\frac{\hbar}{2\pi}\right)^3 \int \hbar\omega f_w \mathbf{v} d^3\mathbf{k} = \frac{1}{4\pi} \int \hbar\omega D(\omega) d\omega \left(\int \mathbf{v} f_w d\Omega\right) \tag{7.175}$$

$$U_w(t, \mathbf{r}) = 3\left(\frac{\hbar}{2\pi}\right)^3 \int \hbar\omega f_w d^3\mathbf{k} = \frac{1}{4\pi} \int \hbar\omega D(\omega) d\omega \left(\int f_w d\Omega\right) \tag{7.176}$$

and similarly for $\mathbf{J}_{q_m}$ and U_m. The governing equations for the ballistic and diffusive parts are

$$\frac{1}{|\mathbf{v}|}\frac{\partial f_w}{\partial t} + \hat{\mathbf{\Omega}} \bullet \nabla_{\mathbf{r}} f_w = -\frac{f_w}{|\mathbf{v}|\tau} + \frac{f_g}{|\mathbf{v}|\tau} \tag{7.177}$$

$$\frac{1}{|\mathbf{v}|}\frac{\partial f_m}{\partial t} + \hat{\mathbf{\Omega}} \bullet \nabla_{\mathbf{r}} f_m = -\frac{f_m - f_0}{|\mathbf{v}|\tau} \tag{7.178}$$

Note here that we have grouped f_g as a source contributing to the ballistic component. A general solution for the ballistic part, eq. (7.177), is (Pomraning, 1973)

$$f_w(t, \mathbf{r}, \hat{\mathbf{\Omega}}) = f_w(t - |\mathbf{r} - \mathbf{r}_w|/|\mathbf{v}|, \mathbf{r}_w, \hat{\mathbf{\Omega}}) \exp\left(-\int_0^{|\mathbf{r}-\mathbf{r}_w|} \frac{ds'}{|\mathbf{v}|\tau}\right) \bullet H(|\mathbf{v}|t - |\mathbf{r} - \mathbf{r}_w|)$$

$$+ \iiint f_g\left(t - \frac{s}{|\mathbf{v}|}, \mathbf{r}', \hat{\mathbf{\Omega}}\right) \exp\left(-\int_0^s \frac{ds'}{|\mathbf{v}|\tau}\right) H(t|\mathbf{v}| - s) \, dV$$

$$+ \iiint f_i(\mathbf{r} - |\mathbf{v}|t\hat{\mathbf{\Omega}}, \hat{\mathbf{\Omega}}) \exp\left(-\int_0^{|\mathbf{v}|t} \frac{ds'}{|\mathbf{v}|\tau}\right) H(s - |\mathbf{v}|t) \, dV \tag{7.179}$$

where f_w is the distribution of the phonons entering the system through the boundary surface, f_g is the distribution of the source inside the domain (denoted by D), f_i represents the initial carrier distribution, and s is the distance between the source ($\mathbf{r}'$) and the point of interest $\mathbf{r}$. The integral equation simply follows the fact that phonons of direction $\hat{\Omega}$ which are at point **r** at time t must have originated at some point $\mathbf{r} - s\hat{\Omega}$ at time $t - s|\mathbf{v}|$ due to the finite speed of the phonon, that is, the time retardation. The exponential in the equation accounts for the out-scattering. The Heaviside or unit step function $H(x)$ is defined as

$$H(x) = \begin{cases} 1 & x \geq 0 \\ 0 & x < 0 \end{cases}$$

indicating that f_i should be taken as zero for negative time and f_w should be taken as zero for points outside the system.

Following a similar derivation to that in the previous section, we obtain the constitutive relation for the diffusive and ballistic components (Chen, 2002; Yang et al., 2002):

$$\frac{\partial \mathbf{J}_{qm}}{\partial t} + \mathbf{J}_{qm} = -\frac{k}{C} \nabla U_m \qquad (7.180)$$

$$\tau \frac{\partial U_w}{\partial t} + \nabla \bullet \mathbf{J}_{qw} = -U_w + \dot{q}_g \qquad (7.181)$$

where k is the thermal conductivity, C is the specific heat capacity, and $\dot{q}_g$ is the internal heat generation responsible for f_g. The final governing equation for the diffusive carriers can be written as

$$\tau \frac{\partial^2 U_m}{\partial t^2} + \frac{\partial U_m}{\partial t} = \nabla \left(\frac{k}{C} \nabla U_m \right) - \nabla \bullet \mathbf{J}_{qw} + \dot{q}_g \qquad (7.182)$$

Equation (7.182) is similar in appearance to the hyperbolic heat conduction equation we discussed in chapter 6, but it is different in nature. First, U_m is only the internal energy of the diffusive carriers. Second, the diffusive carriers are coupled to the ballistic carriers through an additional ballistic term $\nabla \bullet \mathbf{J}_{qw}$. This additional term, together with the appropriate boundary conditions, makes a large difference in the final results in comparison with the hyperbolic heat conduction equation, as will be seen later. We will call eq. (7.182), together with the expressions for the ballistic component, the ballistic–diffusive heat conduction equations, or simply the ballistic–diffusive equations.

In the previous section, we mentioned that whether to treat the reflected carriers as diffusive or ballistic is a point of contention. Chen (2001, 2002) favors considering all the carriers leaving the boundary as ballistic in origin; that is, the boundary does not contribute to the diffusive component. Under this picture, the diffusive heat flux at the boundary is

$$\mathbf{J}_{qm} \bullet \hat{\mathbf{n}} = -\int_{\hat{\Omega} \bullet \hat{\mathbf{n}} < 0} I_{m\omega} \hat{\Omega} \bullet \hat{\mathbf{n}} d\Omega \qquad (7.183)$$

which leads to

$$\mathbf{J}_{qm} \bullet \hat{\mathbf{n}} = -|\mathbf{v}| U_m / 2 \qquad (7.184)$$

We can further eliminate $\mathbf{J}_{q_m}$ in eq. (7.184) on the basis of the constitutive relation, eq. (7.180), to obtain the following boundary conditions for the diffusion components,

$$\tau \frac{\partial U_m}{\partial t} + U_m = \frac{2\Lambda}{3} \nabla U_m \bullet \hat{\mathbf{n}} \quad (7.185)$$

where Λ is the average mean free path of the heat carriers.

We have emphasized in this chapter the consistent use of the temperature definition. When applying the ballistic–diffusive equations or the Boltzmann equations, it is always convenient to specify the distribution of the carriers leaving the boundaries by their temperature. We pointed out before that the temperature of this part of the carriers is not the local temperature that we are familiar with, and thus one must be very careful when using the ballistic–diffusive equations or the Boltzmann equation for phonon transport. The requirement in a consistent temperature definition typically means that the emitted phonon temperature is not known. We take a fixed temperature boundary condition as an example. This boundary temperature is the average local energy density at the boundary. We do not know the distribution function in eq. (7.179) from the given boundary temperature, because this distribution represents only the equivalent phonon temperature leaving the boundaries. The correct way is to use an iterative procedure such that the specific local boundary temperature is satisfied (Yang et al., 2002).

So far, the ballistic–diffusive equations have been tested for heat conduction in thin films by Chen (2001, 2002), and for two-dimensional heat conduction problems with a nanoscale heat source at the boundary or inside the medium (Yang et al., 2002, 2005). Figure 7.16 shows an example of the temperature distribution around a nanoscale heat source at the boundary. Because the heat input region is much smaller than the phonon mean free path in the surrounding, the peak temperature rise is much higher than given by the predictions of the Fourier law, as expected [figure 7.16(b)]. Comparison between the solution of the Boltzmann equation and the ballistic–diffusive equation is favorable, but error increases for the case of an internal nanoscale heat source (Yang et al., 2005). The higher device temperature rise was also predicted by Sverdrup et al. (2001a) on the basis of a solution to the Boltzmann equation. Some experimental results suggest the existence of the predicted effect (Sverdrup et al., 2001b).

7.7 Summary of Chapter 7

Classical size effects can be treated on the basis of the Boltzmann equation. Interfaces and boundaries impose boundary conditions for this equation. We used the relaxation time approximation in this chapter and treated photons, phonons, electrons, and molecules in parallel. Classical size effects have been treated extensively in various different fields, including electronics, photon transport, rarefied gas dynamics, and phonon heat conduction. This chapter is probably the first attempt to bring all the phenomena together to demonstrate the similarities among these vastly different subjects, and also some subtle differences among them.

Sections 7.1 to 7.4 are more or less exact solutions of the Boltzmann equation under various simplifying conditions. For transport along the film plane, it is assumed that the gradient of the distribution function along the transport direction can be approximated by the gradient of the equilibrium distribution function. This leads to analytical expressions for the transport properties. For transport perpendicular to the film plane,

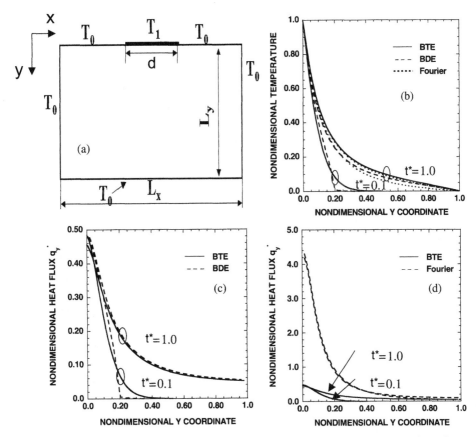

Figure 7.16 (a) Transient heat transfer near a nanoscale heat source at the boundary. Comparison of transient temperature and heat flux distribution at the centerline using the Fourier theory, the Boltzmann equation (BTE), and the ballistic–diffusive equations (BDE) based on thermalized temperature boundary conditions; (b) temperature distribution; (c) and (d) heat flux q_y^* distribution (Yang et al., 2004). All for Knudsen number $Kn = \Lambda/d = 10$. Nondimensional time $t^* = t/\tau$.

we can no longer make the same assumption and the obtained equations determining the temperature, or the Fermi level, are of integral type that typically requires numerical solution. Rarefied gas flow represents a mixture of these two situations.

The Boltzmann equation is difficult to solve. Sections 7.5 and 7.6 introduce two approximations that can greatly simplify the solution of the Boltzmann equation. The diffusion–transmission boundary condition attributes all the size effects to the interface region only. It is easy to use and is a good starting point for estimating the order of magnitude of size effects for transport perpendicular to interfaces. Its accuracy, however, is limited and it is best suited for transport perpendicular to planar interfaces. The ballistic–diffusive treatment should be more accurate but it needs further validation by comparing its predictions with those from the Boltzmann equation. There is room to extend this approach to gas and electron transport.

One important point to keep in mind is that, when size effects are important, the systems are typically highly nonequilibrium. The classical thermodynamic quantities such as temperature and Fermi level no longer have their usual meaning. Nevertheless, we can still feel comfortable in using these equilibrium quantities if we think of them as a representation of local energy density or particle density. It is fair to say that, close to all interfaces and boundaries, all transport processes are highly nonequilibrium. It is important to understand the meaning of conventional thermodynamic variables and to use a definition that is consistent with actual applications. For example, in thermal radiation, the photons coming out of the wall are determined by the local solid temperature, which is not the temperature of the photons inside the solid near the wall. For phonons, a consistent use of the same temperature definition, inside the region as well as at the boundaries, is crucial.

7.8 Nomenclature for Chapter 7

a	thermal or electron diffusivity, $m^2 s^{-1}$	f_0	equilibrium distribution function
A	$= 2RT_0\Pi$, following eq. (7.95)	F_1	function defined by eq. (7.143), m
A_d, A_s	functions given by eqs. (7.24) and (7.25)	F_{ij}	radiation view factor from surface i to point j
A_R	Richardson constant, $A\,m^{-2}\,K^{-2}$	F	external force on the particle, N
c	normalized velocity, defined by eq. (7.64)	g	deviation from equilibrium distribution
C	volumetric specific heat, $J\,m^{-3}\,K^{-1}$	h	Planck constant, J s
d	film thickness, m	$\hbar$	Planck constant divided by 2π, J s
d_a	atomic or molecular diameter, m	H	Heaviside step function
D	density of states per unit volume, m^{-3}	H_n	integral function defined by eq. (7.70)
e	unit charge, C	i	deviation in intensity, $W\,m^{-2}$
e_0	blackbody emissive power, $W\,m^{-2}$	I	intensity, $W\,m^{-2}\,srad^{-1}$
E	energy of one particle, J	I_m	diffusive component of intensity, $W\,m^{-2}\,srad^{-1}$
E_c	conduction band edge, J		
E_f	chemical potential, J	I_{m0}	first term in the spherical harmonics expansion, eq. (7.162), $W\,m^{-2}\,srad^{-1}$
E_n	integral function of order n, defined by eq. (7.17)		
f	one-particle distribution function	$\mathbf{I}_{m1}$	vector in the spherical harmonics expansion, eq. (7.162), $W\,m^{-2}\,srad^{-2}$
f_d	displaced Maxwell distribution		
f_e	distribution of carriers coming toward the interface, i.e., emitted carriers	j_e	electrical current per unit depth, $A\,m^{-1}$

j_m	mass flow rate, kg s^{-1} m^{-1}	α_T	temperature accommodation coefficient
J_q	heat flux, W m^{-2}	α_v	momentum accommodation coefficient
$\mathbf{J}$	flux of heat, W m^{-2}; energy, W m^{-2}; current, A m^{-2}, and particles, s^{-1} m^{-2}	γ	nondimensional temperature, defined by eq. (7.64)
k	thermal conductivity, W m^{-1} K^{-1}	Δ	interface asperity parameter, m
K	defined by eq. (7.72)	ε	emissivity
$\mathbf{k}$	wavevector, m^{-1}	ε_e	electrical permittivity, C^2 N^{-1} m^{-2}
K_e	extinction coefficient, m^{-1}	ς	ratio of constant pressure to constant volume specific heat
Kn_e	electron Knudsen number		
L	channel length or characteristic length, m	η	nondimensional coordinate $= y/(\tau v)$
m	mass, kg	θ	polar angle
M	momentum, kg m s^{-1}	κ_B	Boltzmann constant, J K^{-1}
M_A	molar weight, kg mol^{-1}	λ	wavelength, m
n	particle number density, m^{-3}	Λ	mean free path, m
$\hat{\mathbf{n}}$	surface norm unit vector	μ	directional cosine $= \cos\theta$
N_d	dopant concentration, m^{-3}	μ_e	electron mobility, m^2 V^{-1} s^{-1}
p	interface specularity parameter	ξ	nondimensional film thickness
P	pressure, N m^{-2}	ξ_e	nondimensional film thickness
$\dot{q}_g$	heat generation, W m^{-3}		
$\mathbf{r}$	space coordinate vector	Π	normalized pressure gradient, m^{-1}
R	ideal gas constant, J K^{-1} kg^{-1}	ρ	density, kg m^{-3}
R_u	universal gas constant, J K^{-1} mol^{-1}	σ	electrical conductivity, Ω^{-1} m^{-1}
R_{12}	reflectivity for incident flux from layer 1 toward layer 2	τ	relaxation time, s
S_0	source function	τ_{12}	transmissivity from layer 1 into layer 2
t	time, s	τ_{12}'	transmissivity from layer 1 into layer 2, averaged according to eq. (7.118)
T	temperature, K		
u	average velocity, m s^{-1}		
U	internal energy per unit volume, J m^{-3}	τ_{12}''	transmissivity from layer 1 into layer 2, averaged according to eq. (7.116)
$\mathbf{v}$	velocity, m s^{-1}		
V	external voltage, V		
w_i	weight factor in Gauss–Legendre integration	ϕ	electrostatic potential barrier, V
W	work function, J	φ	azimuthal angle, rad
x, y, z	Cartesian coordinates	Φ	nondimensional deviation function, defined by eq. (7.82)
X	microscopic quantity		
Y_ℓ^m	spherical harmonics function		
Z	function introduced by eq. (7.94)	χ	electron affinity, J

χ_i	fractional thickness of layer i	m	mass; or originating from inside the medium
ψ	angle in cylindrical system, figure 7.11(b)	n	next period
Ψ	nondimensional average velocity, defined in eq. (7.101)	o	inside the medium but adjacent to the wall
		p	previous period
		th	thermal average
ω	angular frequency, Hz	w	originating from wall or boundary
Ω	solid angle, srad		
$\hat{\Omega}$	unit vector along propagation direction	x, y, z	Cartesian component

Subscripts

0	equilibrium
12	from layer 1 to layer 2
1, 2	layer one or layer 2
b	bulk
d	diffuse interface
e	electron; emitted carriers
i	ith layer; initial condition

Superscripts

$'$	deviation from equilibrium
$+$	propagating toward positive y direction, i.e., $\mu > 0$
$-$	propagating toward negative y direction, i.e., $\mu < 0$
$*$	normalized quantity

7.9 References

Asheghi, M., Touzelbaev, M.N., Goodson, K.E., Leung, Y.K., and Wong, S.S., 1998, "Temperature-Dependent Thermal Conductivity of Single-Crystal Silicon Layers in SOI Substrates," *Journal of Heat Transfer*, vol. 120, pp. 31–36.

Berman, R., Foster, E.L., and Ziman, J.M., 1955, "Thermal Conductivity in Artificial Sapphire Crystals at Low Temperatures," *Proceedings of Royal Society A (London)*, vol. 231, pp. 130–144.

Bhatnagar, P.L., Gross, E.P., and Krook, M., 1954, "A Model for Collision Processes in Gases. I. Small Amplitude Processes in Charged and Neutral One-Component Systems," *Physical Review*, vol. 94, pp. 511–525.

Bird, G.A., 1994, *Molecular Gas Dynamics and the Direct Simulation of Gas Flow*, 2nd ed., Oxford University Press, Oxford.

Casimir, H.B.G., 1938, "Note on the Conduction of Heat in Crystals," *Physica*, vol. 5, pp. 495–500.

Cercignani, C., 2000, *Rarefied Gas Dynamics: from Basic Concepts to Actual Calculations*, Cambridge, New York.

Cercignani, C., and Daneri, A., 1963, "Flow of a Rarefied Gas between two Parallel Plates," *Journal of Applied Physics*, vol. 34, pp. 3509–3513.

Chandrasekhar, S., 1960, *Radiative Transfer*, Dover, New York.

Chen, G., 1996a, "Nonlocal and Nonequilibrium Heat Conduction in the Vicinity of Nanoparticles," *Journal of Heat Transfer*, vol. 118, pp. 539–545.

Chen, G., 1996b, "Heat Transport in the Perpendicular Direction of Superlattices and Periodic Thin-Film Structures," Proceedings of the 1996 International Mechanical Engineering Congress, DSC-vol. 59, pp. 13–24.

Chen, G., 1997, "Size and Interface Effects on Thermal Conductivity of Superlattices and Periodic Thin-Film Structures," *Journal of Heat Transfer*, vol. 119, pp. 220–229.

Chen, G., 1998, "Thermal Conductivity and Ballistic Phonon Transport in the Cross-Plane Direction of Superlattices," *Physical Review B*, vol. 57, pp. 14958–14973.

Chen, G., 2001, "Ballistic–Diffusive Heat Conduction Equations," *Physical Review Letters*, vol. 85, pp. 2297–2300.

Chen, G., 2002, "Ballistic–Diffusive Equations for Transient Heat Conduction from Nano- to Macroscales," *Journal of Heat Transfer*, vol. 124, pp. 320–328.

Chen, G., 2003, "Diffusion–Transmission Interface Condition for Electron and Phonon Transport," *Applied Physics Letters*, vol. 82, pp. 991–993.

Chen, G., and Neagu, M., 1997, "Thermal Conductivity and Heat Transfer in Superlattices," *Applied Physics Letters*, vol. 71, pp. 2761–2763.

Chen, G., and Tien, C.L., 1993, "Thermal Conductivity of Quantum Well Structures," *AIAA Journal of Thermophysics and Heat Transfer*, vol. 7, pp. 311–318.

Chen, G., and Zeng, T., 2001, "Nonequilibrium Phonon and Electron Transport in Heterostructures and Superlattices," *Microscale Thermophysical Engineering*, vol. 5, pp. 71–88.

Corrucini, R.J., 1958, "Calculation of Gaseous Heat Conduction in Dewars," *Advances in Cryogenic Engineering*, vol. 3, pp. 353–362.

Davison, B., 1957, *Neutron Transport Theory*, Clarendon Press, Oxford.

Deissler, R.G., 1964, "Diffusion Approximation for Thermal Radiation in Gases with Jump Boundary Conditions," *Journal of Heat Transfer*, vol. 86, pp. 240–246.

Dilliner, U., and Volklein, F., 1990, "Transport Properties of Flash-Evaporated $(Bi_{1-x}Sb_x)_2Te_3$ Films. 2. Theoretical Analysis," *Thin Solid Films*, vol. 187, pp. 263–273.

Dong, W., 1956, "Vacuum Flow of Gases through Channels with Circular, Annular and Rectangular Cross Sections," University of California Radiation Laboratory, Report No. UCRL-3353.

Dufresne, M.A., 1996, "On the Development of a Reynolds Equation for Air Bearings with Contact," Ph.D. thesis, Mechanical Engineering Dept., Carnegie Mellon University.

Eckert, E.R.G., and Drake, R.M., Jr, 1972, *Analysis of Heat and Mass Transfer*, McGraw-Hill, New York, pp. 467–542.

Ferry, D.K., and Goodnick, S.M., 1997, *Transport in Nanostructures*, Cambridge University Press, Cambridge, pp. 127–128.

Flik, M.I., and Tien, C.L., 1990, "Size Effect on Thermal Conductivity of High-T_c Thin-Film Superconductors," *Journal of Heat Transfer*, vol. 112, pp. 872–881.

Fuchs, F., 1938, "The Conductivity of Thin Metallic Films According to the Electron Theory of Metals," *Proceedings of Cambridge Philosophy Society*, vol. 34, pp. 100–108.

Fukui, S., and Kaneko, R. 1998, "Analysis of Ultra-Thin Gas Film Lubrication Based on Linearized Boltzmann Equation: First Report—Derivation of a Generalized Lubrication Equation Including Thermal Creep Flow," *Journal of Tribology*, vol. 110, pp. 253–263.

Goodson, K.E., and Ju, Y.S., 1999, "Heat Conduction in Novel Electronic Films," *Annual Review of Material Science*, vol. 29, pp. 261–293.

Gurevich, V.L., 1986, *Transport in Phonon Systems*, North-Holland, Amsterdam, chapter 2.

Guyer, R.A., and Krumhansl, J.A., 1966, "Thermal Conductivity, Second Sound, and Phonon Hydrodynamic Phenomena in Nonmetallic Crystals," *Physical Review*, vol. 148, pp. 778–788.

Hess, K., 2000, *Advanced Theory of Semiconductor Devices*, IEEE Press, New York.

Hsia, Y.T., and Domoto, G.A., 1983, "An Experimental Investigation of Molecular Rarefaction Effects in Gas Lubricated Bearings at Ultra-Low Clearances," *Journal of Lubrication Technology*, vol. 105, pp. 120–130.

Hyldgaard, P., and Mahan, G.D., 1996, "Phonon Knudsen Flow in GaAs/AlAs Superlattices," in *Thermal Conductivity 23*, ed. Wilkes, K.E., Dinwiddie, R.B., and Graves, R.S., pp. 172–182, Technomic, Lancaster, PA.

Jeans, J.H., 1917, "The Equation of Radiative Transfer of Energy," *Monthly Notices Royal Astronomical Society*, vol. 78, pp. 28–36.

Jen, C.-P., and Chieng, C.-C., 1998, "Microscale Thermal Characterization for Two Adjacent Dielectric Thin Films," *Journal of Thermophysics and Heat Transfer*, vol. 12, pp. 146–152.

Ju, Y.S., and Goodson, K.E., 1999, "Phonon Scattering in Silicon Films with Thickness of Order of 100 nm," *Applied Physics Letters*, vol. 74, pp. 3005–3007.

Karniadakis, G.E., and Beskok, A., 2002, *Micro Flows, Fundamentals and Simulation*, Springer-Verlag, New York.

Kennard, E.H., 1938, *Kinetic Theory of Gases*, McGraw-Hill, New York.

Klitsner, T., VanCleve, J.E., Henry, E.F., and Pohl, R.O., 1988, "Phonon Radiative Heat Transfer and Surface Scattering," *Physical Review B*, vol. 38, pp. 7576–7594.

Knudsen, M., 1909, "Die Gesetze der Molekular Stromung und der Inneren Reibungstromung der Gas durch Rohren," *Annalen der Physik*, vol. 28, pp. 30–75.

Knudsen, M., 1950, *Kinetic Theory of Gases*, Methuen, London.

Kumar, S., and Vradis, G.C., 1994, "Thermal Conductivity of Thin Metallic Films," *Journal of Heat Transfer*, vol. 116, pp. 28–34.

Lee, S.M., Cahill, D.G., and Venkatasubramanian, R., 1997, "Thermal Conductivity of Si–Ge Superlattices," *Applied Physics Letters*, vol. 70, pp. 2957–2959.

Luikov, A.V., 1964, *Heat and Mass Transfer in Capillary-Porous Bodies*, Pergamon, Oxford.

Majumdar, A., 1991 "Effect of Interfacial Roughness on Phonon Radiative Heat Conduction," *Journal of Heat Transfer*, vol. 113, pp. 787–805.

Majumdar, A., 1993, "Microscale Heat Conduction in Dielectric Thin Films," *Journal of Heat Transfer*, vol. 115, pp. 7–16.

Maxwell, J.C., 1952, *Scientific Papers*, vol. 2, Dover, New York.

Maxwell, J.C., 1953, *Collected Works*, Dover Publications, New York.

Mazumder, S., and Majumdar, A., 2001, "Monte Carlo Study of Phonon Transport in Solid Thin Films Including Dispersion and Polarization," *Journal of Heat Transfer*, vol. 123, pp. 749–759.

Millikan, R.A., 1923, "Coefficients of Slip in Gases and the Law of the Reflection of Molecules from the Surfaces of Solids and Liquids," *Physical Review*, vol. 21, pp. 217–238.

Mitsuya, Y., 1993, "Modified Reynolds Equation for Ultra-Thin Film Gas Lubrication Using 1.5-Order Slip-Flow Model and Considering Surface Accommodation Coefficient," *Journal of Tribology*, vol. 115, pp. 289–294.

Modest, M.F., 1989, "The Modified Differential Approximation for Radiative Transfer in General Three-Dimensional Media," *Journal of Thermophysics and Heat Transfer*, vol. 3, pp. 283–288.

Modest, M.F., 1993, *Radiative Heat Transfer*, McGraw-Hill, New York.

Nath, P., and Chopra, K.L., 1973, "Experimental Determination of the Thermal Conductivity of Thin Films," *Thin Solid Films*, vol. 18, pp. 29–37.

Olfe, D.B., 1967a, "A Modification of the Differential Approximation for Radiative Transfer," *AIAA Journal*, vol. 5, pp. 638–643.

Olfe, D.B., 1967b, "Application of a Modified Differential Approximation to Radiative Transfer in a Gray Medium between Concentric Spheres and Cylinders," *Journal of Quantitative Spectroscopy and Radiative Transfer*, vol. 8, pp. 899–907.

Olfe, D.B., 1970, "Radiative Equilibrium of a Gray Medium Bounded by Nonisothermal Walls," *Progress in Astronautics and Aeronautics*, vol. 23, pp. 29–317.

Patterson, G.N., 1971, *Introduction to the Kinetic Theory of Gas Flow*, University of Toronto Press, Toronto, pp. 221–227.

Perlmutter, M., and Howell, J.R., 1964, "Radiant Transfer through a Gray Gas between Concentric Cylinders using Monte Carlo," *Journal of Heat Transfer*, vol. 86, pp. 169–179.

Peterson, R.B., 1994, "Direct Simulation of Phonon-Mediated Heat Transfer in a Debye Crystal," *Journal of Heat Transfer*, vol. 116, pp. 815–822.

Pomraning, G.C., 1973, *The Equation of Radiation Hydrodynamics*, Pergamon Press, Oxford.

Press, W.H., Teukolsky, S.A., Vetterling, W.T., and Flannery, B.P., 1992, *Numerical Recipes in Fortran*, Cambridge University Press, New York.

Probstein, R.F., 1963, "Radiation Slip," *AIAA Journal*, vol. 1, pp. 1022–1024.

Qiu, T.Q., and Tien, C.L., 1993, "Size Effects on Nonequilibrium Laser Heating of Metal Films," *Journal of Heat Transfer*, vol. 115, pp. 842–847.

Ryhming, I.L., 1966, "Radiative Transfer between Two Concentric Spheres Separated by an Absorbing and Emitting Gas," *International Journal of Heat and Mass Transfer*, vol. 9, pp. 315–324.

Schaaf, S.A., and Chambre, P.L., 1961, *Flow of Rarefied Gases*, Princeton University Press, Princeton, NJ.

Schroeder, D., 1992, "The Inflow Moments Method for the Description of Electron Transport at Material Interfaces," *Journal of Applied Physics*, vol. 72, pp. 964–970.

Siegel, R., and Howell, J.R., 1992, *Thermal Radiation Heat Transfer*, 3rd ed., Hemisphere, Washington, DC.

Sondheimer, E.H., 1952, "The Mean Free Path of Electrons in Metals," *Advances in Physics*, vol. 1, pp. 1–42.

Song, D., and Chen, G., 2003, "Monte Carlo Simulation of In-Plane Phonon Transport in Porous Silicon Membranes," *Proceedings of 2003 ASME Heat Transfer Conference*, July 21–23, Las Vegas, Nevada, Paper HT2003-40592.

Sparrow, E.M., Usiskin, C.M., and Hubbard, H.A., 1961, "Radiation Heat Transfer in a Spherical Enclosure Containing a Participating, Heat-Generating Gas," *Journal of Heat Transfer*, vol. 83, pp. 199–206.

Springer, G.S., 1971, "Heat Transfer in Rarefied Gases," *Advances in Heat Transfer*, vol. 7, pp. 163–218.

Stettler, M.A., and Lundstrom, M.S., 1994, "A Detailed Investigation of Heterojunction Transport Using a Rigorous Solution to the Boltzmann Equation," *IEEE Transactions on Electron Devices*, vol. 41, pp. 592–600.

Stoner, R.J., and Maris, H.J., 1993, "Kipitza Conductance and Heat Flow between Solids at Temperatures from 50 to 300 K," *Physical Review B*, vol. 48, pp. 16373–16387.

Sverdrup, P.G., Ju, Y.S., and Goodson, K.E., 2001a, "Sub-Continuum Simulations of Heat Conduction in Silicon–on–Insulator Transistors," *Journal of Heat Transfer*, vol. 123, pp. 130–137.

Sverdrup, P.G., Sinha, S., Asheghi, M., Uma, S., and Goodson, K.E., 2001b, "Measurement of Ballistic Phonon Conduction near Hotspots in Silicon," *Applied Physics Letters*, vol. 78, pp. 3331–3333.

Swartz, E.T., and Pohl, R.O., 1989, "Thermal Boundary Resistance," *Reviews of Modern Physics*, vol. 61, pp. 605–668.

Sze, S.M., 1981, *Physics of Semiconductor Devices*, 2nd ed., Wiley, New York.

Tellier, C.R., and Tosser, A.J., 1982, *Size Effects in Thin Films*, Elsevier, Amsterdam.

Tien, C.L., Majumdar, A., and Gerner, F., 1998, *Microscale Energy Transport*, Taylor and Francis, Washington, DC.

Viskanta, R., and Crosbie, A.L., 1967, "Radiative Transfer through a Spherical Shell of an Absorbing Emitting Gray Medium," *Journal of Quantitative Spectroscopy and Radiative Transfer*, vol. 7, pp. 871–889.

Volklein, F., and Kessler, E., 1986, "Analysis of the Lattice Thermal Conductivity of Thin Films by Means of a Modified Mayadas–Shatzkes Model: the Case of Bismuth Films," *Thin Solid Films*, vol. 142, pp. 169–181.

Volz, S.G., and Chen, G., 1999, "Molecular Dynamics Simulation of Thermal Conductivity of Si Nanowires," *Applied Physics Letters*, vol. 75, pp. 2056–2058.

Walkauskas, S.G., Broido, D.A., Kempa, K., and Reinecke, T.L., 1999, "Lattice Thermal Conductivity of Wires," *Journal of Applied Physics*, vol. 85, pp. 2579–2582.

Weisbuch, C., and Vinter, B., 1991, *Quantum Semiconductor Structures: Fundamentals and Applications*, Academic Press, Boston.

Willis, D.R., 1962, "Heat Transfer in a Rarefied Gas between Parallel Plates at Large Temperature Ratios," Princeton University James Forrestal Research Center, Gas Dynamics Lab., Report No. 615.

Wu, L., and Bogy, D.B., 2001, "A Generalized Compressible Reynolds Lubrication Equation with Bounded Contact Pressure," *Physics of Fluids*, vol. 13, pp. 2237–2244.

Wu, C.-Y., Sutton, W.H., and Love, T.J., 1987, "Successive Improvement of the Modified Differential Approximation in Radiative Heat Transfer," *Journal of Thermophysics and Heat Transfer*, vol. 1, pp. 296–300.

Yang, R.G., Chen, G., and Taur, Y., 2002, "Ballistic–Diffusive Equations for Multidimensional Nanoscale Heat Conduction," *Proceedings of the 2002 International Heat Transfer Conference (IHTC 2002)*, vol. 1, pp. 579–584, Grenoble, France.

Yang, R.G., Laroche, M., Chen, G., and Taur, Y., 2005, "Simulation of Nanoscale Multidimensional Transient Heat Conduction Problems Using Ballistic–Diffusive Equations and Boltzmann Equation," *Journal of Heat Transfer*, in press.

Zeng, T., and Chen, G., 2001, "Phonon Heat Conduction in Thin Films: Impacts of Thermal Boundary Resistance and Internal Heat Generation," *Journal of Heat Transfer*, vol. 123, pp. 340–347.

Ziman, J.M., 1960, *Electrons and Phonons*, Clarendon Press, Oxford, chapter 11.

7.10 Exercises

7.1 *Size effects on the in-plane phonon thermal conductivity of thin films with specular surfaces.* For heat conduction along a freestanding thin film, starting from the Boltzmann equation, prove that if the boundary is specularly reflecting the thin-film thermal conductivity equals the bulk thermal conductivity of the same material.

7.2 *Size effects on the in-plane phonon thermal conductivity of thin films with diffuse surfaces.* The phonon mean free path in silicon is ~2500 Å. Estimate the thermal conductivity of a 1000 Å thick free-standing silicon membrane at room temperature, assuming interfaces scatter phonons diffusely. The room temperature thermal conductivity of silicon is 145 W m^{-1} K^{-1}.

7.3 *Size effects of partially specular and partially diffuse surfaces on the electrical conductivity of thin films.* Derive an expression for the electrical conductivity along a thin film, assuming that the interfaces scatter electrons partially diffusely and partially specularly, with p representing the fraction of specularly scattered electrons. Neglect quantum size effects.

7.4 *Size effects of partially specular and partially diffuse surfaces on the electronic contribution to the thermal conductivity of thin films.* Derive an expression for the electronic contribution to the thermal conductivity along a thin film, assuming that the interfaces scatter electrons partially diffusely and partially specularly, with p representing the fraction of specularly scattered electrons. Neglect quantum size effects.

7.5 *Effects of diffuse interface scattering on the Seebeck coefficient along a thin film.* Derive an expression for the Seebeck coefficient along a thin film, assuming that the interfaces scatter electrons diffusely. Is the Seebeck coefficient higher or lower, due to boundary scattering, compared to that of the bulk material for the same doping concentration?

7.6 *Effects of partially specular and partially diffuse surfaces on the Seebeck coefficient along a thin film.* Derive an expression for the Seebeck coefficient along a thin film, assuming that the interfaces scatter electrons partially diffusely and partially specularly, with p representing the fraction of specularly scattered electrons. Neglect quantum size effects.

7.7 *Size effect on the phonon thermal conductivity along a circular nanowire.* Derive an expression for the effective phonon thermal conductivity of a circular nanowire.

7.8 *Size effect on the phonon thermal conductivity of a square nanowire.* Derive an expression for the effective phonon thermal conductivity of a square nanowire.

7.9 *Phonon transport perpendicular to interfaces of thin films.* A thin film is sandwiched between two identical cladding materials. The parent material of the film has a thermal conductivity of $k_1 = 145$ W m^{-1} K^{-1}, $v_1 = 6400$ m s^{-1}, and $C_1 = 1.66 \times 10^6$ J m^{-3} K^{-1}. The cladding materials have the following properties: $v = 3900$ m s^{-1}, $C = 1.67 \times 10^6$ J m^{-3} K^{-1}. Estimate the effective thermal conductivity in the direction perpendicular to the film as a function of the film thickness for the film thickness in the range of 10–1000 Å, taking the phonon reflection at the two interfaces into consideration.

7.10 *Rarefied gas heat conduction.* Extend the diffusion approximation, together with appropriate boundary conditions for partially accommodating surfaces, for heat conduction through a gas sandwiched between two parallel plates. Derive an expression for the effective thermal conductivity of the gas as a function of the accommodation coefficients on the two surfaces and the Knudsen number.

7.11 *Rarefied Couette flow.* Derive an expression for the velocity distribution using the slip boundary condition, eq. (7.145) for Couette flow, as shown in figure P7.11. The bottom plate is stationary and the top plate is moving at constant speed U_0.

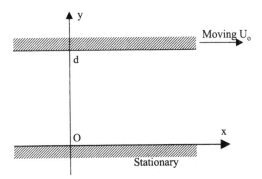

Figure P7.11 Figure for problem 7.11, Couette flow.

7.12 *Electrical conductivity of a double heterojunction structure.* A semiconductor thin film sandwiched between two semiconductor materials has two interfaces and is called a double heterojunction (see figure p7.12). Extend the drift-diffusion approximation, with the diffusion–transmission boundary condition, for electrical conductivity perpendicular to a thin film with a barrier height ϕ at both interfaces. Neglect quantum size effects on the electron energy levels, and the electrostatic potential change due to the depletion or accumulation of electrons in the barrier region.

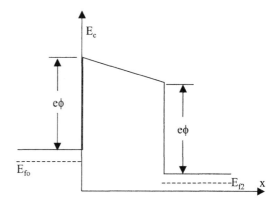

Figure P7.12 Figure for problem 7.12, thermionic emission.

7.13 *Radiative heat transfer between two concentric spheres.* Apply the modified differential approximation to solve the problem of radiative heat transfer between two concentric spheres maintained at two different temperatures T_1 and T_2. Assume each wall is black and the gas between the shells is absorbing, emitting, and isotropically scattering.

7.14 *Radiative heat transfer between two concentric cylinders.* Apply the modified differential approximation to solve the problem of radiative heat transfer between two concentric cylinders maintained at two different temperatures T_1 and T_2. Assume each wall is black and the gas between the cylinders is absorbing, emitting, and isotropically scattering.

7.15 *Size effects of heat conduction across a cylindrical thin shell.* Apply the ballistic–diffusive equations to solve the problem of steady-state phonon heat conduction between a cylindrical thin shell of thickness d. Derive an expression for the temperature distribution in the thin shell. The equivalent equilibrium temperatures at the inner and the outer surfaces are T_i and T_0, respectively. Phonon mean free path in the parent bulk material of the shell is Λ.

7.16 *Size effects on phonon heat conduction across a spherical thin shell.* Apply the ballistic–diffusive equations to solve the problem of steady-state phonon heat conduction between a spherical thin shell of thickness d. The inner and the outer surfaces are maintained at T_i and T_0, respectively. Phonon mean free path in the parent bulk material of the shell is Λ.

7.17 *Size effects on heat conduction with an internal heat source: ballistic–diffusive treatment.* Heat is uniformly generated inside a nanoscale spherical region of diameter d, embedded inside an infinite medium. The heat generation region and the surrounding are of the same material with no boundary; that is the transmissivity is equal to one. Use the ballistic–diffusive approximation to solve for the temperature profile inside the surrounding medium. Phonon mean free path in the surrounding medium is Λ.

7.18 *Size effects on heat conduction with internal heat source: diffusion–transmission interface condition.* Heat is uniformly generated inside a nanoscale spherical region of diameter d, embedded inside an infinite medium. Use the diffusion–transmission boundary condition to derive a simple expression for the temperature distribution inside the sphere as well as in the surrounding medium, assuming that the transmissivity at the interface is equal to one. Bulk phonon mean free path in both media is Λ.

7.19 *Size effects on phonon heat conduction across a spherical thin shell.* Use the diffusion approximation together with the diffusion–transmission boundary condition to solve for phonon heat conduction across a spherical thin shell of thickness d. Derive an expression for the effective thermal conductivity of the shell. Bulk phonon mean free path in the thin shell material is Λ. The inner and outer surfaces of the shell are maintained at T_i and T_0, respectively.

7.20 *Size effects on phonon heat conduction across a cylindrical thin shell.* Use the diffusion approximation together with the diffusion–transmission boundary condition to solve the problem of phonon heat conduction across a cylindrical thin shell of thickness d. Derive an expression for the effective thermal conductivity of the thin shell. Bulk phonon mean free path in the thin shell material is Λ. The inner and the outer surfaces are maintained at T_i and T_0, respectively.

8

Energy Conversion and Coupled Transport Processes

The collision of carriers may cause both energy and momentum transfer among them. Energy conversion processes must involve energy exchange between at least two different carriers. For example, Joule heating is due to the relaxation of energy from electrons to phonons, and thermal radiation from a solid results from the conversion of the energy of phonons or electrons in the solid into photons. The exchange of energy among electrons, phonons, and photons can in principle be determined by evaluating the collision integral in the Boltzmann equation. Discussions in the previous two chapters are based on the relaxation time approximation, which is strictly valid only for elastic scattering processes that do not involve energy exchange among carriers. The relaxation time approximation, however, is often used even for inelastic collision without rigorous justification.

Traditionally, the constitutive equations obtained from the relaxation time approximation are used in combination with the energy conservation law to describe energy exchange processes. For example, in heat conduction processes involving Joule heating, one combines the Fourier law with the first law of thermodynamics to derive the heat conduction equation that determines the temperature distribution. Joule heating is included in the conservation law without consideration of how electrons deliver their energy to the lattice systems. This approach is adequate in most macroscale processes and in systems in which electrons and phonons are close to equilibrium and heat generation occurs over a region much larger than the carrier mean free path, but can fail when these conditions are not satisfied.

In this chapter, we will examine in more detail energy conversion associated with collision processes, starting from the collision integral in the Boltzmann equation.

Carrier scattering, generation, and recombination are discussed in section 8.1, with a detailed treatment of electron–phonon scattering processes and simplified treatments of generation and recombination processes. Terms approximating these processes are added to the Boltzmann equation in addition to the relaxation time term. Such a treatment of carrier collision can be readily combined with the transport processes as represented by the left-hand side of the Boltzmann equation, eq. (6.21). The Boltzmann equation thus obtained can serve as a starting point for analyzing energy conversion and the accompanying transport processes. In section 8.2, we will discuss coupled transport and energy exchange processes, emphasizing the nonequilibrium between electrons and phonons during the transport processes. Section 8.3 examines the effects of recombination on the distribution of heat generation. Section 8.4 introduces two direct energy conversion devices, thermoelectric devices and thermophotovoltaic devices, and discusses the benefits of applying nanostructures to these devices. Many topics discussed in this chapter have not been treated systematically. Interested readers are encouraged to explore these topics further and answer some questions raised either in the text or by themselves.

8.1 Carrier Scattering, Generation, and Recombination

In a scattering process, if the energy of each particle involved is not changed after scattering even though its direction may have changed, the scattering is called *elastic*. On the other hand, if the energy does change, the scattering is called *inelastic*. In section 6.2.1 we gave an integral expression for the scattering term in the Boltzmann equation for two-particle scattering. Other scattering processes can be similarly expressed, although the details are slightly different. In chapters 6 and 7 we used the relaxation time approximation for the scattering term, without much justification. We will see later that the relaxation time approximation is strictly applicable only to elastic scattering processes. Many energy exchange processes, such as energy exchange between electrons and phonons, or between photons and electrons, are intrinsically inelastic. Although, in principle, the integral-differential Boltzmann equation can be solved consistently for the distribution functions, this approach is rarely feasible in practice due to its mathematical complexity. We will see next how we can modify the relaxation time approximation to deal with inelastic scattering processes. The treatment is based on evaluating the scattering matrix in the Fermi golden rule [eqs. (6.13) and (6.14)] and then evaluating the scattering integral [such as eq. (6.16)] in the Boltzmann equation, using approximate expressions for the distribution function. We will illustrate this approach in more detail only for electron scattering by acoustic phonons, and give results for other scattering processes.

8.1.1 Nonequilibrium Electron–Phonon Interactions

Consider the scattering of an electron initially in a state $(\mathbf{k}', E')$ by phonons. The electron can be scattered to another state at wavevector $(\mathbf{k}, E)$ by creating a phonon at $(\mathbf{q}, \hbar\omega)$, or by annihilating (absorbing) a phonon at $(\mathbf{q}, \hbar\omega)$. The conservation of energy and momentum requires

$$E' \pm \hbar\omega = E \quad \mathbf{k}' \pm \mathbf{q} = \mathbf{k} + \mathbf{G} \tag{8.1}$$

where the plus and minus signs mean annihilation and creation of a phonon, respectively, and $\mathbf{G}$ is any reciprocal lattice vector. Umklapp scattering processes ($\mathbf{G} \neq 0$) are usually unimportant for electron–phonon scattering. Thus the rest of our discussion will assume $\mathbf{G} = 0$. This discussion implies that the number of electrons at state $(\mathbf{k}, E)$ can be increased by the scattering of electrons from states $(\mathbf{k}', E')$ by two processes: one is the creation of a phonon (for $E < E'$) and the other is the annihilation of a phonon (for $E > E'$). We will next determine the rate for each process.

The scattering rate can be calculated from the Fermi golden rule [eq. (6.13)]. The scattering matrix element for an electron from state $(\mathbf{k}', E')$ to $(\mathbf{k}, E)$ is

$$M_{\mathbf{k}'\mathbf{k}} = \int \Psi_f^* H' \Psi_i dV = \frac{1}{V} \int e^{-i\mathbf{k}\cdot\mathbf{r}} H' e^{i\mathbf{k}'\cdot\mathbf{r}} dV \tag{8.2}$$

where we have assumed a plane-wave form for the electron wavefunctions, and V is the volume of the crystal. Further evaluation of the above scattering matrix requires an explicit form of the perturbation potential for the electron scattering, H'.

We first consider the perturbation potential of long-wavelength acoustic phonons on the electron scattering. Because the long-wavelength phonons cover many unit cells, the effect of the atomic displacement effectively changes the lattice constant. From the Kronig–Penney model, we know that a lattice constant change creates variation in the bandgap and thus the location of the bottom of the conduction band, E_c, will change. This change corresponds to the perturbation potential of phonons on the electrons. For a one-dimensional lattice with lattice constant a, the perturbation potential is

$$H' = E_c\left(a + \frac{du}{dx}a\right) - E_c(a) = \frac{dE_c}{da}\frac{du}{dx}a$$

where u is the atomic displacement. For a three-dimensional crystal, the above expression can be generalized into

$$H' = V\frac{dE_c}{dV}\nabla \cdot \mathbf{u} = Z_A \nabla \cdot \mathbf{u} \tag{8.3}$$

where $Z_A = V\,dE_c/dV$ is called deformation potential. For semiconductors, Z_A can be estimated on the basis of the temperature dependence of the bandgap and the thermal expansion coefficient (Bardeen and Shockley, 1950). A typical deformation potential value is 8 eV for the conduction bands of semiconductors (Hess, 2000; Lundstrom, 2000). The atomic displacement vector is a superposition of phonon waves of various frequencies,

$$\mathbf{u} = \sum_{\mathbf{q}} \mathbf{u}_\mathbf{q} e^{-i(\omega t - \mathbf{q}\cdot\mathbf{r})} \tag{8.4}$$

where the summation is over all allowable phonon wavevectors **q**. Substituting eq. (8.3) and the time-independent part of eq. (8.4) (the time-dependent part has been included in Fermi's golden rule) into eq. (8.2), we obtain

$$M_{\mathbf{k'k}} = \frac{Z_A}{V} \int e^{-i\mathbf{k}\cdot\mathbf{r}} \nabla \cdot \left(\sum_{\mathbf{q}} \mathbf{u_q} e^{i\mathbf{q}\cdot\mathbf{r}} \right) e^{i\mathbf{k'}\cdot\mathbf{r}} dV$$

$$= \frac{Z_A}{V} \sum_{\mathbf{q}} i\mathbf{q} \cdot \mathbf{u_q} \int e^{i(\mathbf{k'}+\mathbf{q}-\mathbf{k})\cdot\mathbf{r}} dV$$

$$= iZ_A \mathbf{q} \cdot \mathbf{u_q} \delta_{\mathbf{k'}+\mathbf{q},\mathbf{k}} \qquad (8.5)$$

where $\delta_{\mathbf{k'}+\mathbf{q},\mathbf{k}}$ denotes the Kronecker delta function

$$\delta_{\mathbf{k'}+\mathbf{q},\mathbf{k}} = \begin{cases} 1 & \mathbf{k'}+\mathbf{q}=\mathbf{k} \\ 0 & \mathbf{k'}+\mathbf{q} \neq \mathbf{k} \end{cases} \qquad (8.6)$$

This Kronecker delta appears because the integrand in the second equation of (8.5) is rapidly varying between $(-1, 1)$ and its integration is zero except when the exponential equals zero. Equation (8.5) shows that transverse phonons have zero probability of scattering electrons, because $\mathbf{u_q} \cdot \mathbf{q} = 0$. Only longitudinal phonons create instantaneous variations in the lattice constant and scatter electrons.

We can relate the displacement $\mathbf{u_q}$ to the phonon number density from energy conservation considerations. Because eq. (8.4) is a complex representation, the true lattice displacement at mode **q** is given by

$$\tilde{\mathbf{u}}_\mathbf{q} = \mathbf{u_q} e^{-i(\omega t - \mathbf{k}\cdot\mathbf{r})} + \mathbf{u_q^*} e^{i(\omega t - \mathbf{k}\cdot\mathbf{r})} \qquad (8.7)$$

The maximum kinetic energy of the mode over the whole crystal is thus

$$KE_{max} = \frac{1}{2} NM \left(\frac{d\tilde{\mathbf{u}}_\mathbf{q}}{dt} \right)^2 = \frac{1}{2} NM\omega^2 |2\mathbf{u_q}|^2 = 2\rho V \omega^2 |\mathbf{u_q}|^2 \qquad (8.8)$$

where N is the number of atoms in the crystal, M is the mass per atom, and ρ the density. Because the maximum kinetic energy occurs when the potential energy is minimum, it should equal the total energy of the phonons at mode **q**,

$$2\rho V \omega^2 |\mathbf{u_q}|^2 = \hbar \omega n_\mathbf{q} \qquad (8.9)$$

where n_q is the average number of phonons at mode **q** at equilibrium, and is given by the Bose–Einstein distribution. Combining eqs. (8.8) and (8.5), we find that the scattering matrix for an electron scattered from state $\mathbf{k'}$ to $\mathbf{k}$ by annihilating a phonon with wavevector **q** is

$$M_{\mathbf{k'k}} = iZ_A |\mathbf{q}| \sqrt{n_\mathbf{q}} \left(\frac{\hbar}{2\rho V \omega} \right)^{1/2} \delta_{\mathbf{k},\mathbf{k'}+\mathbf{q}} \qquad (8.10)$$

where **q** is the wavevector of the longitudinal acoustic phonons. Substituting eq. (8.10) into eq. (6.13), the transition rate from $\mathbf{k'}$ to $\mathbf{k}$ by absorbing a phonon is obtained. The total rate of increase in the carrier numbers at state **k** due to the phonon annihilation process can be obtained by summing up all the wavevectors $\mathbf{q}$ and $\mathbf{k}'$, and can be expressed as

$$\left(\frac{\partial f}{\partial t}\right)_{c,1} = \frac{2\pi}{\hbar}\sum_{\mathbf{q}}|\mathbf{q}|^2\left(\frac{\hbar Z_A^2}{2\rho V \omega}\right)n_{\mathbf{q}} f_{\mathbf{k}'}'(1-f_{\mathbf{k}})\delta_{\mathbf{k},\mathbf{k}'+\mathbf{q}}\delta(E-E'-\hbar\omega)$$

$$= \sum_{\mathbf{q}}\left(\frac{\pi\omega Z_A^2}{\rho V v_s^2}\right)n_{\mathbf{q}} f_{\mathbf{k}'}'(1-f_{\mathbf{k}})\delta(E_{\mathbf{k}}-E_{\mathbf{k}-\mathbf{q}}-\hbar\omega) \tag{8.11}$$

where $E_{\mathbf{k}}$ and $E_{\mathbf{k}-\mathbf{q}}$ are the energies of the electrons with wavevector $\mathbf{k}$ and $\mathbf{k}-\mathbf{q}$, respectively, $f_{\mathbf{k}}'$ represents the average number of electrons at the starting state $\mathbf{k}'$, $(1-f_{\mathbf{k}})$ gives the number of empty states at $\mathbf{k}$ to receive the electrons, and v_s is the speed of sound. We have assumed a linear phonon dispersion relation, $\omega = v_s|\mathbf{q}|$, to derive the second line of eq. (8.11). We combined the two delta functions into one by imposing the wavevector relationship into the energy relationship. We can further write

$$E_{\mathbf{k}} - E_{\mathbf{k}-\mathbf{q}} - \hbar\omega = E(\mathbf{k}) - E(\mathbf{k}-\mathbf{q}) - \hbar\omega$$

$$\approx \frac{dE}{d\mathbf{k}}\bullet\mathbf{q} - \frac{1}{2}\left(\left|\frac{dE}{d\mathbf{k}}\right|\right)^2|\mathbf{q}|^2 - \hbar\omega$$

$$= \hbar v_e |\mathbf{q}|\cos\theta - \frac{\hbar^2|\mathbf{q}|^2}{2m} - \hbar\omega \tag{8.12}$$

where v_e is the magnitude of the electron group velocity, θ is the angle formed between $\mathbf{k}$ and $\mathbf{q}$, and m is the electron effective mass. Equation (8.12) shows the impact of the momentum conservation rule on the scattering processes. If only energy conservation is imposed, there are many phonons that can satisfy the first of eq. (8.1). However, due to the simultaneous requirement of momentum conservation, the last delta function in eq. (8.11) requires, according to eq. (8.12), that

$$\hbar v_e|\mathbf{q}|\cos\theta - \frac{\hbar^2|\mathbf{q}|^2}{2m} = \hbar\omega \tag{8.13}$$

or

$$\hbar|\mathbf{q}| = 2mv_e\left(\cos\theta - \frac{1}{v_e}\frac{\omega}{|\mathbf{q}|}\right)$$

$$= 2mv_e\left(\cos\theta - \frac{v_s}{v_e}\right)$$

$$\approx 2mv_e\cos\theta \tag{8.14}$$

where the last approximation is based on the fact that the electron group velocity is usually much larger than the speed of sound v_s. This also means that the phonon energy in eq. (8.12) is negligible compared to the electron energy and thus the electron energy change is small during a scattering with a phonon. Equation (8.14) also sets the limit of $\hbar\mathbf{q}$ from zero to $2mv_e$ and $\cos\theta$ to positive values only.

Equation (8.11) only gives the rate of scattering from state $\mathbf{k}'$ to $\mathbf{k}$ due to the phonon annihilation process. The rate for the process that increases the population of electrons at $\mathbf{k}$ through the creation of one phonon can be evaluated by replacing $n_\mathbf{q}$ in eq. (8.11) by $(n_\mathbf{q} + 1)$ and $\delta(E - E' - \hbar\omega)$ by $\delta(E - E' + \hbar\omega)$, because the phonon population increases by one. In addition to the in-scattering mentioned above, the electrons at quantum state $\mathbf{k}$ can also out-scatter to $\mathbf{k}'$ by the creation or annihilation of a phonon. The net balance of the four processes leads to the scattering term in the Boltzmann equation:

$$\left(\frac{\partial f}{\partial t}\right)_c = \sum_\mathbf{q} \left(\frac{\pi Z_A^2 \omega}{\rho V v_s^2}\right) \{f_{\mathbf{k}'}(1 - f_\mathbf{k})[(n_\mathbf{q} + 1)\delta(E_\mathbf{k} - E_{\mathbf{k}+\mathbf{q}} + \hbar\omega)$$

$$+ n_\mathbf{q}\delta(E_\mathbf{k} - E_{\mathbf{k}-\mathbf{q}} - \hbar\omega)] - f_\mathbf{k}(1 - f_{\mathbf{k}'})[(n_\mathbf{q} + 1)\delta(E_\mathbf{k} - E_{\mathbf{k}-\mathbf{q}} - \hbar\omega)$$

$$+ n_\mathbf{q}\delta(E_\mathbf{k} - E_{\mathbf{k}+\mathbf{q}} + \hbar\omega)]\} \tag{8.15}$$

It should be remembered that $\mathbf{k}'$ in $f_{\mathbf{k}'}$ is related to $\mathbf{k}$ and $\mathbf{q}$ through the momentum conservation rule and is thus different for each of the four terms containing the delta function, even though we have written it in the form of a common multiplier. To evaluate the scattering integral, we need to know the distribution functions $f_\mathbf{k}$, $f_{\mathbf{k}'}$, and $n_\mathbf{q}$. As we explained before, the rigorous evaluation of the distribution function requires concurrent solution of eq. (8.15) with the left-hand side of the Boltzmann equation (6.7). This is in general very complicated and not practical for most applications. Our next step is to develop simplified expressions for the scattering in a form similar to the relaxation time. For this purpose, we can decompose the distribution function $f(\mathbf{k})$ into the sum of an even function (f_0) and an odd function (f_1) (Lundstrom, 2000; Hess, 2000),

$$f(\mathbf{k}) = f_0(\mathbf{k}) + f_1(\mathbf{k}) \tag{8.16}$$

Although such a decomposition is always possible for arbitrary f, f_0 and f_1 are in general unknown. Under the relaxation time approximation as expressed in eq. (6.22), f_0 is the equilibrium distribution and f_1 is the deviation from equilibrium. We again assume here that f_1 is much smaller than f_0, an assumption that is well justified for diffusion-dominated transport processes and may be justified even when size effects are important. For the electron system, we further assume that the phonon system is at equilibrium. Substituting eq. (8.16) into eq. (8.15) and neglecting the higher order terms in f_1 leads to the following expressions,

$$\left(\frac{\partial f}{\partial t}\right)_c = \left(\frac{\partial f}{\partial t}\right)_{c,i} + \left(\frac{\partial f}{\partial t}\right)_{c,e} \tag{8.17}$$

$$\left(\frac{\partial f}{\partial t}\right)_{c,i} = \sum_\mathbf{q} \left(\frac{\pi Z_A^2 \omega}{\rho V v_s^2}\right) \{f_{\mathbf{k}',0}(1 - f_{\mathbf{k},0})[(n_\mathbf{q} + 1)\delta(E_\mathbf{k} - E_{\mathbf{k}+\mathbf{q}} + \hbar\omega)$$

$$+ n_\mathbf{q}\delta(E_\mathbf{k} - E_{\mathbf{k}-\mathbf{q}} - \hbar\omega)]$$

$$- f_{\mathbf{k},0}(1 - f_{\mathbf{k}',0})[(n_\mathbf{q} + 1)\delta(E_\mathbf{k} - E_{\mathbf{k}-\mathbf{q}} - \hbar\omega)$$

$$+ n_\mathbf{q}\delta(E_\mathbf{k} - E_{\mathbf{k}+\mathbf{q}} + \hbar\omega)]\} \tag{8.18}$$

$$\left(\frac{\partial f}{\partial t}\right)_{c,e} = -f_{\mathbf{k},1} \sum_{\mathbf{q}} \left(\frac{\pi Z_A^2 \omega}{\rho V v_s^2}\right) \{(1 - f_{\mathbf{k}',0})[(n_\mathbf{q} + 1)\delta(E_\mathbf{k} - E_{\mathbf{k}-\mathbf{q}} - \hbar\omega)$$

$$+ n_\mathbf{q}\delta(E_\mathbf{k} - E_{\mathbf{k}+\mathbf{q}} + \hbar\omega)]\} \tag{8.19}$$

We will see later that $(\partial f/\partial t)_{c,i}$ and $(\partial f/\partial t)_{c,e}$ represent inelastic and elastic scattering respectively. In deriving eq. (8.19), we used the fact that the summation over $\mathbf{q}$ is equivalent to summation over $\mathbf{k}'$ and that $f_{\mathbf{k}',1}$ is an odd function of $\mathbf{k}'$. Equation (8.19) shows that under the approximations we used, this term has the typical *relaxation time* form, with the relaxation time defined as

$$\frac{1}{\tau} = \sum_{\mathbf{q}} \left(\frac{\pi Z_A^2 \omega}{\rho V v_s^2}\right) \{(1 - f_{\mathbf{k}',0})[(n_\mathbf{q} + 1)\delta(E_\mathbf{k} - E_{\mathbf{k}-\mathbf{q}} - \hbar\omega)$$

$$+ n_\mathbf{q}\delta(E_\mathbf{k} - E_{\mathbf{k}+\mathbf{q}} + \hbar\omega)]\} \tag{8.20}$$

To further evaluate eqs. (8.18) and (8.20), we need to assume detailed expressions for $f_{\mathbf{k},0}$ and $n_\mathbf{q}$. We assume that both obey the equilibrium distributions but that the phonons and electrons are at different temperatures, T_p and T_e, respectively,

$$f_{\mathbf{k},0} = \frac{1}{\exp\left(\frac{E-E_f}{\kappa_B T_e}\right) + 1} = f_0(E), \quad n_\mathbf{q} = \frac{1}{\exp\left(\frac{\hbar\omega}{\kappa_B T_p}\right) - 1} \tag{8.21}$$

We first evaluate eq. (8.20) by converting the summation into integration,

$$\frac{1}{\tau} = \left(\frac{1}{8\pi^3}\right) \int_{q_{\min}}^{q_{\max}} q^2 dq \int_0^{2\pi} d\varphi \int_0^{\pi} \sin d\theta \times \left(\frac{\pi Z_A^2 \omega}{\rho v_s^2}\right) (1 - f_{\mathbf{k}',0})$$

$$\times \left[(n_\mathbf{q} + 1)\delta\left(\hbar v_e q \cos\theta - \frac{\hbar^2 q^2}{2m} - \hbar\omega\right) + n_\mathbf{q}\delta\left(-\hbar v_e q \cos\theta - \frac{\hbar^2 q^2}{2m} + \hbar\omega\right)\right] \tag{8.22}$$

where $q = |\mathbf{q}|$ and we have used eq. (8.12). We consider high-temperature situations such that $n_\mathbf{q}$ is large and

$$n_\mathbf{q} + 1 \approx n_\mathbf{q} \approx \frac{\kappa_B T_p}{\hbar\omega} \tag{8.23}$$

To evaluate the integrals of the delta function in eq. (8.22), we will use the following property of the delta function,

$$\int_a^b \delta(x - x_0) dx = \begin{cases} 1 & a \leq x_0 \leq b \\ 0 & x_0 < a \text{ or } x_0 > b \end{cases} \tag{8.24}$$

On the basis of the discussion following eq. (8.14), we see that the $\cos\theta$ in the first delta function of eq. (8.22) is limited to positive values, that is, for θ within $[0, \pi/2)$, and to negative values, that is, for θ within $(\pi/2, \pi]$, in the second delta function. Both these delta functions set the minimum of q, q_{min}, to zero and the maximum of q, q_{max}, to $2mv_e/\hbar$. Equation (8.22) can thus be written as

$$\frac{1}{\tau} = \left(\frac{1}{4\pi^2}\right) \int_0^{2mv_e/\hbar} q^2 dq \times \left(\frac{\pi Z_A^2 \omega}{\rho v_s^2}\right) \left(\frac{\kappa_B T_p}{\hbar \omega}\right) \times (2 - f_{k+q,0} - f_{k-q,0}) \frac{1}{\hbar v_e q} \quad (8.25)$$

For semiconductors that are not heavily doped, the Fermi level is typically inside the bandgap, so that $f_{k+q,0}$ and $f_{k-q,0}$ is much smaller than one and can be neglected. In this case, eq. (8.25) can be explicitly integrated, leading to

$$\frac{1}{\tau} = \left(\frac{Z_A^2}{\pi \rho v_s^2}\right) \left(\frac{\kappa_B T_p}{\hbar^4}\right) m^2 v_e$$

$$= \frac{\sqrt{2} \kappa_B m^{3/2} Z_A^2 T_p}{\pi \rho v_s^2 \hbar^4} E^{1/2} \quad (8.26)$$

This expression is valid for electron scattering with longitudinal acoustic phonons in a semiconductor at high temperature, that is, when the lattice temperature is higher than the Debye temperature. For metals, the relaxation time due to longitudinal acoustic phonons is different from the above expression because (1) $f_{k+q,0}$ and $f_{k-q,0}$ are no longer negligible and (2) the upper limit for q is no longer $2mv_e/\hbar$ since the latter is much larger than the maximum allowable wavevector of the first Brillouin zone. A detailed treatment of relaxation time due to electron and acoustic phonon scattering in metals was given by Wilson (1953), and it is found that

$$\frac{1}{\tau} = \frac{3(6\pi^5)^{1/3}}{16\sqrt{2}} \frac{\kappa_B Z_A^2 T_p}{\sqrt{m} \rho v_s^2 a^4} E^{-3/2} \quad (8.27)$$

where a is the length of the unit cell.

We now turn our attention to inelastic scattering term, eq. (8.18). We follow a similar derivation as for the relaxation time τ and again limit the derivations to the longitudinal acoustic phonon scattering of electrons in semiconductors. Equation (8.18) can be expressed as

$$\left(\frac{\partial f}{\partial t}\right)_{c,i} = \frac{1}{4\pi^2} \int_0^{2mv_e/\hbar} q^2 dq \left(\frac{\pi Z_A^2 \omega}{\rho v_s^2}\right) \frac{1}{\hbar q v_e} n_q f_0(E) \exp\left(\frac{E - E_f}{\kappa_B T_e}\right)$$

$$\times \left\{\left[f_0(E + \hbar\omega) - f_0(E - \hbar\omega) \exp\left(-\frac{\hbar\omega}{\kappa_B T_e}\right)\right]\right.$$

$$\left.\left[\exp\left(\frac{\hbar\omega}{\kappa_B T_p}\right) - \exp\left(\frac{\hbar\omega}{\kappa_B T_e}\right)\right]\right\} \quad (8.28)$$

Assuming that $\kappa_B T_e$ and $\kappa_B T_p$ are much larger than $\hbar\omega$, and also assuming $(E - E_f) \gg \kappa_B T_e$, we can approximate the above equation to

$$\left(\frac{\partial f}{\partial t}\right)_{c,i} \approx f_0 \frac{1}{4\pi^2} \int_0^{2mv_e/\hbar} n_q q\,dq \left(\frac{\pi Z_A^2 \omega}{\hbar v_e \rho v_s^2}\right)\left(\frac{\hbar\omega}{\kappa_B T_p} - \frac{\hbar\omega}{\kappa_B T_e}\right)\left[1 - \exp\left(-\frac{\hbar\omega}{\kappa_B T_e}\right)\right]$$

$$= f_0 \frac{1}{4\pi^2} \int_0^{2mv_e/\hbar} q\,dq \frac{\kappa_B T_p}{\hbar\omega}\left(\frac{\pi Z_A^2 \omega}{\hbar v_e \rho v_s^2}\right)\left(\frac{\hbar\omega}{\kappa_B T_p} - \frac{\hbar\omega}{\kappa_B T_e}\right)\frac{\hbar\omega}{\kappa_B T_e}$$

$$= \frac{2\sqrt{2}}{\pi}\frac{m^{5/2} Z_A^2 E^{3/2}}{\rho \hbar^4 \kappa_B T_e^2} f_0(T_e - T_p) = g(T_e - T_p) \tag{8.29}$$

where the factor g can also be related to the relaxation time in eq. (8.26),

$$g = \frac{2\sqrt{2}}{\pi}\frac{m^{5/2} Z_A^2 E^{3/2}}{\rho \hbar^4 \kappa_B T_e^2} f_0 = \frac{2 m v_s^2 E f_0}{\kappa_B^2 T_e^2 T_p \tau} \tag{8.30}$$

This expression shows that $(\partial f/\partial t)_{c,i}$ represents the inelastic scattering that drives the electrons and phonons to equilibrium. For metals, Qiu and Tien (1993) derived the following expression,

$$\left(\frac{\partial f}{\partial t}\right)_{c,i} = \frac{3\pi^2}{2\sqrt{2}}\left(\frac{3}{4\pi}\right)^{1/3}\frac{\sqrt{m}\,Z_A^2 E^{-1/2}}{\kappa_B \rho a^4} f_0(1-f_0)\left(f_0 - \frac{1}{2}\right)\frac{T_e - T_p}{T_e^2} \tag{8.31}$$

Equation (8.31) describes the energy loss trends in metals. If the electron energy is higher than the Fermi level ($f_0 < \frac{1}{2}$), the term is negative, indicating that electrons are losing high-energy carriers because of their collision with the lattice. If the electron energy is lower than the Fermi level ($f_0 > \frac{1}{2}$) the scattering tends to supply carriers to these states. In contrast, eq. (8.27) represents an elastic scattering process. One can understand this statement by multiplying both sides of the Boltzmann equation by energy and integrating over all allowable energy states. The two scattering terms are

$$\sum_{\mathbf{k}} E \left(\frac{\partial f}{\partial t}\right)_{c,e} = 0 \tag{8.32}$$

$$\sum_{\mathbf{k}} E \left(\frac{\partial f}{\partial t}\right)_{c,i} = G(T_e - T_p) \tag{8.33}$$

where G is called the electron–phonon coupling factor. From eq. (8.29), the electron–phonon coupling factor due to acoustic phonon scattering in nondegenerate semiconductors (such that the Boltzmann statistics is valid) is

$$G = \sum_{\mathbf{k}} E\left(\frac{\partial f}{\partial t}\right)_{c,i} = \int_0^\infty Eg D(E)dE = \frac{n \int_0^\infty Eg D(E)dE}{\int_0^\infty f_0 D(E)dE}$$

$$= \frac{2\sqrt{2}}{\pi}\frac{nm^{5/2} Z_A^2}{\rho \hbar^4 \kappa_B T_e^2}\frac{\int_0^\infty E^3 \exp\left(-\frac{E-E_f}{\kappa_B T_e}\right)dE}{\int_0^\infty E^{1/2}\exp\left(\frac{E-E_f}{\kappa_B T_e}\right)dE}$$

$$= \frac{12\sqrt{2}}{\pi^{3/2}}\frac{nm^{5/2} Z_A^2}{\rho \hbar^4 T_e}(\kappa_B T_e)^{3/2} \tag{8.34}$$

where n is the electron number density and we have used the density of states of a parabolic band for electrons as derived in chapter 3. Substituting typical numbers for electrons, we see that $G \sim 10^{-12} n$ W m^{-3} K^{-1}. From eq. (8.34), the rate of energy loss per electron is thus

$$\frac{de}{dt} = \frac{12\sqrt{2}}{\pi^{3/2}} \frac{m^{5/2} Z_A^2}{\rho \hbar^4 T_e} (\kappa_B T_e)^{3/2} (T_e - T_p) \tag{8.35}$$

The above result differs from that given by Shockley (1951) (see also Conwell, 1967) by a factor of 2/3.

The electron–phonon coupling factor for metals has been evaluated by various researchers (Kaganov et al., 1957; Allen, 1987; Qiu and Tien, 1993). Kaganov et al. (1957) gave the following expression for the *electron–phonon coupling factor* at high temperatures in metals,

$$G = \frac{\pi^2}{6} \frac{m v_s^2 n}{\tau(T_p) T_p} \tag{8.36}$$

where the electron relaxation time is evaluated at the lattice temperature T_p.

We should remind ourselves that in addition to eq. (8.32) and (8.33), the following relation also holds true,

$$\sum_{\mathbf{k}} \left(\frac{\partial f}{\partial t}\right)_{c,i} = 0 \tag{8.37}$$

This can be proven from eq. (8.18), but not from expressions (8.29) and (8.31) due to the approximations invoked in the latter two. Equation (8.37) is important in the derivation of the carrier conservation equation.

The above derivations are based on the expression for the scattering of electrons. We can also carry out similar derivations for the scattering term of phonons. The phonon scattering term due to electron–phonon interaction in the Boltzmann equation is (Ziman, 1960; Srivastava, 1990)

$$\left(\frac{\partial n_q}{\partial t}\right)_c = \frac{4\pi}{\hbar} \sum_{\mathbf{q}} q^2 \left(\frac{\hbar Z_A^2}{2\rho V \omega}\right) f_{\mathbf{k}}(1 - f_{\mathbf{k}'}) [n_{\mathbf{q}} \delta(E_{\mathbf{k}} - E_{\mathbf{k+q}} + \hbar \omega)$$
$$- (n_{\mathbf{q}} + 1) \delta(E_{\mathbf{k}} - E_{\mathbf{k-q}} - \hbar \omega)] \tag{8.38}$$

One should be able to follow a similar strategy to that used above for electrons to express the above expression into a relaxation time term for phonons plus an electron–phonon coupling term. Energy conservation also requires that the energy exchange from electrons to phonons equals the energy received by the phonon system.

Equation (8.20) sums up all possible directions of the electron wavevector $\mathbf{k}'$ that satisfy $\mathbf{k} - \mathbf{k}' = \mathbf{q}$. The relaxation time thus obtained, τ, is also called the single-mode relaxation time. The summation of all $\mathbf{q}$, and thus over all $\mathbf{k}'$, means that τ is an average. Solving the Boltzmann equation using the single-mode relaxation time implies that scattering is isotropic. There exist occasions where such an average is not very good. For example, when the scattering deflects $\mathbf{k}'$ by only a small angle from **k**, the momentum loss along the original **k** direction is small. The single-mode relaxation time, however, does not distinguish a small angle scatter event because it assumes that carriers are scattered isotropically in all directions. A rigorous way to treat small angle scattering is to solve for $f(\mathbf{k})$ directly from the integral-differential Boltzmann equation with eq. (8.15) on the right-hand side. However, this is difficult to do, as already stated. An alternative is to multiply the term inside the summation of eq. (8.20) by $(1 - \mathbf{k} \bullet \mathbf{k}'/|\mathbf{k}||\mathbf{k}'|)$, which represents the fraction of actual momentum lost for each collision. The relaxation time thus obtained is called the momentum relaxation time (Lundstrom, 2000). Similarly, the energy relaxation time is often obtained by multiplying eq. (8.20) by the fraction of energy lost during each collision. The above derivations show that the energy relaxation, however, cannot be expressed in the relaxation form, although the latter is often used in practice and also works reasonably well.

8.1.2 Photon Absorption and Carrier Excitation

As with the electron–phonon scattering discussed in the previous section, one may express the interaction between light and matter by evaluating the scattering integrals in the Boltzmann equation. If the focus is on the photon number balance, as in the photon equation of radiative transfer, we should investigate the scattering integral for photons that includes the absorption, scattering, and emission of photons. We will not discuss these processes here but rather focus on how the photon energy is converted into that of other carriers. For this purpose, we will first consider how photons are absorbed by a material. Such absorption can be through electrons and phonons in crystals, and can also occur for individual atoms and molecules in gases. In the next section, we will turn our attention to the relaxation processes in which the excited material states relax back to equilibrium, during which heat may be generated and/or light can be re-emitted. Rather than using the rigorous approach to evaluate the collision integral, as for electron–phonon scattering presented at the previous section, our discussion will be mostly qualitative in this section. Quantitative treatments can be performed that follow the strategies developed in the previous section (Harrison, 1979).

The absorption coefficient of a material for a given photon at frequency v is proportional to the transition probability from the initial states of the matter to the final states of the matter after absorption of the photon,

$$\alpha(v) = \sum W(i \rightarrow f) n_i P_f \qquad (8.39)$$

where $W(i \rightarrow f)$ is the transition probability from an initial quantum state i to a final quantum state f that can be evaluated on the basis of Fermi's golden rule as given in eq. (6.15), n_i is the occupation number of the initial quantum state, P_f is the probability that the final state of the transition is open, thus allowing such a transition to occur, and the summation is over all allowable initial and final states. For example, if the photon absorption is due to electrons, n_i is given by the Fermi–Dirac distribution corresponding to the initial state f_i and P_f is given by $(1-f_f)$, representing the empty states of the matter that can accept the excited electron. If the photon interacts with phonons, n_i is the Bose–Einstein distribution corresponding to the initial state $n_\mathbf{q}$ while $P_f = 1$

because there is no limit to how many phonons an allowable state can take. During the interaction, energy and momentum should be conserved,

$$E_i + \hbar\omega = E_f \qquad (8.40)$$

$$\mathbf{k_i} + \mathbf{k_p} = \mathbf{k_f} \qquad (8.41)$$

Umklapp scattering is generally not important because the photon wavevector $\mathbf{k}_p$ is very small, due to the long wavelength of photons compared to the wavevectors of electrons and phonons (except X-rays), which means that the initial and final state of the matter must have approximately the same wavevector, $\mathbf{k}_i = \mathbf{k}_f$.

8.1.2.1 Absorption by Gas Molecules

In chapter 2, we discussed the absorption of light by harmonic oscillators, rigid rotors, and atoms according to the energy conservation requirement and selection rules. These absorption processes excite electrons or molecules from a lower energy state into a higher energy state. If the energy of the photons is high enough, it can excite electrons from bound to free, releasing the electron from a state that is confined to an atomic orbital into a free electron. High-energy photons can also break the bonds between molecules. It should be emphasized that absorption does not mean that the absorbed photon energy is converted into heat. Heat conversion occurs when some of these excited states relax back into the lower energy states and release their energy into the random motion of the atoms, molecules, or electrons, as we will discuss in the next section.

8.1.2.2 Absorption by Phonons

A photon can interact with one or several phonons in a crystal. This typically happens in the infrared range when the photon energy becomes comparable to the phonon energy (0.01–0.2 eV). The photon–phonon interaction is caused by the electric dipoles in polar crystals such as GaAs or induced electric dipoles in nonpolar crystals such as silicon (Srivastava, 1990; Pankove, 1971). It is thus anticipated that the absorption coefficients of polar crystals are generally larger than those of nonpolar ones. As the photon represents a transverse electromagnetic wave, it interacts more strongly with transverse optical phonons than with longitudinal ones. Due to the small wavevector of photons, one-phonon absorption processes should occur close to the center of the first Brillouin zone and typically only in polar crystals. In nonpolar crystals such as silicon, the absorption generally involves two or more phonons. The excited phonons usually relax their energy to other phonons through phonon–phonon scattering processes, converting the excess energy into heat during this relaxation process.

8.1.2.3 Absorption by Electrons in Crystals

The absorption of photons by free electrons (or holes) in metals and in semiconductors must be accompanied by emission or absorption of phonons because of momentum conservation. As shown in figure 8.1, the absorption of a photon by a free carrier inside

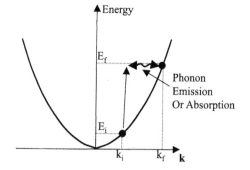

Figure 8.1 Free carrier absorpition involves phonons to help satisfy momentum conservation since the photon wavevector is too small

one band must be accompanied by a large change in the wavevector, which is provided by the emission or absorption of phonons.

Interband absorption can happen in both metals and semiconductors. An electron absorbs a photon and is pumped to a different band as shown in figure 8.2(a). Because of the small photon wavevector, the transition must be nearly vertical at the wavevector **k** such that $E_f - E_i = h\nu$. If the photon energy is smaller than the bandgap, no absorption occurs. Subtle differences occur near the band gap, depending on whether the bandgap is a direct or an indirect gap. Figure 8.2(a) represents a direct gap semiconductor with its valence band maximum aligned to the conduction band minimum. In this case, both energy and momentum conservation rules can be satisfied for photon absorption exactly at the bandgap as long as the photon energy equals that of the bandgap. For an indirect semiconductor, as shown in figure 8.2(b), the absorption of a photon cannot happen at the band edge without the absorption or emission of phonons because photon–electron interaction alone does not satisfy the wavevector (momentum) conservation requirements. Figure 8.3 shows the absorption coefficients of GaAs and silicon near the band edge. GaAs is a direct gap semiconductor and has a very sharp absorption edge, whereas silicon is an indirect one and its absorption edge is not as sharp as that of GaAs.

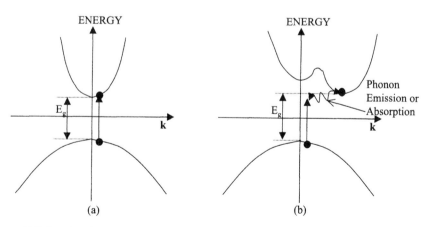

Figure 8.2 Interband absorption of photons near a bandgap in (a) direct and (b) indirect semiconductors. Indirect semiconductors must involve the absorption or emission of phonons to satisfy the momentum conservation rule.

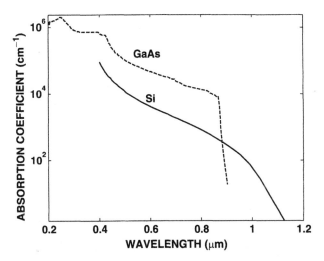

Figure 8.3 Absorption coefficient of Si and GaAs near the band edge, showing the difference between direct bandgap (GaAs) and indirect bandgap (Si) semiconductors (data from Palik, 1985).

The above-discussed absorption mechanisms are all included in the complex optical constants $N = n + i\kappa$ of materials, as we discussed in section 5.1.2. The absorption coefficient can be calculated from the imaginary part of N as [eq. (5.40)]

$$\alpha = \frac{4\pi\kappa}{\lambda} \tag{8.42}$$

where λ is the wavelength in vacuum. From the optical constants and by solving the Maxwell equations (discussed in chapter 5), one can determine the number of photons absorbed locally, and from which the number of electrons, holes, or phonons generated locally can be determined.

Figure 8.4 shows the optical constants for several representative materials. Dielectric crystals distinguish themselves from semiconductors only in that their bandgaps are larger. The extinction coefficient reflects the absorption mechanisms. In the short-wavelength region, semiconductors absorb through band–band excitation of electrons. In the long-wavelength region, semiconductors absorb through phonon excitation and, if the semiconductor is doped, also through free carrier (electrons and holes) absorption. Metals become transparent for short-wavelength photons because electrons cannot respond fast enough to follow the photon frequency. In the long-wavelength region, metals absorb through free carriers and the absorption processes must involve phonon absorption or emission to conserve momentum.

Example 8.1

The solar radiation from the sun can be represented by a blackbody at a temperature of 5600 K. On the earth surface, the incident solar radiation can be treated as parallel with a energy flux of 700 W m^{-2}. Estimate the rate of electron–hole pairs that can be

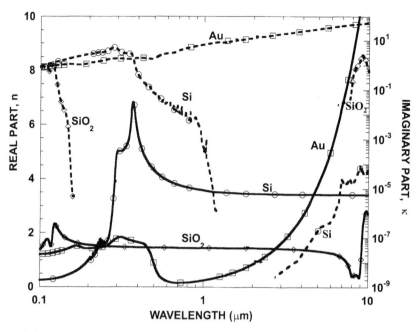

Figure 8.4 Optical constants of representative materials, solid lines real part, dashed lines imaginary part (data from Palik, 1985).

be generated inside a direct gap semiconductor with a bandgap of 1.12 eV when the sun is at normal incidence. The real part of the refractive index is 4 and the semiconductor is semi-infinite.

Solution: Only photons above the bandgap can excite electrons from the valence band into the conduction band, creating electrons in the conduction band and leaving behind holes in the valence band. Generally, one photon only excites one electron–hole pair. The number of electron–hole pairs generated above 1.12 eV per unit area and per unit time is thus

$$\dot{N} = A(1-R) \int_{E_g}^{\infty} cf(E,T)D(E)dE \qquad \text{(E8.1.1)}$$

where A is a constant representing the fraction of solar radiation intercepted by the earth, c is the speed of light, D is the photon density of states [eq. (3.59)], R is the reflectivity of the surface [eq. (5.78) for normal incidence], and f is the Bose–Einstein distribution with $T = 5600$.

$$D(\omega) = \frac{\omega^2}{\pi^2 c^3} \qquad \text{(E8.1.2)}$$

$$R = \left|\frac{n-1}{n+1}\right|^2 = 0.36 \qquad \text{(E8.1.3)}$$

where we have assumed that the imaginary part of the refractive index, κ, is small compared to one and is therefore negligible, which is usually the case near the bandgap. Substituting (E8.1.2) into (E8.1.1), we get

$$\dot{N} = A(1-R) \int_{\omega_g}^{\infty} \frac{\omega^2}{\pi^2 c^2} \frac{1}{\exp\left(\frac{\hbar\omega}{\kappa_B T}\right) - 1} d\omega$$

$$= A(1-R) \frac{\kappa_B^3 T^3}{\hbar^3 \pi^2 c^2} \int_{x_g}^{\infty} \frac{x^3}{e^x - 1} dx \quad \text{(E8.1.4)}$$

where $x_g = E_g/(\kappa_B T) = 1.12 \text{ eV}/0.485 \text{ eV} = 2.58$. A can be evaluated by the ratio of the solar intensity on the surface of the sun to that on the surface of the earth,

$$A = \frac{I_0}{\sigma T^4/\pi} = \frac{3.14 \times 700 \text{ W m}^{-2}}{5.67 \times 10^{-8} (5600)^4 \text{ W m}^{-2}} = 3.94 \times 10^{-5}$$

Numerical integration of the last integral in eq. (E8.1.4) leads to 3.91. The electron–hole pair generation rate is thus

$$\dot{N} = 3.94 \times 10^{-5} (1 - 0.36) \frac{(1.38 \times 10^{-23} \times \text{J K}^{-1})^3 \times (5600 \text{ K})^3}{(1.1 \times 10^{-34} \text{ J s})^3 \pi^2 (3 \times 10^8 \text{ m s}^{-1})^2} \times 3.91$$

$$= 3.8 \times 10^{22} \text{ m}^{-2} \text{ s}^{-1} \quad \text{(E8.1.5)}$$

Note. Because the electromagnetic field decreases exponentially inside the medium, the distribution of these electron–hole pairs inside the solar cell is also exponential (see example 5.1).

8.1.3 Relaxation and Recombination of Excited Carriers

The electrons, phonons, and molecules excited by photon absorption are at nonequilibrium states and tend to relax back to equilibrium by giving off their excess energy. Some of the excess energy will become random internal energy, a process that can be considered as heat generation. The excess energy can also be converted into other forms of energy such as light and electricity. In section 8.1.1, the electron–phonon coupling term as represented by eq. (8.35) expresses the energy exchange rate between electrons and phonons, which can lead to either heat generation or cooling as we will discuss later. The molecules that have become excited due to light absorption can use the excess energy to promote a chemical reaction as, for example, photosynthesis in biological systems. In laser systems, such as the He–Ne laser, argon laser, and so on, the excited states of the molecules emit light. Clearly, not all excitations are converted into heat. We will discuss next, in more detail, the relaxation processes of electrons and holes excited through band-to-band absorption.

The absorption of a photon through band-to-band transition generates an electron and hole pair as shown in figure 8.2. There are different possibilities for this electron–hole pair to relax their excess energy, as shown in figure 8.5. First, the electrons and holes

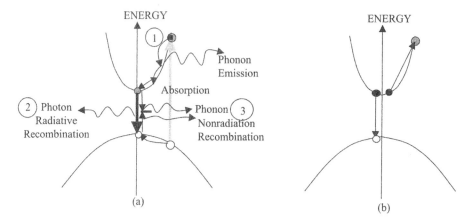

Figure 8.5 Relaxation of an excited state due to photon absorption in a semiconductor. In (a), the electrons and holes first relax back to the bottom of the conduction band and the valence bands by phonon emission, followed by band-to-band recombination, which can be radiative, emitting photons, or nonradiative, emitting phonons. Another form of nonradiative band-to-band recombination, shown in (b), is Auger recombination, in which the electrons and holes release their energy to another electron (or hole), pushing them back to higher energy states in the band, which subsequently releases its excess energy by phonon emission.

can relax to the bottom of the conduction and top of the valence band by emission of phonons, which can be considered as heat generation because these phonons are likely to interact with other phonons and eventually dissipate as random lattice vibration. The electron and hole can also recombine, either directly without relaxing to the bottom of the conduction band and top of the valence band, or after relaxing to these states. The latter is usually the case because electron–phonon interaction ($\sim 10^{-12}$ s) is much faster than band-to-band recombination. There are several band-to-band transition processes: through light emission or phonon emission [figure 8.5(a)], or releasing the excess energy to other electrons or holes (Auger recombination) [figure 8.5(b)]. If light is emitted, the process is called *luminescence*. Depending on how the excited carriers are excited, for example, by other photons or by an electrical field, the luminescence, phenomena are correspondingly called *photoluminescence* and *electroluminescence*, respectively. Band-to-band transition can also generate phonons that are eventually dissipated in other phonons as heat. The light-emission recombination processes are called *radiative* and the heat generation recombination processes are called *nonradiative*. Quantum efficiency is defined as the ratio of the number of photons emitted for a given number of recombined electron–hole pairs,

$$\eta_{\text{quantum}} = \frac{N_{\text{photon}}}{N_{\text{e-h pair}}} \qquad (8.43)$$

which can be measured experimentally and used to estimate the fraction of electrons and holes that are recombined nonradiatively into heat. In addition to recombination, it is also possible to separate the electrons and the holes so that they flow into an external load to do work, as in solar cells or photodiode detectors. In these devices, light absorption generates electrons and holes in the p–n junction region. The built-in electrical field of the p–n junction drives the electrons to the n-region and the holes to the p-region and thus separates the two to prevent radiative or nonradiative recombination. The excess electrons and holes in the n- and p-type regions convert their potential energy difference in an external circuit to do useful work.

Which of the above processes dominates depend on the rate of each process. The relaxation of electrons from excited states to the bottom of the conduction band is generally quick, as determined by electron–electron and electron–phonon relaxation times (10^{-12}–10^{-13} s). The "radiative" or "nonradiative" recombination of electrons and holes through the bandgap depends on whether the bandgap is direct or indirect. In direct bandgap semiconductors such as GaAs, the radiative recombination rate is of the order of 10^{-7}–10^{-9} s. In indirect bandgap semiconductors such as silicon, the recombination is dominated by nonradiative recombination in which phonons are involved. Nonradiative recombination is much slower and ranges from second to microseconds, depending on impurity levels. Because both radiative and nonradiative recombination processes are much slower than electron–electron and electron–phonon scattering processes, excited electrons and holes typically first relax to the bottom of the conduction band and the top of the valence band, releasing the excess energy to phonons as heat. These electrons and holes further recombine through the bandgap by either radiative or nonradiative processes. Semiconductor lasers are always built on direct gap semiconductors such as GaAs because radiative recombination can be made dominant. In detectors, the relevant time scales are the diffusion time of electrons and holes through the space charge region versus the recombination rate. If electrons and holes recombine before they are diverted to the n- and p-type regions, no useful electricity can be generated.

The recombination rate can be evaluated from the collision integral written specifically for the carriers involved in the collision processes (Hess, 2000). For band-to-band recombination, the recombination rate can be expressed as

$$R_r = B \times n \times p \tag{8.44}$$

where n and p are the electron and hole concentrations, respectively, and B is a proportionality constant. When a system is at equilbrium, the recombination will balance the generation of electron–hole pairs due to thermal excitation so that the generation rate can be expressed as

$$R_g = B \times n_0 \times p_0 \tag{8.45}$$

where n_0 and p_0 are the equilibrium concentrations of electrons and holes. When a system is driven out of equilibrium by, for example, photon excitation or injection of current, the net recombination rate is the difference between recombination and excitation,

$$R_n = R_r - R_g \tag{8.46}$$

For electrons, this rate can be expressed approximately,

$$R_n \approx \frac{n - n_0}{\tau_n} \tag{8.47}$$

where τ_n is called the lifetime of the electrons. Understandably, the lifetime depends on the type of semiconductor, the impurity concentration, and the carrier density. For detailed discussion on recombination and generation processes, a book on semiconductor physics should be consulted (e.g., Wang, 1989; Hess, 2000).

Because surfaces often have more impurities and these impurities have energy states different from those in the bulk, they are usually particularly active as nonradiative recombination centers. The surface recombination rate R_s is often described as a surface recombination velocity V_s [m s^{-1}],

$$R_s = V_s(n - n_0) \tag{8.48}$$

Here the units of R_s are m^{-2} s^{-1}.

8.1.4 Boltzmann Equation Revisited

In principle, all of the processes discussed above, including scattering, generation, and recombination, can be derived from the collision integral so that we can write the scattering term into the following generic form,

$$\left(\frac{\partial f}{\partial t}\right)_c = \left(\frac{\partial f}{\partial t}\right)_{\text{scattering}} + \left(\frac{\partial f}{\partial t}\right)_{\text{generation}} + \left(\frac{\partial f}{\partial t}\right)_{\text{recombination}} + \cdots \tag{8.49}$$

For electron–phonon scattering, we have shown that the scattering term can be written as the sum of elastic and inelastic processes,

$$\left(\frac{\partial f}{\partial t}\right)_s = \left(\frac{\partial f}{\partial t}\right)_{s,i} + \left(\frac{\partial f}{\partial t}\right)_{s,e} = g(T_e - T_p) - \frac{f - f_0}{\tau} \tag{8.50}$$

The carrier generation rate term in eq. (8.49) depends on the form of the excitation. For example, in photon-generated band-to-band transition processes, the absorption of one photon generates one electron–hole pair. If a laser beam of intensity I_0 and frequency ν is incident onto the surface of a semiconductor with surface reflectivity R and absorption coefficient α, the intensity distribution inside the semiconductor is given by eq. (E5.1.1)

$$I(x) = (1 - R)I_0 e^{-\alpha x} \tag{8.51}$$

and the local electron–hole generation rate is

$$\left(\frac{\partial f}{\partial t}\right)_{\text{generation}} = \frac{dI/dx}{h\nu} = \frac{(1-R)}{h\nu} I_0 \alpha e^{-\alpha x} \tag{8.52}$$

The generated electrons and holes have energy as determined by the photon energy and the electron band structure.

Conceivably, one may also start from the scattering integral and express electron–hole recombination and generation processes into a relaxation time of the form $-(f - f_0)/\tau_r$. Although such an expression has not been seen in other literature, eq. (8.47) is quite similar to this expression.

In chapters 6 and 7, we solved the Boltzmann equation under the relaxation time approximation

$$\frac{\partial f}{\partial t} + \mathbf{v} \bullet \nabla_{\mathbf{r}} f + \frac{\mathbf{F}}{m} \bullet \nabla_{\mathbf{v}} f = -\frac{f - f_0}{\tau} \qquad (8.53)$$

Such a simple treatment leads to constitutive equations without involving detailed pictures of the energy exchange processes, and should be valid when different carriers are not too far from equilibrium with each other. In the derivation of conservation equations from the Boltzmann equation, as in section 6.4.2, we did take the moments of the collision term and obtained expressions similar to what we discussed above for electron–phonon scattering, generation, and recombination terms. These moment equations, however, are normally coupled with constitutive equations, derived for bulk materials, that may not be valid for nanoscale transport. Since eq. (8.49) and the subsequent expressions for various collision processes also underlie the energy exchange among various carriers, it may be instructive to incorporate these expressions directly into the Boltzmann equation

$$\frac{\partial f}{\partial t} + \mathbf{v} \bullet \nabla_{\mathbf{r}} f + \frac{\mathbf{F}}{m} \bullet \nabla_{\mathbf{v}} f = \left(\frac{\partial f}{\partial t}\right)_{\text{scattering}}$$
$$+ \left(\frac{\partial f}{\partial t}\right)_{\text{generation}} + \left(\frac{\partial f}{\partial t}\right)_{\text{recombination}} + \cdots \qquad (8.54)$$

to examine the energy conversion and transport processes. One can infer that this approach may become particularly necessary when different carriers are at highly nonequilibrium states and when the transport involves nanoscale source terms. The study of nonequilibrium electron–phonon transport during femtosecond laser heating of metals (Qiu and Tien, 1993), the effects of nonlocal heat conduction in MOSFET structures (Sverdrup et al., 2001a), and the nonequilibrium between electron and phonon transport in heterojunction structures (Zeng and Chen, 2002) are a few examples that follow this approach.

8.2 Coupled Nonequilibrium Electron–Phonon Transport without Recombination

In the treatments of chapters 6 and 7, the energy exchange processes between different carriers were ignored in modeling the transport processes, or, at best, added into the energy conservation equations rather than starting with the Boltzmann equation. In this section, we will consider coupled electron and phonon transport, starting with the Boltzmann equation.

We consider transport processes involving only one electron band so that there is no recombination process. The scattering between electrons and phonons can be expressed as eq. (8.50). The Boltzmann equations for electrons and for phonons can be respectively written as

$$\frac{\partial f_e}{\partial t} + \mathbf{v} \bullet \nabla_{\mathbf{r}} f_e + \frac{\mathbf{F}}{m} \bullet \nabla_{\mathbf{v}} f_e = g_e(T_e - T_p) - \frac{f_e - f_{0e}}{\tau_e} \qquad (8.55)$$

$$\frac{\partial f_p}{\partial t} + \mathbf{v} \cdot \nabla_\mathbf{r} f_p = -g_p(T_e - T_p) - \frac{f_p - f_{0p}}{\tau_p} \tag{8.56}$$

where the subscripts e and p denote electrons and phonons, respectively. Multiplying eq. (8.55) by the electron kinetic energy E, and integrating over all wavevectors $\mathbf{k}$, we obtain

$$\frac{\partial u_e}{\partial t} + \nabla_\mathbf{r} \cdot \sum_\mathbf{k} E \mathbf{v} f_e - \frac{e\mathscr{E}}{m} \cdot \sum_\mathbf{k} E \nabla_\mathbf{v} f_e = -G(T_e - T_p) \tag{8.57}$$

where $\mathscr{E}$ is the electric field, and

$$u_e = \sum_\mathbf{k} E f_e \tag{8.58}$$

Here, we have used the convention of figure 6.9(b), with E_c as the reference level for E and E_f such that E and E_f represent the kinetic energy and chemical potential, respectively. The last term in eq. (8.55) drops out because it is odd with respect to $\mathbf{k}$, as one can see from eq. (8.32). The third term on the left-hand side of eq. (8.57) is

$$-\frac{e\mathscr{E}}{\hbar} \sum_\mathbf{k} [\nabla_\mathbf{k}(E f_e) - f_e \nabla_\mathbf{k} E] = e\mathscr{E} \sum_\mathbf{k} f_e \mathbf{v} = -\mathscr{E} \cdot \mathbf{J}_e \tag{8.59}$$

where $\mathbf{J}_e (= -e \sum_\mathbf{k} f_e \mathbf{v})$ is the current density of electrons. In the second step of eq. (8.59), we have used the fact that the distribution of f_e at large wavevectors is zero. Thus the third term on the left-hand side of eq. (8.57) represents the gain in electron energy under an electric field. The second term on the left-hand side of eq. (8.57) can be written as

$$\nabla_\mathbf{r} \cdot \left\{ \sum_\mathbf{k} (E - E_f) \mathbf{v} f_e \right\} + \nabla_\mathbf{r} \cdot \left(E_f \sum_\mathbf{k} \mathbf{v} f_e \right)$$
$$= \nabla \cdot \mathbf{Q}_e + \mathbf{J}_e \cdot \left(-\frac{1}{e} \nabla E_f \right) \tag{8.60}$$

where $\mathbf{Q}_e$ is the heat flux carried by electrons, including both heat conduction flux and the Peltier heat flux as dealt with in chapter 6 [eq. (6.100)]. We have assumed $\mathbf{J}_e$ is continuous in eq. (8.60). Substituting eqs. (8.58–8.60) back into eq. (8.57), we obtain the energy conservation equation for electrons as

$$\frac{\partial u_e}{\partial t} + \nabla \cdot \mathbf{Q}_e - G(T_e - T_p) = (\mathscr{E} + \nabla E_f/e) \cdot \mathbf{J}_e \tag{8.61}$$

where

$$\mathscr{E} + \nabla E_f/e = -\nabla \varphi_e + \nabla E_f/e = -\nabla(\varphi_e - E_f/e) = -\nabla \Phi \tag{8.62}$$

and Φ is the electrochemical potential [eq. (6.73a)]. Normally, the right-hand side is considered as Joule heating and thus a heat source term. Equation (8.61) indicates that it actually represents the gain of electron energy under an electrochemical potential gradient. This gain is balanced by the energy storage as internal energy, the energy loss from electrons to phonons, and the energy loss due to the gradient in the heat current carried by electrons. Similarly, we can derive the energy conservation equation for phonons as

$$\frac{\partial u_p}{\partial t} + \nabla \cdot \mathbf{Q}_p + G(T_e - T_p) = 0 \quad (8.63)$$

where u_p is the phonon internal energy and $\mathbf{Q}_p$ is the phonon heat flux.

The next question we ask is: what are the appropriate constitutive equations for the electron and phonon heat fluxes? Under the relaxation time approximation and the assumption of small deviation from spatial equlibrium, we derived the classical laws for the heat flux of electrons [eq. (6.100)] and for phonons. The phonon heat flux is the normal Fourier law but the electron heat flux includes both the heat conduction part and the coupling with the electric field [eq. (6.100)],

$$\mathbf{Q}_e = L_{21}(-\nabla \Phi) + L_{22}(-\nabla T) \quad (8.64)$$

If we start from eq. (8.55) and again make the same assumption that the spatial deviation from equilibrium is small (local equilibrium assumption) as we did in chapter 6, we can show that the same expression for the electron heat flux holds true, except that we should use the electron temperature T_e in all the calculations. The extra term $-g_e(T_e - T_p)$ does not enter into the constitutive equation for heat flux because this term is an even function of the wavevector, meaning that there is an equal amount in the forward and backward directions, thus not contributing to the heat flux.

In chapter 7, we dealt with classical size effects and showed that in this case the constitutive equations for bulk materials are no longer applicable. All the considerations in chapter 7 are based on the simple relaxation time approximation. The existence of nonequilibrium among carriers in small scales may mandate a re-examination of the topics considered in chapter 7. There are only very few studies addressing this issue, which will be discussed here.

8.2.1 Hot Electron Effects in Short Pulse Laser Heating of Metals

In short pulse laser heating of metals, electrons absorb the photon energy and transfer it to phonons. If the laser pulse is very short, the electrons will be driven to a temperature much higher than that of the phonons. Using the Fourier law expression for the electron and phonon heat flux, eqs. (8.61) and (8.63) can be written as

$$C_e \frac{\partial T_e}{\partial t} = \nabla(k_e \nabla T_e) - G(T_e - T_p) \quad (8.65)$$

$$C_p \frac{\partial T_p}{\partial t} = \nabla(k_p \nabla T_p) + G(T_e - T_p) \quad (8.66)$$

In the above equations, we have neglected the thermoelectric effects, particularly the Thomson effect for the open-circuit condition. Also neglected is the potential existence of a transient electrical current. Typically, the phonon thermal conductivity is small and can be neglected. These equations were originally derived by Kaganov

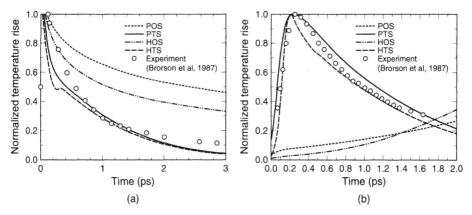

Figure 8.6 Nonequilibrium electron temperature rise during femtosecond pulsed laser heating of metals (Qiu and Tien, 1993; courtesy of ASME). Dots reflect electron temperature. (a) Front side, and (b) back side of the film. POS denotes the parabolic one-step heat conduction model based on the Fourier law without considering the electron and phonon temperature difference. PTS denotes the parabolic two-step heat conduction model based on the Fourier law, considering electron and phonon temperature difference. HOS (and HTS) denote hyperbolic one-step (and two-step) heat conduction models.

et al. (1957). Qiu and Tien (1992, 1993) investigated experimental results of femtosecond laser heating on various metallic thin films, using these equations. One example is given in figure 8.6, which demonstrates good comparison between model results and experimental data. Qiu and Tien called eqs. (8.64) and (8.65) the parabolic two-step model to differentiate it from the hyperbolic two-step model they developed. The latter includes the Cantaneo type of constitutive equations for electrons and phonons, the validity of which is questionable as discussed in chapter 6. In fact, because the metallic films under consideration are typically of the order of hundreds to thousands of angstroms, comparable to the electron and phonon mean free path, the validity of the diffusion type of equations, be it the Fourier or the Cantaneo type, should also be subject to question. One can conceivably solve the coupled Boltzmann equation and examine whether the ballistic transport will have any significant impact on the explanation of experimental data. The thermoelectric effect on the transport process is also interesting to consider due to the large electron temperature gradient existing in the film. However, judging from the reasonably good agreement between two-step parabolic and experimental data (Qiu and Tien, 1993), it is likely that the two-step parabolic heat conduction model captures the dominant processes while other effects discussed above are of less importance.

8.2.2 Hot Electron and Hot Phonon Effects in Semiconductor Devices

Metal-oxide-semiconductor field effect transistors (MOSFETs), as shown in figure 1.1(a), are the workhorses of integrated circuits. In MOSFET devices, electrons go through a short channel underneath the gate oxide as shown in that figure. The electron transport can be nonlocal or ballistic. In this case, the external electrical field accelerates the electrons to the drain side, where the energy gained by the electrons is dumped to phonons as seen from the Monte Carlo simulation shown in figure 1.1(b). Because most phonons are heated in a small region that is much shorter than the phonon mean free path in the bulk substrate, phonon transport is also nonlocal, as figure 7.15 shows. In fact, phonon heating can also be a cascade process. Electrons may first release their energy to optical phonons, which subsequently pass their energy to acoustic phonons that conduct heat through the substrate to the ambient. Lai and Majumdar (1996) considered the nonequilibrium between optical and acoustic phonons on the basis of diffusion theories. Ju and Goodson (1999) further suggested splitting the acoustic phonons into two groups of different velocities due to the large dispersion of the phonons, and introduced a two-fluid model to distinguish fast and slow propagating acoustic phonons, similar to that between electrons and phonons. Sverdrup et al. (2001b) reported some initial data suggesting that both nonlocal and nonequilibrium effects are important with acoustic phonons. More experimental validations are needed.

We can quickly evaluate the difference between electron and phonon temperatures under an electric field using a simple one-dimensional model. Considering electrons driven by an external field along a conductor and neglecting thermoelectric effects, the energy balance equation for electrons and phonons can be written as

$$k_e \frac{d^2 T_e}{dx^2} - G(T_e - T_p) = J_e \left(-\frac{d\Phi}{dx} \right) \tag{8.67}$$

$$k_p \frac{d^2 T_p}{dx^2} + G(T_e - T_p) = 0 \tag{8.68}$$

From eqs. (8.67) and (8.68) we obtain

$$\frac{d^2(T_e - T_p)}{dx^2} - G \left(\frac{1}{k_e} + \frac{1}{k_p} \right)(T_e - T_p) = \frac{1}{k_e} J_e \left(-\frac{d\Phi}{dx} \right) \tag{8.69}$$

A general solution of the above equation is

$$T_e(x) - T_p(x) = \frac{J_e}{G(1 + k_e/k_p)} \left(-\frac{d\Phi}{dx} \right) + C_1 e^{-\beta x} + C_2 e^{\beta x} \tag{8.70}$$

where

$$\beta = \sqrt{\frac{G}{k_e} + \frac{G}{k_p}} \tag{8.71}$$

Let us consider the following boundary conditions

$$x = 0, \quad T_e - T_p = 0 \tag{8.72}$$

$$x \to \infty, \quad T_e - T_p \text{ is finite} \tag{8.73}$$

With these two boundary conditions, we can determine the temperature distribution as

$$T_e(x) - T_p(x) = \frac{J_e}{G(1 + k_e/k_p)} \left(-\frac{d\Phi}{dx}\right)(1 - e^{-\beta x}) \tag{8.74}$$

and the corresponding heat generation distribution is

$$\dot{q}(x) = \frac{J_e}{(1 + k_e/k_p)} \left(-\frac{1}{e}\frac{d\Phi}{dx}\right)(1 - e^{-\beta x}) \tag{8.75}$$

The first term in eq. (8.70) is a measure of the temperature difference between electrons and phonons in a bulk medium and the last two terms are determined by the boundary conditions. The temperature difference between electrons and phonons in a bulk medium is of the order of

$$\Delta T_{ep} \approx \frac{J_e}{G(1 + k_e/k_p)} \left(-\frac{d\Phi}{dx}\right) \tag{8.76}$$

In metals, the thermal conductivity is controlled by the electronic thermal conductivity. G is typically $\sim 10^{16}$ W m^{-3} K^{-1} (Qiu and Tien, 1993), k_e/k_p is typically ~ 10–100, and the electrical resistivity is $\sim 10^{-8}$ Ω m. Consider a conductor 1 mm in diameter carrying 1 A current, leading to a current density of $\sim 10^6$ A m^{-2}; the estimated temperature difference between electrons and phonons is $\sim 10^{-13}$ K. This temperature difference is clearly negligible, so electrons and phonons are essentially at equilibrium with each other.

In modern semiconductor devices, the situation is different. The electric field in some devices such as the MOSFET easily reaches 10^6 V m^{-1}. The current density can be estimated from eq. (6.82) as

$$J_e \approx en\mu_e \mathcal{E} \tag{8.77}$$

where μ_e is the electron mobility. We have estimated that $G \approx 10^{-12}n$ [W m^{-3} K^{-1}] for semiconductors following eq. (8.34). So we can eliminate the carrier density, n, and estimate the temperature difference between electrons and phonons from eq. (8.76). The electronic thermal conductivity will be much smaller than the phonon thermal conductivity, so k_e/k_p in the denominator of eq. (8.74) can be neglected. For a typical carrier concentration $n \sim 10^{17}$ cm^{-3}, and a field of 10^6–10^7 V m^{-1}, the temperature difference between electrons and phonons would be $\sim 10^2$–10^3 K. In this case, the electrons are much hotter than the lattice. Simulations of hot-electron effects in MOSFET devices often show electron temperatures in the range of 10^4 K (Lundstrom, 2000).

Equation (8.75) also reveals that electric heating in a conductor is not necessarily uniform but determined by the boundary conditions. If the temperatures of electrons and phonons are equal at $x = 0$, the heat generation close to $x = 0$ is not uniform but increases exponentially as shown in eq. (8.75). The distance it takes for the temperature difference between electrons and phonons to reach 63% of the saturated value is

$$L_e = \frac{1}{\sqrt{\frac{G}{k_e} + \frac{G}{k_p}}} \tag{8.78}$$

and this distance is called the energy coupling length. Using a typical value for metal, this length scale is $\sim 10^{-7}$ m. For semiconductors, this value could be as large as several microns. This means that Joule heating can be made nonuniform, an effect that can be exploited to make potentially better thermoelectric devices (Zakordonets and Logvinov, 1997; Shakouri and coauthors, 1997, 1998; Zeng and Chen, 2000).

8.2.3 Cold and Hot Phonons in Energy Conversion Devices

The electron temperature can also fall below the phonon temperature. This must be the case if cooling of the lattice occurs, as in the Peltier effect at the junction of two dissimilar materials when the current flows in the correct direction. Surprisingly, the detailed picture for the electron and phonon energy conversion processes at the junction between two materials is not very clear. There have been a few studies in the past, starting on the basis of the drift-diffusion equation (Stratton, 1962) or nonequilibrium electron–phonon transport models as represented by eqs. (8.60) and (8.62) (Zakordonets and Logvinov, 1997; Anatychuk and Balut, 1998; Mahan, 2001). There are several questions that remain to be further clarified. One is the ballistic electron and phonon transport effect near the boundary, which may render the diffusion type of equations nonapplicable. Studies of electron transport near interfaces, based on the Boltzmann equation, can be found in the literature (Stettler and Lundstrom, 1994), but their implications for energy conversion are not clear. The other key question is on the boundary conditions. The diffusion-transmission boundary condition discussed in chapter 7 (Chen, 2003), for example, may be applied in conjunction with the nonequilibrium electron–phonon transport equations. Zeng and Chen (2002) considered concurrent ballistic electron and phonon transport inside a thin film. They used the diffusion approximation coupled with the discontinuity type of boundary conditions in the film side but not on the electrodes side. Their simulations show that close to the cold region the electron temperature is higher, whereas near the hot region the electron temperature is lower than that of the phonons.

8.3 Energy Exchange in Semiconductor Devices with Recombination

The discussion in the previous section is limited to a single type of charged carriers. In semiconductor devices, both electrons and holes can be present. When transport processes involve both electrons and holes, the electrons and holes can recombine and release their energy as either heat or light. Determining where this recombination occurs and whether the energy is released as heat or light, is important for device thermal modeling. In this section, we will first discuss the formulation of the energy source term, then follow with two examples. One is energy conversion in a p–n junction and the other is the heat generated when a semiconductor is subjected to laser heating.

8.3.1 Energy Source Formulation

The energy source term in semiconductor devices was discussed in several papers by Wachutka (1990) and Lindefelt (1994). Wachutka (1990) started from the energy conservation equations whereas Lindefelt (1994) started from the Boltzmann equation. Since we have seen how the constitutive equations can be derived from the Boltzmann equation in chapter 6, we will follow Wachutka's approach. We should first point out that these treatments assume that electrons, holes, and phonons are at equilibrium with each other and neglect the nonlocal transport effects. We have seen in previous chapters that such an assumption is usually not valid in nanodevices. Thus, the content of this section should be viewed with caution when applied to nanostructures. This section can serve as a starting point for further development on related topics.

The general philosophy behind both Wachutka's and Lindefelt's treatments is the energy conservation equation

$$\frac{\partial u}{\partial t} + \nabla \bullet \mathbf{J}_u = 0 \tag{8.79}$$

where u is the total system energy per unit volume, and $\mathbf{J}_u$ is the flux of the total system energy. In this expression, it is implicitly assumed that there is no local external energy input such as a radiation source. The external electrical field will be included in the energy flux term. The total internal energy and the energy flux include contributions from electrons, holes, and phonons. Our goal is to cast eq. (8.79) into the typical heat conduction equation form,

$$C\frac{\partial T}{\partial t} = \nabla \bullet (k\nabla T) + \dot{q} \tag{8.80}$$

so that the $\dot{q}$ term can be identified as a source term. Here C is the volumetric specific heat. This source term was referred to as heat generation by Wachutka (1990) and Lindefelt (1994). Calling it heat generation, however, is misleading since this term can include reversible thermoelectric effects, discussed in chapter 6, and the energy gain from an external field, as the right-hand side of eq. (8.60) represents. In addition, this term can include radiative recombination that releases the energy as light. Thus we will call it the *energy source term*. Again, since we have discussed extensively in previous chapters, particularly in chapter 7, circumstances where the above equation is no longer valid in nanostructures, we emphasize that neither Wachutka's nor Lindefelt's formulation of the heat source includes nonlocal transport effects and thus their applicability to nanoscale devices should be questioned. The extension to include nonlocal transport and nonequilibrium between carriers, however, should not be particularly difficult.

In chapter 6, we obtained expressions for the energy flux carried by electrons. We need an expression for the rate of internal energy change in eq. (8.79). Taking electrons as an example, the first law of thermodynamics can be expressed as

$$dU_e = T\,dS - P\,dV + E_{fe}dN \tag{8.81}$$

where U_e, S, V, and N are respectively the internal energy, the entropy, the volume, and the number of electrons. These variables depend on the size of the system and are called extensive variables. The other three variables, T, P, E_{fe}, are the temperature, pressure, and chemical potential of the electrons, and are intensive variables because they are independent of system size.

We will now show that the three intensive variables are not independent of each other. From thermodynamics, we can also express the energy as

$$U_e = TS - PV + E_{fe}N \tag{8.82}$$

which is one form of the Euler equation (Callen, 1985). Combining eqs. (8.81) and (8.82) leads to

$$S\,dT - V\,dP + N\,dE_{fe} = 0 \tag{8.83}$$

This relation, called the Gibbs–Duhem relation, shows that the three intensive variables are not independent of each other and thus we can choose two of them in formulating a problem. When dealing with steam engines, for example, we often use temperature and pressure. For electrons, since the total volume occupied by the electrons is constant, we will choose temperature and chemical potential. If we normalize all extensive variables by the volume of the electrons, we can obtain from eqs. (8.81) and (8.82)

$$du_e = T\,ds + E_{fe}\,dn \tag{8.84}$$

where $u_e = U_e/V$, $s = S/V$, and $n = N/V$. Taking the entropy s as a function of (T, n), we have

$$ds = \left(\frac{\partial s}{\partial T}\right)_n dT + \left(\frac{\partial s}{\partial n}\right)_T dn \tag{8.85}$$

where the variable in the subscript is held fixed during the partial differentiation. The partial derivatives in eq. (8.84) can be further expressed as

$$\left(\frac{\partial s}{\partial T}\right)_n = \left(\frac{\partial s}{\partial u}\right)_n \left(\frac{\partial u}{\partial T}\right)_n = \frac{C_e}{T} \tag{8.86}$$

$$\left(\frac{\partial s}{\partial n}\right)_T = -\left(\frac{\partial E_{fe}}{\partial T}\right)_n \tag{8.87}$$

where C_e is the volumetric specific heat of electrons. The last equation can be derived from the Helmholtz free energy $F(T, V, N) = U - TS$ (section 4.1.2) and

$$dF = -S\,dT - P\,dV + E_{fe}dN \tag{8.88}$$

Substituting eqs. (8.85–8.87) into eq. (8.84) leads to

$$du_e = C_e dT + \left[E_{fe} - T\left(\frac{\partial E_{fe}}{\partial T}\right)_n\right] dn \tag{8.89}$$

In the above derivations, we have not considered the existence of an external electric field. In such a situation, the internal energy of electrons, eq. (8.82), should be modified to include the electrostatic potential φ_e,

$$U_e = TS - PV + E_{fe}N - e\varphi_e N \tag{8.90}$$

If we include the electrostatic potential the rate of internal energy change can be expressed as

$$du_e = C_e dT + \left[\psi_e - T\left(\frac{\partial \psi_e}{\partial T}\right)_n\right] dn \qquad (8.91)$$

where $\psi_e = -e\Phi = E_{fe} - e\varphi_e$ is the combined Fermi energy and electrostatic potential energy of electrons. When electrons, holes, and phonons are all included, the rate of the internal energy change can be expressed as

$$\frac{\partial u}{\partial t} = C\frac{\partial T}{\partial t} + \left[\psi_e - T\left(\frac{\partial \psi_e}{\partial T}\right)_n\right]\frac{\partial n}{\partial t} - \left[\psi_h - T\left(\frac{\partial \psi_h}{\partial T}\right)_p\right]\frac{\partial p}{\partial t} \qquad (8.92)$$

where C is the volumetric specific heat of electrons, holes, and phonons, p is the concentration of holes, and ψ_h is combined Fermi energy and electrostatic potential energy of holes. The negative sign for the hole contribution arises from the convention that the hole energy increases as the electron energy decreases in a typical band diagram. Here it has been assumed that electrons and holes may have different Fermi levels, while being at the same temperature. This assumption is often made in the treatment of transport processes in semiconductors, as we will see more in the next section.

Now we deal with the second term in eq. (8.79). The energy flux of the electrons $\mathbf{J}_{u,e}$ is composed of the heat flux and the flux of electrochemical potential,

$$\mathbf{J}_{u,e} = \mathbf{Q}_e - \frac{\psi_e}{e}\mathbf{J}_e = (\Pi_e - \psi_e/e)\mathbf{J}_e - k_e \nabla T \qquad (8.93)$$

where Π_e is the Peltier coefficient, $\mathbf{J}_e$ is the electron current flux vector, and k_e the electron thermal conductivity. In this notation, we consider that the actual electron flow direction is opposite to the current vector $\mathbf{J}_e$ (the Peltier coefficient for electrons is negative). In eq. (6.105), we derived an expression for internal heat generation for the case when $\mathbf{J}_e$ is constant in space. Here we will not make this assumption because, when recombination occurs, the electron current may be a function of the spatial coordinate.

A similar expression can be written for the energy flux of holes. Substituting eqs. (8.92) and (8.93) into eq. (8.79), we arrive at eq. (8.80) with the following expression for the energy source term,

$$\dot{q} = -\nabla \bullet \left[(\Pi_e - \psi_e/e)\mathbf{J}_e + (\Pi_h - \psi_h/e)\mathbf{J}_h\right]$$

$$- \left[\psi_e - T\left(\frac{\partial \psi_e}{\partial T}\right)_n\right]\frac{\partial n}{\partial t} + \left[\psi_h - T\left(\frac{\partial \psi_h}{\partial T}\right)_p\right]\frac{\partial p}{\partial t} \qquad (8.94)$$

where $\mathbf{J}_h$ is the hole current density. The divergence term includes the energy gain from the external field, the Joule heating, the Thomson effect, and the heat generation by recombination. The time-dependent terms represent the loss of carrier energy during transient processes.

8.3.2 Energy Conversion in a p–n Junction

Semiconductor p–n junctions are elementary structures for electronic devices. To identify the distribution of the energy sources in a p–n junction according to eq. (8.94), we

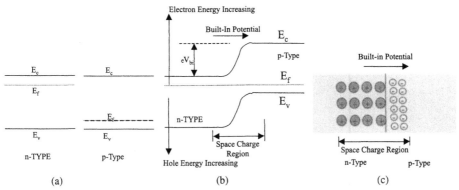

Figure 8.7 Formation of a p–n junction: (a) conduction and valence band edges and Fermi levels before the n-type and p-type materials are joined; (b) when the p–n junction forms, the Fermi level is continuous at equilibrium, forcing the band edge to bend, and a built-in potential is created across the space charge region; (c) this can also be understood as meaning that the electrons in the n-type region diffuse into the p-type region, leaving immobile positive charges (donors) behind, and holes from the p-type region diffuse into the n-type region, leaving immobile negative charges behind. The positive and negative charge regions near the interface form an electrostatic potential to resist further diffusion, and thus establish an equilibrium state as in (b).

need to determine how the electron and the hole currents, J_e and J_h, vary in the p–n junction. We will thus first discuss the principle of a p–n junction and the mathematical formulation for the electron and hole current densities, and then the heat source distribution.

8.3.2.1 Current Flow in a p–n Junction

When two semiconductors (see appendix B) with different doping are joined together, they form a p–n junction. The band structure of a p–n junction is as shown in figure 8.7(b). Before the two semiconductors are joined together, the electrochemical potential levels in each material are not related [figure 8.7(a)]. When they are connected together, the electrochemical potential levels must be equal in both sides because of the requirement of thermodynamic equilibrium, just as the temperatures must be equal [figure 8.7(b)]. This happens because electrons from the n-type material diffuse into the p-type material, leaving positively charged immobile ions behind [figure 8.7(c)]. Conversely, holes from the p-type material diffuse into the n-type material, leaving negatively charged ions behind. At the interface, the positively charged ions in the n-side and the negatively charged ions in the p-side form an electrostatic potential barrier that resists further diffusion to establish an equilibrium state for the whole structure. This is reflected in the band diagram [figure 8.7(b)]. The drawing of the built-in potential may seem a little strange in figure 8.7 since it points from the lower edge of the conduction band to the higher edge, but keep in mind that the larger the electron energy, the more negative is its electrostatic potential. The region near the interface where negatively and positively charged ions are not neutralized by electrons and holes is called the space charge region. The concentrations of free electrons or holes in this space charge region are very low compared to the number of electrons and holes in the bulk material. The

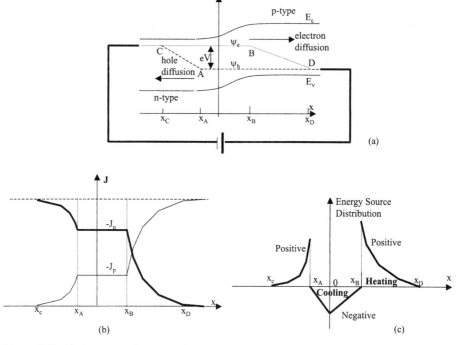

Figure 8.8 (a) A p–n junction under forward bias; (b) current distribution due to electrons and holes; (c) energy source distribution.

built-in potential, V_{bi}, and the p–n junction width, w, are given by (Sze, 1981; see also appendix B)

$$V_{bi} = \frac{\kappa_B T}{e} \ln\left(\frac{n_{n0} p_{p0}}{n_i^2}\right) \qquad (8.95)$$

$$w = \sqrt{\frac{2\varepsilon_s}{e}\left(\frac{N_A + N_D}{N_A N_D}\right) V_{bi}} \qquad (8.96)$$

where n_{n0} and p_{p0} are the equilibrium electron and hole concentrations in the bulk region (away from the space charge region) of the n-type and p-type semiconductors, respectively, N_A and N_D are the concentrations of p-type and n-type dopants in the p-type and n-type regions, respectively, n_i is the intrinsic carrier concentration, and ε_s is the electrical permittivity of the semiconductor (chapter 5).

Now let's consider the current–voltage characteristics of a p–n diode under an externally applied positive voltage (forward bias), as shown in figure 8.8. In this case, excess electrons will move from the n-type region into the p-type region and holes from the p-type region into the n-type region, breaking the equilibrium established when there is no external bias. The electrons and holes are no longer in equilibrium with each other and we cannot use a Fermi level to describe both carriers. We can, however, consider that electrons as one entity at local equilibrium with chemical potential, E_{fe}, and regard holes as another entity with a chemical potential E_{fh}. Both of these chemical potentials are functions of location and even time for transient cases. Assuming that the Boltzmann statistics are still valid, we can express the local electron and hole concentrations as [eq. (4.64)]

$$n = N_c \exp\left(-\frac{E_c - \Psi_e}{\kappa_B T}\right) \quad (8.97)$$

$$p = N_v \exp\left(-\frac{\Psi_h - E_v}{\kappa_B T}\right) \quad (8.98)$$

where N_v and N_c are the effective densities-of-states in the conduction and the valence bands, respectively (appendix A). Here we have written the conduction and valence band edges E_c and E_v explicitly, rather than taking E_c as the reference level. This corresponds to the reference system in figure 6.9(a).* Consequently, Ψ_e and Ψ_h include both chemical potential and electrostatic potential, and are called quasi-Fermi levels.

To describe the motion of electrons or holes, we can use the drift-diffusion equation introduced in eq. (6.82). For electrons

$$J_e = e\mu_e n \mathscr{E} + ea_e \frac{dn}{dx} = e\mu_e \left(n\mathscr{E} + \frac{\kappa_B T}{e}\frac{dn}{dx}\right) \quad (8.99)$$

where we dropped the vector sign for one-dimensional problems and used the Einstein relation between the diffusivity and the mobility [eq. (6.86)]. The electrical field can be related to the conduction band edge as

$$\mathscr{E} = \frac{1}{e}\frac{dE_c}{dx} \quad (8.100)$$

Substituting eqs. (8.97) and (8.100) into eq. (8.99), we have

$$J_e = \mu_e n \frac{d\Psi_e}{dx} = -e\mu_e n \frac{d\Phi_e}{dx} \quad (8.101)$$

and similarly

$$J_h = \mu_h p \frac{d\Psi_h}{dx} = e\mu_h p \frac{d\Phi_h}{dx} \quad (8.102)$$

which means that the electrochemical potentials, Φ_e and Φ_h, $\Phi_e(= -\Psi_e/e)$ and $\Phi_h(= \Psi_h/e)$, or quasi-Fermi levels, for electrons and holes, are the driving forces for charge flow. The changes of Ψ_e and Ψ_h in the p–n junction as a function of space are schematically illustrated in figure 8.8(a). In the bulk regions (beyond points C and D), the changes in the quasi-Fermi levels can be neglected because the resistances of the bulk regions are small compared to those of the junction region. It can be further proven that in the space charge region the variation of the quasi-Fermi level is also small (Shockley, 1949; Sze, 1981). Thus most of the quasi-Fermi level drops occur in transition regions

---

*Because both E_c and E_v vary as a function of spatial coordinate, the reference system as shown in figure 6.9(a) is more convenient here.

starting from the edges of the space charge region and linking with the bulk regions, that is, B–D for electrons in the p-type region and A–C for holes in the n-type region. We will see that heat is also generated in these transition regions. To find the heat generation, we need first to determine the current distribution due to electron transport in B–D and hole transport in A–C. Since the external electric field in the transition regions is weak, the drift term in eq. (8.99) can be neglected. We can write down the charge conservation equation for electrons in region B–D as [eq. (6.152)]

$$a_e \frac{d^2 n}{dx^2} - \frac{n - n_{p0}}{\tau_n} = 0 \qquad (8.103)$$

where the second term represents the loss of electrons due to recombination and n_{p0} is the equilibrium electron concentration in the bulk region of the p-type semiconductor, given by (see appendix A)

$$n_{p0} = \frac{n_i^2}{p_{p0}} \approx \frac{n_i^2}{N_A} \qquad (8.104)$$

where N_A is the acceptor concentration in the p-region. The appropriate boundary conditions for eq. (8.103) are

$$x = x_B : n = n_{p0} \exp\left(\frac{eV}{\kappa_B T}\right) \qquad (8.105)$$

and

$$x \to \infty : n = n_{p0} \qquad (8.106)$$

where V is the externally applied voltage. In eq. (8.105), we have used eq. (8.97) and the fact that ψ_p is approximately constant between A and D. Equation (8.103) can be solved with the above boundary conditions, giving the n-type carrier distribution in the p-type region as (B–D)

$$n - n_{p0} = n_{p0} \left[\exp\left(\frac{eV}{\kappa_B T}\right) - 1\right] \exp\left(-\frac{x - x_B}{L_e}\right) \qquad (8.107)$$

where

$$L_e = \sqrt{a_e \tau_e} \qquad (8.108)$$

is the electron diffusion length before it is recombined with a hole. The diffusion current of the n-type carrier at $x = x_B$ is

$$J_e = ea_e \left.\frac{dn}{dx}\right|_{x=x_B} = -\frac{ea_e n_{p0}}{L_e}[e^{eV/k_B T} - 1] \qquad (8.109)$$

where the negative sign means that the current flows into the negative x-direction because the electron diffuses along the positive x-direction. This current must be supplied by the diffusion of holes from the bulk p-type region into region B–D. Similarly, at $x = A$, the hole diffusion current into the n-type region is

$$J_h = -\frac{ea_h p_{n0}}{L_h}[e^{eV/k_B T} - 1] \qquad (8.110)$$

which must be supplied by electron diffusion from the bulk of the n-region. The total current flowing through the p–n junction is thus

$$J = J_h + J_e = -J_s(e^{eV/k_B T} - 1) \tag{8.111}$$

The negative sign is not normally explicitly written and is only specific to the coordinate system we chose in figure 8.8(a). J_s is called the saturation current, which can be expressed as

$$\begin{aligned} J_s &= e\left(p_{n0}\sqrt{\frac{a_h}{\tau_h}} + n_{p0}\sqrt{\frac{a_e}{\tau_e}}\right) \\ &= e\left(\frac{n_i^2}{N_A}\sqrt{\frac{a_h}{\tau_h}} + \frac{n_i^2}{N_D}\sqrt{\frac{a_e}{\tau_e}}\right) \\ &= eN_c N_v \left(\frac{1}{N_A}\sqrt{\frac{a_h}{\tau_h}} + \frac{1}{N_D}\sqrt{\frac{a_e}{\tau_e}}\right)\exp\left(-\frac{E_G}{\kappa_B T}\right) \end{aligned} \tag{8.112}$$

Equation (8.111) is the celebrated Shockley equation (Shockley, 1949), which is a description of ideal p–n junctions at low current densities. The factors that make real devices deviate from this equation are (1) the finite lengths of the bulk regions of the semiconductor such that the boundary condition, eq. (8.106), is no longer valid, and (2) the recombination that occurs inside the depletion region. We will not discuss these nonideal factors any further. Figure 8.8(b) shows the current distributions carried by electrons and holes in the junction region. From A to C, the hole (minority carrier) current diminishes to zero as the holes diffuse and recombine with electrons. The electron current at x_A, $J_n(x_A)$, supplies the electron diffusion current between B and D. Another component of the electron current in A–C is drawn from the bulk n-type region through recombination with holes. The hole current in B–D consists similarly of two parts, one to supply the diffusion current in A–C and the other drawn by recombination with electrons in B–D.

8.3.2.2 Energy Sources in a p–n Junction

From the above discussion, we can see that in a p–n junction it is the diffusion of minority carriers (holes in the n-type region or electrons in the p-type region) and their recombination with the majority carrier (electrons in the n-type region or holes in the p-type region) that control the current flow. Because of these characteristics, p–n diodes are also called minority carrier devices. During recombination, heat is released. We now evaluate the energy source distribution. Starting from eq. (8.94), since we are dealing with a steady-state problem, the time derivative terms drop out and only the first two terms contribute to the energy source:

$$\dot{q} = -\frac{d}{dx}[(\Pi_e - \psi_e/e)J_e + (\Pi_h - \psi_h/e)J_h] \tag{8.113}$$

Now we apply eq. (8.113) to three separate sections in figure 8.8(a), C–A, A–B, and B–D. Consider the region between B and D first. In this region, it is the diffusion of electrons that drives the current flow. For nondegenerate semiconductors, that is, semiconductors with electron and hole distributions that can be approximated by the Boltzmann distribution, we derived in example 6.1 an expression for the Seebeck coefficient, from which the Peltier coefficient can be expressed as

$$\Pi_n = \frac{T\kappa_B}{e}\left[-\frac{\psi_e - E_c}{\kappa_B T} - \left(\gamma_e + \frac{5}{2}\right)\right]$$

$$= -\frac{E_{fe}}{e} - \left(\gamma_e + \frac{5}{2}\right)\frac{\kappa_B T}{e} \tag{8.114}$$

where γ_e is the parameter in the energy dependance of the electron scattering ($\sim E^{\gamma_e}$).

In region B–D, E_c is a constant. If we further consider an isothermal condition, consistent with Shockley's derivation, the first term in eq. (8.113) can be expressed as

$$\dot{q}_n(x_B \le x \le x_D) = -\frac{[E_c + (\gamma_e + 5/2)\kappa_B T]}{e}\frac{dJ_e}{dx}$$

$$= \left[E_c + \left(\gamma_e + \frac{5}{2}\right)\kappa_B T\right]\frac{n - n_{p0}}{\tau_e}$$

$$= \left[E_c + \left(\gamma_e + \frac{5}{2}\right)\kappa_B T\right]\frac{n_{p0}}{\tau_e}\left[\exp\left[\frac{eV}{\kappa_B T}\right] - 1\right]\exp\left(-\frac{x - x_B}{L_e}\right) \tag{8.115}$$

To evaluate the second term in eq. (8.113), we start from the fact that the total electron and hole current at each location must be a constant, and thus

$$\nabla J_e(x) = -\nabla J_p(x) \tag{8.116}$$

Thus the second term in eq. (8.113) is

$$\dot{q}_p(x_B \le x \le x_D) = -(\Pi_h - \psi_h/e)\frac{dJ_e}{dx}$$

$$= \left[-E_v + \left(\gamma_h + \frac{5}{2}\right)\kappa_B T\right]\frac{n_{p0}}{\tau_e}\left[\exp\left(\frac{eV}{\kappa_B T}\right) - 1\right]$$

$$\exp\left(-\frac{x - x_B}{L_e}\right) \tag{8.117}$$

where Π_h is the Peltier coefficient for holes and γ_h is the parameter in the energy dependence of the hole scattering. The total energy source term in the region $x_B < x < x_D$ is therefore

$$\dot{q}(x) = \left[E_c - E_v + \left(\gamma_h + \frac{5}{2}\right)\kappa_B T\right.$$

$$\left. + \left(\gamma_e + \frac{5}{2}\right)\kappa_B T\right]\frac{n_{p0}}{\tau_e}\left[\exp\left(\frac{eV}{\kappa_B T}\right) - 1\right]\exp\left(-\frac{x - x_B}{L_e}\right)$$

$$= \left[E_c - E_v + \left(\gamma_h + \frac{5}{2}\right)\kappa_B T + \left(\gamma_e + \frac{5}{2}\right)\kappa_B T\right]\frac{n - n_{p0}}{\tau_e} \tag{8.118}$$

This is the energy released by the electron–hole recombination. The term E_c-E_v inside the square brackets of the last equation is the energy across the bandgap E_g, while the rest of the terms are the thermal energy of the electron and hole currents. This energy release is often treated as heat generation, but it should be remembered that some of the energy can be converted into light, as in luminescence, particularly in direct gap semiconductors. For indirect semiconductors such as silicon-based devices, most of the recombination is nonradiative and thus can be treated as heat.

We can integrate eq. (8.118) from x_B to infinity to obtain the total energy released in this section,

$$\dot{Q}_{BD} = \left[E_g + \left(\gamma_h + \frac{5}{2}\right)\kappa_B T + \left(\gamma_e + \frac{5}{2}\right)\kappa_B T\right]\frac{n_{p0}L_e}{\tau_e}\left[\exp\left(\frac{eV}{\kappa_B T}\right) - 1\right]$$

$$= -\left[E_g + \left(\gamma_h + \frac{5}{2}\right)\kappa_B T + \left(\gamma_e + \frac{5}{2}\right)\kappa_B T\right]\frac{J_e(x_B)}{e} \quad (8.119)$$

where the negative sign is because J_e in eq. (8.109) is negative. Similarly, the total energy released in region x_C to x_A is

$$\dot{Q}_{CA} = -\left[E_g + \left(\gamma_h + \frac{5}{2}\right)\kappa_B T + \left(\gamma_e + \frac{5}{2}\right)\kappa_B T\right]\frac{J_h(x_A)}{e} \quad (8.120)$$

The total energy released, due to the recombination of electrons and holes, in regions C–A and B–D is then

$$\dot{Q}_{CA-BD} = -\left[E_g + \left(\gamma_h + \frac{5}{2}\right)\kappa_B T + \left(\gamma_e + \frac{5}{2}\right)\kappa_B T\right]\frac{J}{e} \quad (8.121)$$

This energy released in regions AC and BD, however, is larger than the total energy supplied by the external source, which should be $-V \times J (J < 0$ here). Where does the extra energy come from? We find the answer in region A–B, the space charge region. In this region, the recombination current can be neglected and thus J_e and J_p are constant. However, E_c and E_v vary. Applying eq. (8.113) to this region, we have

$$\dot{q}(x_A \leq x \leq x_B) = \frac{J_e}{e}\frac{d}{dx}\left[E_c + \left(\gamma_e + \frac{5}{2}\right)\kappa_B T\right]$$
$$- \frac{J_h}{e}\frac{d}{dx}\left[-E_v + \left(\gamma_h + \frac{5}{2}\right)\kappa_B T\right] \quad (8.122)$$

The profile of E_c and E_v can be obtained from solving the Poisson equation (Sze, 1981, appendix B). The solutions are not listed here since we are interested in finding whether the energy source term in this region can help the energy balance of the whole device. Integrating eq. (8.122) leads to

$$\dot{Q}_{AB} = [J_e(x_B)E_c(x_B) - J_e(x_B)E_c(x_A)$$
$$+ J_h(x_B)E_v(x_B) - J_h(x_B)E_v(x_A)]/e \quad (8.123)$$

Adding eqs. (8.121) and (8.123), we can write the total energy dissipated in the p–n junction as

$$\dot{Q}_{CD} = -[E_g - (E_c - \psi_e)_{x_B} - (\psi_h - E_v)_{x_B}]J_e(x_B)/e$$
$$- [E_g - (E_c - \psi_e)_{x_A} - (\psi_h - E_v)_{x_A}]J_h(x_A)/e$$
$$+ \frac{[J_e(x_A) + J_h(x_A)]}{e}\left[(\psi_e - E_c) - \left(\gamma_e + \frac{5}{2}\right)\kappa_B T\right]_{x_A}$$
$$- \frac{[J_e(x_B) + J_h(x_B)]}{e}\left[(\psi_h - E_v) + \left(\gamma_h + \frac{5}{2}\right)\kappa_B T\right]_{x_B} \quad (8.124)$$

Referring to figure 8.8(a), we see that $[E_g - (E_c - \psi_e)_{x_B} - (\psi_h - E_v)_{x_B}]$ is equal to the difference between Ψ_e and Ψ_h and thus the externally applied voltage. The first two terms in the above equation give $-JV$, which is what we would like to see. The third term can be written as

$$J\left[(\psi_e - E_c) - \left(\gamma_e + \frac{5}{2}\right)\kappa_B T\right]_{x_c} = J\Pi_e(x_c) \quad (8.125)$$

because the electrochemical and the electrostatic potential do not vary between C and A. Equation (8.125) represents the energy carried into the region by electrons (both J and Π_e are negative). Similarly, the last term in eq. (8.124) represents the energy carried by holes into this region. The thermoelectric energy represented by these two terms, which is small for typical p–n junctions, is not treated in Shockley's model. We should also point out that the above derivation of the energy source distribution did not include the resistance of the majority carriers, that is, electrons in the n-type region and holes in the p-type region, as is the case in the Shockley model. For real devices, these factors should be taken into account.

On the basis of the above discussion and from eq. (8.123), we see that $\dot{Q}_{AB}$ is negative because electrons driven by an external voltage extract energy from the built-in field and release the energy in other regions of the device. The energy source distribution for the p–n junction under forward bias is sketched in Figure 8.8(c). The negative energy source means that energy is extracted out of the region. Local cooling of the p–n junction is thus possible. This issue was discussed by Stratton (1962), who indeed showed possible cooling at a p–n junction. Surprisingly, there does not exist much discussion on the energy source distribution in a p–n junction. Recently, there has been discussion in the literature on using the internal cooling of heterojunctions for semiconductor devices, particularly semiconductor lasers, based on a simplified treatment of the energy source distributions (Pipe et al., 2002). This section provides an example of how one can treat such problems more rigorously. Treatments of heterojunctions, and the inclusion of nonequilibrium electron–phonon interactions and ballistic electron transport, are potentially fruitful directions to explore for the benefit of electronic devices and photonic devices, or energy conversion devices.

8.3.3 Radiation Heating of Semiconductors

Another example of heat generation due to recombination is photon absorption in semiconductors. Although all the light (photons) may be absorbed very close to the surface,

ENERGY CONVERSION AND COUPLED TRANSPORT PROCESSES   385

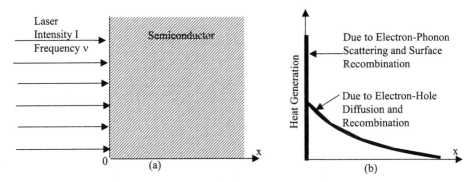

Figure 8.9 Heat source distribution due to optical absorption. Although all the photons may be absorbed very near the surface, the heat generation distribution still spreads out due to electron–hole diffusion.

heat generation is not necessarily limited to this region. We consider the one-dimensional steady-state laser heating of a piece of semiconductor material as shown in figure 8.9. Under the assumption that absorption occurs at the surface only, the carrier density distribution n is determined by the diffusion equation

$$a\frac{d^2n}{dx^2} - \frac{n - n_0}{\tau} = 0 \tag{8.126}$$

where the diffusivity and the recombination time are the average values for electrons and holes. Equation (8.126) is subject to the following boundary conditions:

$$x = 0, \quad -a\frac{dn}{dx} + V_s(n - n_0) = \frac{I(1-R)}{h\nu} \tag{8.127}$$

$$x \to \infty, \quad n \to n_0 \tag{8.128}$$

where h is the Planck constant, I the radiation intensity, R the surface reflectance, V_s the surface recombination velocity, and ν the photon frequency. The boundary condition at $x = 0$ implies that the carriers generated at the surface due to absorption either recombine at the surface or diffuse away from the surface. The solution of eq. (8.126) is

$$n(x) - n_0 = \frac{I(1-R)}{h\nu(V_s + a/L)} e^{-x/L} \tag{8.129}$$

where $L = (a\tau)^{1/2}$ is the recombination length, the average distance the electrons and holes diffuse before they recombine. On the further assumption that the recombination is all nonradiative, the heat generation distribution due to electron–hole recombination can be calculated as

$$\dot{q}(x) = -\frac{d}{dx}[(\Pi_e - \psi_e/e)J_e + (\Pi_h - \psi_h/e)J_h]$$

$$= \frac{d}{dx}\left\{\left[E_g + \left(\gamma_e + \frac{5}{2}\right)\kappa_B T + \left(\gamma_h + \frac{5}{2}\right)\kappa_B T\right]J_e\right\}$$

$$= \left[E_g + \left(\gamma_e + \frac{5}{2}\right)\kappa_B T + \left(\gamma_h + \frac{5}{2}\right)\kappa_B T\right]\frac{I(1-R)}{h\nu\tau(V_s + a/L)}e^{-x/L} \tag{8.130}$$

where we have have calclulated J_e from the concentration distribution and used $J_h = -J_e$. The contribution of thermoelectric energy is often neglected in simulation, which can be justified if the band gap is larger than the average thermal energy. The surface heat flux is made of the following two parts,

$$q'' = \left\{ h\nu - \left[E_g + \left(\gamma_e + \frac{5}{2}\right)\kappa_B T + \left(\gamma_h + \frac{5}{2}\right)\kappa_B T \right] \right\} \frac{I(1-R)}{h\nu}$$

$$+ V_s \left[E_g + \left(\gamma_e + \frac{5}{2}\right)\kappa_B T + \left(\gamma_h + \frac{5}{2}\right)\kappa_B T \right] \frac{I(1-R)}{h\nu (V_s + a/L)} \quad (8.131)$$

The first term on the right-hand side of eq. (8.131) is the photon energy above the bandgap. The relaxation of the excess carrier energy above the bandgap is a fast electron–phonon scattering process similar to that in metals. The carrier diffusion during this period is negligible except in extremely thin films. Heat generated due to the conversion of this excess part of the carrier energy can be treated as surface heating. We should point out, however, that the use of $(\gamma_e + 5/2)\kappa_B T$ and $(\gamma_h + 5/2)\kappa_B T$ in this term is somewhat arbitrary since the thermal energy is $3/2\,\kappa_B T$. This choice is justified on the basis of energy balance considerations. The second term in eq. (8.131) is due to nonradiative surface recombination. Equation (8.130) shows that, even for the steady-state surface-absorption process the heat source spreads to a distance on the order of L. Using typical values of $a = 12.3$ cm^2 s^{-1} and $\tau = 50 \times 10^{-6}$ s in silicon yields $L = 248$ μm. This heat-spreading length is much larger than the typical film thickness used in electronic and photonic devices. It demonstrates that a surface absorption can generate a volumetric heat source distribution.

8.4 Nanostructures for Energy Conversion

Most, if not all, energy conversion devices rely on some transport processes to move energy from one place to another. Transport processes often incur large entropy generation and degrade the efficiency of energy conversion. Some of the size effects in nanostructures, both quantum and classical, can be utilized for improving the efficiency of energy conversion devices. In this section, we will examine two examples based on solid-state energy conversion. One is thermoelectric refrigeration and power generation, and the other is the thermophotovoltaic power generation. Because the basic transport phenomena have already been discussed in chapters 5–7, our discussion in this section will emphasize energy conversion devices and how nanoscale size effects can be exploited for improving the device efficiency or the power density.

8.4.1 Thermoelectric Devices

In section 6.3.4, we discussed three thermoelectric effects: the Seebeck effect, the Peltier effect, and the Thomson effect. The Peltier effect is exploited for thermoelectric refrigerators and the Seebeck effect is the basis of thermoelectric power generators (Ioffe, 1957; Goldsmid, 1964). Thermoelectric devices are typically made of multiple p-type and n-type semiconductor elements and connected such that the current flow

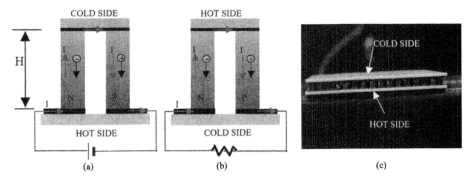

Figure 8.10 Thermoelectric devices: (a) cooling mode; (b) power generation mode; (c) an actual device made of many pairs of p-n legs electrically in series and thermally in parallel.

is in series while the heat flow is in parallel, as shown in figure 8.10. The reason for the use of both p- and n-type elements (legs) is because the Seebeck and Peltier coefficients in these two types of materials are usually of opposite sign, so that both types of element contribute to the desired thermoelectric effect. The reason that the legs are electrically in series is due to the small electrical resistance of each element. Putting many legs in series increases the total resistance, which simplifies the power source [figure 8.10(c)].

8.4.1.1 Thermoelectric Refrigeration

We consider a pair of thermoelectric legs used for cooling, as shown in figure 8.10(a). Current is passed from the n-type leg to the p-type leg so that both electrons in the n-type leg and holes in the p-type leg move away from the cold junction, thus carrying thermal energy out of the cold junction. The Peltier cooling is proportional to $(\Pi_2 - \Pi_1)I_e$, where I_e is the current. In addition to the heat taken out by the current flow, there is also reverse heat flow from the hot side to the cold side due to heat conduction by the solid. Some of the Joule heat generated inside the thermoelectric element is also conducted back to the cold junction. The net cooling power of the couple is thus

$$Q_C = (S_h - S_n)I_e T_C - K(T_H - T_C) - \frac{1}{2}I_e^2 R_e \quad (8.132)$$

where S is the Seebeck coefficient [eq. (6.93)], TS equals the Peltier coefficient [eq. (6.104)], and

$$K = \frac{k_h A_h}{H_h} + \frac{k_e A_e}{H_e} \text{ and } R_e = \frac{H_h}{\sigma_h A_h} + \frac{H_e}{\sigma_e A_e} \quad (8.133)$$

are the thermal conductance and electrical resistance of the two branches, respectively. Here H and A are the height and cross-sectional area of the thermoelectric elements, and ρ the electrical resistivity. The one-half factor in eq. (8.132) appears because half of the Joule heat conducts back to the cold side. The electrical power W_e consumed by the pair of legs is

$$W_e = (S_h - S_e)I_e(T_H - T_C) + I_e^2 R_e \quad (8.134)$$

where the first term on the right-hand side accounts for the additional reverse voltage developed over the device due to the Seebeck effect and the second term is power dissipation due to Joule heating. The coefficient of performance used to describe the efficiency of refrigeration is

$$\phi = \frac{Q_C}{W_e} = \frac{(S_h - S_e)I_e T_C - K(T_H - T_C) - \frac{1}{2}I_e^2 R_e}{(S_h - S_e)I_e(T_H - T_C) + I_e^2 R_e} \quad (8.135)$$

At a given temperature difference, the coefficient of performance is dependent on the current I_e. For the refrigeration application, two special cases are of interest. One is the maximum cooling power. The other is the maximum coefficient of performance at a given temperature difference. For the first case, the current I_{eq} can be determined by solving $dQ_C/dI_e = 0$, which gives

$$I_{eq} = \frac{(S_h - S_e)T_C}{R_e} \quad (8.136)$$

$$\phi_q = \frac{0.5 Z T_C^2 - (T_H - T_C)}{Z T_H T_C} \quad (8.137)$$

where $Z = (S_h - S_e)^2/(KR_e)$ is called the figure of merit of the device. The product KR_e is a minimum when $(H_e A_h)/(H_h A_e) = [(\rho_h k_e)/(\rho_e k_h)]^{1/2}$, which gives the maximum Z as

$$Z = \frac{(S_h - S_e)^2}{[(k_h \rho_h)^{1/2} + (k_e \rho_e)^{1/2}]^2} \quad (8.138)$$

The device figure of merit thus depends on the properties of both legs. The figure of merit of individual materials

$$Z = S^2 \sigma / k \quad (8.139)$$

is nevertheless a useful measure of the material's potential for making a useful device. Because Z has units of inverse Kelvin, the nondimensional figure of merit ZT is often used. To increase the coefficient of performance, one needs to identify materials with a large Z or ZT.

The second case is to maximize the coefficient of performance by setting $d\phi/dI_e = 0$, which leads to

$$I_\phi = \frac{(S_h - S_e)(T_H - T_C)}{R_e(\sqrt{1 + ZT_M} - 1)} \quad (8.140)$$

$$\phi_{max} = \frac{T_C(\sqrt{1 + ZT_M} - T_H/T_C)}{(T_H - T_C)(\sqrt{1 + ZT_M} + 1)} \quad (8.141)$$

where $T_M = (T_H + T_C)/2$ is the average temperature of the hot and cold junctions. It is clear that as Z goes to infinity, ϕ_{max} approaches the coefficient of performance of a Carnot cycle.

Another parameter that is useful for refrigeration performance is the maximum temperature difference that a given system can ideally reach. The maximum temperature difference is reached when there is no net cooling power taken from the source. Therefore, the coefficient of performance is zero at the maximum temperature difference. The corresponding maximum temperature difference can be obtained from eq. (8.132) as

$$(T_H - T_C)_{max} = \frac{1}{2}ZT_C^2 \tag{8.142}$$

8.4.1.2 Thermoelectric Power Generation

The thermoelectric power generation mode can be analyzed similarly. In a power generation mode, as shown in figure 8.10(b), the heat supplied to the hot side should be

$$Q_H = (S_h - S_e)I_e T_H + K(T_H - T_C) - \frac{1}{2}I_e^2 R_e \tag{8.143}$$

and the power output is

$$W_e = I_e^2 R_L \tag{8.144}$$

where R_L is the external load resistance of the output circuit. The output current is given by

$$I_e = \frac{(S_h - S_e)(T_H - T_C)}{R_e + R_L} \tag{8.145}$$

Therefore, the thermal efficiency is

$$\eta = \frac{W_e}{Q_H} = \frac{I_e^2 R_L}{(S_h - S_e)I_e T_H + K(T_H - T_C) - \frac{1}{2}I_e^2 R_e} \tag{8.146}$$

It can be shown that the maximum power is obtained with a matched load $R_L = R_e$. The maximum efficiency, which does not have to occur at the maximum power, is determined by setting $d\eta/dR_L = 0$,

$$\eta_{max} = \frac{(T_H - T_C)(\sqrt{1 + ZT_M} + 1)}{T_H(\sqrt{1 + ZT_M} + T_c/T_H)} \tag{8.147}$$

Equations (8.141) and (8.147) show that one could make thermoelectric refrigerators and power generators close to the Carnot efficiency if materials with large ZT could be identified. The search for high ZT materials, however, has proven to be a very difficult path. The best ZT materials are found in heavily doped semiconductors. Insulators have poor electrical conductivities. Metals have relatively low Seebeck coefficients. In addition, the thermal conductivity of a metal, which is dominated by electrons, is proportional to the electrical conductivity, as dictated by the Wiedmann–Franz law [eq. (6.88)]. It is thus hard to realize high ZT in metals. In semiconductors, the thermal conductivity consists of contributions from electrons (k_e) and phonons (k_p), with the majority coming from phonons. The phonon thermal conductivity can be reduced without causing too

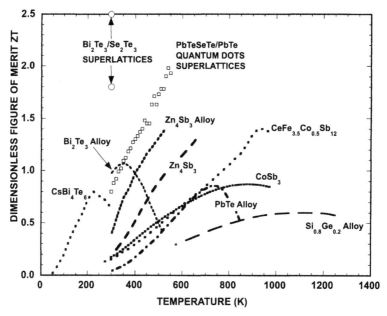

Figure 8.11 State of the art of thermoelectric materials. Lines are for bulk materials and dots are for nanostructured materials. A room-temperature ZT of 1.3 has recently been reported on PbTe-based superlattices by Harman et al. (2002) (from Chen et al., 2003).

much reduction in the electrical conductivity. A proven approach to reduce the phonon thermal conductivity is through alloying (Ioffe, 1957). The mass difference scattering in an alloy reduces the lattice thermal conductivity significantly without much degradation of the electrical conductivity. The traditional cooling materials are alloys of Bi_2Te_3 with Sb_2Te_3 (such as $Bi_{0.5}Sb_{1.5}Te_3$, p-type) and Bi_2Te_3 with Bi_2Se_3 (such as $Bi_2Te_{2.7}Se_{0.3}$, n-type), with a ZT at room temperature approximately equal to one (Goldsmid, 1964). A typical power generation material is an alloy of silicon and germanium, with a ZT of ~0.6 at 700 °C.

8.4.1.3 Benefits of Nanostructures

Before the 1990s, the search for high ZT materials was mostly limited to bulk materials. The possibilities of using nanostructures for improving ZT started to attract attention in the 1990s (Hicks and Dresselhaus, 1993). Compared to the research in bulk materials that emphasizes reduction in thermal conductivity, nanostructures offer a chance of improving both the electron and phonon transport through the use of quantum and classical size and interface effects. Several directions have been explored, such as quantum size effects for electrons (Hicks and Dresselhaus, 1993; Dresselhaus et al., 2001) (see exercises 6.11 and 6.12) and interface scattering of phonons (Chen et al., 2003; Venkatasubramanian, et al. 2001) (see chapter 7). Interface scattering and filtering of electrons and nonequilibrium between electrons and phonons have both been explored (Shakouri and Bowers, 1997; Moyzhes and Nemchinsky, 1998; Zeng and Chen 2002).

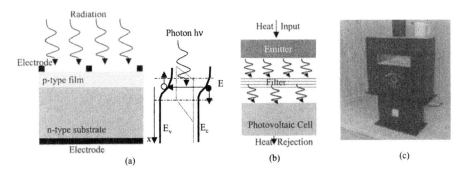

Figure 8.12 (a) The principle of a photovoltaic cell; (b) principle of thermophotovoltaic power generation; (c) a prototype thermophotovoltaic converter powered by a stove (Fraas et al., 2003; courtesy of Dr. L.M. Fraas).

Impressive ZT values have been reported in some superlattice structures based on both an enhancement of the electron performance (Harman et al., 2002) and a reduction in thermal conductivity (Harman et al., 2002; Venkatasubramanian, et al., 2001), with biggest benefit coming from the thermal conductivity reduction. Figure 8.11 shows a snapshot of the ZT of state-of-the art materials (Chen et al., 2003). The ZT of many nanostructures has surpassed that of bulk materials. However, these nanostructured materials are difficult to synthesize and useful devices have yet to be made from them. More information on nanostructured thermoelectric materials can be found in several reviews (Tritt, 2001; Chen and Shakouri, 2002; Chen et al., 2003).

8.4.2 Solar Cells and Thermophotovoltaic Power Conversion

8.4.2.1 Basic Principles

Solar cells (and, more generally, photovoltaic cells) absorb the photon energy from the sun and convert it into electricity (Chapin et al., 1954; Sze, 1981). The working principle of a solar cell comprising a p–n junction is sketched in figure 8.12(a). Photons from the sun generate electron–hole pairs in the space charge region. The electrostatic field in this region pulls holes to the p-type region and electrons to the n-type region. The accumulation of electrons and holes in these two regions generates a voltage. Thermophotovoltaic power generators are similar to photovoltaic cells but use a heat source to generate photons rather than solar energy (Coutts, 1999). A regular thermophotovoltaic device usually consists of the following parts: a heat source, an emitter to emit photons, a filter to reflect unwanted photons, photovoltaic cells to generate electricity, and a thermal management system to keep the photovoltaic cell cool, as shown in figure 8.12(b). Figure 8.12(c) is an example of a commercial thermophotovoltaic power generator powered by a furnace.

The p–n junction current as given by eq. (8.111) is balanced by the electron–hole generation rate

$$J_e = -J_s(e^{eV/k_BT} - 1) + J_G \tag{8.148}$$

where J_s is also called the dark current for diodes operated as photon detectors, and J_G is the current source due to photon absorption, which we will discuss later.

Under an open-circuit condition, eq. (8.148) gives the open-circuit voltage of the photovoltaic cell as

$$V_0 = \frac{\kappa_B T}{e} \ln\left(\frac{J_G}{J_s} + 1\right) \qquad (8.149)$$

As eq. (8.112) shows, the dark current is dependent on the bandgap,

$$J_s = A \exp\left(-\frac{E_G}{\kappa_B T}\right) \qquad (8.150)$$

where the coefficient A can be derived from eq. (8.112) for the Shockley ideal diode model. In Shockley's model, the dark current is due to nonradiative recombination outside the space charge region. A more fundamental limit is the radiative recombination (Shockley and Queisser, 1961; Henry, 1980) that must exist on the basis of the Kirchoff law in radiation; that is, the absorption must balance the emission for a system in equilibrium. This fundamental limit leads to the following expression for A (Henry, 1980),

$$A \approx \frac{e(n^2 + 1)E_g^2 \kappa_B T}{4\pi^2 \hbar^3 c^2} \qquad (8.151)$$

where n is the refractive index of the photovoltaic cell and c is the speed of light in vacuum. This A is typically much smaller than that due to nonradiative recombination as given by eq. (8.112). The dark current caused by radiative recombination is often used to estimate the maximum efficiency of a solar cell.

Substituting eq. (8.150) into (8.149), the open-circuit voltage ($J_e = 0$) can be expressed as

$$V_0 \approx \frac{E_G}{e} - \frac{\kappa_B T}{e} \ln\left(\frac{A}{J_G}\right) \qquad (8.152)$$

which implies that the output voltage depends on the bandgap. The power output per unit area from a solar cell is

$$W_e = J_e V = -J_s V(e^{eV/k_B T} - 1) + J_G V \qquad (8.153)$$

and the maximum power output can be obtained from $dW_e/dV = 0$. This mathematical operation leads to the following expression for the optimum current and voltage,

$$J_{opt} = J_s \frac{eV_{opt}}{\kappa_B T} \exp\left[\frac{eV_{opt}}{\kappa_B T}\right] \approx J_G \left[1 - \frac{eV_{opt}}{\kappa_B T}\right] \qquad (8.154)$$

$$V_{opt} = \frac{\kappa_B T}{e} \ln\left(\frac{J_G/J_s + 1}{1 + eV_{opt}/(\kappa_B T)}\right) \approx V_0 - \frac{\kappa_B T}{e} \ln\left(1 + \frac{eV_{opt}}{\kappa_B T}\right) \qquad (8.155)$$

We now determine J_G, the current source due to photon excitation of electrons and holes. The radiation source is at temperature T_s. The quantity of photons entering the photovoltaic cell depends on the emissivity of the emitter ε_ω, the transmissivity of the medium between the emitter and the photovoltaic cell τ_ω, the reflectivity of the

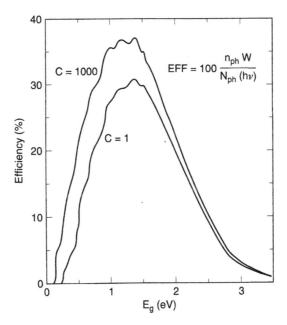

Figure 8.13 Maximum efficiency of a solar cell as a function of the bandgap at 1 and 1000 times (as represented by C) solar radiation intensity (Henry, 1980; courtesy of the American Institute of Physics).

photovoltaic cell itself R_ω, and the radiation view factor. We further assume that each photon entering the photovoltaic cell generates one electron–hole pair and neglect the multiple reflection effect. The photon-generated electron–hole current density is then

$$J_G = 2e \int_{E_G/\hbar}^{\infty} \varepsilon_\omega \tau_\omega (1 - R_\omega) \frac{I_{b\omega}(T,\omega)}{\hbar \omega} d\omega \tag{8.156}$$

where $I_{b\omega}/(\hbar\omega)$ is the photon flux. The efficiency of a solar cell is defined as the amount of electrical power generated, divided by the incident solar power,

$$\eta = \frac{J_e V}{I_s} \tag{8.157}$$

The maximum efficiency of a solar cell as a function of the bandgap is shown in figure 8.13 (Henry, 1980). For solar radiation, an optimum bandgap exists. If the bandgap is too large, there are few photons that can be used for electron–hole pair generation. If the bandgap is too small, the voltage output of the photovoltaic cell is too small.

For a thermophotovoltaic power generator, the efficiency is defined on the basis of the net power leaving the radiator,

$$\eta = \frac{J_{opt} V_{opt}}{\int_0^\infty \varepsilon_\omega I_{b\omega} d\omega} \tag{8.158}$$

To maximize the efficiency, the ideal emissivity of the emitter should be unity above the bandgap and zero below the bandgap. The maximum efficiency of thermophotovoltatic systems is also addressed in the literature (Baldasaro et al., 2001).

8.4.2.2 Benefits of Nanostructures

Nanostructures can benefit a thermophotovoltaic power generation system in different ways. A conventional photovoltaic cell used in thermophotovoltaic systems usually has multilayer thin film coatings as filters to selectively transmit only photons above the bandgap and reflect photons below the bandgap, using interference effects (section 5.3.1) (Demichelis et al., 1982). Selective emitters based on photonic crystals (section 3.5.2) are also being considered to control the emission spectrum of the emitter (Fleming et al., 2002; Narayanaswamy and Chen, 2004). These spectral control technologies are beneficial for improving the efficiency of thermophotovoltaic systems. Increased radiation transfer across small gaps due to tunneling (section 5.4.2) can be used to increase the power density of thermophotovoltaic systems (Pan et al., 2000; DiMatteo et al., 2001; Whale and Cravalho, 2002). The tunneling of bulk evanescent waves, however, is broadband and favors long-wavelength photons below the bandgap. More recently, surface waves, such as surface plasmons and surface photon-polaritons, have drawn significant attention (Greffet et al., 2002). These surface waves exist only in a narrow frequency range in which the dielectric constant of the material is negative, and they decay exponentially on both sides of an interface. The wave can tunnel into a third material, just like the tunneling of evanescent wave resulting from total internal reflection. Figure 8.14 shows an example of radiative heat transfer between two parallel plates, one of them made of boron nitride that supports a surface wave (Narayanaswamy and Chen, 2003). When the separation between the two plates is large, boron nitride does not emit in the surface wave frequency range. As the distance between the two plates shrinks into the nanometer regime, surface wave tunneling occurs, leading to orders of magnitude increases in the photon flux in a narrow frequency range. These surface waves can also be coupled into the far field by gratings on the surface (Greffet et al., 2002). Because the surface plasmons and photon-polaritons have a large mean free path and thus long-range correlation, the thermal emission from these surface waves has good coherent properties.

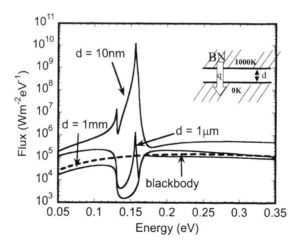

Figure 8.14 Radiation heat flux between two parallel plates at different separations. The dashed line is blackbody radiation. When the distance is large, BN does not emit in the frequency range of 0.15 eV. As the gap becomes small, surface phonon-polaritons tunnel through the gap and create an increase of orders of magnitude in the radiation heat flux (Narayanaswamy and Chen, 2003).

For solar cells, we have no control on the radiation source. Thus any improvement of solar cells mainly comes from the photovoltaic cell itself. Various quantum structures can be used to improve the photon collection and thus the cell efficiency. A key driver for solar cell research, however, is the cost. Thin-film-based quantum structures are not cost competitive. Electrolytic solar cells based on photosensitive dyes in conjunction with nanostructures have recently made impressive progress (O'Regan and Grätzel, 1991; Huynh et al., 2002). In such solar cells, photosensitive dyes are used to convert photons into electrons while inorganic nanostructures are used to collect the electrons. Efficiency higher than 10% has been reached with such photovoltaic cells (O'Regan and Grätzel, 1991).

8.5 Summary of Chapter 8

The attention of this chapter is focused on how energy is exchanged from one type of carrier to another. This chapter also serves to illustrate that energy conversion is closely coupled to energy transport.

Energy exchange between different carriers results from their collisions. The collision rates depend on the scattering matrix elements and are usually calculated on the basis of the Fermi golden rule. During collision, momentum and energy conservation rules are obeyed. In section 8.1.1, we treat electron–phonon scattering in detail and show that the collision integral in the Boltzmann equation cannot in general be simply approximated by the relaxation time approximation, which represents only the elastic scattering process. The energy exchange between electrons and phonons is proportional to the temperature difference between them. The proportionality constant is called the electron–phonon coupling factor. The temperature differences between electrons and phonons are usually small. However, there are also many cases in which electrons and phonons are at highly nonequilibrium states, such as during short pulse laser heating of metals and semiconductors (section 8.2.1) and electron transport under a high field (section 8.2.2). Photon absorption and the subsequent relaxation of excited carriers are treated in sections 8.1.2 and 8.1.3. These processes can also be evaluated in principle on the basis of the scattering matrix, details of which are not given here. Emphasis has been placed on pointing out that not necessarily all radiation absorption is heat. Heat generation occurs only if the excited carriers relax their excess energy into the random lattice vibration. Electron–hole recombination can lead to light emission instead of heat generation. In thermal modeling, it is thus important to distinguish where and how much heat is generated.

From the treatment of the scattering processes, it becomes clear that the collision integral usually includes more than just the relaxation time term. The previous two chapters (6 and 7) are based exclusively on the relaxation time approximation of the Boltzmann equation. The new additional source terms in the Boltzmann equation and their impacts on heat transfer and energy conversion in nanostructures await further exploration.

Section 8.2 emphasizes the coupling of energy conversion with the transport processes. With the appropriate form of the collision integral, the coupled energy conversion and transport processes can in principle be treated on the basis of the Boltzmann equation. This section focuses on a single type of charged carrier so that recombination can be neglected, and emphasizes the nonequilibrium between electrons and phonons.

The equations derived, however, do not include ballistic transport and their validities for nanostructures should be further examined. Several application examples are given in this chapter, including short pulse laser interaction with metals, hot electron effects in MOSFET devices, and hot/cold electrons and phonons in thermoelectric devices.

In semiconductors, with both electrons and holes used as energy carriers, recombination processes must be dealt with carefully. In section 8.3, a general formulation of the energy source term is given, again based on the diffusion approximation that signals room for improvement with regard to applications to nanostructures. The general formulation is applied to energy conversion inside a semiconductor p–n junction, and shows that heat generation does not occur inside the space charge region where the electrostatic field changes significantly. Heat is actually extracted out of the space charge region and deposited in its near surroundings through electron–hole recombination. In a different example, it is shown that the absorption of photons from a laser beam at the surface actually creates a heat source, distributed beneath the surface, due to electron–hole diffusion and their recombination.

The last section of this chapter explains several solid-state energy conversion devices, thermoelectric devices, solar cells, and thermophotovoltaic devices, with a brief discussion on how nanostructures can potentially improve the efficiency and power density of these devices. Although the contents of this section do not necessarily integrate well with the previous three sections, it is presented here in hope that interested readers will explore the connection between the microscopic pictures and the actual device operation processes. The benefits of nanostructures for these devices have been mentioned in various places in the preceding chapters. The discussion on devices brings them together.

Through this chapter, we have been constantly reminding the readers that many of the derivations are based on assumptions that may not be valid for nanostructures. It is hoped that these reminders will arouse the interest of some readers to pursue some of the topics further.

8.6 Nomenclature for Chapter 8

a	lattice constant, m	$\mathscr{E}$	magnitude of electrostatic field, V m^{-1}
A	cross-sectional area, m^2	f	distribution function
B	constant in eq. (8.44), m^3 s^{-1}	f_0	equilibrium distribution function
C	volumetric specific heat, J m^{-3} K^{-1}	F	external force, N
		g	defined by eq. (8.29), K^{-1} s^{-1}
D	density of states per unit volume, m^{-3} J^{-1}	G	electron–phonon coupling factor, W m^{-3} K^{-1}
e	unit charge, C	G	reciprocal lattice vector, m^{-1}
E	energy of one particle, J	h	Planck constant, J s
E_c	conduction band edge, J	$\hbar$	Planck constant divided by 2π, J s
E_f	chemical potential, J		
E_g	bandgap, J	H	height of thermoelectric element, m
E_v	valence band edge, J		

H'	perturbation Hamiltonian operator, J	R_s	surface recombination rate, m^{-2} s^{-1}
I	intensity, W m^{-2} srad^{-1}	s	entropy per unit volume, J K^{-1} m^{-3}
I_e	current, A	S	entropy, J K^{-1}; Seebeck coefficient, V K^{-1}
J_G	current density due to electron–hole pair generation, A m^{-2}	t	time, s
J_s	saturation current density, also called dark current in photodiode, Am^{-2}	T	temperature, K
		T_e	electron temperature, K
		T_p	phonon temperature, K
$\mathbf{J}$	current flux, A m^{-2}	$\mathbf{u}$	atom displacement, m
k	thermal conductivity, W m^{-1} K^{-1}	u	internal energy per unit volume, J m^{-3}
$\mathbf{k}$	wavevector, m^{-1}	U	Internal energy, J
K	thermal conductance, W K^{-1}	v_e	magnitude of the electron group velocity, m s^{-1}
L	recombination length, m		
L_e	electron–phonon energy coupling length, defined in eq. (8.78), m	v_s	speed of sound for longitudinal phonons, m s^{-1}
m	electron effective mass, kg	V	crystal volume, m^3; voltage drop, V
M	scattering matrix, J		
n	electron number density, m^{-3}	V_{bi}	built-in potential, V
N_A	p-type dopant (acceptor) concentration, m^{-3}	V_s	surface recombination velocity, m s^{-1}
		w	width of p–n junction, m
N_D	n-type dopant (donor) concentration, m^{-3}	W_e	electrical power input, W
		W_{ij}	transition rate from quantum state i to quantum state j, s^{-1}
p	hole concentration, m^{-3}		
P	pressure, N m^{-2}	x	Cartesian coordinate
P_f	probability that the final quantum state can accept carrier	Z_A	deformation potential introduced in eq. (8.3), eV
		α	absorption coefficient, m^{-1}
q	magnitude of phonon wavevector, m^{-1}	β	quantity defined by eq. (8.71), m^{-1}
$\dot{q}$	energy source term, W m^{-3}		
$\mathbf{q}$	phonon wavevector, m^{-1}	γ	parameter in the energy dependence of the relaxation time
$\mathbf{Q}_e$	electron heat flux, W m^{-2}		
$\mathbf{Q}_p$	phonon heat flux, W m^{-2}		
R	reflectivity	δ	delta function
R_e	electrical resistance, Ω	ε_e	electrical permittivity, C^2 N^{-1} m^{-2}
R_g	generation rate, m^{-3} s^{-1}		
R_L	external load resistance, Ω	ε_ω	spectral-dependent emissivity
R_n	net recombination rate, m^{-3} s^{-1}	η	quantum efficiency; power generation efficiency
R_r	recombination rate, m^{-3}s^{-1}	θ	polar angle

κ	imaginary part of complex refractive index	\multicolumn{2}{l}{**Subscripts**}

κ imaginary part of complex refractive index
κ_B Boltzmann constant, J K^{-1}
λ wavelength in vacuum, m
μ_e electron mobility, m^2 V^{-1} s^{-1}
ν frequency, Hz
Π Peltier coefficient, V
ρ density, kg m^{-3}
σ electrical conductivity, Ω^{-1} m^{-1}
τ relaxation time, s
τ_n lifetime, s
ϕ coefficient of performance
φ azimuthal angle
φ_e electrostatic potential, V
Φ electrochemical potential, V
ψ quasi-Fermi level, J
Ψ wave function

Subscripts

0   equilibrium
ω angular frequency, Hz.rad
Ω solid angle, srad
C   cold side
e   electron
f   final state
h   hole
H   hot side
i   initial state
opt   optimal
p   phonon
l   at state *l*

Superscripts

k   particle with wavevector **k**
q   particles with wavevector **q**
*   complex conjugate

8.7 References

Allen, P.B., 1987, "Theory of Thermal Relaxation of Electrons in Metals," *Physical Review Letters*, vol. 59, pp. 1460–1463.

Anatychuk, L.I., and Bulat, L.P., 1998, "New Nonlinear Thermoelectric, Heat-Conducting and Thermomagnetic Effects and their Classification," *Journal of Thermoelectricity*, vol. 7, pp. 41–55.

Baldasaro, P.F., Raynolds, J.E., Charache, G.W., DePoy, D.M., Ballinger, C.T., Donovan, T., and Borrego, J.M., 2001, "Thermodynamic Analysis of Thermophotovoltaic Efficiency and Power Density Tradeoffs," *Journal of Applied Physics*, vol. 89, pp. 3319–3327.

Bardeen, J., and Shockley, W., 1950, "Deformation Potentials and Mobilities in Nonpolar Crystals," *Physical Review*, vol. 80, pp. 72–80.

Brorson, S.D., Fujimoto, J.G., and Ippen, E.P., 1987, "Femtosecond Electronic Heat-Transport Dynamics in Thin Gold-Films," *Physical Review Letters*, vol. 59, pp. 1962–1965.

Callen, H.B., 1985, *Thermodynamics and an Introduction to Thermostatistics*, 2nd ed., Wiley, New York.

Chapin, D.M., Fuller, C.S., and Pearson, G.L., 1954, "A New Silicon p–n Junction Photocell for Converting Solar Radiation into Electrical Power," *Journal of Applied Physics*, vol. 25, pp. 676–677.

Chen, G., 2003, "Diffusion-Transmission Condition for Transport at Interfaces and Boundaries," *Applied Physics Letters*, vol. 82, pp. 991–993.

Chen, G., and Shakouri, A., 2002, "Heat Transfer in Nanostructures for Solid-State Energy Conversion," *Journal of Heat Transfer*, vol. 124, pp. 242–252.

Chen, G., Dresselhaus, M.S., Dresselhaus, G., Fleurial, J.P., and Caillat, T., 2003, "Recent Developments in Thermoelectric Materials," *International Materials Review*, vol. 48, pp. 45–66.

Conwell, E.M., 1967, "High Field Transport in Semiconductors," in *Solid-State Physics*, ed. F. Seitz, D. Turnbull, and H. Ehrenreich, Supplement 9, Academic Press, New York, pp. 119–127.

Coutts, T.J., 1999, "A Review of Progress in Thermophotovoltaic Generation of Electricity," *Renewable and Sustainable Energy Reviews*, vol. 3, pp. 77–184.

Demichelis, F., Minetti-Mezzetti, E., Agnello, M., and Perotto, V., 1982, "Bandpass Filters for Thermophotovoltaic Conversion Systems," *Solar Cells*, vol. 5, pp. 135–141.

DiMatteo, R.S., Greiff, P., Finberg, S.L., Young-Waithe, K., Choy, H.K.H., Masaki, M.M., and Fonstad, C.G., 2001, "Enhanced Photogeneration of Carriers in a Semiconductor via Coupling across a Nonisothermal Nanoscale Vacuum Gap," *Applied Physics Letters*, vol. 79, pp. 1894–1896.

Dresselhaus, M.S., Lin, Y.M., Cronin, S.B., Rabin, O., Black, M.R., Dresselhaus, G., and Koga, T., 2001, "Quantum Wells and Quantum Wires for Potential Thermoelectric Applications," *Semiconductors and Semimetals*, vol. 71, pp. 1–121.

Fleming, J.G., Lin, S.Y., El-Kady, I., Biswas, R., and Ho, K.M., 2002, "All-Metallic Three-Dimensional Photonic Crystals with a Large Infrared Gap," *Nature*, vol. 417, pp. 52–55.

Fraas, L.M., Avery, J.E., Huang, H.X., and Martinelli, R.U., 2003, "Thermophotovoltaic System Configurations and Spectral Control," *Semiconductor Science and Technology*, vol. 18, pp. S165–S173.

Goldsmid, H.J., 1964, *Thermoelectric Refrigeration*, Plenum Press, New York.

Greffet, J.J., Carminati, R., Joulain, K., Mulet, J.P., Mainguy, S., and Chen, Y., 2002, "Coherent Emission of Light by Thermal Sources," *Nature*, vol. 416, pp. 61–64.

Harman, T.C., Taylor, P.J., Walsh, M.P., and LaForge, B.E., 2002, "Quantum Dot Superlattice Thermoelectric Materials and Devices," *Science*, vol. 297, pp. 2229–2232.

Harrison, W.A., 1979, *Solid State Theory*, Dover, New York.

Henry, C.H., 1980, "Limiting Efficiencies of Ideal Single and Multiple Energy Gap Terrestrial Solar Cells," *Journal of Applied Physics*, vol. 51, pp. 4494–4450.

Hess, C., 2000, *Advanced Theory of Semiconductor Devices*, IEEE Press, New York, chapter 8.

Hicks, L.D., and Dresselhaus, M.S., 1993, "Effect of Quantum-Well Structures on the Thermoelectric Figure of Merit," *Physical Review B*, vol. 47, pp. 12727–12731.

Huynh, W.U., Dittmer, J.J., and Alivisatos, A.P., 2002, "Hybrid Nanorod–Polymer Solar Cells," *Science*, vol. 295, pp. 2425–2427.

Ioffe, A.F., 1957, *Semiconductor Thermoelements and Thermoelectric Cooling*, Infosearch, London.

Ju, Y.S., and Goodson, K.E., 1999, "Phonon Scattering in Silicon Films of Thickness below 100 nm," *Applied Physics Letters*, vol. 74, pp. 3005–3007.

Kaganov, M.I., Lifshitz, I.M., and Tanatarov, L.V., 1957, "Relaxation between Electrons and Crystalline Lattices," *Soviet Physics JETP*, vol. 4, pp. 173–178.

Lai, J., and Majumdar, A., 1996, "Concurrent Thermal and Electrical Modeling of Sub-Micrometer Silicon Devices," *Journal of Applied Physics*, vol. 79, pp. 7353–7361.

Lindefelt, U., 1994, "Heat Generation in Semiconductor Devices," *Journal of Applied Physics*, vol. 75, pp. 942–957.

Lundstrom, M., 2000, *Fundamentals of Carrier Transport*, 2nd ed., Cambridge University Press, Cambridge, UK, chapter 3.

Mahan, G.D., 2001, "Thermionic Refrigeration," in *Semiconductors and Semimetals*, vol. 71, pp. 157–174.

Moyzhes, B.Y., and Nemchinsky, V., 1998, "Thermoelectric Figure of Merit of Metal-Semiconductor Barrier Structure based on Energy Relaxation Length," *Applied Physics Letters*, vol. 73, pp. 1895–1897.

Narayanaswamy, A., and Chen, G., 2003, "Surface Modes for Near-Field Thermophotovoltaics," *Applied Physics Letters*, vol. 82, pp. 3544–3546.

Narayanaswamy, A., and Chen, G., 2004, "Thermal Radiation Control with 1D Metallo-Dielectric Photonic Crystals," *Physical Review B*, vol. 70, pp. 125101/1–4.

O'Regan, B., and Grätzel, M.A., 1991, "A Low-Cost, High-Efficiency Solar Cell Based on Dye-Sensitized TiO$_2$ Films," *Nature*, vol. 335, pp. 737–739.

Palik, E.D., 1985, 1991, *Handbook of Optical Constants of Solids*, Academic Press, Orlando, vol. 1 (1985), vol. 2 (1991).

Pan, J.L., Choy, H.K.H., and Fonstad, C.G., 2000, "Very Large Radiative Transfer over Small Distances from a Black Body for Thermophotovoltaic Applications," *IEEE Transactions on Electron Devices*, vol. 47, pp. 241–249.

Pankove, J.I., 1971, *Optical Processes in Semiconductors*, Dover, New York, pp. 76–80.

Pipe, K.P., Ram, R.J., and Shakouri, A., 2002, "Bias-Dependent Peltier Coefficient and Internal Cooling in Bipolar Devices," *Physical Review B*, vol. 66, pp. 1–11.

Qiu, T.Q., and Tien, C.L., 1992, "Short Pulse Laser Heating on Metals," *International Journal of Heat and Mass Transfer*, vol. 35, pp. 719–726.

Qiu, T.Q., and Tien, C.L., 1993, "Heat Transfer Mechanisms during Short-Pulse Laser Heating of Metals," *Journal of Heat Transfer*, vol. 115, pp. 835–841.

Shakouri, A., and Bowers, J.E., 1997, "Heterostructure Integrated Thermionic Coolers," *Applied Physics Letters*, vol. 71, pp. 1234–1236.

Shakouri, A., Lee, E.Y., Smith, D.L., Narayanamurti, V., and Bowers, J.E., 1998, "Thermoelectric Effects in Submicron Heterostructure Barriers," *Microscale Thermophysical Engineering*, vol. 2, pp. 37–42.

Shockley, W., 1949, "The Theory of p–n Junctions in Semiconductors and p–n Junction Transistors," *Bell System Technical Journal*, vol. 28, pp. 435–489.

Shockley, W., 1951, "Hot Electrons in Germanium and Ohm's Law," *Bell System Technical Journal*, vol. 30, pp. 990–1034.

Shockley, W., and Queisser, H.J., 1961, "Detailed Balance Limit of Efficiency of p–n Junction Solar Cells," *Journal of Applied Physics*, vol. 32, pp. 510–519.

Srivastava, G.P., 1990, *The Physics of Phonons*, Adam Hilger, Bristol, pp. 200–226.

Stettler, M.A., and Lundstrom, M.S., 1994, "A Detailed Investigation of Heterojunction Transport Using a Rigorous Solution to the Boltzmann Equation," *IEEE Transactions on Electron Devices*, vol. 41, pp. 592–600.

Stratton, R., 1962, "Diffusion of Hot and Cold Electrons in Semiconductor Barriers," *Physical Review*, vol. 126, pp. 2002–2014.

Sverdrup, P.G., Ju, Y.S., and Goodson, K.E., 2001a, "Sub-Continuum Simulations of Heat Conduction in Silicon-on-Insulator Transistors," *Journal of Heat Transfer*, vol. 123, pp. 130–137.

Sverdrup, P. G., Sinha, S., Uma, S., Asheghi, M., and Goodson, K. E., 2001b, "Measurement of Ballistic Phonon Conduction Near Hotspots in Silicon," *Applied Physics Letters*, vol. 78, pp. 3331–3333.

Sze, S.M., 1981, *Physics of Semiconductor Devices*, 2nd ed., Wiley, New York.

Tritt, T.M., ed., 2001, Recent Trends in Thermoelectric Materials Research, *Semiconductors and Semimetals*, vols. 69–71, Academic Press, San Diego.

Venkatasubramanian, R., 2001, "Thin-film Thermoelectric Devices with High Room-Temperature Figures of Merit," *Nature*, vol. 413, pp. 597–602.

Wachutka, G.K., 1990, "Rigorous Thermodynamic Treatment of Heat Generation and Conduction in Semiconductor Device Modeling," *IEEE Transactions on Computer-Aided Design*, vol. 9, pp. 1141–1149.

Wang, S., 1989, *Fundamentals of Semiconductor Theory and Device Physics*, Prentice Hall, Englewood Cliffs, NJ, p. 467.

Whale, M.D., and Cravalho, E.G., 2002, "Modeling and Performance of Microscale Thermophotovoltaic Energy Conversion Devices," *IEEE Transactions on Energy Conversion*, vol. 17, pp. 130–142.

Wilson, A.H., 1953, *Theory of Metals*, Cambridge University Press, Cambridge, UK.

Zakordonets, V.S., and Logvinov, G.N., 1997, "Thermoelectric Figure of Merit of Monopolar Semiconductors with Finite Dimensions," *Semiconductors*, vol. 31, pp. 265–267.

Zeng, T., and Chen, G., 2000, "Energy Conversion in Heterostructures for Thermionic Cooling," *Microscale Thermophysical Engineering*, vol. 4, pp. 39–50.

Zeng, T., and Chen, G., 2002, "Interplay between Thermoelectric and Thermionic Effects in Heterostructures," *Journal of Applied Physics*, vol. 92, pp. 3152–3161.

Ziman, J.M., 1960, *Electrons and Phonons*, Clarendon Press, Oxford, chapter 5.

8.8 Exercises

8.1 *Electron mobility in semiconductors* (Bardeen and Shockley, 1950). On the basis of eqs. (8.26) and (6.85),
   (a) derive an expression for the electron mobility of a nondegenerate semiconductor due to acoustic phonon scattering;
   (b) show that the mobility depends on temperature through $T^{-3/2}$;
   (c) estimate the average momentum relaxation time in silicon due to acoustic phonon scattering.

8.2 *Electron–phonon coupling factor of semiconductors*. Estimate the value of the electron–phonon coupling factor in silicon (due to acoustic phonon scattering) as a function of the electron temperature for a carrier concentration of 10^{17} cm^{-3}. Take a deformation potential value of 5 eV.

8.3 *Relationship between mobility and electron–phonon coupling factor*. Derive a relationship between the electron mobility and the electron–phonon coupling factor for semiconductors due to acoustic phonon scattering.

8.4 *Classical collision model* (Shockley, 1951; Wang, 1989). In an intuitive model of electron–phonon interaction, we treat the phonon as a particle having mass M. The phonon mass can be approximately modeled as the mass of a single atom. Consider collinear collision of an electron having mass m and momentum $p_{e,i}$ with a phonon of mass M initially at rest. Since the phonon mass is much larger than the electron mass, the electron will bounce back with a momentum $p_{e,f}$ while the phonon gains a momentum of $P_{p,f}$.
   (a) On the basis of energy and momentum conservation, show that the amount of energy transport per collision from electron to phonon is

$$(\delta E)_{e \to p} = \frac{P_{p,f}^2}{2M} = \frac{(2p_{e,i})^2}{2m} = 2mv_e^2 = \kappa_B T_e$$

where v_e is the electron random velocity and we have used the relation $mv_e^2/2 = \kappa_B T_e/2$ (since we are dealing with one degree of freedom in a collinear system).

   (b) Similarly, consider a phonon with an initial momentum $P_{p,i}$ colliding with an electron initially at rest, and show that the energy exchange per collision is

$$(\delta E)_{p \to e} = 2Mu^2 = \kappa_B T_p$$

   (c) On the basis of the above results, show that the energy exchange rate between electron and phonon can be expressed as

$$\left(\frac{dE}{dt}\right)_c \approx G(T_e - T_p)$$

with $G = \kappa_B/\tau$, where τ is the time interval between each collision, or the relaxation time.

8.5 *Electron–phonon coupling factor of metals.* Estimate the electron–phonon coupling factor in gold.

8.6 *Electron–hole pair generation due to light absorption.* Gallium antimonide (GaSb) is used as a photovoltaic cell material in thermophotovoltaic energy conversion. It is a direct gap semiconductor with a gap of 0.72 eV. The real part of the refractive index is 3.8. The absorption coefficient above the bandgap is $\sim 10^4$ cm^{-1}. Determine the distribution of the electron–hole pairs for thermal radiation at normal incidence from a blackbody source of 1500 K.

8.7 *Hot electrons under a high field.* A voltage V is applied to an n-type semiconductor of length L between two electrodes, as shown in figure P8.7. The electron conductivity is σ and the electron–phonon coupling factor is G. We further assume that at the cathode, $x = 0$, the electrons and phonons are at the same room temperature. At the anode, the phonons are maintained at room temperature but the electrons' boundary condition is close to adiabatic (this is equivalent to assuming that the thermal conductivity and the Peltier coefficient of electrons in the anode are close to zero). The electron and phonon thermal conductivities of the semiconductor are k_e and k_p, respectively. Neglect thermoelectric effects.

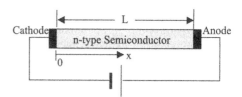

Figure P8.7 Figure for Problem 8.7.

Determine
(a) the temperature distributions of electrons and phonons;
(b) the heat source distribution in the semiconductor.

8.8 *Heat generation in a p–n junction.* Consider a silicon p–n junction. The dopant concentrations for both the p and the n sides are 10^{18} cm^{-3}. The mobilities of electrons and holes are 170 cm^2 V^{-1} s^{-1} and 60 cm^2 V^{-1} s^{-1}, respectively. The lifetimes of both electrons and holes are $\sim 10^{-10}$ s. The intrinsic carrier concentration n_i is 1.1×10^{10} cm^{-3}. The bandgap of silicon is 1.12 eV.
(a) Estimate the electron and the hole diffusion lengths.
(b) Estimate the saturation current density.
(c) Plot the heat source distribution due to electron–hole recombination in the n-type region.

8.9 *Heat source distribution under laser irradiation.* A laser beam with a wavelength of 1.55 μm and with an intensity of 10,000 W m^{-2} is incident on a semiconductor with a bandgap of 0.66 eV and a complex refractive index of (4, 0.01). The electron–hole mobility is 1000 cm^2 V^{-1} s^{-1} and the electron–hole recombination lifetime is 1 ms. Assuming that all recombination is nonradiative, determine the heat source distribution at steady state.

8.10 *Thermoelectric cooler.* Bismuth telluride (Bi$_2$Te$_3$) is a common thermoelectric material. State-of-the-art bulk n-type Bi$_2$Te$_3$ has the following properties: Seebeck coefficient, –240 mV K^{-1}; electrical resistivity, 10 μΩ m; thermal conductivity, 2.2 W m^{-1} K^{-1}.

(a) Calculate the figure of merit of this material.

(b) On the basis of the Wiedmann–Franz law and the Lorentz number, separate the thermal conductivity contributions due to electrons and to phonons.

(c) If, through the use of nanostructures, the phonon thermal conductivity can be reduced to $0.25 \text{ W m}^{-1}\text{ K}^{-1}$ without degrading the Seebeck coefficient or the electrical conductivity, what figure of merit can one get?

(d) If a p-type material with identical properties can also be obtained, calculate the maximum temperature difference that can be generated with a thermoelectric device made of the state-of-the-art material and the nanostructured material. Assume that all properties are temperature independent.

8.11 *Thermophotovoltaic generator.* Assuming that the photovoltaic cell of a thermophotovoltaic converter has a refractive index of 4 and that the filter is ideal, such that all photons above the bandgap are absorbed and all below the bandgap are reflected back to the heat source, determine the optimum bandgap and the maximum efficiency of the generator as a function of the emitter temperature.

8.12 *Dielectric-coupled thermophotovoltaic generator.* One idea to increase the power output of a thermophotovoltaic generator is to place a dielectric material between the emitter and the photovoltaic cell, as shown in figure P8.12. For simplicity, we assume that the refractive indices of all three media are matched above the bandgap of the photovoltaic cell and are equal to 4. We further assume that the photovoltaic cell has a built-in filter that reflects all radiation below the bandgap back to the emitter.

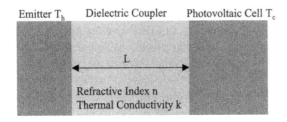

Figure P8.12 Figure for Problem 8.12.

(a) Show that the blackbody radiation heat flux is n^2 higher in the dielectric coupler than in vacuum, where n is the refractive index.

(b) Assuming that the dielectric coupler has zero thermal conductivity, determine the optimum bandgap as a function of the emitter temperature and evaluate the maximum efficiency of the thermophotovoltaic system.

(c) If the dielectric coupler has a thermal conductivity $k = 1 \text{ W m}^{-1}\text{ K}^{-1}$ but no absorption, evaluate the maximum efficiency for a heat source at 1000°C as a function of the thickness of the coupler.

Elaborate how absorption inside the dielectric coupler will affect the result for (c). Assume that the absorption coefficient above the bandgap is α and there is no absorption below the bandgap; also neglect the re-emission of the dielectric coupler.

9

Liquids and Their Interfaces

So far we have considered photons, electrons, phonons, and gas molecules. The transport processes of these energy carriers have many common characteristics and thus we have treated them in parallel. Transport in liquid is considerably more difficult to deal with. Compared to gases, liquids have molecules are closely packed and have short-range interactions, while compared to crystalline solids, liquids lack the periodicity of crystal structures. For these reasons, we cannot develop a parallel treatment for transport in liquids as we have done in previous chapters for other energy carriers. This chapter provides a brief description of transport processes in liquids and near the interfaces between liquids and their surrounding media, such as liquid–solid, liquid–liquid, and liquid–vapor interfaces.

We will start with a brief introduction to the methods used to deal with transport in bulk liquids. Historically, some of the earliest approaches were attempts to modify kinetic theory, particularly the Boltzmann equation, to include, for example, the finite size of liquid molecules and the potential interaction among molecules. The success of modified kinetic theories, however, is rather limited. Another line of development was pioneered by Einstein (1905) in his studies of the Brownian motion of particles in a liquid. Brownian motion was generalized by Langevin, and further developed by others in the linear response theory (Kubo et al., 1998). With the development of computational tools, the study of liquids has gradually shifted to computer simulations based on the linear response theory. In section 9.1, we will discuss the modification of the Boltzmann equation by Enskog, Einstein's Brownian motion theory, and the Langevin equation. The linear response theory will be discussed in chapter 10.

From the discussion on transport in bulk liquids, we will see that the transport processes involve not only kinetic energy exchange but also potential energy exchange. For nanoscale liquid transport, the interfacial force and the interfacial potential may be very different from those in bulk liquids, and this can impact the transport processes. We thus spend a large fraction of this chapter discussing the interfacial forces and interface potential between liquids and their surroundings. Regretfully, since most studies of transport between liquids and their interfaces are based on computer simulations of small domains, general information on how interfacial forces and potentials impact the liquid transport processes remains scarce in the literature.

At the end of this chapter, we will discuss some size effects on the thermodynamic properties at liquid–vapor interfaces. These effects are obtained from thermodynamics and are, in general, well understood. With the increasing attention to nanotechnology, these effects have new applications. Computational simulations also provide some new insights into these phenomena.

9.1 Bulk Liquids and Their Transport Properties

9.1.1 Radial Distribution Function and van der Waals Equation of State

We often regard liquids as structureless. This is not exactly true! Surrounding every liquid molecule are other molecules and their positions cannot be completely random because (1) the finite size of the liquid molecule determines that there are only a few other molecules in its immediate surrounding, and (2) these few molecules are more or less stabilized by the interatomic potential. One measure of the structure of a liquid is the radial distribution function, $g(r)$, which is defined as

$$4\pi r^2 n g(r) dr = \begin{array}{l}\text{the number of molecules with centers between} \\ r \text{ and } r + dr \text{ measured relative to a specific molecule}\end{array} \quad (9.1)$$

where $n (= N/V)$ is the average particle number density. Typical distributions of $g(r)$ for liquids, crystalline solids, and gases are shown in figure 9.1. The first peak in $g(r)$ for a liquid represents the coordination shell of nearest neighbors, the second peak represents the next nearest neighbors, and so on. The radial distribution function is a measure of the correlation of the atom under consideration with the surrounding atoms. Figure 9.1 indicates that after a few intermolecular distances the correlation between liquid atoms totally disappears. For example, in liquid argon, this occurs at around 20 Å (Boon and Yip, 1980). In contrast, the radial distribution function $g(r)$ is periodic for crystalline solids and is equal to one for ideal gases.

One use of the radial distribution function $g(r)$ is in the construction of the equation of states (Carey, 1999). From chapter 4, eqs. (4.14) and (4.15), we see that the equation of states can be constructed from the canonical partition function Z, which depends on the energy states of the system. For a liquid system, we can divide its energy into the sum of its kinetic and potential energy. It is easy to write down the kinetic energy expression for every molecule in the liquid system. The radial distribution can be used to construct an expression of the potential energy. For example, if the average potential energy between

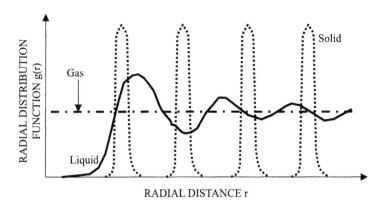

Figure 9.1 Typical behavior of the radial distribution function of liquids, crystalline solids (one-dimensional), and gases.

two molecules as a function of their separation r is $\phi(r)$, the total potential energy of an N-particle system is then

$$\Phi_N = \frac{N}{2} \int_0^\infty 4\pi r^2 \phi(r) n g(r) dr \qquad (9.2)$$

The integral represents the potential interaction of one particle with the rest of the particles. $N/2$ includes all the particles in the system, where the factor $\frac{1}{2}$ accounts for the sharing of ϕ between two particles. One example of the potential distribution function is the Lennard–Jones potential between two molecules [see eq. (3.7)],

$$\phi(r) = 4\varepsilon \left[\left(\frac{\sigma}{r}\right)^{12} - \left(\frac{\sigma}{r}\right)^6 \right] \qquad (9.3)$$

where ε and σ are the Lennard–Jones potential parameters (table 3.1). We make a further assumption on the radial distribution function,

$$g(r) = \begin{cases} 0 & r < D \\ 1 & r \geq D \end{cases} \qquad (9.4)$$

which means that there is no other molecule within a distance D surrounding the molecule, and outside D the other molecules are not correlated to the molecule. This is essentially a hard-sphere model for a molecule of diameter D. Under this model, eq. (9.2) can be expressed as

$$\Phi_N = -\frac{aN^2}{V} \qquad (9.5)$$

where

$$a = -2\pi \int_D^\infty r^2 \phi(r) dr \qquad (9.6)$$

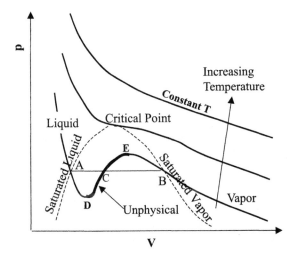

Figure 9.2 Isotherms predicted by the van der Waals equation. In region D–E, the pressure increases with increasing volume, which is unlikely to occur. Real systems avoid this region by a phase transition directly from A to B.

the negative sign ensuring that a is positive since the attractive part of the Lennard–Jones potential is negative. With eq. (9.5) for the potential energy, the canonical partition function can be derived for such a system (see, for example, Carey, 1999; Kittel and Kroemer, 1980). We will skip the details of the derivation and give the final canonical distribution function (the Helmholtz free energy) as

$$F(T, V, N) = -N\kappa_B T \left[\ln(V - BN) - \frac{3}{2} \ln \left(\frac{h^2}{2\pi m \kappa_B T} \right) \right]$$
$$+ \kappa_B T (N \ln N - N) - \frac{N^2 a}{V} \qquad (9.7)$$

The above expression is an extension of eq. (4.28) for an ideal gas to a hard-sphere fluid. The extension includes two parts. One is the addition of the potential energy term we have just obtained. The other is to replace the volume in eq. (4.28) by $(V - BN)$, where B is the volume of an individual molecule. This correction accounts for the finite size of the molecules. From the Helmholtz free energy, the equation of states can be derived from $p = -(\partial F/\partial V)_{T,N}$ as

$$p = \frac{N\kappa_B T}{V - BN} - \frac{aN^2}{V^2} \qquad (9.8)$$

The above equation is the celebrated van der Waals equation of states. Compared to the ideal gas equation of states, we see that BN accounts for the volume of the molecules and that the last term on the right-hand side represents the potential energy contribution to pressure.

The van der Waals equation of states is an idealized model that can describe both the vapor and the liquid phases (Goodstein, 1985). Figure 9.2 shows the isotherms of the equation of states on a p–V diagram. There exists a critical temperature below which the p–V curves have a local minimum (point D) and a local maximum (point E). Between the two extrema (DE), the pressure increases as the volume expands. This is an unphysical result. Real systems overcome this by a sudden change in volume to go from point A to point B of the curve, which corresponds to the evaporation of liquid into vapor. Other familiar curves, such as the saturated liquid and saturated vapor lines, as well as the critical point, are also marked in the figure.

9.1.2 Kinetic Theories of Liquids

Given the great success of the Boltzmann kinetic theory for gases and, as we have seen, for electrons and phonons, it was natural to modify the Boltzmann equation for liquid transport (Enskog, 1922; Chapman and Cowling, 1953; Rice and Gray, 1965; Kohler, 1972). The Boltzmann equation applies only to dilute systems of particles, assuming that these particles occupy no volume and that their interactions are limited to the instant of their collision, which is much shorter than the time spent by the particles moving freely before and after the collision. Liquid molecules clearly violate these assumptions. The molecules in a liquid are closely packed. While the repulsion force between the molecules is strong and similar to that between gas molecules, the long-range attraction force can no longer be neglected because such a force is the very reason that holds the molecules together in the liquid.

In chapter 6, we explained that the Boltzmann equation is a one-particle distribution function approximation to the general Liouville equation. In this section, we continue to use f to denote this one-particle distribution function and $f^{(2)}$ to denote the two-particle distribution function. The one-particle distribution function is an average of the N-particle distribution function $f^{(N)}(t, \mathbf{r}_i, \mathbf{p}_i)$ in the Liouville equation over the rest of the $(N-1)$ particles, eq. (6.4), which is repeated here,

$$f(t, \mathbf{r}_1, \mathbf{p}_1) = \frac{N!}{(N-1)!} \int \cdots \int f^{(N)}(t, \mathbf{r}_i, \mathbf{p}_i) d\mathbf{r}_2 \ldots d\mathbf{r}_N d\mathbf{p}_2 \ldots d\mathbf{p}_N \quad (9.9)$$

We will drop the subscript 1 from here on. One can also define a two-particle distribution function

$$f^{(2)}(t, \mathbf{r}, \mathbf{p}, \mathbf{r}_2, \mathbf{p}_2) = \frac{N!}{(N-2)!} \int \cdots \int f^{(N)}(t, \mathbf{r}_i, \mathbf{p}_i) d\mathbf{r}_3 \ldots d\mathbf{r}_N d\mathbf{p}_3 \ldots d\mathbf{p}_N \quad (9.10)$$

This two-particle distribution function describes the joint probability distribution of finding particles 1 at $(\mathbf{r}, \mathbf{p})$ and particle 2 at $(\mathbf{r}_2, \mathbf{p}_2)$. Similarly, higher order distribution functions can also be defined. From the Liouville equation, a hierarchy of equations for each of the distribution functions can be derived (Liboff, 1998). The equation of the lower order will involve also the distribution of higher orders. For example, the governing equation for the first order distribution function in its general form, is

$$\frac{\partial f}{\partial t} + \mathbf{v} \bullet \nabla_\mathbf{r} f + \frac{\mathbf{F}}{m} \bullet \nabla_\mathbf{v} f = S(f, f^{(2)}) \quad (9.11)$$

The left-hand side comprises the familiar terms in the Boltzmann equation and the right-hand side is a generalized scattering term, which is a function of f and $f^{(2)}$. To close this equation, $f^{(2)}$ must be related to f. Similarly, governing equations for higher order terms must be truncated by introducing closure relations, as in turbulence modeling.

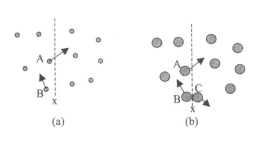

Figure 9.3 The difference between the Boltzmann kinetic formulation (a) and the Enskog formulation (b). In the Boltzmann formulation, heat flows across the interface at x only when the molecules (such as A) go across the interface. Molecules such as B do not contribute momentum and heat flux across x. In the Enskog model, even though B does not go across x, it collides with C and momentum and energy are transferred across the interface. This transfer occurs at a distance D and can be considered as due to the potential interaction between B and C.

The scattering integral in the Boltzmann equation, eq. (6.20), is based on the following assumption,

$$f^{(2)}(t, \mathbf{r}, \mathbf{p}, \mathbf{r}_2, \mathbf{p}_2) = f(t, \mathbf{r}, \mathbf{p}) f(t, \mathbf{r}_2, \mathbf{p}_2) \qquad (9.12)$$

This expression means that no correlation exists between the position and the momentum of the colliding particles; this is called the molecular chaos assumption. For liquid, the long-range attraction force between molecules creates correlations among molecules and thus the above assumption is no longer valid. One approach to address this problem, taken by Rice and Allnatt (1961), is to derive a governing equation for $f^{(2)}$. The Rice–Allnatt equations and their solutions for even simple cases are very complex and will not be elaborated on here (see Rice and Gray, 1965). An earlier approach, taken by Enskog (Enskog, 1922; Chapman and Cowling, 1953, Velarde, 1974), is more intuitive and will be briefly discussed here. The Enskog model assumes that the liquid molecules are hard spheres of diameter D, and neglects long-range attracting forces. However, such long-range interactions are implicitly included in the Enskog model. Referring to figure 9.3, in the Boltzmann equation model, momentum and energy transfer across an imaginary plane x occurs only when a molecule, such as A, goes across the plane. In the Enskog model, however, momentum and energy transfer across plane x occurs in a collision between molecules B and C, even if molecule B does not go across this plane. Because of the finite diameter of the molecules, this latter process occurs when the two molecules are within a distance D in the Enskog hard sphere model. Enskog assumed that the two-particle distribution is related to the one-particle distribution function through the radial distribution function g:

$$f^{(2)}(t, \mathbf{r}, \mathbf{p}, \mathbf{r}_2, \mathbf{p}_2) = f(t, \mathbf{r}, \mathbf{p}) f(t, \mathbf{r}_2, \mathbf{p}_2) g(|\mathbf{R}_{12}|) \qquad (9.13)$$

where $\mathbf{R}_{12} = \mathbf{r}_1 - \mathbf{r}_2$. In addition, due to the finite size of the particles, they do not occupy the same location at the moment of collision but are separated by a distance D. The Enskog equation may be written as (Chapman and Cowling, 1953; Kohler, 1972; Ferziger and Kaper, 1972; Velarde, 1974)

$$\frac{\partial f}{\partial t} + \mathbf{v} \cdot \nabla_{\mathbf{r}} f + \frac{\mathbf{F}}{m} \cdot \nabla_{\mathbf{v}} f = S \qquad (9.14)$$

with the following expression for the scattering term,

$$S = \iint D^2\hat{\boldsymbol{\Omega}} \cdot (\mathbf{v}_2 - \mathbf{v}) d\Omega\, d^3\mathbf{p}_2$$
$$\times \left[g\left(\left|\mathbf{r} + \frac{1}{2}D\hat{\boldsymbol{\Omega}}\right|\right) f(t,\mathbf{r},\mathbf{p}') f(t,\mathbf{r}+D\hat{\boldsymbol{\Omega}},\mathbf{p}_2') \right.$$
$$\left. - g\left(\left|\mathbf{r} - \frac{1}{2}D\hat{\boldsymbol{\Omega}}\right|\right) f(t,\mathbf{r},\mathbf{p}) f(t,\mathbf{r}-D\hat{\boldsymbol{\Omega}},\mathbf{p}_2) \right] \quad (9.15)$$

where the superscript ′ represents properties after collision, $\hat{\boldsymbol{\Omega}}$ is the unit vector connecting the centers of the two colliding spherical molecules and the integration of $d\Omega$ means over the entire solid angle formed between the two molecules. The first term inside the square brackets represents the in coming scattering into $(\mathbf{r},\mathbf{p})$ and the second term is the out going scattering. From momentum conservation, $\mathbf{p}'$ and $\mathbf{p}_2'$ are related to $\mathbf{p}$ and $\mathbf{p}_2$ by

$$\mathbf{p}' = \mathbf{p} + \hat{\boldsymbol{\Omega}}[(\mathbf{p}_2 - \mathbf{p}) \bullet \hat{\boldsymbol{\Omega}}] \quad \mathbf{p}_2' = \mathbf{p}_2 - \hat{\boldsymbol{\Omega}}[(\mathbf{p}_2 - \mathbf{p}) \bullet \hat{\boldsymbol{\Omega}}] \quad (9.16)$$

Solution of the Enskog equation leads to the following expressions for the viscosity and thermal conductivity of liquids (Kohler, 1972; Ferziger and Kaper, 1972; Velarde, 1974):

$$\mu_E = \frac{\mu_0}{g(D)}\left[1 + \frac{4}{15}\pi n D^3 g(D)\right]^2 + \frac{3}{5}\Pi \quad (9.17)$$

$$k_E = \frac{k_0}{g(D)}\left[1 + \frac{2}{5}\pi n D^3 g(D)\right]^2 + \frac{3\kappa_B}{2m}\Pi \quad (9.18)$$

$$\mu_0 = \frac{5}{16}\frac{\sqrt{\pi m \kappa_B T}}{D^2 \pi} \quad (9.19)$$

$$k_0 = \frac{25}{32}\frac{\sqrt{\pi m \kappa_B T}}{D^2 \pi}\frac{3\kappa_B}{2m} \quad (9.20)$$

$$\Pi = \frac{4}{9}\sqrt{\pi m \kappa_B T}\, n^2 D^4 g(D) \quad (9.21)$$

$$g = \frac{(1 - 11nb/9)}{1 - 2nb} \quad (9.22)$$

$$b = \frac{2}{3}\pi D^3 \quad (9.23)$$

where the subscript E represents Enskog results, and k_0 and μ_0 are the kinetic theory results for dilute gases [eq. (6.68a)]. These expressions can be written in normalized form as

$$\frac{\mu_E}{\mu_0} = bn\left(\frac{1}{y} + 0.8 + 0.76y\right) \tag{9.24}$$

$$\frac{k_E}{k_0} = bn\left(\frac{1}{y} + 1.2 + 0.76y\right) \tag{9.25}$$

where $y = nbg(D)$. In using the above expressions to calculate the transport coefficient, the key is to calculate an effective diameter D for the gas molecules. This can be done, for example, on the basis of experimental data on dilute gases and from eq. (9.19) or (9.20). Once the diameter is obtained, μ_E and k_E can be calculated as a function of the mass density (or the molecular number density n) (Sengers, 1965).

Both the viscosity and the thermal conductivity expressions can be interpreted as the result of a kinetic contribution, a collisional contribution, which is equivalent to potential contribution, and a cross-coupling term between the kinetic and collisional processes. Taking the thermal conductivity as an example, these two contributions can be expressed as

$$\frac{k_E}{k_0 nb} = \frac{1}{y} + 1.2 + 0.76y = \text{kinetic} + \text{cross} + \text{collisional} \tag{9.26}$$

which shows that the importance of the kinetic contribution decreases as the density increases, whereas the collisional contribution increases with density. Figure 9.4 compares the results of molecular dynamics simulation with the results from the Enskog equation for each of the three terms (Alder et al., 1970), showing good agreement for $V \geq 5V_0$ or $n \leq 0.2n_0$, where V_0 and n_0 are the volume and number density when the molecules are close packed. At higher densities, significant deviations occur. Near solidification, for example, the Enskog theory underestimates the shear viscosity by a factor of two.

Despite the work of Enskog and many others, the success of the kinetic theory for liquids, is limited. With the advance of computational power, direct simulation of liquid molecules, that is, molecular dynamics, and the analysis of the simulation using linear response theory, have largely replaced the kinetic theory approach for the study of liquid transport properties. In chapter 10, we will discuss in more detail some molecular dynamics simulation techniques and the linear response theory. In the following, we will discuss Brownian motion. Einstein (1905) pioneered an entirely different approach from that of the Boltzmann equation to study Brownian motion in liquids, and this eventually led to the development of the linear response theory.

9.1.3 Brownian Motion and the Langevin Equation

Einstein is best known for his relativity theory. Few people are familiar with his work on Brownian motion, which was a key step in the development of the atomic theory of matter. Research on Brownian motion was carried out by Einstein (1905, 1906a, 1906b, 1956) for his doctoral dissertation when he was an engineer in a patent office in Bern,

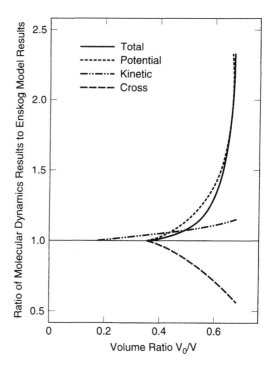

Figure 9.4 Shear viscosity of a hard-sphere fluid, calculated from molecular dynamics, relative to its Enskog values for contributions from the kinetic, the potential, and cross terms, where $V_0 = ND^3/\sqrt{2}$ is the volume at close packing (after Alder et al., 1970).

Switzerland. In fact, when he wrote his first paper on the Brownian motion in 1905, he was unaware that Brownian motion had been observed by botanist Brown in 1827. His motivation was to determine the size of molecules. Prior to his work, the well-established Boltzmann kinetic theory was applicable only to gases. He intended to develop a kinetic theory for liquids and, from the theory, to determine the size of the molecules.

Einstein considered the mass transfer of dilute solutes in a solvent. He first proved, from statistical thermodynamics, that the solute generates an additional pressure, called the osmotic pressure, that can be expressed as

$$p = \frac{1}{V} N \kappa_B T = n \kappa_B T \qquad (9.27)$$

where N is the number of solute particles in the total solution volume V and $n = N/V$ is the solute concentration per unit volume. This expression is similar to the ideal gas law and is valid only when the solute is dilute, similar to the condition of an ideal gas. This osmotic pressure can be measured by an osmometer, which employs a semi-permeable membrane that permits the crossing of only the solvent, not the solute (Hiemenz, 1986). Einstein considered next the mass diffusion of solute particles under a concentration gradient of the particles (Einstein, 1908). The osmotic pressure drives the diffusion of the solute particles. When the solute particles traverse the solvent, they experience a drag that can be modeled by the well-established Stokes law in continuum fluid mechanics,

$$\mathbf{F}_D = 3\pi \mu D \mathbf{u} \qquad (9.28)$$

where μ is the dynamic viscosity, D the particle diameter, and **u** the particle velocity. At steady state, the osmotic pressure force is balanced by the frictional force on the particles, leading to a steady drift velocity that is determined by the following balance equation,

$$A_c[p(x) - p(x+dx)] - dN(3\pi D\mu)u_x = 0 \tag{9.29}$$

for particles drifting in the x-direction, where dN is the number of solute particles in the volume $dV = A_c dx$. Substituting eqs. (9.27) into (9.29) leads to

$$nu_x = -\frac{\kappa_B T}{3\pi D\mu}\frac{dn}{dx} \tag{9.30}$$

The left-hand side is the volume flux of the solute particles. Thus the above equation is just Fick's law of diffusion,

$$J_x = -a\frac{dn}{dx} \tag{9.31}$$

where the diffusivity $a\,[\mathrm{m}^2\,\mathrm{s}^{-1}]$ is related to the viscosity by

$$a = \frac{\kappa_B T}{3\pi D\mu} \tag{9.32}$$

Equation (9.32) is the celebrated Einstein relation. If both a and μ are known, the diameter of the Brownian particle can be estimated, as was Einstein's original intent. This expression is similar to eq. (6.86), which is also called the Einstein relation. Equation (6.86) is an electron analogy to what Einstein actually derived for Brownian particles, except that in eq. (6.86) the electron mobility is a measure of how mobile the charge is; that is, the higher the mobility, the larger the diffusivity and conductivity of electrons. By contrast, the viscosity in eq. (9.32) is a measure of the resistance to the particle flow and thus is in the denominator rather than the numerator.* The Einstein relation shows an intrinsic relationship between transport properties (diffusivity) and the internal friction (viscosity), and is one example of the general fluctuation–dissipation theorem that we will discuss in the next chapter.

Equation (9.32) reflects Einstein's motivation of determining the diameter of the solute molecules, which are assumed to be much larger than those of the solvent, from the diffusivity of the solute in the solvent. Another question is how to determine the solute diffusivity. Einstein further showed, by solving the mass balance equation, that the mean displacement of the solute particles, for transport along the x-direction only, is

$$\langle \Delta x \rangle = \langle (x(t) - x(0))^2 \rangle^{1/2} = \sqrt{2at} \tag{9.33}$$

while for motion in a three-dimensional space the mean displacement is given by

$$\langle \Delta \mathbf{r} \rangle = \langle (|\mathbf{r}(t) - \mathbf{r}(0)|)^2 \rangle^{1/2} = \sqrt{6at} \tag{9.34}$$

---

*We have used μ to present both the mobility of electrons in eq. (6.86) and the viscosity of a fluid, as is customary in electronics and fluid mechanics, despite the opposite meaning of these two quantities.

where $x(t)$ and $r(t)$ are the instantaneous position of the particle. Thus, from measuring the mean displacement of the solute, the diffusivity can be determined.

In addition to the above approach, Einstein established another method to determine the diameter of solute particles. He proposed to measure the viscosity of the solvent and of the solution, μ_0, and μ, respectively, and derived, again assuming dilute solute particles, the following relationship between the two viscosities,

$$\frac{\mu}{\mu_0} = 1 + 2.5\varphi = 1 + 2.5n\frac{1}{6}\pi D^3 \qquad (9.35)$$

where φ is the volumetric concentration of the solute particles. We will not repeat Einstein's derivation but instead refer the reader to his original work (1906a; 1956). This result again applies only to dilute solutes. Many studies have been done to extend his results to higher volumetric concentrations (Hiemenz, 1986). These works should be a good starting point to examine recent claims on the novel properties of nanoparticle-seeded fluids, also referred to as nanofluids (Choi et al., 2001).

The Einstein relation can also be derived from the stochastic approach developed by Langevin to treat Brownian motion of particles much larger than those of the surrounding medium. The key idea of the Langevin equation is to assume that the motion of a Brownian particle is subject to a frictional force that is linearly proportional to its velocity, as in the Stokes law [eq. (9.28)], and a random driving force, $\mathbf{R}(t)$, imparted by the random motion of the molecules in the bath. The requirement that the Brownian particle is much larger in size than the molecules in the bath implies that the collision time of the bath molecules with the Brownian particle is much shorter than the relaxation time of the Brownian particle from its initial velocity, and hence there is no time correlation between the Brownian particle velocity and the molecular velocity. In the absence of an external force, the Langevin equation that governs the instantaneous velocity of the Brownian particle can be written as

$$m\frac{d\mathbf{u}}{dt} = -m\eta\mathbf{u} + \mathbf{R}(t) \qquad (9.36)$$

where η is the friction coefficient for Brownian particles in a fluid. The Stokes law gives $\eta = 3\pi D\mu/m$. The random driving force $\mathbf{R}(t)$ has the following characteristics:

$$\langle \mathbf{R}(t) \rangle = 0 \qquad (9.37)$$

$$\langle \mathbf{R}(t) \cdot \mathbf{u}(t) \rangle = 0 \qquad (9.38)$$

$$\langle \mathbf{R}(t+s) \cdot \mathbf{R}(s) \rangle = 2\pi R_0 \delta(t) \qquad (9.39)$$

where the bracket $\langle \rangle$ represents the ensemble average, a concept we discussed in chapter 4. Equation (9.37) indicates that the random driving force averages to zero because it acts in all directions. Equation (9.38) states that the random driving force is not correlated to the velocity of the Brownian particle. This can be justified if the Brownian particle size is large and its velocity relaxation time is much longer than the characteristic fluctuation time of the random driving force. Equation (9.39) implies that the autocorrelation of the random driving force is infinitely short.

Now, we show how to derive the Einstein relation from the Langevin equation. Taking the inner product of both sides of eq. (9.36) with $\mathbf{r}(t)$, the instantaneous position of the Brownian particle, and utilizing the relation

$$\mathbf{r} \cdot \frac{d\mathbf{u}}{dt} = \mathbf{r} \cdot \frac{d^2\mathbf{r}}{dt^2} = \frac{1}{2}\frac{d^2|\mathbf{r}^2|}{dt^2} - |\mathbf{u}|^2 \tag{9.40}$$

we obtain

$$\frac{1}{2}m\frac{d^2|\mathbf{r}^2|}{dt^2} + \frac{1}{2}m\eta\frac{d|\mathbf{r}^2|}{dt} = m|\mathbf{u}^2| + \mathbf{r}(t) \cdot \mathbf{R}(t) \tag{9.41}$$

By ensemble averaging, the last term on the right-hand side of the above equation drops out because there is no correlation between the particle instantaneous position and the random driving force. Applying the equipartition theorem,

$$\langle |\mathbf{u}(t)|^2 \rangle = \frac{3\kappa_B T}{m} \tag{9.42}$$

we obtain from eq. (9.41)

$$\frac{1}{2}m\frac{d^2\langle |\mathbf{r}^2| \rangle}{dt^2} + \frac{1}{2}m\eta\frac{d\langle |\mathbf{r}^2| \rangle}{dt} = 3\kappa_B T \tag{9.43}$$

The initial conditions for the above differential equation are

$$\langle |\mathbf{r}(0)|^2 \rangle = 0 \tag{9.44}$$

$$\frac{d}{dt}\langle |\mathbf{r}(0)|^2 \rangle = 2\langle \mathbf{r}(0) \cdot \mathbf{u}(0) \rangle = 0 \tag{9.45}$$

Equation (9.43) can be readily solved with the above initial conditions, leading to

$$\langle |\mathbf{r}(t)|^2 \rangle = \left(\frac{6\kappa_B T}{\eta m}\right)\left(t - \frac{1}{\eta} + \frac{1}{\eta}e^{-\eta t}\right) \tag{9.46}$$

At large times such that $\eta t \gg 1$, eq. (9.46) becomes

$$\langle |\mathbf{r}(t)|^2 \rangle = \left(\frac{6\kappa_B T}{\eta m}\right)t \tag{9.47}$$

Combining eq. (9.47) with eq. (9.34) leads to the Einstein relation

$$a = \frac{\kappa_B T}{\eta m} \tag{9.48}$$

This is identical to eq. (9.32), if we substitute $\eta = 3\pi D\mu/m$.

From the Langevin equation, one can also derive another way to calculate the friction coefficient. We start by integrating eq. (9.36) directly for a solution of the velocity. Strictly speaking, because the driving force is random, direct integration of eq. (9.36) is problematic from a mathematical point of view. This difficulty can be overcome by spectral analysis of the equation. Here, however, we will put aside the mathematical rigor and integrate eq. (9.36) directly as if **R**(t) were a continuous function. The end results are the same as those from a rigorous mathematical treatment. Integration of eq. (9.36) leads to

$$m\mathbf{u}(t) = m\mathbf{u}(0)e^{-\eta t} + e^{-\eta t}\int_0^t e^{\eta s}\mathbf{R}(s)ds \quad (9.49)$$

Taking the dot product of eq. (9.49) with **u**(0), we have

$$m\mathbf{u}(t) \cdot \mathbf{u}(0) = m\mathbf{u}(0) \cdot \mathbf{u}(0)e^{-\eta t} + e^{-\eta t}\int_0^t e^{\eta s}\mathbf{R}(s) \cdot \mathbf{u}(0)ds \quad (9.50)$$

Ensemble-averaging the above equation yields

$$\langle \mathbf{u}(t) \cdot \mathbf{u}(0)\rangle = \langle \mathbf{u}(0) \cdot \mathbf{u}(0)\rangle e^{-\eta t} = \frac{3\kappa_B T}{m}e^{-\eta t} \quad (9.51)$$

The ensemble average of the product of the same time-varying function at two different times $\langle A(t)A(t')\rangle$ is called the autocorrelation function, which again will be defined more carefully in chapter 10. In eq. (9.51), the left-hand side is the velocity autocorrelation function. By integrating both sides of the above equation from $t = 0$ to $t \to \infty$, the following expression for the friction coefficient can be obtained,

$$\int_0^\infty \langle \mathbf{u}(t) \cdot \mathbf{u}(0)\rangle dt = \frac{3\kappa_B T}{\eta m} = 3a \quad (9.52)$$

Equation (9.52) shows that the friction coefficient and thus the diffusivity can be calculated from the velocity autocorrelation function. This approach of calculating the transport properties from the auto correlation functions has gained popularity with increasing computational power, because the history of individual particles can be monitored through molecular dynamics simulation, as we will see in the next chapter. Modern treatments of liquids rely heavily on computational simulations (Alder and Wainwright, 1967; Boon and Yip, 1980; Hansen and McDonald, 1986). For example, the molecular dynamic calculations of Alder and Wainwright (1967) showed that the velocity autocorrelation decays much slowly than the exponential function suggested by eq. (9.46) and by the Enskog theory. Figure 9.4 shows another example where molecular dynamics is used to examine the validity of the Enskog equation. We will leave more discussion on the molecular dynamics simulations to the next chapter.

9.2 Forces and Potentials between Particles and Surfaces

The discussion in the previous section shows that for transport in liquids the intermolecular potential plays a direct role in energy and momentum exchange. For liquid transport in nanostructures, we naturally expect that the forces and potentials between the liquid molecules and their surroundings may become important. In our previous consideration of boundary effects on the transport of dilute particles, based on the Boltzmann equation or on wave propagation, the boundary impacts the transport only at the point of particle trajectory or wavefunction overlapping with the boundary because potential energy does not directly enter into the transport picture. On the other hand, because dilute particles have relatively long mean free paths, a collision that occurs at the boundary can affect the distribution quite far from the interface. Liquid molecules, however, have short mean free paths, suggesting that boundary effects will most likely be limited to the region where the interfacial potential changes significantly from that in the bulk fluid. Despite this relatively straightforward argument, however, few studies exist on how the interface potential impacts the transport properties. Most investigations have so far been based on nonequilibrium molecular dynamics methods (Koplik and Banavar, 1995; Thompson and Troian, 1997). On the other hand, there exists quite a large literature on the forces and potentials between liquids and their interfaces, arising from studies on surface tension, colloids, and complex fluids. Understanding such potentials and forces is an important step in appreciating the interfacial transport processes and incorporating them into either modeling or direct molecular dynamic simulations. In this section, we will briefly summarize the interfacial interactions. We will start from the intermolecular potentials to build expressions for interactions among surfaces, expressed in terms of forces or interaction potentials. These surface interactions include van der Waals interactions, which are typically attractive, and electrostatic interactions that are typically repulsive, and other forces arising from the structure of the molecules and the interfaces. We will talk about forces and potentials interchangeably, with the understanding that force is related to the potential through

$$\mathbf{F} = -\nabla \Phi \tag{9.53}$$

9.2.1 Intermolecular Potentials

Fundamentally, all the interatomic and intermolecular potentials are due to electrostatic interactions. In chapter 3, we discussed bonding forces in crystals, such as van der Waals bonding, ionic bonding, covalent bonding, and metallic bonding. In solids, the force interactions are mainly due to electrostatic interactions among atoms. In liquids, a greater variety of force interactions exists because liquids are made of molecules and have more degrees of freedom. Figure 9.5 summarizes common types of interactions between atoms, ions and molecules in vacuum and the corresponding interaction potential Φ (relative to the energy of the system in vacuum when the two parts are far apart) (Israelachvili, 1992). The basic building block for all these potentials is the Coulomb potential between two single charged particles, from which one can derive other types of interaction potential based on the charge configurations in atoms and molecules.

Example 9.1 *Charge–dipole interaction*

A dipole consists of two oppositely charged particles separated by a distance d. Derive an expression for the potential between a charge and a fixed dipole in vacuum, that is, the third line in figure 9.5.

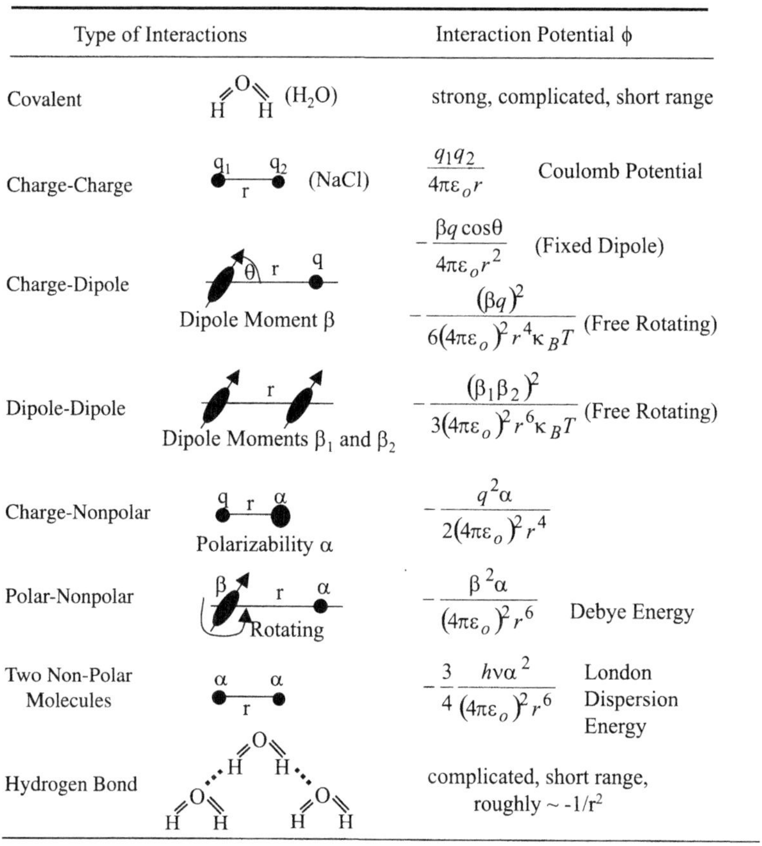

418 NANOSCALE ENERGY TRANSPORT AND CONVERSION

Type of Interactions		Interaction Potential ϕ	
Covalent	(H_2O)	strong, complicated, short range	
Charge-Charge	(NaCl)	$\dfrac{q_1 q_2}{4\pi\varepsilon_o r}$	Coulomb Potential
Charge-Dipole	Dipole Moment β	$-\dfrac{\beta q \cos\theta}{4\pi\varepsilon_o r^2}$	(Fixed Dipole)
		$-\dfrac{(\beta q)^2}{6(4\pi\varepsilon_o)^2 r^4 \kappa_B T}$	(Free Rotating)
Dipole-Dipole	Dipole Moments β_1 and β_2	$-\dfrac{(\beta_1 \beta_2)^2}{3(4\pi\varepsilon_o)^2 r^6 \kappa_B T}$	(Free Rotating)
Charge-Nonpolar	Polarizability α	$-\dfrac{q^2 \alpha}{2(4\pi\varepsilon_o)^2 r^4}$	
Polar-Nonpolar	Rotating	$-\dfrac{\beta^2 \alpha}{(4\pi\varepsilon_o)^2 r^6}$	Debye Energy
Two Non-Polar Molecules		$-\dfrac{3}{4}\dfrac{h\nu\alpha^2}{(4\pi\varepsilon_o)^2 r^6}$	London Dispersion Energy
Hydrogen Bond		complicated, short range, roughly $\sim -1/r^2$	

Figure 9.5 Common types of interaction between atoms, ions, and molecules in vacuum (after Israelachvili, 1992).

Solution: Consider a charge q placed in the field of the dipole, as shown in figure E9.1. The potential energy between the charge and the dipole can be thought of as the superposition of the potential energy of charge q interacting with Q and $-Q$, respectively,

$$\phi = -\frac{Qq}{4\pi\varepsilon_0}\left[\frac{1}{AB} - \frac{1}{AC}\right] \tag{E9.1.1}$$

where

$$AB = \left[\left(r - \frac{d}{2}\cos\theta\right)^2 + \left(\frac{d}{2}\sin\theta\right)^2\right]^{1/2} \tag{E9.1.2}$$

$$AC = \left[\left(r + \frac{d}{2}\cos\theta\right)^2 + \left(\frac{d}{2}\sin\theta\right)^2\right]^{1/2} \tag{E9.1.3}$$

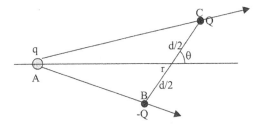

Figure E9.1 Figure for example 9.1.

In writing eq. (E9.1.1), we did not include the potential energy of the positive and negative charge of the dipole itself, which is included in the dipole self-energy. Substituting eqs. (E9.1.1) and (E9.1.2) into eq. (E9.1.3) and making use of the approximation that $r \gg d$, we obtain

$$\phi(r, \theta) = -\frac{q\beta \cos \theta}{4\pi \varepsilon_0 r^2} \tag{E9.1.4}$$

where $\beta = Qd$ is the dipole moment.

Comment. If the dipole is freely rotating, we can obtain the angle-averaged potential based on the Boltzmann factor

$$e^{-\bar{\phi}(r)/kT} = \frac{1}{4\pi} \int e^{-\phi(r,\theta)/\kappa_B T} d\Omega \tag{E9.1.5}$$

where the integration is over the entire solid angle. Using Taylor expansion for the exponential factor, one can show (Israelachvili, 1992)

$$\bar{\phi}(r) \approx -\frac{q^2 \beta^2}{6(4\pi \varepsilon_0)^2 \kappa_B T r^4} \tag{E9.1.6}$$

9.2.2 Van der Waals Potential and Force between Surfaces

On the basis of the elementary potential interactions discussed in the previous section, the force interaction between particles and surfaces can be obtained by summing up the interactions between the atoms or molecules involved. Such a summation often leads to qualitatively different behavior compared to the elementary forces between charge and atoms given in figure 9.5.

Starting from the attractive van der Waals potential between two atoms, $\phi(r) = -C/r^6$, one may sum the interaction energies of all the atoms in one body that interact with all the atoms in the other to obtain the interaction potential between the two bodies for a variety of geometries that are listed in figure 9.6 (Israelachvili, 1992). In this figure, A is called, the Hamaker constant (Hamaker, 1937)

$$A = \pi^2 C n_1 n_2 \tag{9.54}$$

where n_1 and n_2 are the number densities of molecules of the two interacting media. A typical values of C is 10^{-77} J m^6 and $n \approx 3 \times 10^{28}$ m^{-3}, leading to $A \approx 10^{-19}$ J. For example, for water, $A = 1.5 \times 10^{-19}$ J and for CCl$_4$, $A = 0.5 \times 10^{-19}$ J.

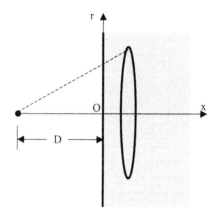

Figure E9.2 Figure for example 9.2.

Example 9.2 Derive an expression for the van der Waals potential between one atom and a surface.

Solution: We take a coordinate system as shown in figure E9.2. The atom is at equal distance from all parts of a differential ring inside the surface and thus all the atoms on the ring experience the same potential. The van der Waals attracting potential between the atom and this differential volume of $2\pi r^2 dr dx$ is

$$d\Phi = -\frac{C}{\left(\sqrt{(D+x)^2 + r^2}\right)^6} n 2\pi r\, dr\, dx \tag{E9.2.1}$$

Integrating the above expression for x from D to ∞ and for r from 0 to ∞, we obtain

$$\Phi = -\frac{\pi C n}{6D^3} \tag{E9.2.2}$$

which is identical to the formula in figure 9.6(b).

In example 9.2, and similarly for all cases in figure 9.6, we neglected the influence of the atoms inside the same solid and assumed that the medium between the two surfaces is vacuum. This treatment always leads to a positive Hamaker constant and thus an attractive potential between the two bodies. An alternative approach is the Lifshitz (1956) theory, which neglects the atomic structure and treats the objects as continuous media. The final expressions are similar to those listed in figure 9.6 but the Hamaker constant between surfaces 1 and 2, separated by a medium 3, can be expressed in terms of the dielectric constants of the media. If all three media are dielectrics and the electronic excitation frequencies are the same (ν_e), the Hamaker constant from the Lifshitz theory can be approximated as

$$A \approx \frac{3}{4}\kappa_B T \left(\frac{\varepsilon_{r1} - \varepsilon_{r3}}{\varepsilon_{r1} + \varepsilon_{r3}}\right)\left(\frac{\varepsilon_{r2} - \varepsilon_{r3}}{\varepsilon_{r2} + \varepsilon_{r3}}\right) + \frac{3h\nu_e}{8\sqrt{2}} \frac{(n_{r1}^2 - n_{r3}^2)(n_{r2}^2 - n_{r3}^2)}{\sqrt{(n_{r1}^2 + n_{r3}^2)(n_{r2}^2 + n_{r3}^2)}\left[\sqrt{n_{r1}^2 + n_{r3}^2} + \sqrt{n_{r2}^2 + n_{r3}^2}\right]} \tag{9.55}$$

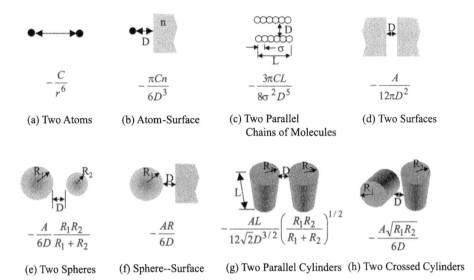

Figure 9.6 Van der Waals interaction free energies between two bodies. The Hamaker constant $A = \pi^2 C n_1 n_2$, where n_1 and n_2 are the atom number densities in the two media and C is the coefficient in the atom–atom pair potential (after Israelachvili, 1992).

where ε_r is the static dielectric constant (at zero frequency), n_r is the refractive index in the visible frequency range, and the number in the subscript represents the medium. The purpose for showing the above formula is to demonstrate the possibility of negative Hamaker constant values and thus a repulsive van der Waals force between two macroscopic objects, separated by a third medium, depending on the relative magnitudes of the dielectric constants of the media involved. Such negative Hamaker constant values have indeed been observed, for example, between fused quartz and air, separated by a water layer, and between CaF_2 and helium vapor, separated by liquid helium.

The van der Waals force between surfaces is also called the London force or dispersion force. This potential is universal among all surfaces because it arises from the induced dipoles among atoms.

9.2.3 Electric Double Layer Potential and Force at Interfaces

Surfaces immersed in liquids are usually charged because of the ionization or dissociation of surface groups or the adsorption of ions from the solution onto a previously uncharged surface [figure 9.7(a)]. The charges accumulated at the surface are balanced by an equal but oppositely charged region of counterions. Some of these counterions are also bounded to the surface, forming a so-called Stern or Helmholtz layer, which is usually very thin (a few angstroms). The remaining counterions distribute near the surfaces but are free to move, forming a diffuse electric double layer. This electric double layer is of fundamental importance for a wide range of technologies such as batteries, fuel cells, colloids, and in biochemistry and biotechnology.

We first determine the magnitude of the potential developed on the solid–liquid interface. This potential can be easily measured, using the solid as an electrode. Under the

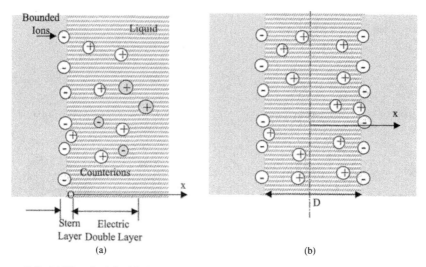

Figure 9.7 (a) Electrical double layer near the interfacial region. The charge on the surface can be due to dissociation or ionization of the surface materials or adsorption from the liquid. The charges in the liquid can be of multiple species. (b) A repulsive potential develops between two closely spaced surfaces when the charges in the liquids are of only one type, due, for example, to ionization or dissociation from the surface.

condition of local equilibrium, the ion density on the solid surface obeys the Boltzmann distribution

$$c_s = c_{zp} e^{-Ze\psi_s/\kappa_B T} \qquad (9.56)$$

where Z is the number of charges per ions and e is the unit charge ($-e$ for an electron), ψ_s is the electrostatic potential of the solid surface, and c_{zp} is the ion density at zero surface electrostatic potential. Equation (9.55) can be written as

$$\psi_s = -\frac{\kappa_B T}{e} \ln\left(\frac{c_s}{c_{zp}}\right) = \frac{\kappa_B T}{e} \ln\left(\frac{n}{n_{zp}}\right) \qquad (9.57)$$

where n and n_{zp} are the counterion densities in the solution corresponding to c_s and c_{zp} which are much easier to measure than the charge densities on the solid side. Equation (9.57) is called the Nernst equation. This is very similar to the Seebeck coefficient expression we derived in example 6.1.

As an example, consider a saturated solution of AgI in pure water with a solid AgI electrode. The saturated solution has equal amounts of 8.7×10^{-9} moles per liter (mol L^{-1}) of Ag$^+$ and I$^-$ ions at 25°C. It is found experimentally that the AgI electrode is negatively charged in this situation, meaning that more I$^-$ ions are adsorbed on the surface. At zero potential, the Ag$^+$ concentration is 3×10^{-6} mol L^{-1}. Using eq. (9.57), the Nernst potential is $\psi_s = -150$ mV.

To find the ion distribution near the surface, we need to solve the following equation governing the distribution of electrostatic potential ψ that can be derived from eq. (5.13),

$$-\varepsilon_0 \varepsilon_r \nabla^2 \psi = \rho_n \qquad (9.58)$$

where ε_r is the dielectric constant and ρ_n is the net charge number density. The charge distribution in the solution is also given by the Boltzmann distribution, leading to the Poisson–Boltzmann equation

$$-\varepsilon_0\varepsilon_r \nabla^2 \psi = \sum_i Z_i e n_{0i} \exp\left(-\frac{Z_i e \psi}{\kappa_B T}\right) \quad (9.59)$$

where n_{0i} is the ion concentration far from the surface and the summation is over all the ions in the solution. No simple analytical solution exists for the Poisson–Boltzmann equation. The Debye–Hückel theory of the electric double layer considers the limit when $Z_i e \psi \ll \kappa_B T$ such that the Poisson–Boltzmann equation can be linearized as

$$-\varepsilon_0\varepsilon_r \nabla^2 \psi = \sum_i Z_i e n_{0i} \left(1 - \frac{Z_i e \psi}{\kappa_B T}\right) \quad (9.60)$$

Far away from the surface, the liquid has no net charge. This requires $\sum_i Z_i e n_{0i} = 0$. We further consider a planar geometry as shown in figure 9.7(a) such that eq. (9.60) can be written as

$$\frac{d^2 \psi}{dx^2} = \psi \sum_i \frac{Z_i^2 e^2 n_{0i}}{\varepsilon_0 \varepsilon_r \kappa_B T} \quad (9.61)$$

with the boundary conditions

$$x = 0, \psi = \psi_s \text{ and } x \to \infty, \psi \to 0 \quad (9.62)$$

The solution for the potential distribution is then

$$\psi(x) = \psi_s e^{-x/\delta} \quad (9.63)$$

where δ is called the Debye length,

$$\frac{1}{\delta} = \sqrt{\sum_i \frac{Z_i^2 e^2 n_{oi}}{\varepsilon_0 \varepsilon_r \kappa_B T}} \quad (9.64)$$

which is very similar to the p–n junction width given by eq. (8.96). In fact, the development of p–n junction theory also relies on solving the Poisson–Boltzmann equation and exploited extensively the Debye–Hückel theory for the electric double layer (Shockley, 1949). The Debye length is of the order of a few nanometers for typical electrolytes, but can extend to hundreds of nanometers, depending on the dielectric constant and the ion concentration.

Now we consider the force balance inside the liquid. Because the liquid is stationary, the electrostatic force on the liquid must balance the pressure force. For the one-dimensional geometry in figure 9.7(a), this leads to

$$-\frac{dp}{dx} + \rho_n \left(-\frac{d\psi}{dx}\right) = 0 \quad (9.65)$$

where $(-d\psi/dx)$ gives the electric field and $\rho_n(-d\psi/dx)$ gives the electrostatic force. Again, substituting in the Boltzmann distribution for charge, we can write the above equation as

$$dp = -d\psi \sum_i Z_i e n_{0i} \exp\left(-\frac{Z_i e \psi}{\kappa_B T}\right) \quad (9.66)$$

The above equation can be integrated, from infinity where $p = p_\infty$ and $\psi = 0$, leading to

$$p(x) - p_\infty = \sum_i n_{0i} \kappa_B T \left[\exp\left(-\frac{Z_i e \psi(x)}{\kappa_B T}\right) - 1\right]$$

$$= \sum_i \kappa_B T \left[n_i(x) - n_{0i}\right] \quad (9.67)$$

The right-hand side of eq. (9.67) is always positive and thus the pressure inside the electric double layer is higher than that inside the bulk liquid at the equilibrium state. When the surface potential is negative, the anion concentration in the liquid near the surface is in excess of its equilibrium distribution far away from the surface and the cation concentration is smaller than its equilibrium distribution. The net effect is that the electric double layer creates an attraction force between the ions on the solid surface and the counterions in the liquid. This attractive electrostatic force is balanced by the positive pressure in the liquid.

Hence, when two solid surfaces are brought close to each other as shown in figure 9.7(b), a repulsive force develops between the two surfaces because the electrostatic force between the liquid and the solid surfaces no longer balances the positive pressure inside the liquid. A detailed exact solution for the symmetric surface case with only one type of counterions in the liquid has been obtained without invoking the Debye–Hückel approximation (Israelachvili, 1992). In this case, the potential distribution and the repulsive pressure between the two surfaces are given by

$$\exp\left(-\frac{Ze\psi}{\kappa_B T}\right) = \frac{1}{\cos^2 Kx} \quad (9.68)$$

$$p(D) = \kappa_B T n_0 (D) = 2\varepsilon_0 \varepsilon_r \left(\frac{\kappa_B T}{Ze}\right)^2 K^2 \quad (9.69)$$

where n_0 is the counterion number density at the middle plane when the two surfaces are separated by a distance D, and $1/K$ is of the same order as the Debye length. K and n_0 are determined by the surface charge density c_s,

$$-\frac{2\kappa_B T K}{Ze} \tan\left(\frac{KD}{2}\right) = \frac{c_s}{\varepsilon_0 \varepsilon_r} \quad (9.70)$$

$$K^2 = \frac{(Ze)^2 n_0}{2\varepsilon_0 \varepsilon_r \kappa_B T} \quad (9.71)$$

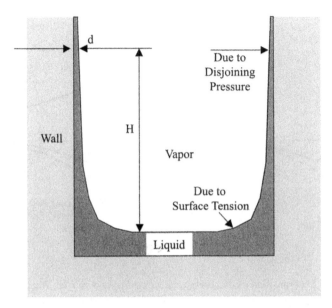

Figure 9.8 A repulsive disjoining pressure in a liquid film can raise the liquid film much higher than can surface tension.

As an example, consider two charged surfaces with $c_s = 0.2\,\text{C}\,\text{m}^{-2}$ (which is one charge per 0.6 nm^2) separated by $D = 2$ nm of water. Assuming monovalent counterions, that is, $Z = 1$, eq. (9.70) gives $K = 1.34 \times 10^9$ m^{-1} and eq. (9.69) gives $p(D) = 1.7 \times 10^6$ N m^{-2} or 17 atm. On the other hand, the van der Waals attraction force between the two surfaces, based on figure 8.5, is only $A/(12\pi D^3) \approx 3 \times 10^4$ N m^{-2} for a typical Hamaker constant of 10^{-20} J, which is much smaller than the repulsive force of the electrical double layer.

The repulsive force inside the liquid, due to the electric double layer for the case discussed here and also due to a van der Waals force (when the Hamaker constant is negative), leads to the concept of disjoining pressure (Derjaguin et al., 1987). The repulsive force means that medium 3, in between media 1 and 2, experiences an expansion, or a negative pressure, which is superimposed onto normal compressive pressure. This disjoining pressure can result in a variety of consequences that may affect thin-film spreading and phase-change processes (Israelachvili, 1992; Wayner, 1998).

Figure 9.8 shows an example where the disjoining pressure plays an important role. The vapor–liquid interface can be thought of as the plane of symmetry in figure 9.7(b) if the additional liquid–vapor interface charge adsorption is neglected. The corresponding repulsive pressure (disjoining pressure) inside the liquid layer due to the electric double layer between the liquid–solid interface can be calculated from eq. (9.69) by taking $D = 2d$, where d is the liquid film thickness. In the limit of high surface charge density c_s, eq. (9.70) leads to the solution $K \to \pi/D$ and thus

$$p_d(d) = 2\varepsilon_0 \varepsilon_r \left(\frac{\pi \kappa_B T}{2 Z e d} \right)^2 \qquad (9.72)$$

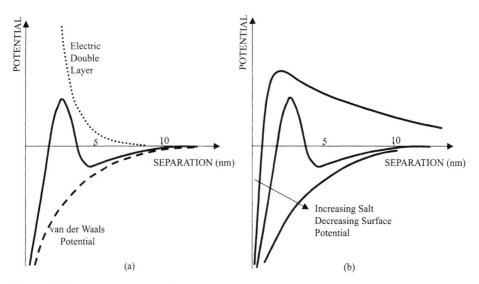

Figure 9.9 (a) Superposition of the double layer potential and van der Waals potential (DLVO theory). (b) Combined potential under varying salt concentration and surface potential.

This is known as the Langmuir equation. This repulsive pressure sucks the liquid film to a height H. At steady state, this liquid film is balanced by the gravitational force, $\rho_d g H$, which leads to

$$H = \frac{2\varepsilon_0 \varepsilon_r}{\rho_d g} \left(\frac{\pi \kappa_B T}{2 Z e d} \right)^2 \tag{9.73}$$

where ρ_d is the density of the liquid.

Combining the discussion in this section with that in the previous one, we see that the interaction potential between two close surfaces separated by a liquid layer experiences both electrostatic and van der Waals force,

$$\psi = \psi(\text{van der Waals}) + \psi(\text{electric double layer}) \tag{9.74}$$

The van der Waals force is usually attractive, although repulsive force can occur for some combinations of surfaces. The electric double layer generates a repulsive force. One possible combined potential profile between two surfaces is illustrated in figure 9.9(a). Whether the maximum or minimum in the figure appears or not apparently depends on the strength of each potential component. Unlike the interatomic potential, which has a short-range electrostatic repulsive force and a long-range van der Waals force, the electrostatic repulsive force due to a double layer is fairly long range. If the van der Waals potential between surfaces is attractive, it tends to pull the surfaces toward each other, whereas the repulsive electric double layer force prefers the separation of the solid surfaces.

To make the discussion more concrete, we consider a particulate solution. The balance between these two forces determines whether the particles form a stable solution (a colloid) or aggregate. For highly charged particles in a dilute electrolyte, there is a strong long-range repulsion that peaks at some distance (1–4 nm), creating an energy

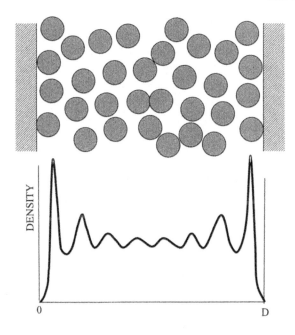

Figure 9.10
Illustration of density profile of liquids confined between two walls.

barrier [figure 9.9(b)]. In more concentrated electrolytic solutions, a secondary minimum forms before the energy barrier due to the decreased Debye layer thickness. This secondary minimum, although only marginally thermodynamically favorable, forms a metastable state of the particles, or a colloid. If the surface charge density is low, the repulsive potential is small and no energy barrier forms. In this case, the van der Waals force dominates and pulls the particles together. This aggregation process is also called coagulation or flocculation. This picture of colloids is called the DLVO theory, after Derjaguin and Landau (1941) and Verwey and Overbeek (1948).

9.2.4 Surface Forces and Potentials Due to Molecular Structures

The above analysis of the van der Waals force and the electric double layers are essentially a continuum approach that ignores the detailed molecular structure at the interfaces or the structures of the interfaces themselves. Within a few molecular layers of the interface, the molecular and interfacial structures alter the behavior of the interfacial potential.

Close to solid–liquid surfaces, the liquid molecules are more regularly arranged because of the constraints of the immobile solid atoms. The density profile oscillates as shown in figure 9.10. This density variation causes an oscillatory pressure variation, called the solvation force, that is superimposed onto the van der Waals and electric double layer forces based on the continuum analysis. As indicated in figure 9.10, the structural variations occur only within a few ($\sim$3) molecular layers near the surface, as does the oscillatory solvation force. Consequently, as two surfaces are brought close together, the pressure between the two surfaces varies (Israelachvili, 1992; Koplik and Banavar, 1995). In addition to the oscillatory solvation force, a repulsive force exists between two hydrophilic surfaces and is called the hydration force. The force decays exponentially from the surface at a much faster rate than that of the electric double layer and is active within a range of 1–5 nm. The origins of this hydration force are either the repulsive nature of the similarly charged hydrogen-bonding surface groups that modify the structure of the water near the surface (their effective range is about 3–5 nm), or the repulsion of the thermally excited molecular chains protruding from one medium into the other (their range is 1–2 nm). Similarly, between hydrophobic surfaces, an attractive hydrophobic force exists and this force has a range of $\sim$10 nm and can be much stronger than that predicted for the van der Waals attraction force.

It is assumed in the above discussions that the interacting surfaces are well defined, rigid, and smooth. At liquid–liquid, and liquid–vapor interfaces, thermal fluctuation changes the interface constantly, and this can be considered as a roughness of the interface. Similarly, liquid–polymer and polymer–polymer interfaces are also "rough" to the size of the polymers. As two such surfaces come together, a repulsive force develops, due to either the confinement of the motion of each interface (fluctuation force) or the interaction of the molecules (steric force). These repulsive forces are often understood from the configurational entropy perspective. For a more detailed discussion on these forces, refer to the excellent textbook by Israelachvili (1992).

9.2.5 Surface Tension

At a liquid–vapor interface, the atoms experience a different potential from those atoms deep inside the liquid, and a similar argument can also be made for solid–vapor and solid–liquid interfaces. The liquid molecules at the interface have the tendency to escape to the vapor side and the intermolecular spacing at the interfaces is larger than deep inside the liquid. This causes an effective density variation, as illustrated in figure 9.11. The larger intermolecular distance also means that the molecules at the interfacial region have a higher potential energy than those inside. The excess energy needed to bring the liquid molecules from inside the liquid to the interfacial region, per unit area of the interface, is the surface tension γ, or the energy "cost" of creating the surface. The surface tension units are [J m^{-2}], or [N m^{-1}] in the more familiar force unit. This force unit is also related to the conventional understanding of surface tension as the tangential force along the interface per unit length. Under this force picture, the work required to stretch a liquid membrane as shown in figure 9.12 is

$$W = \int F\, dx = \int 2\gamma L\, dx = 2\gamma L\, \Delta x = \gamma \Delta A_s \qquad (9.75)$$

where ΔA_s is the total surface area increase and the factor of two arises from the existence of two interfaces of the membrane. During this stretching, the liquid molecules inside the membrane are pushed toward the interfaces. In thermodynamics, work is a path dependent variable. However, the surface tension is a thermodynamic variable, analogous to the pressure in bulk fluids (Defay et al., 1966; Hiemenz, 1986). Equation (9.75) is similar to the $p\Delta V$ work in bulk fluids due to volume expansion. In fact, the analogy between surface layer and bulk fluid extends far beyond surface tension and pressure. Surface states are important topics in colloidal and surface chemistry.

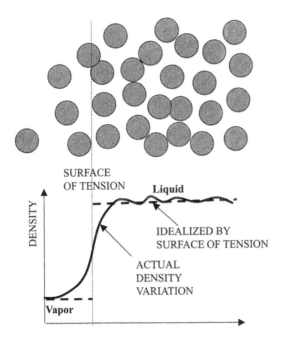

Figure 9.11 Density variation near a liquid–vapor interface. The surface of tension is an idealized plane that divides the interfacial region into the vapor side and the liquid side. On each side, bulk properties of liquid and vapor apply up to the surface of tension.

The bulk equations of states and fluid flow all have analogies for two-dimensional surfaces, which will not be discussed further (Hiemenz, 1986).

The surface tension concept approximates the interface as a mathematical plane (figure 9.11)—the surface where the tension acts, which sharply divides the liquid and the vapor. On each side of the surface of tension, the properties are assumed to be uniform and equal to that of the bulk values on the same side. The physical interface between liquid and vapor phases, however, is not sharp, as figure 9.11 indicates, with the density varying from that of the liquid to that of the vapor over a narrow range. The exact location of the surface of tension, however, depends on definition. Young (1972) defines the surface of tension on the basis of the mechanical force balance, whereas the Gibbs (1928) treatment, based on thermodynamics, defines the surface (Gibbs surface) as the location where the liquid and the vapor are of equal molar concentration. Here,

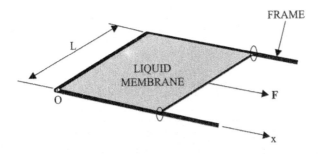

Figure 9.12 Stretching of a liquid membrane.

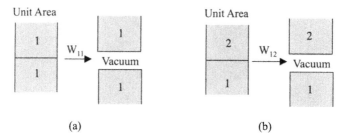

Figure 9.13 (a) Work of cohesion W_{11} is the energy needed to create two surfaces of unit area from the same matter. (b) Work of adhesion W_{12} is the energy needed per unit area to separate two different surfaces in contact.

we will not get into details of these discussions but refer interested readers to references on this topic (Defay et al., 1966; Hiemenz, 1986).

Theoretically, the surface tension can be determined from the interfacial potential discussed earlier. We consider first the separation of a solid into two parts, separated by a vacuum, as shown in figure 9.13(a). The energy needed per unit area, or the work done to separate the solid, is called the work of cohesion W_{11}. After separation, two surfaces are formed. The energy input during the separation of the solid is stored at the two surfaces, and thus the surface tension is

$$\gamma_1 = \frac{1}{2} W_{11} \quad (9.76)$$

For liquids, we cannot directly use this process to create two surfaces physically. However, we can increase the area of an existing surface easily. The energy needed per unit increase in the surface area is the surface tension, which can be similarly calculated from eq. (9.76) if we imagine that liquid can be similarly separated. To evaluate W_{11}, we can refer to figure 9.6 for the van der Waals force interaction between atoms in the medium. The work done in separating two parallel plates from a separation of D to infinity is

$$W_{11}^d = \frac{A}{12\pi D^2} \quad (9.77)$$

where we have used the superscript d to denote that this is due to van der Waals, or the London dispersion potential. When the two media are in contact, the effective separation $D \approx \sigma/2.5$. For a typical value of $\sigma = 0.4$ nm, $D \approx 0.16$ nm. Substituting this value into eqs. (9.77) and (9.76), we obtain

$$\gamma_1 \approx \frac{A}{24\pi} (0.16 \text{ nm})^{-2} \quad (9.78)$$

This simple estimation based on $D = 0.16$ nm actually gives very good values of surface energy for a wide variety of solid and liquid surfaces, as shown in Table 9.1.

Now consider the separation of two immiscible liquids in contact into two stand-alone parts at the interface [figure 9.13(b)]. After separation, the interfacial energy on each

LIQUIDS AND THEIR INTERFACES

Table 9.1 Hamaker constant and surface tension of typical fluids

Material	Theoretical Hamaker Constant (10^{-20} J)	Surface Tension (mN m^{-1})	
		Eq. (9.78)	Experimental
Liquid helium	0.057	0.28	0.12–0.35
n-pentane	3.75	18.3	16.1
n-octane	4.5	21.9	21.6
CCl$_4$	5.5	26.8	29.7
Acetone	4.1	20	23.7
Ethanol	4.2	20.5	22.8
Methanol	3.6	18	23
Glycol	5.6	28	48
Glycerol	6.7	33	63
Water	3.7	18	73

Source: Israelachvili, 1992.

surface is γ_1 and γ_2. The energy difference between the surface energy after separation and the interfacial tension γ_{12} before separation is called the work of adhesion,

$$W_{12} = \gamma_1 + \gamma_2 - \gamma_{12} \tag{9.79}$$

The above is the Dupré equation, which leads to a way of calculating the surface tension γ_{12}. The work of adhesion can be approximately estimated from the work of cohesion,

$$W_{12} = \sqrt{W_{11d} W_{22d}} = 2\sqrt{\gamma_{1d} \gamma_{2d}} \tag{9.80}$$

Equations (9.79) and (9.80) lead to

$$\gamma_{12} = \gamma_1 + \gamma_2 - 2\sqrt{\gamma_{1d} \gamma_{2d}} \tag{9.81}$$

Equation (9.81) is often called the Girifalco–Good–Fowkes equation. The square root term is due to the London dispersion force (van der Waals force), while ($\gamma_1 + \gamma_2$) includes all mechanisms contributing to the interfacial potential, on the basis of the argument that only the van der Waals potential operates across the interface.

Consider now a curved interface, figure 9.14. Because the surface tension is tangential to the surface, it has a component pointing toward the concave side of the interface, which must be balanced by a pressure difference across the surface. The normal force component due to the pressure difference is

$$(p'' - p')\Delta x \Delta y$$

where Δx and Δy are along two orthogonal directions. The surface tension force along the two lines perpendicular to length Δx is $\gamma \Delta x$, and its normal force component is

$$2\gamma \Delta x \sin(\Delta \theta_y) = 2\gamma \Delta x \times \Delta \theta_y = 2\gamma \Delta x \frac{\Delta y}{2r_y} \tag{9.82}$$

where r_y is the radius of curvature for the curve Δy. A similar expression exists for the force component along the two lines perpendicular to Δy. Equating the normal

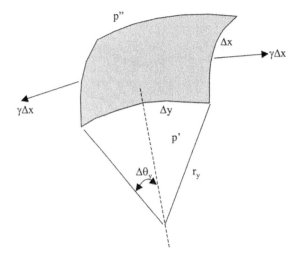

Figure 9.14 Derivation of Laplace equation for pressure difference created due to surface tension on a curved surface.

pressure force to the projection of the surface tension force, we obtain the Laplace equation

$$p'' - p' = \gamma \left(\frac{1}{r_x} + \frac{1}{r_y} \right) \tag{9.83}$$

where r_x and r_y are the two local radii of curvature in two orthogonal directions, usually taken along the principal directions of the surface such that r_x and r_y are the principal radii of curvature. For a spherical surface, $r_x = r_y = r$, the Laplace equation becomes

$$p'' - p' = \frac{2\gamma}{r} \tag{9.84}$$

When a liquid condenses on a solid surface, there are three phases and potentially three interfaces, the liquid–vapor, the liquid–solid, and the solid–vapor. Knowing the interfacial tension for each interface, we can use a simple force balance to derive the static contact angle between a droplet and a surface as shown in figure 9.15,

$$\gamma_{13} = \gamma_{23} \cos\theta + \gamma_{12} \tag{9.85}$$

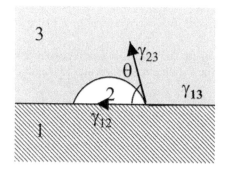

Figure 9.15 Derivation of the Young equation.

which is known as the Young equation. Depending on the values of surface tension at the three interfaces, a solution for θ may exist. In this case, the liquid forms droplets on the surface. When there exists no solution for θ, the liquid film spreads out to wet the surface.

9.3 Size Effects on Single-Phase Flow and Convection

The above discussion suggests that in the range of a few to tens of nanometers, and maybe even hundreds of nanometers in the case of the electric double layer (depending on the Debye length), the intermolecular potential between liquid molecules and liquid–solid surfaces may be modified. These modifications can potentially impact the fluid flow and heat transfer characteristics in micro- and nanostructures. Up to now, however, there exists no simple way to estimate how the interfacial potential affects fluid flow and heat transfer characteristics. The experimental data on pressure-driven fluid flow in microchannels scatter widely and lead to different interpretations, and not much data are available for nanochannels. In this section, we will first comment on pressure-driven flow in micro- and nanoscale channels, then follow with a brief introduction to electrokinetic and electrophoretic flows. The latter is much better understood and widely used in biotechnology.

9.3.1 Pressure-Driven Flow and Heat Transfer in Micro- and Nanochannels

Developments in micro-electro-mechanical systems (MEMS) have attracted strong interest in fluid flow and heat transfer in microchannels (Ho and Tai, 1998). Quite a large amount of experimental work has been performed on fluid flow in microchannels, particularly in relation to applications in biotechnology. Theoretically, one question that has been debated is whether fluid flow in microchannels deviates from the laws used for macrochannels. Unfortunately, experimental work so far has not been conclusive. The earliest experiments on liquid flow in microchannels were carried out by Poiseuille [see Sutera and Skalak, (1993) for an interesting historical review]. Poiseuille was interested in blood flow in the arterioles and venules but realized that controlled experiments using simple liquids would give him clearer formulations of the laws governing blood flow. His experiments employed glass tubes with diameters in the range of 15–600 μm and lengths of 6.77–100.5 mm. Many fluids were tested, including water, aqueous salt solutions, teas, wines and spirits, extracts of plants and roots, ethers, alcohols, and solutions of ammonia. On the basis of these experiments, he derived the following expression for the volumetric flow rate,

$$Q_V = \frac{\pi D^4 \Delta P}{128 \mu L} \tag{9.86}$$

which we now call the Poiseuille law. In eq. (9.86), D is the tube diameter, L is the length of the tube, and ΔP is the total pressure drop.

At the time of Poiseuille's experiments, the concept of viscosity was just being developed and was not used in his work. Yet the viscosity values deduced from his experimental data are within 0.1% of the accepted values! These facts strongly suggest that liquid flow in microchannels of comparable diameter should not deviate significantly from predictions of the classical laws. However, some recent experiments in microchannels of comparable or even larger diameters indicate deviations from Poiseuille's results. What are the possible reasons for the deviation observed in recent experiments? Examining these experiments, we can infer the following causes:

1. *Entrance and exit region effects.* In Poiseuille's experiments, the entrance region was carefully shaped and the exit was situated in water to eliminate the effects of surface tension at the exit. Long tubes were used such that the entrance region was much shorter than the fully developed flow. More recent experiments using silicon or stainless steel microchannels usually do not allow similar precautions to be taken for the entrance and exit effects.
2. *Surface roughness effects.* Stainless steel tubes and micromachined channels typically have a surface roughness larger than that of blown glass tubes. There exist observations of early transition to turbulence and deviation from the Moody friction factor chart in microchannels. Although the simple scaling law based on relative surface roughness to the diameter seems to be sufficient for the modification of the friction factor in macrochannels, one may argue that this may not be the most appropriate parameter in a microchannel. For example, with decreasing diameter the volume-to-surface ratio of the liquid in contact with the wall decreases. How does such a ratio influence the friction factor for microchannel flow is an open question.
3. *Effects of surface forces.* Because the surfaces of silicon and stainless steel are different from that of glass, the surface forces discussed in section 9.2 may play a role in determining the friction factor or heat transfer characteristics. However, there is no modeling or simulation to quantify the effects of surface forces on single-phase flow.
4. *Experimental error.* Because fluid flow in microchannels typically requires high pressure and the flow rate is very small, it takes a long time to reach steady state. An accurate characterization of the channel diameter is critical because the flow rate is proportional to $\sim D^4$.

Given the existing large variations that exist in reported fluid flow characteristics in microchannels, it is not strange to see a similarly large variation in the experimental data on heat transfer characteristics in such channels (Obot, 2000; Sobhan and Garimella, 2000). Despite these large variations, consensus is gradually emerging that these variations are more likely to be due to causes similar to those listed above rather than the breakdown of the continuum approximation.

Moving further down in scale, fluid flow in nanoscale becomes of interest to the understanding of a wide range of phenomena, such as self-assembly, DNA dynamics in liquids, and nanofabrication. The extreme of very narrow channels is amenable to molecular dynamics simulation, as reviewed by Koplik and Banavar (1995). One question on which molecular dynamics can provide considerable insight is that of boundary conditions. Koplik et al. (1989) concluded that, in a simple liquid undergoing Poiseuille or Couette flow, for all practical purposes the average velocity vanishes at the wall and thus the non-slip boundary conditions are valid. Thompson and Robbins (1990) showed that in-plane ordering is a key factor in being able to transmit shear stress across the fluid–solid interface. By varying the temperature, the wall–fluid commensurability, and the wall–fluid couplings, the range of behaviors from complete slip to non-slip was observed in their simulations. For simple spherical molecules, a direct correlation between the extent of in-plane ordering and the degree of slip was found. A general boundary condition was proposed, based on molecular dynamics simulation of Couette flow using the Lennard–Jones potential (Thompson and Troian, 1997). So far, there have not been many studies on convective heat transfer in nanochannels, despite the fact that molecular dynamics has been widely used to study phase change and heat conduction problems.

Example 9.3 *Order of magnitude for slip to occur*

To estimate the order of magnitude for the slip to occur, we consider a simple model as shown in figure E9.3(a). A molecule sits on top of a plane of molecules forming a solid surface. Estimate how much shear force is needed to slide the molecule on the surfaces.

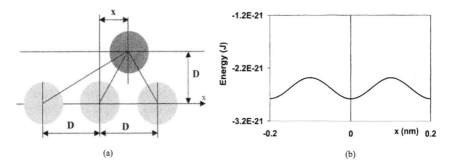

Figure E9.3 Figure for example 9.3.

Solution: We consider the van der Waals attraction potential between the molecule of interest and the solid wall. Assuming that the distance between the molecule and the surface is D, the interaction potential as a function coordinate x is then

$$\Phi(x, y) = -\sum_{i=-\infty}^{\infty} \frac{C}{[D^2 + (iD - x)^2 + (jD - y)^2]^3} \quad \text{(E9.3.1)}$$

where y is the coordinate direction perpendicular to the page. In figure E9.3(b), we plotted the above expression using typical values of $D (= 0.2$ nm$)$ and C (10^{-77} J m^6) as a function of x for $y = 0$. From the potential distribution, the attraction force acting on the molecule can be estimated, which is $\sim 10^{-11}$ N. Considering that there is approximately one molecular per $(0.2$ nm$)^2$ of surface area, the shear stress needed to cause the layer of molecules to slip is then $\sim 10^{10}$ N m^{-2}. To put this value into perspective, the shear stress generated for a velocity change of 1 m s^{-1} over 1 μm is 10^3 N m^{-2} using the viscosity of water. These numbers strongly suggest that the slip flow is unlikely to occur.

Comments. The above model is very crude. It does not include the interaction of the molecules with the other molecules and does not include the thermal motion of the molecules. We encourage interested readers to develop more rigorous models along similar lines of reasoning.

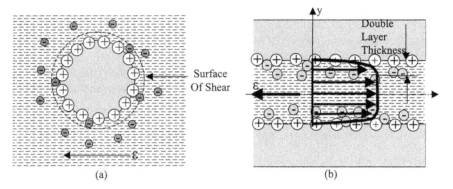

Figure 9.16 Electrokinetic phenomena: (a) electrophoresis; (b) electro-osmosis.

9.3.2 Electrokinetic Flows

The electrical double layer discussed in section 9.2.3 can be exploited to create various kinds of flows that have become increasingly important for micro- and nanofluidic devices. Under an external electrical field, positively and negatively charged particles experience forces in opposite directions, which may set either the liquid or the solid into motion, depending on their relative mobility. For example, if a field is applied along a stationary dielectric solid wall with an electric double layer, the ions in the liquid side experience an electrostatic force and are set into motion. These ions will drag the bulk liquid into motion through viscous force (Rice and Whitehead, 1965). Such fluid flow phenomena are known as electro-osmosis [figure 9.16(b)]. In other situations, such as when particles are suspended inside an electrolyte, an external electric field can set the charged particles into motion, a phenomenon called electrophoresis [figure 9.16(a)].

9.3.2.1 Electrophoretic Motion

As an example of electrophoresis, we consider the motion of a charged sphere inside an electrolytic solution under an external electric field $\mathscr{E}$, as shown in figure 9.16(a). The particle experiences a force and will be set into motion. A layer of liquid molecules in the immediate vicinity of the particle will also move at the same velocity as the particle due to the strong molecular bonding between the solid and the liquid molecules. This layer is typically only a few atomic layers thick but it affects the net charge and thus the electrostatic force on the particle [similar to the Stern layer in figure 9.7(a)]. The boundary between this immobile liquid layer and the mobile liquid molecules is called the surface of shear and its location is generally difficult to determine exactly. The balance of the electrostatic force with the viscous force leads to

$$3\pi\mu D\mathbf{u} = q\mathscr{E} \text{ or } \mathbf{u} = \frac{q\mathscr{E}}{3\pi\mu D} \quad (9.87)$$

where $\mathbf{u}$ is the drift velocity of the particle relative to the solution. The charge on a particle q should be considered as the net charge within the surface of shear. It can be related to the electrostatic potential at the surface of shear, which is also called the zeta potential (ζ), by solving the Poisson–Boltzmann equation for a spherical coordinate. When the Debye layer thickness δ is much larger than the particle diameter, the zeta potential is (Hiemenz, 1986)

$$\zeta = \frac{q}{2\pi\varepsilon_0\varepsilon_r D}\frac{\delta}{\delta + D/2} \approx \frac{q}{2\pi\varepsilon_0\varepsilon_r D} \tag{9.88}$$

Substituting eq (9.88) into (9.87) leads to an expression of the particle velocity in terms of the zeta potential,

$$\mathbf{u} = \frac{2\zeta\varepsilon_0\varepsilon_r \mathscr{E}}{3\mu} \tag{9.89}$$

The above analysis can be applied to biomolecules. Different biomolecules (DNAs and proteins) usually have different charge and effective diameters and thus will have different terminal velocities. Under the same electric field, they have different drift velocities and thus travel different distances, which means that different biomolecules can be separated. This is the basis of gel-electrophoresis, which is widely used in biology to separate biomolecules (Manchenko, 2003).

9.3.2.2 Electro-Osmotic Flow

We consider now the electrostatically driven osmotic fluid flow in microchannels formed between two parallel plates as shown in figure 9.16(b). Taking a differential control volume of the fluids, the balance of the viscous force and the electrostatic force gives

$$\mu\frac{d^2 u}{dy^2} = \rho_n \mathscr{E}_x = \mathscr{E}_x \left(-\varepsilon_0\varepsilon_r \frac{d^2\psi}{dy^2}\right) \tag{9.90}$$

where $\mathscr{E}_x$ is the x component of the electric field. This equation should be coupled to the Poisson–Boltzmann equation to obtain the potential profile. Here we assume that the thickness of the electric double layer δ is small compared to the plate spacing; that is, $\delta \ll D$. Outside the electric double layer, $\psi = 0$ and thus eq. (9.90) becomes

$$\mu\frac{d^2 u}{dy^2} = 0 \text{ for } -\left(\frac{D}{2} - \delta\right) \leq y \leq \left(\frac{D}{2} - \delta\right) \tag{9.91}$$

The above equation can be integrated once to give

$$\mu\frac{du}{dy} = C_1 \tag{9.92}$$

The symmetry requirement at $y = 0$ leads to $C_1 = 0$. Thus the velocity distribution must be a constant outside the electric double layer,

$$u(y) = u_0 \text{ for } -\left(\frac{D}{2} - \delta\right) \leq y \leq \left(\frac{D}{2} - \delta\right) \tag{9.93}$$

Now consider transport inside the electric double layer. Integrating eq. (9.90) once leads to

$$\mu \frac{du}{dy} = -\varepsilon_0 \varepsilon_r \mathscr{E}_x \left(\frac{d\psi}{dy} \right) + C_2 \tag{9.94}$$

At the edge of the electric double layer, $du/dy = 0$ and $d\psi/dy = 0$, thus $C_2 = 0$. Integrating eq. (9.94) again, from the wall to the edge of the electric double layer, we have

$$\mu[u_0 - u(y = -D/2)] = -\varepsilon_0 \varepsilon_r \mathscr{E}_x [\psi(y = -D/2 + \delta) - \zeta] \tag{9.95}$$

where ζ is the zeta potential on the surface of the shear (we have neglected the thickness of the surface of the shear). At the edge of the electric double layer, $y = -D/2 + \delta$, the potential ψ is zero. On the surface of the shear, the velocity is zero. Thus, eq. (9.95) gives the velocity of the fluid in the center region of the channel as

$$u_0 = \frac{\varepsilon_0 \varepsilon_r \mathscr{E}_x \zeta}{\mu} \tag{9.96}$$

The velocity profile of an electro-osmotic flow is sketched in figure 16(b). Because the electric double layer thickness is much smaller than the plate separation, the flow is essentially a plug flow with uniform velocity. Within the electric double layer, the velocity decreases continuously to zero at the surface of the shear.

The above derivation of electro-osmotic flow does not consider the fluid structure near the surfaces such as that sketched in figure 9.10. A recent molecular dynamics study (Freund, 2002) found that ions are more attracted to the wall than the Poisson–Boltzmann equation predicts for electro-osmotic flow in nanochannels, suggesting that discrete nature of ions can be important for flow in nanostructures.

9.4 Size Effects on Phase Transition

Size has profound effects on phase change processes. One can easily appreciate this from the Laplace equation, eq. (9.84), that shows pressure dependence on curvature. Since pressure is related to other thermodynamic properties, it is reasonable to anticipate that certain thermodynamic properties will be influenced by size. Examples are surface tension, phase transition pressure and temperature, and so on. In-depth discussion of the thermodynamics of small systems can be found in the work of Hill (1963, 1964). Here we focus on the effects of curvature. We further limit our discussion to the phase transition of a pure substance; in other words, only one material exists in the system so that only two phases are present.

The starting point in analyzing curvature effects on thermodynamic properties is based on the Laplace equation and thermodynamic relations. We consider a spherical geometry (droplet or bubble). The inner pressure is p″ and the surrounding fluid pressure is p′. When the system goes from one equilibrium state to another, the Laplace equation (9.84) leads to

$$dp'' - dp' = d\left(\frac{2\gamma}{r}\right) \tag{9.97}$$

Next, we eliminate some variables so that we can solve the above equation. Because each phase is in thermal equilibrium, we can use the Gibbs–Duhem equation [eq. (8.83)] for each of the bulk phases

$$s'dT - v'dp' + d\mu' = 0 \tag{9.98}$$

$$s''dT - v''dp'' + d\mu'' = 0 \tag{9.99}$$

where s and v are the entropy and volume per mole, respectively, and μ is the chemical potential. A similar equation, called the Gibbs equation, exists for the interface,

$$d\gamma = -s_i dT - \Gamma d\mu_i \tag{9.100}$$

where Γ is the number density of molecules per unit area at the surface of tension, s_i is the entropy per unit area, and μ_i is the chemical potential at the interface. Equations (9.97)–(9.100) form the basis for analysing the effects of curvature on thermodynamic properties. Which of the variables we choose to eliminate depends on whether the liquid or the vapor is inside the sphere, and what are the system constraints, that is, constant pressure or constant temperature. We discuss a few cases below.

9.4.1 Curvature Effect on Vapor Pressure of Droplets

First we consider a droplet system at constant temperature so that p'' is the pressure inside the liquid droplet. At equilibrium, since $\mu' = \mu'' = \mu_i = \mu$, eqs. (9.98) and (9.99) lead to

$$v'dp' = v''dp'' \tag{9.101}$$

Substituting eq. (9.101) into (9.97) and eliminating p'' yields

$$d\left(\frac{2\sigma}{r}\right) = \frac{v' - v''}{v''} dp' \tag{9.102}$$

If we further assume $v' \gg v''$, the ideal gas law for the vapor phase, and that v'' (liquid) is independent of pressure, the above equation can be integrated, leading to the Kelvin equation

$$\ln\left(\frac{p'}{p_0}\right) = \frac{2\sigma}{r} \frac{v''}{RT} \tag{9.103}$$

where R is the universal gas constant and p_0 is the normal vapor pressure when the interface is flat ($r \to \infty$). This equation shows that the equilibrium vapor pressure increases as the liquid droplet radius decreases. For a given vapor pressure, smaller droplets tend to evaporate. Thus, in a mist of droplets of pure substance, the large droplets will grow at the expense of the small droplets since they have a lower vapor pressure.

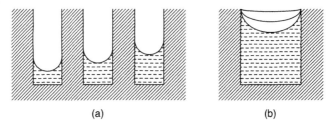

Figure 9.17 Capillary condensation inside nanopores. (a) Before the pores are filled, the vapor pressure inside the pores is lower than in bulk form; even superheated vapor may condense if eq. (9.104) is satisfied. (b) Once the meniscus reaches the mouth of the pore, the vapor pressure starts to increase as the radius of curvature decreases.

If, instead of droplets, bubbles are formed in a liquid ambient, a similar analysis leads to the Kelvin equation for the vapor pressure p'' in the bubble,

$$\ln\left(\frac{p''}{p_0}\right) = -\frac{2\gamma}{r}\frac{v'}{RT} \qquad (9.104)$$

which shows that the equilibrium vapor pressure inside a bubble is lower than its corresponding equilibrium pressure when a flat interface exists.

We can use the Kelvin relation to understand the condensation of a vapor in a porous medium [figure 9.17(a) and (b)]. Let us first consider an idealized problem in which all the pores are cylinders of the same radius. We suppose that the pores are partially filled with a liquid in contact with its own vapor and assume furthermore that the walls of the pores are completely wetted by the liquid. So long as the menisci are away from the mouths of the pores, all the menisci will have hemispherical surfaces of radius r. The vapor pressure in the pores is given by eq. (9.104) and is less than p_0. Consequently, liquid can exist in a porous medium in equilibrium with superheated vapor. If the vapor pressure is increased slightly, condensation will occur in all pores in which the meniscus has not yet reached the mouth of the pore; further condensation would result in an increase of the radius of curvature of the surface. Condensation in this pore therefore ceases when the radius of curvature reaches the equilibrium value corresponding to the vapor pressure. Thus condensation will proceed in the partially filled pores and be halted in the filled pores, until a point is reached at which all the pores are similarly filled and the liquid in them has everywhere the radius of curvature corresponding to the vapor pressure. Further increase in the vapor pressure results in condensation in all the pores and the flattening of the menisci, which become plane when $p'' = p_0$. The vapor is now saturated and any further increase in p'' is immediately offset by condensation of bulk liquid.

This reduced vapor pressure when the vapor is on the concave side and the increased vapor pressure when the vapor is on the convex side may be related to some technologies being used in nanowire growth. One method developed, for example, is to condense vapors of materials onto a template (Heremans et al., 2000), such as anodized alumina, that can have channels with a diameter in the range of 10–200 nm. In this case, the Kelvin relation suggests that the vapor pressure inside the channels may be lower than that of the bulk saturation pressure and thus even superheated vapors may condense in very small channels. Another example is the vapor–liquid–solid growth of nanowires in free space (Morales and Lieber, 1998). In this case, vapor condenses on convex surfaces which may have a higher vapor pressure that favors the growth of wires with bigger diameters. These possibilities, however, have not been studied experimentally, nor fully exploited in the control of nanowire growth.

9.4.2 Curvature Effect on Equilibrium Phase Transition Temperature

We now examine the size dependence of the equilibrium phase transition temperature. For this purpose, we will assume that the external pressure is constant, that is, $dp' = 0$ in eq. (9.98). In the case of droplets, we can subtract eq. (9.98) from (9.99) to obtain

$$\frac{\Delta h}{T}dT + v''dp'' = 0 \qquad (9.105)$$

where $\Delta h = T(s' - s'')$ is the latent heat. Substituting eq. (9.105) into (9.97) leads to

$$\frac{dT}{T} = -\frac{v''}{\Delta h}d\left(\frac{2\gamma}{r}\right) \qquad (9.106)$$

Integrating the above equation from $r \to \infty$ ($T = T_0$) to r, we obtain

$$\ln\frac{T}{T_0} = -\frac{2\gamma}{r}\frac{v''}{\Delta h} \qquad (9.107)$$

This shows that the equilibrium temperature of small droplets is lower than that of a flat interface.

If, instead of droplets, bubbles are formed inside liquid under a constant liquid pressure, a slightly more complicated derivation, due to the compressibility of the vapor phase, leads to (Defay et al., 1956)

$$\overline{\Delta h}\left(\frac{1}{T_0} - \frac{1}{T}\right) = R\ln\frac{2\sigma/r + p'}{p'} \qquad (9.108)$$

where $\overline{\Delta h}$ is the average latent heat in the range between T_0 and T. It should also be pointed out that the latent heat can also be size dependent and could be analyzed on the basis of the same sets of equations [see Defay et al. (1956) for details]. Equation (9.108) shows that the equilibrium temperature of small vapor bubbles must be higher than the normal phase transition temperature T_0, which explains the existence of superheated liquid.

9.4.3 Extension to Solid Particles

A similar analysis can be extended to small solid particles by replacing the liquid with the solid properties. Such an extension leads to the following conclusions:

1. The vapor pressure of small crystals is greater than that of large crystals. In the presence of vapor, large crystals will grow at the expense of smaller crystals.
2. Small crystals melt at a temperature lower than the bulk melting point. The melting point T of a small crystal is given by

$$\ln\frac{T}{T_0} = -\frac{2\gamma^{s\ell}}{r}\frac{v_s}{\Delta h_{s\ell}} \qquad (9.109)$$

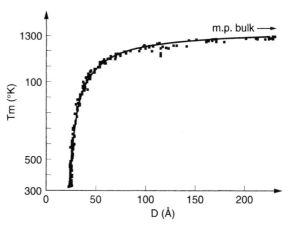

Figure 9.18 Experimental results on the melting point of gold nanoparticles as a function of the particle diameter (Buffat and Borel, 1976; courtesy of APS Associate Publisher).

where T_0 is the normal melting point at the same external pressure, $\gamma^{s\ell}$ is the interfacial tension at the solid–liquid interface, v_s is the molar volume of the solid, and $\Delta h_{s\ell}$ is the heat of fusion. A well-established example is the lowering of the melting point of gold nanoparticles, as shown in figure 9.18 (Buffat and Borel, 1976).

3. The melting point of a substance solidified in the pores of an inert material will depend on the size of the pores. For a more detailed discussion, see Defay et al. (1966).
4. Small crystals may have a heat of fusion and a heat of sublimation smaller than the value for bulk solid.

9.4.4 Curvature Effect on Surface Tension

The size dependence of surface tension has been under investigation since Gibbs (1928) but still remains a topic of debate. Tolman's (1949) work is a classic in this field. The Tolman theory treats a single-component system with two phases and uses the same set of equations (9.97)–(9.100). Considering a droplet at constant temperature, we have

$$d\left(\frac{2\gamma}{r}\right) = \frac{v' - v''}{v''} dp'' \tag{9.110}$$

or

$$\frac{2}{r} d\gamma + 2\gamma d\left(\frac{1}{r}\right) = \frac{v' - v''}{v''} dp'' \tag{9.111}$$

To eliminate dp'', we use eq. (9.100),

$$d\gamma = -\Gamma d\mu_i = -\Gamma v'' dp'' \tag{9.112}$$

From eqs. (9.111) and (9.112), we obtain

$$\frac{d\gamma}{\gamma} = -\frac{2\Gamma}{(2\Gamma/r) + \rho_d'' - \rho_d'} d\left(\frac{1}{r}\right) \tag{9.113}$$

where ρ_d'' and ρ_d' are the densities of the two phases.

For a liquid droplet, ρ'_d (the vapor side) is negligible and eq. (9.113) can be integrated to give

$$\frac{\gamma(r)}{\gamma_0} \approx \frac{1}{1 + (2\Gamma/\rho''_d)/r} \quad (9.114)$$

where γ_0 is the surface tension for a flat surface. Equation (9.114) shows that the surface tension of a droplet decreases with decreasing droplet diameter. For a vapor bubble inside a liquid, ρ''_d (the vapor side) is negligible and eq. (9.113) leads to

$$\frac{\gamma(r)}{\gamma_0} \approx \frac{1}{1 - (2\Gamma/\rho'_d)/r} \quad (9.115)$$

and thus the surface tension increases with decreasing bubble diameter. The above expression can also be generalized as

$$\frac{\gamma(r)}{\gamma_0} \approx \frac{1}{1 + 2\delta_T/r} \quad (9.116)$$

where $\delta_T \approx \Gamma/(\rho''_d - \rho'_d)$ is called the Tolman length. It is interpreted as the distance between equimolar surface, where the densities of the liquid phase and the vapor phase are equal, and the surface of the tension. A more accurate expression was given by Tolman (1949).

Typically, the Tolman length is short. Molecular dynamics simulation by Haye and Bruin (1994) found that for the Lennard–Jones potential, $\delta_T/\sigma = 0.2 \pm 0.05$. Since in the Lennard–Jones potential σ is only a few angstroms (3.542 Å for argon), δ_T is very small, so that the effect of radius on surface tension is negligible in most situations. The size dependence of surface tension is of great interest to nucleation theory and has received most attention in this field (Laaksonen et al., 1995). Typically, it is found that the Tolman length is a function of radius and temperature and thus the original theory by Tolman is often modified and revisited (Kalikmanov, 1997; Granasy, 1998).

9.5 Summary of Chapter 9

Because the liquid molecules are closely packed and lack long-range correlations as in crystalline solids, conventional kinetic theory, which is based on the assumption of dilute particles and infrequent interaction among the particles, is not applicable to liquid systems. This difficulty has prevented us from pursuing a treatment of the transport processes in liquid parallel to what we have done for gaseous molecules, electrons, phonons, and photons in previous chapters. This chapter attempts to provide an overview of theoretical tools for studying transport in liquids and key aspects of the interfaces between liquids and their surroundings.

We started with a review of theoretical tools that treat transport in bulk liquids. The radial distribution function provides a useful description of the structure of a liquid and we used it to derive the van der Waals equation of state, which captures the vapor, the liquid, and the liquid–vapor coexisting regions, at least qualitatively. For transport in liquid, one effort is the extension of Boltzmann's kinetic theory for gases to higher densities. The Enskog equation is an example. This equation includes the effects of the finite size of the molecules and, implicitly, the potential energy exchange. The solutions of the Enskog equation for dense gas, however, are not valid at liquid densities. Einstein started a different line of investigation—Brownian motion in liquids. The major result of his investigation was to relate viscosity to the diffusivity of Brownian particles, which provides a means of estimating the size of Brownian particles (and big molecules). Einstein's work laid the foundation for the later development of linear response theory, which has a much broader impact than just liquid transport. The linear response theory applies equally well to all other transport processes we have discussed in this book so far. We have reserved the linear response theory for the next chapter and, instead, discussed only the Langevin equation. This equation generalizes Brownian motion by splitting the forces acting on a Brownian particle into a resistive term that is proportional to the particle velocity and a random force term that arises from the particle interacting with the surrounding. Although Brownian motion is discussed in the context of particle transport in a liquid, it is a very general phenomenon and has found applications in many different fields. We noted that many studies on transport in fluids are based on linear response theory, which also rely heavily on direct computer simulation; hence the reason that we delay further discussion to the next chapter.

Because potential interaction, with its associated momentum and energy exchange, is an important part of transport in liquid, we next turned our attention to the potentials at the interfaces of the liquid and its surroundings: liquid–solid, liquid–vapor, and liquid–liquid interfaces. These interfacial potentials play a central role in surface chemistry and colloids, and have been extensively studied. However, their impacts on momentum and energy transport processes have yet to be fully explored. The interfacial potentials between surfaces can be constructed from the individual atomic and molecular potentials. These interfacial potentials were discussed under three categories:

1. The van der Waals potential (also called London potential or dispersion potential), which usually leads to an attractive force between surfaces separated by a vacuum. The van der Waals potential can also be repulsive, depending on the medium between the two surfaces, and is characterized by the Hamaker constant.
2. The electric double layer potential, formed as a result of the existence of surface charges and leading to a repulsive force between two surfaces separated by the liquid.
3. The potentials and forces arising from molecular structures near the interface, such as the oscillatory solvation force due to the regular molecular arrangement near a solid surface, the hydration force between hydrophilic surfaces (repulsive) and hydrophobic surfaces (attractive), the steric force due to the overlapping of long molecules (important in polymer systems), and the fluctuation force due to molecular protrusion at mobile surfaces and random thermal fluctuation.

The van der Waals force usually decays as D^{-n} (table 9.1), where D is the separation between surfaces. The electric double layer decays exponentially and its characteristic length is the Debye length, which also depends on ion concentration and thus can be controlled, while the molecular structure forces are typically active within a few molecular diameters of the interface (a few nanometers). From the interfacial potential, one can easily appreciate the concept of surface tension, a key phenomenon for micro- and nanosystems, where the inertial force is normally unimportant and the surface force dominates as the surface-to-volume ratio increases with decreasing size. The Laplace equation determines the shape of an interface and the Young equation determines the contact angle between a liquid droplet and a solid surface.

In section 9.3, we discussed the effects of size on single-phase liquid transport. There have been many studies on fluid flow and heat transfer in microchannels, but the results have been controversial. Size-dependent viscosity has been reported. Similarly, there exist reports on the deviation of heat transfer in microchannels from the continuum theorem. We caution that these data must be viewed critically. The monumental experimental work of Poiseuille, on fluid flow, which is almost 180 years old, was performed in glass tubes with inner diameter 15 μm and above. Yet his data can be explained by the continuum theory. This fact strongly indicates that discrepancies from more recent experiments in microchannels of comparable diameter, when compared with continuum theory, may not be due to the breakdown of the continuum theory but to the differences of experimental conditions and models used. We also note that, at the other extreme, there have been some studies on liquid transport in nanostructures, mostly based on molecular dynamics simulations. Because the interfacial forces discussed in section 4.2 have an active range from a few nanometers to tens or even hundreds of nanometers, these forces may impact the heat transfer and fluid flow processes in nanostructures. This is a direction that calls for more exploration, perhaps with a combination of simulation and modeling tools. In comparison with pressure-driven flow, however, electrokinetic flow, which is due to the existence of the electric double layer, has been well studied. Two basic motion configurations were discussed: electrophoretic flow due to the motion of the particles, and electro-osmotic flow due to the motion of the ionic solution. Such electrokinetic driven flows are widely used in biotechnology and have also found increasing use in a variety of microfluidic devices.

In the last section, we discussed the size dependence of thermodynamic properties of small particles, such as the phase transition pressure, temperature, and surface tension. The results have been well established but are finding new applications in nanotechnology.

9.6 Nomenclature for Chapter 9

a	constant, eq. (9.6), J m^3	e	unit charge, C
A	Hamaker constant, J	$\mathscr{E}$	electric field, V m^{-1}
A_s	surface area, m^2	f	one-particle distribution function for molecules, m^{-3}
b	volume defined by eq. (9.23), m^3	$f^{(2)}$	two-particle distribution function, m^{-6}
B	volume per molecule, m^3	$f^{(N)}$	N-particle distribution function
c	charge concentration on surface, C m^{-2}		
C	coefficient in the atom–atom pair potential, J m^6	F	Helmholtz free energy, J; force, N
D	effective diameter of molecule or spacing between surfaces, m	$\mathbf{F}_D$	drag force, N
		g	radial distribution function, eq. (9.1)

h	Planck constant, J s	ε	Lennard–Jones potential parameter, J
H	height, m		
J	particle number flux, $m^{-2} s^{-1}$	ε_0	dielectric permittivity of vacuum, $C^2 N^{-1} m^{-2}$
k	thermal conductivity, $W m^{-1} K^{-1}$		
		ε_r	dielectric constant
m	mass, kg	ζ	zeta potential, V
n	particle or atom number density, m^{-3}	η	friction coefficient, s^{-1}
		κ_B	Boltzmann constant, $J K^{-1}$
n_r	refractive index	μ	dynamic viscosity, $N s m^{-2}$
N	total number of particles		
p	pressure, $N m^{-2}$	Π	quantity defined by eq. (9.21), $kg m^{-3} s^{-1}$
p	particle momentum, $kg m s^{-1}$		
q	charge, C	ρ_d	mass density, $kg m^{-3}$
Q	charge, C	ρ_n	net charge density, $C m^{-3}$
Q_V	volumetric flow rate, $m^3 s^{-1}$	σ	Lennard–Jones potential parameter, m
r	radius, radial coordinate, or radial separation between particles, m		
		φ	volumetric fraction of Brownian particles in a solution
r	position vector, m		
R	random driving force in Brownian motion, N	ϕ	potential energy between particles, J
		Φ	total potential energy, J
s	entropy per mole, $J K^{-1} mol^{-1}$	Ψ	electrostatic potential, $J C^{-1}$
t	time, s		
T	temperature, K	Ω	solid angle, srad
u	Brownian particle velocity or drift velocity, $m s^{-1}$	$\hat{\Omega}$	unit vector connecting two colliding molecules
v	volume per mole, $m^3 mol^{-1}$		
W	work, J		Subscripts
W_{11}	cohesion energy between two identical surfaces, $J m^{-2}$		
		0	flat surface, far away
W_{12}	adhesion energy between two different surfaces, $J m^{-2}$	d	dispersion force
		E	Enskog model
Z	canonical partition function, or number of net elementary charge per ion	s	surface
		x	x-component
		zp	when surface potential is zero
α	electric polarizability, $C^2 m N^{-1}$		Superscripts
β	electric dipole moment, C m		
γ	surface tension, $J m^{-2}$ or $N m^{-1}$	$'$	outside the spherical surface, convex side
Γ	number density of molecules at the Gibbs surface, m^{-2}	$''$	inside the spherical surface, concave side
δ	Debye length, m		
δ_T	Tolmann length, m		Symbol
Δh	enthalpy or latent heat, $J mol^{-1}$	$\langle \rangle$	ensemble or time averaging

9.7 References

Alder, B.J., and Wainwright, T.E., 1967, "Velocity Autocorrelations for Hard Spheres," *Physical Review Letters*, vol. 18, pp. 988–990.

Alder, B.J., Gass, D.M., and Wainwright, T.E., 1970, "Studies in Molecular Dynamics. VIII. The Transport Coefficients for a Hard-Sphere Fluid," *Journal of Chemical Physics*, vol. 53, pp. 3813–3826.

Boon, J.P., and Yip, S., 1980, *Molecular Hydrodynamics*, Dover, New York, p. 26.

Buffat, Ph., and Borel, J.-P., 1976, "Size Effect on the Melting Temperature of Gold Particles," *Physical Review A*, vol. 13, pp. 2287–2298.

Carey, V.P., 1999, *Statistical Thermodynamics and Microscale Thermophysics*, Cambridge University Press, Cambridge, UK, chapter 6.

Chapman, S., and Cowling, T.G., 1953, *The Mathematical Theory of Non-Uniform Gases*, Cambridge University Press, Cambridge, UK.

Choi, S.U.S., Zhang, Z.G., Yu, W., Lockwood, F.E., and Grulke, E.A., 2001, "Anomalous Thermal Conductivity Enhancement in Nanotube Suspensions," *Applied Physics Letters*, vol. 79, pp. 2252–2254.

Defay, R., Prigogine, I., Bellemans, A., and Everett, D.H., 1966, *Surface Tension and Adsorption*, Longmans, London.

Derjaguin, B.V., Churaev, N.V., and Muller, V.M., 1987, *Surface Forces* (translated from the Russian by Kisin, V.I.), New York, Consultants Bureau.

Derjaguin, B.V., and Landau, L., 1941, *Acta Physicochim*, USSR, vol. 14, pp. 633–662.

Einstein, A., 1905, "On the Movement of Small Particles Suspended in a Stationary Liquid Demanded by the Molecular Kinetic Theory of Heat," *Annalen der Physik*, vol. 17, p. 549.

Einstein, A., 1906a, "A New Determination of the Molecular Dimensions," *Annalen der Physik*, vol. 19, pp. 289–306; correction, ibid., vol. 34, pp. 591–592 (1911).

Einstein, A., 1906b, "On the Theory of the Brownian Movement," *Annalen der Physik*, vol. 19, pp. 371–381.

Einstein, A., 1908, "The Elementary Theory of the Brownian Motion," *Zeitschrift für Elektrochemie*, vol. 14, pp. 235–239.

Einstein, A., 1956, *Investigations on the Theory of the Brownian Movement*, ed. R. Furth, Dover, New York.

Enskog, D., 1922, "Kungliga Svenska Vatenskapsakademiens Handlingar," *Ny Föld*, vol. 63, no. 4 (1922). For English translation, see Brush, S.G., 1972, *Kinetic Theory*, vol. 3, Pergamon Press, Oxford.

Ferziger, J.H., and Kaper, H.G., 1972, *Mathematical Theory of Transport Processes in Gases*, North-Holland, Amsterdam, chapter 12.

Freund, J.B., 2002, "Electro-Osmosis in a Nanometer Scale Channel Studied by Molecular Dynamics," *Journal of Chemical Physics*, vol. 116, pp. 2194–2200.

Gibbs, J.W., 1928, *Collected Works*, vol. I, Longmans Green, New York, p. 219.

Goodstein, D.L., 1985, *States of Matter*, Dover, New York, chapter 4.

Granasy, L., 1998, "Semiempirical van der Waals/Cahn–Hillar Theory: Size Dependence of the Tolman Length," *Journal of Chemical Physics*, vol. 109, pp. 9660–9663.

Hamaker, H.C., 1937, "The London–van der Waals Attraction between Spherical Particles," *Physica*, vol. 4, pp. 1058–1072.

Hansen, J.P., and McDonald, I.R., 1986, *Theory of Simple Liquids*, Academic Press, London, chapter 7.

Haye, M.J., and Bruin, C., 1994, "Molecular Dynamics Study of the Curvature Correction to the Surface Tension," *Journal of Chemical Physics*, vol. 100, pp. 556–559.

Heremans, J., Thrush, C.M., Lin, Y.-M., Cronin, S., Zhang, Z., Dresselhaus, M.S., and Mansfield, J.F., 2000, "Bismuth Nanowire Arrays: Synthesis and Galvanomagnetic Properties," *Physical Review B*, vol. 61, pp. 2921–2130.

Hiemenz, P.C., 1986, *Principles of Colloid and Surface Chemistry*, 2nd ed., Marcel Dekker, New York.

Hill, T.L., 1963, *Thermodynamics of Small Systems, Part I*, Benjamin, New York.
Hill, T.L., 1964, *Thermodynamics of Small Systems, Part II*, Benjamin, New York.
Ho, C.-M., and Tai, Y.-C., 1998, "Micro-Electro-Mechanical Systems (MEMS) and Fluid Flow," *Annual Review of Fluid Mechanics*, vol. 30, pp. 579–612.
Israelachvili, J.N., 1992, *Intermolecular and Surface Forces*, 2nd ed., Academic Press, London.
Kalikmanov, V.I., 1997, "Semiphenomenological Theory of the Tolman Length," *Physical Review E*, vol. 55, pp. 3068–3071.
Kittel, C., and Kroemer, H., 1980, *Thermal Physics*, 2nd ed., Freeman and Company, New York.
Kohler, F., 1972, *Liquid State*, Verlag Chemie, Weinheim, chapters 8 and 9.
Koplik, J., and Banavar, J.R., 1995, "Continuum Deductions from Molecular Dynamics," *Annual Review of Fluid Mechanics*, vol. 27, pp. 257–292.
Koplik, J., Banavar, J.R., and Willemsen, J.F., 1989, "Molecular Dynamics of Poiseuille Flow at Solid Surface," *Physics of Fluids A*, vol. 1, pp. 781–794.
Kubo, R., Toda, M., and Hashitsume, N., 1998, *Statistical Physics II*, 2nd ed., Springer, Berlin.
Laaksonen, A., Talanquer, V., and Oxtoby, D.W., 1995, "Nucleation—Measurements, Theory, and Atmospheric Applications," *Annual Review of Physical Chemistry*, vol. 46, pp. 489–524.
Liboff, R.L., 1998, *Kinetic Theory*, Wiley, New York.
Lifshitz, E.M., 1956, "The Theory of Molecular Attractive Force between Solids," *Soviet Physics, JETP* (English Translation), vol. 2, pp. 72–83.
Manchenko, G.P., 2003, *Handbook of Detection of Enzymes on Electrophoretic Gels*, CRC Press, Boca Raton, FL.
Morales, A.M., and Lieber, C.M., 1998, "A Laser Ablation Method for the Synthesis of Crystalline Semiconductor Nanowires," *Science*, vol. 279, pp. 208–211.
Obot, N.T., 2000, "Toward a Better Understanding of Friction and Heat/Mass Transfer in Microchannels—A Literature Review," in *Proceedings of the International Conference on Heat Transfer and Transport Phenomena in Microscale*, ed. G.P. Celata, Begell House, New York, pp. 72–79.
Rice, C.L., and Whitehead, R., 1965, "Electrokinetic Flow in a Narrow Cylindrical Capillary," *Journal of Physical Chemistry*, vol. 69, pp. 4017–4023.
Rice, S.A., and Allnatt, A.R., 1961, "On Kinetic Theory of Dense Fluids. VI. Singlet Distribution Function for Rigid Spheres with an Attractive Potential," *Journal of Chemical Physics*, vol. 34, p. 2144.
Rice, S.A., and Gray, P., 1965, *The Statistical Mechanics of Simple Liquids*, Wiley, New York.
Sengers, J.V., 1965, "Thermal Conductivity and Viscosity of a Moderately Dense Gas," *International Journal of Heat and Mass Transfer*, vol. 8, pp. 1103–1116.
Shockley, W., 1949, "The Theory of p–n Junctions in Semiconductors and p–n Junction Transistors," *Bell System Technical Journal*, vol. 28, pp. 435–489.
Sobhan, C.B., and Garimella, S.V., 2000, "A Comparative Analysis of Studies on Heat Transfer and Fluid Flow in Microchannels," in *Proceedings of the International Conference on Heat Transfer and Transport Phenomena in Microscale*, ed. G.P. Celata, Begell House, New York, pp. 80–92.
Sutera, S.P., and Skalak, R., 1993, "The History of Poiseuille's Law," *Annual Review of Fluid Mechanics*, vol. 25, pp. 1–19.
Thompson, P.A., and Robbins, M.O., 1990, "Shear Flow Near Solids: Epitaxial Order and Flow Boundary Conditions," *Physical Review A*, vol. 41, pp. 6830–6837.
Thompson, P.A., and Troian, S.M., 1997, "A General Boundary Condition for Liquid Flow at Solid Surfaces," *Nature*, vol. 389, pp. 360–362.
Tolman, R.C., 1949, "The Effect of Droplet Size on Surface Tension," *Journal of Chemical Physics*, vol. 17, pp. 333–337.
Velarde, M.G., 1974, "On the Enskog Hard-Sphere Kinetic Equation and the Transport Phenomena of Dense Simple Gases," in *Transport Phenomena*, ed. G. Kirczenow and J. Marro, Springer-Verlag, Berlin, pp. 289–336.

LIQUIDS AND THEIR INTERFACES 449

Verwey, E.J.W., and Overbeek, J. Th. G., 1948, *Theory of Stability of Lyophobic Colloids*, Elsevier, Amsterdam.

Wayner, P.C., Jr, 1998, "Interfacial Forces and Phase Change in Thin Liquid Films," in *Microscale Energy Transport*, ed. C.L. Tien, A. Majumdar, and F.M. Gerner, Taylor & Francis, Washington, DC.

9.8 Exercises

9.1 *van der Waals equation of states—critical point and the corresponding states.* The critical point of a liquid is where the D and E points in figure 9.2 coincide with each other. This is an inflection point (point where both the first-order and the second-order derivatives are zero) on the constant T curve in a p–V diagram.

(a) Start from the van der Waals equation, and demonstrate the critical temperature T_c and pressure p_c

$$T_c = \frac{8a}{27 B \kappa_B} \text{ and } P_c = \frac{a}{27 B^2}$$

(b) Show that the van der Waals equation can be written as

$$p_r = \frac{8 T_r}{3 v_r - 1} - \frac{3}{v_r^2}$$

where $p_r = p/p_c$, $T_r = T/T_c$, and $v_r = V/V_c$. Thus, in this normalized form, the equation of states is independent of the actual fluids. This is called the law of corresponding states.

9.2 *Saturation pressure.* The saturation pressure, line A and B in figure 9.2 can be found by requiring that the chemical potentials of the liquid and the vapor phases are equal to each other. Derive a relation between the saturation pressure and temperature for a van der Waals liquid.

9.3 *Enskog equation.* Argon gas at low density and 300 K has a thermal conductivity of 0.018 W m^{-1} K^{-1}. Estimate its thermal conductivity as a function of density at higher density. In what range do you expect this estimation to be valid?

9.4 *Einstein relation.* The viscosity of water at 300 K is 4×10^{-4} N s m^{-2} and the diffusivity of monodisperse Brownian particles in water is 1.1×10^{-11} m^2 s^{-1}. Estimate the diameter of the particle.

9.5 *Viscosity of nanofluids.*

(a) The viscosity of water is 4×10^{-4} N s m^{-2}. Nanoparticles are seeded into water with a volume concentration of 5%. What is the viscosity of the nanoparticle-loaded fluid according to the Einstein theory?

(b) The Einstein theory is based on the assumption of dilute particles such that interparticle interactions can be neglected. Estimate the interparticle distance for a 5% volume loading as a function of the nanoparticle diameter, assuming that nanoparticles are monodisperse.

9.6 *Van der Waals potential between two nanowires.* Carbon nanotubes and nanowires can be grown into aligned and closely spaced dense arrays [see figure 1.4(c)]. Estimate the van der Waals potential and the attractive force per unit length between two parallel silicon nanowires of equal diameter (10 nm) with a center-to-center spacing of 20 nm. Assume $C = 5 \times 10^{-78}$ J m^6. The silicon lattice constant is 5.2 Å.

9.7 *Debye length.* Estimate the Debye length in water containing 0.01 mole of NaCl. The dielectric constant of water is 78.54.

9.8 *Liquid helium and disjoining pressure.* It is known that if liquid helium is placed in a beaker, it rapidly climbs up the walls and down the other side, and eventually leaves the container. This is caused by a negative Hamaker constant between the helium vapor and the container wall.

(a) Show that the liquid helium film varies as a function of its thickness,

$$D = \left(-\frac{A}{6\pi \rho g H}\right)^{1/3}$$

(b) The Hamaker constant between helium vapor and the container, made of CaF_2, is -0.59×10^{-20} J. Estimate the liquid helium film height at a thickness of $D = 2$ nm. The density of liquid helium is 125 kg m^{-3}.

9.9 *Capillary rise of liquid in a tube.* In a small tube inserted into a liquid bath, the liquid rises above the height of the bath surface due to the surface tension if the contact angle is less than 90° (figure P9.9). Show that the height of the liquid column is

$$H = \frac{2\sigma \cos\theta}{(\rho_l - \rho_v) g r_i}$$

where ρ_l and ρ_v are the density of the liquid and its vapor and r_i is the inner radius of the capillary tube. For glass tubes with $r_i = 10$ μm, 100 μm, and 1 mm, estimate the heights of water inside the tube ($\gamma = 72.8$ mN m^{-1})

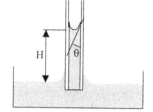

Figure P9.9 Figure for exercise 9.9.

9.10 *Electrokinetic flow.* Consider fully developed electro-osmotic flow between two parallel plates, assuming that the Debye thickness is much smaller than the separation of the two plates. Use the Hückel–Debye approximation to find the electric double layer potential distribution,

(a) Develop an expression for the velocity distribution within the electric double layer.

(b) Assuming that a constant heat flux is applied to the fluid on both surfaces and the thermal profile is fully developed, derive an expression for the Nusselt number.

9.11 *Effects of radius on water droplet surface tension and saturation vapor pressure.* For water, taking 9.6 Å^2 as the surface area occupied by a water molecule on the surface of tension, half of the monolayer concentration is $\Gamma = 0.9 \times 10^{-9}$ mol cm^{-2}. The liquid phase density is $\rho'' = 5.55 \times 10^{-2}$ mol cm^{-3}. Calculate the surface tension and the saturation vapor pressure of water droplets as a function of the diameter in the range of $r = 10^{-9}$–10^{-6} m.

9.12 *Effects of radius on water vapor bubble surface tension and saturation vapor pressure.* Using the same data as in exercise 9.11, calculate the surface tension and the saturation vapor pressure of a water bubble inside water as a function of its diameter in the range of $r = 10^{-9}$–10^{-6} m.

9.13 *Melting temperature of Au nanoparticles.* The surface tension between liquid and solid gold is 0.27 N m^{-1}, and its latent heat is 6.27×10^4 J kg^{-1}. Estimate the melting temperature of gold nanoparticles as a function of their radius. The melting point of bulk gold is 1064.43°C.

9.14 *Bismuth condensation into anodized alumina template.* Consider the condensation of bismuth vapor onto an anodized alumina template with cylindrical channels. The channel diameter is between 5 and 20 nm. Comment on the filling process of superheated bismuth vapor into the channel. The surface tension of bismuth is 378×10^{-3} N m^{-1} at melting point 271°C.

10

Molecular Dynamics Simulation

So far most of our discussion of energy storage and energy transport has been built on the reciprocal space: the dispersion relations between wavevectors and frequencies. We discarded the history of the motion of individual particles (electrons and atoms) and focused on their collective modal behavior. However, the approximate trajectory of individual particles can be traced if the computational power is sufficient, and from the calculated trajectory of all the particles we can evaluate the desired macroscopic properties or examine the microscopic processes in real space. When the particles are individual molecules or atoms, the approach is typically called molecular dynamics simulation. In classical molecular dynamics, the equations of motion for each individual atom in the system are established on the basis of an empirical interatomic force (or potential) and Newton's second law. These equations for all the atoms in the system are coupled through the interatomic potential and solved numerically. A quantum molecular dynamics simulation solves the coupled time-dependent Schrödinger equations for all the particles in the system. Exact direct numerical solution of the Schrödinger equations for a system comprising a large number of particles is impractical, and so various approximations have been used. For example, Car and Parrinello (1985) combined the classical atomistic simulation for atomic ions with the density function theory for electrons. Both classical and quantum molecular dynamic simulations require extensive computation but are becoming increasingly useful as computers become faster. The simulation results are usually analyzed on the basis of statistical mechanics principles. In this chapter, we focus on classical molecular dynamic simulations, since quantum molecular dynamics is still limited to a small number of atoms. Molecular dynamic simulation methods have been used as a basic tool in a wide range of fields. A single chapter will not be able to cover even a small number of potential applications. There are many textbooks and research monographs published on molecular dynamic simulation (Cicotti et al., 1987; Allen and Tildesley, 1987; Haile, 1992; Rapaport, 1995; Frenkel and Smit, 1996). Interested readers should consult these books for numerical details. Review articles on molecular dynamic simulations of thermal transport are also a good source on past work (Chou et al., 1999; Maruyama, 2000).

Our discussion will focus on the basic principles underlying molecular dynamic simulations and the computation of thermodynamic and transport properties. We start with the equations of motion that determine the time evolution of the simulated system, including the Newtonian, the Lagrangian, and the Hamiltonian descriptions. The interatomic potentials that enter into the equations of motion have been discussed in chapters 3 and 9 but will be further elaborated in section 10.2 to include potentials commonly used in molecular dynamic simulations. While solving the equations of motion is conceptually simple, analyzing the trajectories of atoms to obtain the desired physical properties or to gain insight on the detailed transport processes is more challenging. Statistical mechanics, and particularly the linear response theory, is the cornerstone for analyzing molecular dynamics simulation results. It will be introduced in section 10.3. We should emphasize, however, that linear response theory itself has much wider applications than just in molecular dynamics simulations. In section 10.4, numerical integration methods for the equations of motion will be introduced. The simulation of thermal transport properties and processes will be discussed in section 10.5.

10.1 The Equations of Motion

Classical molecular dynamics is based on solving Newton's equations of motion for all the molecules in the system. For the ith particle with a mass m_i, its trajectory $\mathbf{r}_i(t)$ is governed by

$$m_i \frac{d^2 \mathbf{r}_i}{dt^2} = \sum_{\substack{j=1 \\ j \neq i}}^{N} \mathbf{F}_{ij} \quad (i = 1, 2, \ldots, N) \tag{10.1}$$

where N is the number of particles in the system and $\mathbf{F}_{ij}$ denotes the force exerted on particle i by particle j. The summation is carried out over all other particles in the system. Equation (10.1) represents a total of $3N$ equations due to the three Cartesian components of $\mathbf{r}_i$ (if only translational motion is considered). The major tasks of a molecular dynamic simulation are to solve these $3N$ equations, also known as the many-body problem, and then to analyze the simulation results to obtain the information of interest.

The Newtonian equations of motion, though always correct in classical mechanics, have limitations, however, when describing different kinds of motion. For example, they are based on Cartesian coordinates and thus angular rotation and their associated coordinates cannot be easily incorporated. To describe the rotation of an object by Newton's equation of motion, one needs to use the angular momentum equations rather than eq. (10.1) directly. Such drawbacks can be overcome by using other equations of motion, particularly (1) Lagrange's equations of motion or (2) Hamilton's equations of motion. We will briefly introduce these equations for later use when dealing with the molecular dynamics simulation of systems at constant temperature. Both the Lagrange and the Hamilton equations of motion are based on a set of generalized coordinates that describe all degrees of freedom in a system. For example, if a system has N diatomic molecules, each molecule has three degrees of freedom in translation for its center of mass, two degrees of freedom in rotation, and one other degree of freedom in the relative vibration of the molecule; the total coordinates required to specify the state of the system are then $n = 6N$. We can, conceivably, treat the problem as if there were $2N$ individual atoms in the system and use Newton's equation of motion for each atom, which also leads to $6N$ equations. Such a treatment, however, separates the diatomic molecule into two independent atoms and thus adds unnecessary complications to the problem. In the Lagrangian or Hamiltonian approach, we use generalized coordinates $r_1, r_2, \ldots, r_n$ that include position, angle, and so on, and define the generalized particle coordinate vector

$$\mathbf{r} = (r_1, r_2, \ldots, r_n) \tag{10.2}$$

and the generalized particle velocity

$$\dot{\mathbf{r}} = (\dot{r}_1, \dot{r}_2, \ldots, \dot{r}_n) \tag{10.3}$$

In a conservative force field, the potential energy of the system is a function of its position only, $U(\mathbf{r})$, and its kinetic energy K may depend on both $\mathbf{r}$ and $\dot{\mathbf{r}}$; thus we have $K(\mathbf{r},\dot{\mathbf{r}})$. The Lagrangian of the system is defined as

$$L(t,\mathbf{r},\dot{\mathbf{r}}) = K(\mathbf{r},\dot{\mathbf{r}}) - U(\mathbf{r}) \tag{10.4}$$

The Lagrangian depends on time because both $\mathbf{r}$ and $\dot{\mathbf{r}}$ are functions of time. The Lagrangian itself does not give the equations of motion. The Lagrange equation of motion is derived from Hamilton's principle, which states that the motion of a system between two fixed points in space, which corresponds to two states of the system, renders the action integral

$$S = \int_1^2 L(\mathbf{r},\dot{\mathbf{r}})dt \tag{10.5}$$

an extremum. Using variational calculus, one can prove from this principle that S will be an extremum when

$$\frac{d}{dt}\left(\frac{\partial L}{\partial \dot{r}_i}\right) - \frac{\partial L}{\partial r_i} = 0 \quad (i = 1, 2, \ldots, n) \tag{10.6}$$

This constitutes the Lagrange equations of motion which describe the evolution of the system. The advantage of the Lagrange equations over the Newtonian equations lies in the generalized coordinates of the former compared to the simple Cartesian coordinates used in the latter. Thus, rotational and vibrational motion can be described consistently, using one set of equations. Later, when dealing with molecular dynamics simulation in a constant-temperature ensemble, we will introduce new additional coordinates that mimic the functions of the surroundings. If r_i denotes a component of the Cartesian coordinates for one of the atoms in a monatomic system (assuming identical mass), the Lagrangian of the system is then

$$L(t,\mathbf{r},\dot{\mathbf{r}}) = \frac{1}{2}m\sum_{i=1}^{n}\dot{r}_i^2 - U(r_1, r_2, \ldots, r_n) \tag{10.7}$$

Substituting eq. (10.7) into (10.6), we obtain

$$m\frac{d\dot{r}_i}{dt} = -\frac{\partial U}{\partial r_i} = F_i \tag{10.8}$$

which in Cartesian coordinates is nothing but Newton's equations of motion.

Although Lagrange's equations of motion extend the Cartesian coordinates of Newton's equations to generalized coordinates, the equations are not symmetric with interchange of particle coordinates and velocities, as is clear in eq. (10.6). The Hamilton equations of motion to be derived are equivalent to the Lagrange equations of motion but are in a form that is symmetric with their variables. The Hamilton equations of motion are obtained by replacing the generalized velocities $\dot{\mathbf{r}}$ in the Lagrangian, with a generalized momentum defined as

$$p_i \equiv \frac{\partial L}{\partial \dot{r}_i} \tag{10.9}$$

Using this variable, Lagrange's equations, eq. (10.6) become

$$\frac{dp_i}{dt} - \frac{\partial L}{\partial r_i} = 0 \text{ or } \dot{p}_i = \frac{\partial L}{\partial r_i} \tag{10.10}$$

The equivalence to the Lagrangian in the new variables system $(\mathbf{r},\mathbf{p})$, is obtained through the so-called Legendre transformation

$$H(t,\mathbf{r},\mathbf{p}) = \sum_i \frac{\partial L}{\partial \dot{r}_i}\dot{r}_i - L = \sum_i p_i\dot{r}_i - L \tag{10.11}$$

and the obtained function $H(t,\mathbf{r},\mathbf{p})$ is called the Hamiltonian. To derive the corresponding equations of motion, we first recognize the identity

$$dH = \sum_i \left(\frac{\partial H}{\partial r_i}dr_i + \frac{\partial H}{\partial p_i}dp_i\right) + \frac{\partial H}{\partial t}dt \tag{10.12}$$

From eq. (10.11), we have,

$$dH(t,\mathbf{r},\mathbf{p}) = \sum_i (\dot{r}_i dp_i + p_i d\dot{r}_i) - dL$$

$$= \sum_i (\dot{r}_i dp_i + p_i d\dot{r}_i) - \sum_i \left(\frac{\partial L}{\partial r_i}dr_i + \frac{\partial L}{\partial \dot{r}_i}d\dot{r}_i\right) - \frac{\partial L}{\partial t}dt$$

$$= \sum_i \left(\dot{r}_i dp_i - \frac{\partial L}{\partial r_i}dr_i\right) - \frac{\partial L}{\partial t}dt$$

$$= \sum_i (\dot{r}_i dp_i - \dot{p}_i dr_i) - \frac{\partial L}{\partial t}dt \tag{10.13}$$

where in the third step we have used eq. (10.9) and in the fourth step we used eqs. (10.10) and (10.11). Comparing eq. (10.13) with (10.12), we obtain the Hamilton equations of motion

$$\dot{p}_i = -\frac{\partial H}{\partial r_i} \quad \dot{r}_i = \frac{\partial H}{\partial p_i} \quad (i = 1, 2, \ldots, n) \tag{10.14}$$

These are first-order differential equations. Their beauty lies in the fact that the coordinates and momenta are symmetric and are of equal status in the Hamilton equations of motion. Because of their similar status, (r,p) can be considered as a new set of coordinates and they form an (r,p) phase space, which we discussed in chapters 4 and 6. The solution of Hamilton's equations gives the time history of a system in phase space, called a trajectory.

As an example, we again consider the simple case of Cartesian coordinates with the Lagrangian given by eq. (10.7). From eq. (10.9), the generalized momentum is then $p_i \equiv m\dot{r}_i$, that is, the linear momentum of the particles. According to eq. (10.11), the Hamiltonian is

$$H(\mathbf{r},\mathbf{p}) = \frac{1}{2m}\sum_{i=1}^{n} p_i^2 + U(\mathbf{r}) = K + U \tag{10.15}$$

which is the total energy of the system and is constant for an isolated system,

$$\frac{dH}{dt} = 0 \tag{10.16}$$

Substituting eq. (10.15) into the first of eq. (10.14) leads to the Newton equations of motion.

The three ways of describing the motion, called the Newtonian mechanics, the Lagrangian mechanics, and the Hamiltonian mechanics, are all equivalent. The difference mainly lies in the mathematical convenience. For some situations, it is easier to start with, for example, the Hamiltonian of a system to derive the equations of motion. We will see such examples when dealing with molecular dynamics simulations of systems at constant temperatures.

Example 10.1 *Equivalence of Newtonian, Lagrangian, and Hamiltonian equations of motion*

A ball of mass m is thrown vertically into the air with an initial velocity v_0. Derive the equations of motion for the ball using Newtonian, Lagrangian, and Hamiltonian mechanics. Neglect air friction.

Solution: The time-dependent height of the ball is $x(t)$. Under the Newtonian mechanics, the equation of motion is

$$m\frac{d^2x}{dt^2} = -mg \tag{E10.1.1}$$

where g is the gravitational acceleration and the x-coordinate points up, opposite to gravity. The above equation is subject to initial conditions

$$t = 0: \quad x = 0 \text{ and } dx/dt = v_0 \tag{E10.1.2}$$

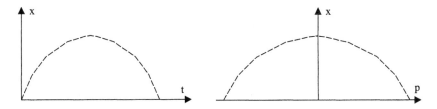

Figure E10.1 (a) Time history of the ball height. (b) Trajectory of the ball in phase space.

and its solution is

$$x(t) = -\frac{g}{2}t^2 + v_0 t \quad (E10.1.3)$$

The potential energy of the ball is $U = mgx$. The Lagrangian of the ball is, according to eq. (10.4),

$$L(x, \dot{x}) = \frac{1}{2}m\dot{x}^2 - mgx \quad (E10.1.4)$$

From eq. (10.6), the Lagrange equation of motion is

$$\frac{d}{dt}\left(\frac{\partial L}{\partial \dot{x}}\right) - \frac{\partial L}{\partial x} = 0 \quad (E10.1.5)$$

or

$$\frac{d}{dt}(m\dot{x}) + mg = 0 \quad (E10.1.6)$$

which is the same as eq. (E10.1.1) since $\dot{x} = dx/dt$.

The generalized momentum, according to eq. (10.9), is

$$p = \frac{\partial L}{\partial \dot{x}} = m\dot{x} \quad (E10.1.7)$$

which is also the momentum that we are familiar with. The Hamiltonian of the system, according to eq. (10.11), is

$$H(x, p) = p\frac{p}{m} - \frac{1}{2}m\left(\frac{p}{m}\right)^2 + mgx = \frac{p^2}{2m} + mgx \quad (E10.1.8)$$

From eq. (10.14), the Hamilton equations of motion are

$$\frac{dp}{dt} = -\frac{\partial H}{\partial x} = -mg \quad \text{and} \quad \frac{dx}{dt} = \frac{\partial H}{\partial p} = \frac{p}{m} \quad (E10.1.9)$$

The first of the Hamiltonian equations is identical to the Newton equation of motion. The second is consistent with eq. (E10.1.5).

Comment. Figure E10.1(a) shows the time history of the height of the ball. The phase space is formed by (x, p), and the Hamiltonian in the phase space is a parabolic line, determined by

$$\frac{p^2}{2m} + mgx = \frac{mv_0^2}{2} \quad (E10.1.10)$$

Each point on the trajectory represents the position and momentum of the ball at a specific time. In comparison, the Lagrangian as given by eq. (E10.1.4) can be best described by a two-dimensional surface in a three-dimensional $(x, \dot{x}, L)$ space. The Hamilton principle states that the actual trajectory of the ball is the one that leads to an extremum of eq. (10.5) among all possible trajectories described by this surface.

10.2 Interatomic Potential

Choosing an interatomic potential is a crucial step in a molecular dynamics simulation in that it determines the realism of the simulated system. Even if all numerical simulation and the subsequent analysis are performed perfectly, an inaccurate potential can lead to inaccurate conclusions about the phenomenon of interest. However, there exists no easy yet rigorous way to compute the interatomic potentials. Direct quantum mechanical molecular dynamics, or even a hybrid of quantum mechanics for electrons and classical molecular dynamics (Car and Parrinello, 1985; Lee and Chang, 1994), is a possible approach in theory but the lack of computational power has limited the number of atoms to less than ~100. Most molecular dynamics simulations are based on empirical or semi-empirical potentials. These potentials are established on the basis of various models of the force interactions among atoms. The models have unknown coefficients that are determined by fitting the molecular dynamics simulation results with experimentally determined physical parameters such as the lattice constant, melting point, elastic constants, and so on. A large number of empirical potentials have been proposed, even for the same material system. We briefly explain a few commonly used potentials.

For an N-body system of particles, the potential function can be resolved into one-body, two-body, three-body, and so on, as follows (Stillinger and Weber, 1985),

$$U(1, 2, \ldots, N) = \sum_i u_1(\mathbf{r}_i) + \sum_i \sum_{j>i} u_2(\mathbf{r}_i, \mathbf{r}_j)$$
$$+ \sum_i \sum_{j>i} \sum_{k>j>i} u_3(\mathbf{r}_i, \mathbf{r}_j, \mathbf{r}_k)$$
$$+ \cdots + u_N(\mathbf{r}_i, \mathbf{r}_j, \mathbf{r}_k, \ldots, \mathbf{r}_N) \qquad (10.17)$$

where $\mathbf{r}_i$ is the position of the ith particle, and the inequality signs for the summations are to avoid repeated counting of individual atoms. The single particle potential u_1 usually represents external potentials such as those due to walls or an external electric field. In many cases, this term is absent. The two-body potential, also called the pair potential, can have different expressions, as we discussed in chapter 3. The best-known two-body potential is the Lennard–Jones (LJ) potential

$$u_2(r_{ij}) = 4\varepsilon \left[\left(\frac{\sigma}{r_{ij}} \right)^{12} - \left(\frac{\sigma}{r_{ij}} \right)^{6} \right] \qquad (10.18)$$

where r_{ij} is the distance between particles, ε is the energy parameter, and σ is the length parameter. As explained in chapter 3, the first term in the Lennard–Jones potential models the strong repulsive force resulting from the overlapping the inner-shell electrons or ions, and the second term is the attractive electrostatic force between the instantaneous dipole of one atom and the induced instantaneous dipole of the other. The Lennard–Jones potential has a tail at large interatomic separations that requires a large simulation time if all such long-range interactions are included. In practice, the tail is truncated at an arbitrarily chosen cut-off radius a that depends on the density of the material. A typical value is $a = 2.5\,\sigma$. However, the truncation of the potential leaves a discontinuity in the potential and its derivative at $r = a$. One remedy is to modify the truncated potential, for example, into (Rapaport, 1995)

$$u_{t2}(r_{ij}) = \begin{cases} u_2(r_{ij}) - u_2(a) - (du_2/dr)_{r=a}(r_{ij} - a) & \text{if } r_{ij} < a \\ 0 & \text{otherwise} \end{cases} \quad (10.19)$$

which ensures the continuity of the energy and its first derivative (force) at $r = a$. Such a modification of the original interatomic potential, however, changes the total energy of the original system and thus it should be used carefully.

The Lennard–Jones potential is a realistic representation of molecular crystals such as argon and has been widely used for liquids and various solids as a qualitative model. Real material systems are more complicated and typically cannot be adequately described by the Lennard–Jones potential or other forms of the pair potential. The two-body potential, for example, cannot stabilize tetrahedral semiconductors such as silicon. An empirical potential for silicon is the Stillinger–Weber (1985) potential, which has a pair potential $u_2(r_{ij})$ and a three-body potential u_3,

$$u_2(r_{ij}) = \begin{cases} \varepsilon A_{sw}\left[B_{sw}\left(\frac{\sigma}{r_{ij}}\right)^4 - 1\right]\exp\left[\left(\frac{r_{ij}-a}{\sigma}\right)^{-1}\right] & r_{ij} < a \\ 0 & r_{ij} \geq a \end{cases} \quad (10.20)$$

and

$$u_3(r_i, r_j, r_k) = \varepsilon[h(r_{ij}, r_{ik}, \theta_{jik}) + h(r_{ji}, r_{jk}, \theta_{ijk}) + h(r_{ki}, r_{kj}, \theta_{ikj})] \quad (10.21)$$

with

$$h(r_{ij}, r_{ik}, \theta_{jik})$$
$$= \lambda \exp\left[\gamma\left(\frac{r_{ij}-a}{\sigma}\right)^{-1} + \gamma\left(\frac{r_{ik}-a}{\sigma}\right)^{-1}\right]\left(\cos\theta_{jik} + \frac{1}{3}\right)^2 \quad (10.22)$$

where θ_{jik} is the angle between $\mathbf{r}_{ij}$ and $\mathbf{r}_{ik}$. The three-body potential stabilizes the tetrahedral bond angle of the silicon crystals. Table 10.1 shows the recommended parameters for silicon crystals. Values for the energy ε and distance σ are deduced from the observed atomic energy and lattice spacing at 0 K in silicon. Other parameters are obtained by fitting the simulation with experimental data for the melting point and liquid structure.

Although the Stillinger–Weber potential is widely used, it has limitations. For example, Biswas and Hamann (1985) noted that the potential does not describe qualitatively the behavior of nontetrahedral polytypes of silicon, and proposed a separate form

Table 10.1 Recommended parameters in the Stillinger–Weber potential for silicon

| $\varepsilon = 3.4723 \times 10^{-19}$ J | $\sigma = 0.209\,51$ nm | $A_{sw} = 7.049\,556\,277$ | |
| $B_{sw} = 0.602\,224\,5584$ | $a = 1.80\,\sigma$ | $\lambda = 21.0$ | $\gamma = 1.20$ |

of the three-body potential. Among the various other potentials proposed for silicon, the Tersoff potential (Tersoff, 1988a) is also often used. The Tersoff potential expresses the total energy of the system as

$$U = \frac{1}{2}\sum_{i=1}^{N}\sum_{\substack{j=1\\i\neq j}}^{N} u_{ij} = \frac{1}{2}\sum_{i=1}^{N}\sum_{\substack{j=1\\i\neq j}}^{N} f_c[a_{ij}f_R(r_{ij}) - b_{ij}f_A(r_{ij})] \quad (10.23)$$

where

$$f_R(r) = A_t \exp(-\lambda_1 r) \quad (10.24)$$
$$f_A(r) = B_t \exp(-\lambda_2 r) \quad (10.25)$$

and f_c is a cutoff function

$$f_c(r) = \begin{cases} 1, & r < R_t - D_t \\ \frac{1}{2} - \frac{1}{2}\sin\left[\frac{\pi}{2}\frac{(r-R_t)}{D_t}\right], & R_t - D_t < r < R_t + D_t \\ 0 & r > R_t + D_t \end{cases} \quad (10.26)$$

Although u_{ij} looks like a two-body potential and f_A and f_R in eq. (10.23) resemble the attractive and repulsive terms, the key departure from the normal two-body potential lies in a_{ij} and particularly, b_{ij}. The latter is a measure of the bond order and is assumed to be a monotonically decreasing function of the coordination of atoms i and j. Tersoff (1988a) proposed the following form for b_{ij},

$$b_{ij} = (1 + b_t^{n_t}\xi_{ij}^{n_t})^{-1/2n_t} \quad (10.27)$$

where ξ_{ij} is given by

$$\xi_{ij} = \sum_{\substack{k=1\\k\neq i,j}}^{N} f_c(r_{ik})g(\theta_{ijk})\exp[\lambda_3^3(r_{ij} - r_{ik})^3] \quad (10.28)$$

$$g(\theta) = 1 + \frac{c_t^2}{d_t^2} - \frac{c_t^2}{d_t^2 + (l_t - \cos\theta)^2} \quad (10.29)$$

He also suggested the following form for a_{ij},

$$a_{ij} = (1 + \alpha_t^{n_t}\eta_{ij}^{n_t})^{-1/2n_t} \quad (10.30)$$

where η_{ij} is given by

$$\eta_{ij} = \sum_{\substack{k=1\\k\neq i,j}}^{N} f_c(r_{ik})\exp[\lambda_3^3(r_{ij} - r_{ik})^3] \quad (10.31)$$

but argued that α_t is generally small, such that $a_{ij} = 1$.

Table 10.2 Suggested values for silicon for Tersoff potential (Tersoff, 1988)

$A_t = 3264.7$ eV	$B_t = 95.373$ eV	$\lambda_1 = 3.2394$ Å^{-1}
$\alpha_\tau = 0$	$b_t = 0.33675$	$n_t = 22.956$
$c_t = 4.8381$	$d_t = 2.0417$	$\ell_t = 0.0000$
$\lambda_2 = \lambda_3 = 1.3258$ Å^{-1}	$R_t = 3.0$ Å	$D_t = 0.2$ Å

The parameters recommended by Tersoff for silicon are given in table 10.2. Comparisons of the various silicon potentials have been made by Balamane et al. (1992), Cook and Clancy (1993), and Halicioglu et al. (1988). The same potentials have also been widely applied to other material systems such as germanium (Ding and Andersen, 1986; Tersoff, 1989) and carbon (Tersoff, 1988b).

The Stillinger–Weber potential and the Tersoff potential have been mainly used for covalent bonding systems. For metals, a more common approach to constructing the potential function is the embedded-atom method developed by Daw and Baskes (1983, 1984). The starting point of this method is the observation that the total electron density in a metal is reasonably approximated by the linear superposition of contributions from the individual atoms. The electron density in the vicinity of each atom can then be expressed as a sum of the electron density contributed by the atom in question plus that contributed by all the surrounding atoms. This latter contribution to the electron density is a slowly varying function of position and can be approximated as constant. The potential energy of this atom is then the energy associated with the electrons of the atom plus a contribution due to the constant background electron density. This sum defines an embedding energy of an atom as a function of the background electron density and the atomic species. In addition, there is a repulsive electrostatic energy component due to core–core overlap. Under this picture, the embedded-atom method gives the total energy in the form

$$U = \sum_{i=1}^{N} u_i[\rho(\mathbf{r}_i)] + \frac{1}{2} \sum_{i=1}^{N} \sum_{\substack{j=1 \\ j \neq i}} u_{ij}(\mathbf{r}_{ij}) \qquad (10.32)$$

where ρ is the host electron density at atom i due to the remaining atoms of the system, u_i is the energy needed to embed atom i into the background electron density ρ, and u_{ij} is the core–core pair repulsion between atoms i and j separated by the distance $\mathbf{r}_{ij}$. The electron density at an atomic site is a superposition of the electrons of all other atoms in the host

$$\rho(\mathbf{r}_i) = \sum_{j \neq i} \rho_a(r_{ij}) \qquad (10.33)$$

Daw and Baskes (1983, 1984) discussed how to obtain empirically embedding energies u_i and pairwise potential u_{ij}. Please refer to their work for details.

For water, the hydrogen bond (figure 9.5) between molecules due to the imbalance of the charge distribution within an H_2O molecule should be included. Many empirical potentials for water have been developed (Guillot, 2002). The individual H_2O molecule is usually treated as rigid, with fixed interatomic distances and angles. Commonly used

Table 10.3 Parameters for water potentials (Jorgensen et al., 1983; Maruyama, 2000)

Parameter	Unit	ST2	SPC/E	TIP4P
r_{OH}	nm	0.100	0.100	0.09572
HOH angle	°	109.47	109.47	104.52
σ_{OO}	nm	0.310	0.3166	0.3154
$\sigma_{OO} \times 10^{-21}$	J	0.52605	1.0797	1.0772
r_{OM}	nm	0.08	0	0.015
Q_H	e	0.2357	0.4238	0.52
Q_M	e	−0.2357	−0.8476	−1.04

strategies are to treat the intermolecular potential between two water molecules as a sum of one pair of Lennard–Jones potentials between the two oxygen atoms and electrostatic potentials between charges of the two molecules (Jorgensen et al., 1983)

$$u_2(r_{ij}) = 4\varepsilon_{OO}\left[\left(\frac{\sigma_{OO}}{r_{ij}}\right)^{12} - \left(\frac{\sigma_{OO}}{r_{ij}}\right)^{6}\right] + \sum_i \sum_j \frac{Q_i Q_j}{4\pi \varepsilon R_{ij}} \quad (10.34)$$

where r_{ij} is the distance of oxygen atoms, and ε_{OO} and σ_{OO} are the Lennard–Jones potential parameters for the oxygen atoms; Q_i and Q_j are the effective charge of each atom in the H$_2$O molecule and R_{ij} their distance, and ε is the electric permittivity of water. In some empirical potentials for H$_2$O, charges are located at each atomic site, such as the SPC model (Jorgensen et al., 1983), while in others, charges are also attached to some fictitious sites where no atoms exist, such as the TIP4P (Jorgensen et al., 1983) and ST2 (Stillinger and Rahman 1974) models. The summation of the Coulomb potentials is over all the pairs of point charges between two H$_2$O molecules. Table 10.3 gives the parameters used in these potentials (Jorgensen et al., 1983; Maruyama, 2000). In this table, M represents a fictitous charge site lying along the bisect of the HOH angle.

10.3 Statistical Foundation for Molecular Dynamic Simulations

Before moving further into solving the equations of motion, we will first discuss the statistical tools needed to analyze the molecular dynamics simulation results. This approach is taken because the equations of motion can be modified so that molecular dynamics simulation results correspond to different ensembles. From chapter 4, we have seen that the equilibrium distributions differ among different ensembles. Thus, how the molecular dynamics simulation data is going to be analyzed determines in turn what equations of motion should be solved.

10.3.1 Time Average versus Ensemble Average

A molecular dynamic simulation gives the time history of the coordinates and momenta of all particles in the system, $\mathbf{r}(t)$ and $\mathbf{p}(t)$. If a quantity A is an explicit function of the coordinates and momenta of the system, $A(\mathbf{r}(t),\mathbf{p}(t))$, that is, a phase space variable, such as the system kinetic energy, the time average of A is

$$\bar{A} = \lim_{\Delta t \to \infty} \frac{1}{\Delta t} \int_{t_0}^{t_0+\Delta t} A\left(\mathbf{r}(t), \mathbf{p}(t)\right) dt \tag{10.35}$$

In this chapter, we use an overbar "–" to denote time average and "⟨ ⟩" for ensemble average. For an equilibrium system, the time interval Δt can be taken as large as possible since $\bar{A}$ is time independent. For a transient process, Δt is usually taken much smaller than the characteristic time of the transient but large enough for there to be enough statistical variations of system in this interval, although this is not always possible.

While the time average is usually most convenient for a molecular dynamic simulation, the kinetic theories we have introduced earlier were based on the ensemble average. As explained in chapter 4, an ensemble consists of all microscopic systems that satisfy the same macroscopic constraints. Each system in the ensemble can be represented by a point in phase space, and the whole ensemble forms a cluster of points in the phase space. The distribution function $f^{(N)}(t,\mathbf{r},\mathbf{p})$ measures the number of these systems at $(\mathbf{r},\mathbf{p})$ per unit volume of the phase space at time t. In chapter 4, we discussed the distribution function for equilibrium systems and, in chapter 6, the time evolution of the distribution function was given, as represented by the Liouville and Boltzmann equations. The ensemble average of a phase variable $A(\mathbf{r},\mathbf{p})$, that is, a function that depends on $(\mathbf{r},\mathbf{p})$ only, is then

$$\langle A(t) \rangle = \iint A(\mathbf{r},\mathbf{p}) f^{(N)}(t,\mathbf{r},\mathbf{p}) d\mathbf{r} d\mathbf{p} \tag{10.36}$$

where $A(\mathbf{r},\mathbf{p})$ is the value of A at time t, $d\mathbf{r} = dr_1 dr_2 \ldots dr_n$, and $d\mathbf{p} = dp_1 dp_2 \ldots dp_n$. If a system is ergodic, as in most cases, its time average equals the ensemble average.

There are, however, different ensembles. In chapter 4, we introduced the microcanonical ensemble with fixed (N,E,V), the canonical ensemble with fixed (N,T,V), and the grand canonical ensemble with fixed (μ,T,V), where N is the number of particles in the system, E the system energy, V the system volume, and μ the chemical potential. A valid question is then: which ensemble average does the time average correspond to? Most molecular dynamic simulations maintain the system energy constant, for fixed particles and fixed volume, hence the time average of quantities obtained from such a simulation corresponds to the microcanonical ensemble average. There are situations where it is desirable to impose a constant temperature on the system rather than a constant energy. This is very common in practice and in thermodynamic analysis, but turns out to be more difficult to realize in molecular dynamics simulations. Strategies have been developed to realize such systems (Anderson, 1980; Nosé, 1984; Hoover, 1985), as will be discussed later.

In analyzing molecular dynamics simulation results we often deal with the time correlation between two phase variables $A[\mathbf{r}(t + \Delta t), \mathbf{p}(t + \Delta t)] = A(t + \Delta t)$

and $B[\mathbf{r}(t),\mathbf{p}(t)] = B(t)$ of an equilibrium system at two different times, $t + \Delta t$ and t,

$$\overline{A(t + \Delta t)B(t)} = \lim_{\tau \to \infty} \frac{1}{\tau} \int_0^\tau A(t + \Delta t) B(t) dt \tag{10.37}$$

Note that the above integration is not over the time delay Δt, but over time origin t. The integration means that the correlation function samples all possible different time origins. When $B = A$, the correction function $\overline{A(t + \Delta t) A(t)}$ is called the autocorrelation function. If the system is ergodic, this time average is equal to the ensemble average,

$$\langle A(t + \Delta t)B(t) \rangle$$
$$= \iint A[\mathbf{r}(t + \Delta t),\mathbf{p}(t + \Delta t)]B[\mathbf{r}(t),\mathbf{p}(t)]f_0^{(N)}(t,\mathbf{r},\mathbf{p})d\mathbf{r}d\mathbf{p} \tag{10.38}$$

where $f_0^{(N)}$ is the equilibrium distribution function of the system.

10.3.2 Response Function and Kramers–Kronig Relations

Our previous treatment of transport has been mostly based on the kinetic theory of dilute particles that predicts the transport from the distribution functions of reduced order, such as the one-particle distribution function in the Boltzmann equation, which is an average of the nth-order distribution function in the Liouville equation, discussed in section 6.1, over the $(n - 1)$ momentum and spatial coordinates. We have already pointed out some limitations of this kinetic theory approach, particularly when applied to a dense medium such as liquid. A different approach is the linear response theory, developed in the 1950s by Kubo (Kubo, 1957; Kubo et al., 1957, 1998) and others, which starts from the Liouville equation and investigates the response of the system to small external disturbances from equilibrium. This approach leads to general results for the system response functions and transport coefficients, such as viscosity, thermal conductivity, and electrical conductivity, under the approximation of the linearized Liouville equation. Linear response theory is widely used to analyze molecular dynamics simulation results. Here, we provide a brief introduction to the basic ideas underlying it. Although this introduction is embedded in a chapter dealing with molecular dynamics, the theory itself has much wider implications, as will become apparent.

Consider a system that is initially at equilibrium with a Hamiltonian H_0. The system is then subjected to a small spatially homogeneous external force $F(t)$, which causes a perturbation to the system Hamiltonian, $H'(t)$. The assumption of a spatially homogeneous external force is not essential and can be relaxed to include spatial inhomogeneity (Kohler, 1972; Hansen and McDonald, 1986). The perturbation Hamiltonian $H'(t)$ can be related to the external force $F(t)$ by

$$H'(t) = -AVF(t) \tag{10.39}$$

where $A(\mathbf{r},\mathbf{p})$ is a phase variable (per unit volume), that is, a function of the positions and momenta of particles in the system, and V is the volume of the system. We will discuss the form of A and $F(t)$ in more detail later. We are interested in the response of another phase variable, $B(\mathbf{r},\mathbf{p})$, to this external force disturbance. In the linear response theory, this response will be linear to the force through an "after-effect function" $\Phi_{BA}(t)$,

$$\langle B(t)\rangle = \int_{-\infty}^{t} \phi_{BA}(t-t')F(t')dt' = \int_{0}^{\infty} \phi_{BA}(\tau)F(t-\tau)d\tau \qquad (10.40)$$

where we also assume that for the unperturbed case $\langle B_0 \rangle = 0$. The case when $\langle B_0 \rangle \neq 0$ (such as for internal energy and pressure) is only a simple extension. Equation (10.40) simply says that the response at time t is related to the previously applied force, as is required for all natural processes. This requirement is sometimes called causality. Since $F(t)$ and the response $\langle B \rangle$ are both real functions, the after-effect function $\phi_{BA}(t)$ must also be real. Our task is to derive expressions for $\phi_{BA}(t)$ that relate the response to the disturbance. This task will be accomplished in the next section. Here we will examine the properties of $\phi_{BA}(t)$ itself.

First, we introduce the temporal Fourier transformation of a function $F(t)$ and its inverse transformation,

$$F(t) = \int_{-\infty}^{\infty} \tilde{F}(\omega)e^{i\omega t}d\omega, \quad \tilde{F}(\omega) = \frac{1}{2\pi}\int_{-\infty}^{\infty} F(t)e^{-i\omega t}dt \qquad (10.41)$$

Expressing the integrand of eq. (10.40) as the inverse of its Fourier transform, we can write

$$\langle B(t)\rangle = \int_{0}^{\infty} \phi_{BA}(\tau)d\tau \left(\int_{-\infty}^{\infty} \left[\frac{1}{2\pi}\int_{-\infty}^{\infty} F(t''-\tau)e^{-i\omega t''}dt'' \right] e^{i\omega t}d\omega \right)$$

$$= \int_{0}^{\infty} \phi_{BA}(\tau) \left[\int_{-\infty}^{\infty} \tilde{F}(\omega)e^{i\omega t}e^{-i\omega \tau}d\omega \right]$$

$$= \int_{-\infty}^{\infty} d\omega \tilde{F}(\omega)e^{i\omega t}\int_{0}^{\infty} \phi_{BA}(\tau)e^{-i\omega \tau}d\tau$$

$$= \int_{-\infty}^{\infty} \chi_{BA}(\omega)\tilde{F}(\omega)e^{i\omega t}d\omega \qquad (10.42)$$

The function $\chi_{BA}(\omega)$ is called the dynamic susceptibility or the response function,

$$\chi_{BA}(\omega) = \int_{0}^{\infty} \phi_{BA}(\tau)e^{-i\omega \tau}d\tau \qquad (10.43)$$

and is generally a complex valued function,

$$\chi_{BA}(\omega) = \chi'_{BA}(\omega) + i\chi''_{BA}(\omega) \qquad (10.44)$$

The real and imaginary parts of the dynamic susceptibility are related to each other. One can see this from eq. (10.43) since the real and imaginary parts are the cosine and (−sine) Fourier transformation of $\phi_{BA}(t)$, respectively. If one of these is known over the whole spectrum, $\phi_{BA}(t)$ can be obtained from the inverse transformation and thus the other part can also be calculated. The relationship between χ' and χ'', without going into detail, is called the Kramers–Kronig relation,

$$\chi''_{BA}(\omega) = -P\left[\frac{1}{\pi}\int_{-\infty}^{\infty}\frac{\chi'_{BA}(\omega')}{\omega'-\omega}d\omega'\right]$$

$$= -\frac{1}{\pi}\left[\lim_{\omega'\to\omega_-}\int_{-\infty}^{\omega_-}\frac{\chi'_{BA}}{\omega'-\omega}d\omega' + \lim_{\omega'\to\omega_+}\int_{\omega_+}^{\infty}\frac{\chi'_{BA}}{\omega'-\omega}d\omega'\right] \quad (10.45)$$

where P represents the Cauchy principal value since the denominator goes to zero at $\omega' = \omega$. The calculation of the principal value is given by the second equation in which ω_- and ω_+ mean that the limit is taken for ω' to approach ω from $\omega' < \omega$ and $\omega' > \omega$, respectively. Similarly, if the imaginary part is known, the real part can be calculated from

$$\chi'_{BA}(\omega) = P\left[\frac{1}{\pi}\int_{-\infty}^{\infty}\frac{\chi''_{BA}(\omega')}{\omega'-\omega}d\omega'\right] \quad (10.46)$$

Despite the mathematical simplicity, the major difficulty in the application of the Kramers–Kronig relation lies in the fact that either the real or the imaginary part over the whole spectrum must be known.

As a consequence of the causality, the Kramers–Kronig relations are general and apply to various susceptibilities. The real and imaginary parts of the electric susceptibility defined in eq. (5.7) obey the Kramers–Kronig relations, from which the corresponding forms of the Kramer–Kronig relations for the real and imaginary parts of the dielectric constant can be obtained. The Fresnel coefficient for reflection at an interface is the response function for an incident electromagnetic field, and its real and imaginary part also obeys the Kramers–Kronig relation.

It can also be shown that the imaginary part of the dynamic susceptibility $\chi_{AA}(\omega)$ ($B = A$) is related to the rate at which energy is dissipated in the system due to the external disturbance $F(\omega)$ (Kohler, 1972)

$$\frac{dU}{dt} = \left\langle\frac{dH'(t)}{dt}\right\rangle = -\frac{\omega}{2}V\chi''_{AA}(\omega)|F(\omega)|^2 \quad (10.47)$$

The real part, χ'_{AA}, is a measure of the magnitude of the response. We will use eq. (10.47) in the next section to explain the fluctuation–dissipation theorem.

10.3.3 Linear Response Theory

After the above general discussion on dynamic susceptibility, we move on to show how it can be calculated under the linear response theory. As mentioned before, the linear response theory starts from the Liouville equation, eq. (6.3), which is recast as

$$\frac{\partial f^{(N)}}{\partial t} + \sum_{i=1}^{n} \frac{\partial H}{\partial p_i} \times \frac{\partial f^{(N)}}{\partial r_i} - \sum_{i=1}^{n} \frac{\partial H}{\partial r_i} \times \frac{\partial f^{(N)}}{\partial p_i} = 0 \qquad (10.48)$$

where $f(N)$ is the N-particle distribution function and we have used eq. (10.14) to replace $\dot{r}_i$ and $\dot{p}_i$. In chapter 6, we used $r^{(i)}$ and $p^{(i)}$ to distinguish the coordinates of N particles in phase space and reserved $\mathbf{r}$ and $\mathbf{p}$ for the coordinates of one particle. Here, since we are always dealing with N particles, it is more convenient to use $(\mathbf{r},\mathbf{p})$ for the generalized coordinates of N particles and (r_i, p_i) as the components of the generalized coordinates. We will also use the following two notations for the Liouville equation to simplify the mathematical operations,

$$\frac{\partial f^{(N)}}{\partial t} = \{H, f^{(N)}\} \qquad (10.49)$$

and

$$\frac{\partial f^{(N)}}{\partial t} = -i\Gamma f^{(N)} \qquad (10.50)$$

The $\{A, B\}$ denotes the Poisson bracket that obeys the following operational rules,

$$\{A, B\} \equiv \sum_{i=1}^{n} \left(\frac{\partial A}{\partial r_i} \frac{\partial B}{\partial p_i} - \frac{\partial A}{\partial p_i} \frac{\partial B}{\partial r_i} \right) \qquad (10.51)$$

and Γ is the Liouville operator, defined as

$$\Gamma \equiv i\{H,\} \qquad (10.52)$$

Using the definitions in eqs. (10.51) and (10.52), it is easy to show that eqs. (10.49) and (10.50) are identical eq. (10.48). Any quantity that follows Γ will be placed into the Poisson bracket, following H. Using the Liouville operator notation, a formal solution of eq. (10.50) is obtained by treating Γ as an ordinary function and using the method of separation of variables,

$$f^{(N)}(t,\mathbf{r},\mathbf{p}) = \exp(-it\,\Gamma) f^{(N)}(0,\mathbf{r},\mathbf{p}) \qquad (10.53)$$

where we have assumed that Γ does not depend explicitly on time. The exponential of the operator, $\exp(-it\,\Gamma)$, should be understood as another operator that acts on the function following it, which is $f^{(N)}(0,\mathbf{r},\mathbf{p})$ in this case. The exponential of an operator can be expressed as a power series of the operator, using its Taylor expansion (Evans and Morriss, 1990),

$$\exp(-it\Gamma) = \sum_{n=0}^{\infty} \frac{(t)^n}{n!}(-i\Gamma)^n = \sum_{n=0}^{\infty} \frac{t^n}{n!}\frac{d^n}{dt^n} \qquad (10.54)$$

It also follows that if A is a phase function $A(\mathbf{r},\mathbf{p})$, then

$$\dot{A} = \frac{dA}{dt} = \sum_{i=1}^{n}\left(\frac{\partial A}{\partial r_i}\frac{\partial r_i}{\partial t} + \frac{\partial A}{\partial p_i}\frac{\partial p_i}{\partial t}\right)$$

$$= \sum_{i=1}^{n}\left(\frac{\partial A}{\partial r_i}\frac{\partial H}{\partial p_i} - \frac{\partial A}{\partial p_i}\frac{\partial H}{\partial r_i}\right)$$

$$= \{A, H\} = i\Gamma A \tag{10.55}$$

To investigate how a disturbance to the Hamiltonian, eq. (10.39), affects the distribution function, we assume that the small external disturbance causes a small deviation $f_1^{(N)}$ to the general distribution function $f^{(N)}(\mathbf{r},\mathbf{p})$, which was originally at an equilibrium state $f_0^{(N)}$,

$$f^{(N)} = f_0^{(N)} + f_1^{(N)} \tag{10.56}$$

The Liouville equation under the disturbance is

$$\frac{\partial f_0^{(N)}}{\partial t} + \frac{\partial f_1^{(N)}}{\partial t} = \{H_0 + H', f_0^{(N)} + f_1^{(N)}\} \tag{10.57}$$

Following the definition of the Poisson bracket, eq. (10.57) can be written as

$$\frac{\partial f_0^{(N)}}{\partial t} + \frac{\partial f_1^{(N)}}{\partial t} = \{H_0, f_0^{(N)}\} + \{H_0, f_1^{(N)}\} + \{H', f_0^{(N)}\} + \{H', f_1^{(N)}\} \tag{10.58}$$

The first terms on the left- and right-hand sides in eq. (10.58) are equal to each other, from the Liouville equation for an undisturbed system. The approximation made in the linear response theory is to neglect the last term on the right-hand side of eq. (10.58), because both H' and f_1^N are asumed to be small, so that the equation becomes

$$\frac{\partial f_1^{(N)}}{\partial t} = \{H_0, f_1^{(N)}\} + \{H', f_0^{(N)}\} \tag{10.59}$$

Since the undisturbed system is at equilibrium, its distribution function is given by the Boltzmann distribution

$$f_0^{(N)} = \text{const} \bullet e^{-H_0/\kappa_B T} \tag{10.60}$$

where we assume that the system is conservative such that its Hamiltonian is also its total energy. The use of the above distribution function implies dealing with a canonical ensemble. Using eqs. (10.39), (10.50), and (10.60), eq. (10.59) becomes

$$\frac{\partial f_1^{(N)}}{\partial t} = -i\Gamma_0 f_1^{(N)} - \{A, f_0^{(N)}\}VF(t)$$

$$= -i\Gamma_0 f_1^{(N)} + \frac{f_0^{(N)}}{\kappa_B T}\{A, H_0\}VF(t)$$

$$= -i\Gamma_0 f_1^{(N)} + \frac{f_0^{(N)}}{\kappa_B T}\dot{A}VF(t) \tag{10.61}$$

where Γ_0 is the Liouville operator for undisturbed potential H_0, and $\dot{A}$ is the time derivative of the phase variable in the original equilibrium system. In the above derivation, we have also used eq. (10.55). A formal solution for eq. (10.61) is

$$f_1^{(N)}(t,\mathbf{r},\mathbf{p}) = \frac{V}{\kappa_B T} \int_{-\infty}^{t} e^{-i(t-t')\Gamma_0} \{\dot{A}[\mathbf{r}(t'),\mathbf{p}(t')]F(t')f_0^{(N)}\} dt' \qquad (10.62)$$

where we have written the time dependence of A through $(\mathbf{r},\mathbf{p})$ to assist our future discussion. With $f_1^{(N)}$ determined, we can calculate the ensemble average of any dynamic variable B that responds to the disturbance,

$$\langle B(t) \rangle = \iint B[\mathbf{r}(t),\mathbf{p}(t)] f^{(N)}(t,\mathbf{r},\mathbf{p}) d\mathbf{r} d\mathbf{p}$$

$$= \iint B[\mathbf{r}(t),\mathbf{p}(t)] f_1^{(N)}(t,\mathbf{r},\mathbf{p}) d\mathbf{r} d\mathbf{p}$$

$$= \frac{V}{\kappa_B T} \int_{-\infty}^{t} dt' F(t') \iint f_0^{(N)} B[\mathbf{r}(t),\mathbf{p}(t)] e^{-i(t-t')\Gamma_0} \dot{A}[\mathbf{r}(t'),\mathbf{p}(t')] d\mathbf{r} d\mathbf{p}$$

$$= \int_{-\infty}^{t} \phi_{BA}(t-t') F(t') dt' \qquad (10.63)$$

Again, we have assumed that the ensemble average of B at equilibrium is zero. Comparing eq. (10.63) with eq. (10.40), we see that the after-effect function can be expressed as

$$\phi_{BA}(t-t') = \frac{V}{\kappa_B T} \iint f_0^{(N)} B[\mathbf{r}(t),\mathbf{p}(t)] e^{-i(t-t')\Gamma_0} \dot{A}[\mathbf{r}(t'),\mathbf{p}(t')] d\mathbf{r} d\mathbf{p}$$

$$= \frac{V}{\kappa_B T} \iint f_0^{(N)} \dot{A}[\mathbf{r}(t'),\mathbf{p}(t')] e^{i(t-t')\Gamma_0} B[\mathbf{r}(t),\mathbf{p}(t)] d\mathbf{r} d\mathbf{p}$$

$$= \frac{V}{\kappa_B T} \iint f_0^{(N)} \dot{A}[\mathbf{r}(t'),\mathbf{p}(t')] B[\mathbf{r}(t-t'),\mathbf{p}(t-t')] d\mathbf{r} d\mathbf{p}$$

$$= \frac{V}{\kappa_B T} \langle B(t-t') \dot{A}(t') \rangle \qquad (10.64)$$

In the second step of the above derivation, we have used a special property of the Liouville operator on phase variables (Evans and Morriss, 1990),

$$\iint B e^{-i(t-t')\Gamma_0} A d\mathbf{r} d\mathbf{p} = \iint A e^{i(t-t')\Gamma_0} B d\mathbf{r} d\mathbf{p} \qquad (10.65)$$

In the third step of eq. (10.64), we used the solution of eq. (10.55) to express $e^{i(t-t')\Gamma_0} B$ at time t in terms of value of B at time $(t-t')$. Because the correlation function of an equilibrium system does not depend on the origin of time, we can set $t' = 0$ in eq. (10.64) to write the after-effect function as

$$\phi_{BA}(t) = \frac{V}{\kappa_B T} \langle B(t)\dot{A}(0)\rangle = -\frac{V}{\kappa_B T} \langle \dot{B}(t)A(0)\rangle$$

$$= \frac{V}{\kappa_B T} \overline{B(t'+t)\dot{A}(t')} \qquad (10.66)$$

In writing the above expressions, we have taken advantage of the fact that the correlation function for an equilibrium system does not depend on the origin of time. The last expression in eq. (10.66) relates the ensemble average to the time average, again for ergodic systems, and thus connects the molecular dynamic simulation to the ensemble statistics.

The above derivation is based on the existence of external disturbance. In the limit $F(t)$ is very small (approaching zero), the phase space trajectory $[\mathbf{r}(t), \mathbf{p}(t)]$ is not affected by $F(t)$, and $A(\mathbf{r},\mathbf{p})$ and $B(\mathbf{r},\mathbf{p})$ are just the phase functions of the system in equilibrium. Although the time averages of phase functions are zero for an equilibrium system, their correlations can have nonzero values. From eqs. (10.43) and (10.66), the susceptibility can be written as

$$\chi_{BA}(\omega) = \frac{V}{\kappa_B T} \int_0^\infty \langle B(t)\dot{A}(0)\rangle e^{-i\omega t} dt \qquad (10.67)$$

Equation (10.67) relates the susceptibility to the correlation functions of an equilibrium system. Thus, by monitoring the time history of A and B and their correlation in an equilibrium system, their response under nonequilibrium conditions can be obtained. This expression leads to various forms of the Green–Kubo formula for transport properties, as we will see later (Green, 1952; Kubo, 1957).

We use now the above expression for the susceptibility to derive the fluctuation–dissipation theorem. Equation (10.47) shows that the dissipation in a system is related to the imaginary part of $\chi_{AA}(\omega)$, which can be computed from

$$\chi''_{AA}(\omega) = \frac{1}{2i}[\chi_{AA}(\omega) - \chi^*_{AA}(\omega)] \qquad (10.68)$$

where the complex conjugate $\chi^*_{AA}(\omega)$ can be calculated from

$$\chi^*_{AA}(\omega) = \frac{V}{\kappa_B T}\int_0^\infty \langle A(\tau)\dot{A}(0)\rangle e^{i\omega \tau} d\tau = -\frac{V}{\kappa_B T}\int_{-\infty}^0 e^{-i\omega t}\langle A(t)\dot{A}(0)\rangle dt \qquad (10.69)$$

where we have used the identity

$$\langle A(0)\dot{A}(-t)\rangle = \frac{d}{dt}\langle A(0)A(-t)\rangle = -\frac{d}{dt}\langle A(0)A(t)\rangle = -\langle A(0)\dot{A}(t)\rangle \qquad (10.70)$$

From eqs. (10.68) and (10.69), we obtain

$$\chi''_{AA}(\omega) = \frac{V}{2i\kappa_B T}\int_{-\infty}^\infty e^{-i\omega t}\langle A(t)\dot{A}(0)\rangle dt = \frac{V\omega}{2\kappa_B T}\int_{-\infty}^\infty e^{-i\omega t}\langle A(t)A(0)\rangle dt \qquad (10.71)$$

Because the imaginary part is also related to the dissipation of energy, as eq. (10.47) suggests, the above relation is also called the fluctuation–dissipation theorem, which was first derived by Callen and Welton (1951). The fluctuation theory relates the time correlation of a dynamic variable A to the dissipation caused by the same variable through the imaginary part of the susceptibility. The Einstein relation, eq. (9.52), is a special case of this fluctuation–dissipation theorem. If quantum correction is considered, eq. (10.71) should be modified to

$$\chi''_{AA}(\omega) = \frac{V}{\hbar \coth[\hbar\omega/(2k_B T)]} \int_{-\infty}^{\infty} e^{-i\omega t} \langle A(t)A(0)\rangle dt \qquad (10.72)$$

where coth is the hyperbolic cotangent function

$$\frac{1}{2}\coth\left(\frac{\hbar\omega}{2k_B T}\right) = \frac{1}{2} + \left[\exp\left(\frac{\hbar\omega}{\kappa_B T}\right) - 1\right]^{-1} \qquad (10.73)$$

Although this expression resembles the Bose–Einstein distribution, its physical origin is not due to Bose–Einstein statistics but arises from the special characteristics of quantum-mechanical operators. Hence, eq. (10.72) also applies to the fluctuation of electron systems. This form of fluctuation–dissipation theorem is often used in dealing with thermal emission (Rytov et al., 1987; Polder and van Hove, 1971; Greffet et al., 2002).

In the above treatment, the disturbance $F(t)$ is assumed to be spatially homogeneous. If the disturbance is also space dependent, a spatially dependent response function can be introduced such that the response of the dynamical variable can be expressed as (Kohler, 1972, Hansen and McDonald, 1986; Kubo et al., 1998)

$$\langle B(\mathbf{r}, t)\rangle = \int_{-\infty}^{t} \int_V \phi_{BA}(\mathbf{r} - \mathbf{r}', t - t') F(\mathbf{r}', t') d\mathbf{r}' dt' \qquad (10.74)$$

Through both time and space Fourier transforms, a frequency and wavevector-dependent susceptibility can be defined

$$\chi_{BA}(\mathbf{k}, \omega) = \int_0^{\infty} \int_V \phi_{BA}(\mathbf{r}, t) e^{i(\mathbf{k}\cdot\mathbf{r} - \omega t)} d\mathbf{r} dt \qquad (10.75)$$

and the fluctuation–dissipation theorem is generalized to

$$\chi''_{AA}(\mathbf{k}, \omega) = \frac{\omega}{2V\kappa_B T} \int_{-\infty}^{\infty} e^{-i\omega t} \langle A(\mathbf{k}, t)A(-\mathbf{k}, 0)\rangle dt \qquad (10.76)$$

where $A(\mathbf{k}, t)$ is the spatial Fourier component of $A(\mathbf{r}, t)$,

$$A(\mathbf{k}, t) = \int_V A(\mathbf{r}, t) e^{i\mathbf{k}\cdot\mathbf{r}} d\mathbf{r} \qquad (10.77)$$

Based on the linear response theory, expressions for the transport coefficient can be derived. These expressions are often called the Green–Kubo formula. Let's consider a simple example first.

Example 10.2 *Green–Kubo formula for electrical conductivity*

Consider a system of charged particles. Suppose that a spatially uniform, time-dependent electric field $\mathbf{E}(t)$ is applied to the particles. This field creates a disturbance to the system Hamiltonian

$$H'(t) = \sum_{i=1}^{N}(-Z_i e \mathbf{r}_i) \bullet \mathbf{E}(t) \tag{E10.2.1}$$

so that

$$A(\mathbf{r}) = -\frac{1}{V}\sum_{i=1}^{N}(Z_i e \mathbf{r}_i) \tag{E10.2.2}$$

where Z_i is the number of charges per particle and e is the positive unit charge, so that the negative sign gives the electron charge. We can see that A depends on $\mathbf{r}$ and is thus a phase function. The term in the bracket is the dipole moment. The corresponding current can be expressed in terms of instantaneous velocity of all the charged particles,

$$\mathbf{J}(t) = \frac{1}{V}\sum_{i=1}^{N}\left(-Z_i e \frac{d\mathbf{r}_i}{dt}\right) \tag{E10.2.3}$$

Comparing eq. (E10.2.2) with (E10.2.3), we see that

$$\dot{A}(t) = \mathbf{J}(t) \tag{E10.2.4}$$

The ensemble average of the current, in the form of eq. (10.40), is

$$\langle \mathbf{J}(t) \rangle = \int_{-\infty}^{t} \phi(t-t')\mathbf{E}(t')dt' \tag{E10.2.5}$$

and the response function is given, according to eq. (10.66), by

$$\phi_{BA}(t) = \frac{V}{\kappa_B T}\langle \mathbf{J}(t,\mathbf{r})\dot{A}(0,\mathbf{r})\rangle = \frac{V}{\kappa_B T}\langle \mathbf{J}(t,\mathbf{r})\,\mathbf{J}(0,\mathbf{r})\rangle \tag{E10.2.6}$$

where the two-vector product $\mathbf{J}(\tau,\mathbf{r})\mathbf{J}(0,\mathbf{r})$ should be understood as the product of a (3×1) matrix, (J_x, J_y, J_z), with a (1×3) matrix, that is, the transpose of $\mathbf{J}$, leading to a (3×3) matrix. The Fourier transform of eq. (E10.2.5) is

$$\langle \mathbf{J}(\omega) \rangle = \bar{\bar{\sigma}}(\omega)\mathbf{E}(\omega) \tag{E10.2.7}$$

which is Ohm's law with a (3×3) electrical conductivity tensor. For an isotropic electrical conductor, the conductivity is given by

$$\sigma(\omega) = \frac{V}{3\kappa_B T}\int_{0}^{\infty}\langle J_x(t)J_x(0) + J_y(t)J_y(0) + J_z(t)J_z(0)\rangle e^{-i\omega t}dt \tag{E10.2.8}$$

where the factor of 3 arises from our averaging the autocorrelation functions in three directions. In a molecular dynamics simulation, eq. (E10.2.3) can be used to evaluate the instantaneous current flux in the system.

Comment. The above expression for electrical conductivity is one form of the Green–Kubo formula, which relates the transport coefficient to the atomic level fluctuation, originally derived by Kubo (1957). The usual static electrical conductivity is given by

$$\sigma = \lim_{\omega \to 0} \sigma(\omega) \qquad (E10.2.9)$$

The lengthy derivations in sections 10.3.1–10.3.3 lead to three major results: (1) the Kramer–Kronig relations, eqs. (10.45) and (10.46), that are a requirement of causality, in other words, the event cannot happen before the cause; (2) the susceptibility relation, eq. (10.66), that relates the susceptibility to the correlation function and eventually to the familiar Green–Kubo formula for transport properties; and (3) the fluctuation–dissipation theorem, eqs. (10.71), (10.72), and (10.76).

10.3.4 Linear Response to Internal Thermal Disturbance

The above description of the linear response theory is based on the perturbation of an external force to the system Hamiltonian, that is, eq. (10.39). Given an external disturbance $F(t)$, the phase function $A(\mathbf{r},\mathbf{p})$ that couples $F(t)$ to the system Hamiltonian can be determined. Example 10.1 shows a clear case of how an external electric field perturbs the system Hamiltonian. For thermal transport and viscous momentum transport in fluids, however, the driving forces are the temperature and pressure gradients, which are themselves of statistical nature and are internal to the system. The treatment of such internal thermal disturbance is not as clear-cut as the linear response theory developed in the previous section and is particularly difficult for heat conduction. In fact, many different approaches toward deriving a Green–Kubo formula for thermal conductivity have been developed. Kubo et al. (1957) used the generalized thermodynamic forces that drive the entropy flux. Green's (1954) original derivation was based on the Fokker–Planck equation for random processes. Luttinger (1964) used the equivalence between energy and mass under a gravitational field, and employed the latter as the external potential to derive the Green–Kubo formula. A mathematically elegant wavevector-based analysis is often used (Hansen and McDonald, 1986; Evans and Morriss, 1990). Zwanzig (1965) gave a critical review of various derivations of the Green–Kubo formula for heat conduction. Here we will adapt an approach first developed by Mori (1958) and recast by Allen and Feldman (1993). This derivation is not the most rigorous but it is more intuitive.

Consider a canonical system at equilibrium temperature T which is subject to a small temperature disturbance, δT. We assume that δT is stationary and its gradient, $\nabla(\delta T)$, is a constant. Since the final expression for the thermal conductivity is expressed in terms of the fluctuations in the original equilibrium system, the end result is not subject to this assumption. We further assume that the system is at local equilibrium after the disturbance. The local probability distribution in a canonical system is

$$f = C \exp\left[-\frac{h\Delta V}{\kappa_B (T + \delta T)}\right]$$

$$\approx C \exp\left[-\frac{h\Delta V}{\kappa_B T}\left(1 - \frac{\delta T}{T}\right)\right] \quad (10.78)$$

where h is the local energy density and ΔV is the local volume. The above expression shows that the local temperature disturbance is equal to a local energy disturbance of $\sim -h\Delta V \delta T/T$, and thus the total disturbed Hamiltonian of the system is

$$H'(t) \approx -\int h(\mathbf{r}(t))\frac{\delta T(\mathbf{r})}{T}dV \quad (10.79)$$

where $h(\mathbf{r}(t))$ should be understood as the local internal energy density when the system is at equilibrium and the integration is over the system volume. Using this perturbed Hamiltonian, we can write eq. (10.59) as

$$\frac{\partial f_1^{(N)}}{\partial t} = \{H_0, f_1^{(N)}\} + \{H', f_0^{(N)}\}$$

$$= -i\Gamma_0 f_1^{(N)} + \frac{f_0^{(N)}}{\kappa_B T}\dot{H}'$$

$$= -i\Gamma_0 f_1^{(N)} - \frac{f_0^{(N)}}{\kappa_B T^2}\int \frac{\partial h(\mathbf{r}(t))}{\partial t}\delta T dV$$

$$= -i\Gamma_0 f_1^{(N)} - \frac{f_0^{(N)}}{\kappa_B T^2}\int (-\nabla \bullet \mathbf{j}_Q)\delta T dV$$

$$= -i\Gamma_0 f_1^{(N)} - \frac{f_0^{(N)}}{\kappa_B T^2}\int [-\nabla \bullet (\mathbf{j}_Q \delta T) + \mathbf{j}_Q \bullet \nabla \delta T]dV$$

$$= -i\Gamma_0 f_1^{(N)} - \frac{f_0^{(N)}}{\kappa_B T^2}\nabla \delta T \bullet \int \mathbf{j}_Q dV$$

$$= -i\Gamma_0 f_1^{(N)} - \frac{V f_0^{(N)}}{\kappa_B T^2}\nabla \delta T \bullet \mathbf{J}_Q \quad (10.80)$$

where $\mathbf{j}_Q$ is the instantaneous local heat flux and $\mathbf{J}_Q$ the instantaneous total heat flux of the system per unit volume. From the third to the fourth step in eq. (10.80), we have used the energy conservation principle

$$\frac{\partial h}{\partial t} + \nabla \bullet \mathbf{j}_Q = 0 \quad (10.81)$$

and from the fifth to the sixth step in eq. (10.80) we assumed that $\mathbf{j}_Q \delta T$ on the system boundary is zero. The formal solution of eq. (10.80) is

$$f_1^{(N)}(t, \mathbf{r}, \mathbf{p}) = -\frac{V}{\kappa_B T^2}\nabla \delta T \bullet \int_{-\infty}^{t} e^{-i(t-t')\Gamma_0}\mathbf{J}_Q dt' \quad (10.82)$$

We now calculate the ensemble average of the system total heat flux $\mathbf{J}_Q(t)$ per unit volume. We examine the x-component, J_{Qx}, as an example,

$$\langle J_{Qx}(t)\rangle = \iint J_{Qx}[\mathbf{r}(t),\mathbf{p}(t)]f^{(N)}(t,\mathbf{r},\mathbf{p})d\mathbf{r}d\mathbf{p}$$

$$= -\frac{V}{\kappa_B T^2}\int_{-\infty}^{t}dt'\iint f_0^{(N)} J_{Qx}[\mathbf{r}(t),\mathbf{p}(t)]e^{-i(t-t')\Gamma_0}$$

$$\times \{\nabla\delta T \bullet \mathbf{J}_Q[\mathbf{r}(t'),\mathbf{p}(t')]\}d\mathbf{r}d\mathbf{p}$$

$$= -\frac{V}{\kappa_B T^2}\int_{-\infty}^{t}dt\langle J_{Qx}(t)\mathbf{J}_Q(t-t')\rangle \bullet \nabla\delta T \quad (10.83)$$

This expression is similar to eq. (10.63). In addition, the equation is identical in form to Fourier's heat conduction law. Following steps similar to those between eqs. (10.63) and (10.67), we can write the thermal conductivity tensor (2nd order) as

$$k_{ij}(\omega) = \frac{V}{\kappa_B T^2}\int_0^\infty \langle J_{Qi}(\tau)J_{Qj}(0)\rangle e^{-i\omega\tau}d\tau \quad (10.84)$$

where $i,j(=x,y,z)$ represent coordinate components. Equation (10.84) again expresses the thermal transport coefficient in terms of the equilibrium heat flux autocorrelation function, similar to the expression of the electrical conductivity in terms of the equilibrium autocorrelation function of the current fluctuation, and is called the Green–Kubo formula for thermal conductivity. For an isotropic medium, the thermal conductivity can be expressed as

$$k(\omega) = \frac{V}{3\kappa_B T^2}\int_0^\infty [\langle J_{Qx}(\tau)J_{Qx}(0)\rangle + \langle J_{Qy}(\tau)J_{Qy}(0)\rangle$$

$$+ \langle J_{Qz}(\tau)J_{Qz}(0)\rangle]e^{-i\omega\tau}d\tau \quad (10.85)$$

where the factor of three appears because the average of heat flux autocorrelation function in three directions is taken. The above expression is a general one, showing that the thermal conductivity is a function of the frequency of the disturbance. At high frequency, that is, for fast transport processes, the effective thermal conductivity can deviate significantly from the steady-state thermal conductivity (Volz, 2001). However, we should note that the above derivation still assumes local equilibrium, and thus the discussion in chapter 6 on the validity of the Cattaneo equation in fast processes should be kept in mind. The static thermal conductivity is often quoted as the Green–Kubo formula and used in molecular dynamics simulation,

$$k = \frac{V}{3\kappa_B T^2}\int_0^\infty [\langle J_{Qx}(\tau)J_{Qx}(0)\rangle + \langle J_{Qy}(\tau)J_{Qy}(0)\rangle + \langle J_{Qz}(\tau)J_{Qz}(0)\rangle]d\tau \quad (10.86)$$

Similarly, the Green–Kubo expression for the shear viscosity can be expressed as

$$\mu = \frac{V}{\kappa_B T} \int_0^\infty \langle \tau_{xz}(\tau)\tau_{xz}(0)\rangle d\tau \qquad (10.87)$$

where τ_{xz} is the off-diagonal component of the stress tensor on the constant z-plane along the x-direction.

10.3.5 Microscopic Expressions of Thermodynamic and Transport Properties

We have yet to express the heat flux, shear stress, and other macroscopic quantities in terms of the trajectories of individual particles. The easiest example is the kinetic energy of the system,

$$K = \sum_{i=1}^{n} \frac{p_i^2}{2m} \qquad (10.88)$$

Similarly, the total system energy is the sum of the kinetic energy and the potential energy,

$$H = \sum_{i=1}^{n} h_i = \sum_{i=1}^{n} \left(\frac{p_i^2}{2m} + u_i \right) \qquad (10.89)$$

Even this simple expression, however, encounters some ambiguity because the potentials are generally between particles and it is hard to define the potential of an individual particle. This can be resolved if we simply sum up all the interaction potentials. For example, if the interaction potential can be expressed as pair potential u_{ij}, the total energy can be written as

$$H = \sum_{i=1}^{n} \frac{p_i^2}{2m} + \frac{1}{2}\sum_{i=1}^{n}\sum_{j \neq i=1}^{n} u_{ij}(r_{ij}) \qquad (10.90)$$

with

$$h_i = \frac{p_i^2}{2m} + \frac{1}{2}\sum_{j=1}^{n} u_{ij}(r_{ij}) \qquad (10.91)$$

where $\frac{1}{2}$ in the potential energy expression accounts for counting each particle twice in the double summation. The thermodynamic properties are often obtained from basic thermodynamic relations. For example, the temperature can be related to the kinetic energy of the system,

$$\langle K \rangle = \frac{3}{2}N\kappa_B T = \overline{\sum_{i=1}^{n} \frac{p_i^2}{2m}} \qquad (10.92)$$

where, again, the bar means time average. This expression only counts for the translational kinetic energy. If other degrees of freedom such as rotation and vibration are involved, they should also be included.

A microscopic expression for pressure can be obtained from the virial theorem in thermodynamics (Toda et al., 1995)

$$3PV = \sum_{i=1}^{n} \left\langle p_i \frac{\partial H}{\partial p_i} \right\rangle - \sum_{i=1}^{n} \left\langle r_i \frac{\partial H}{\partial r_i} \right\rangle \tag{10.93}$$

The first term on the right-hand side represents the kinetic energy contribution and equals $3N\kappa_B T$, and the second is the potential energy contribution. If the potential interacts as a two-body potential as in eq. (10.90), eq. (10.93) becomes

$$PV = N\kappa_B T - \frac{1}{6} \sum_{i=1}^{N} \sum_{j=1}^{N} \left\langle r_{ij} \frac{\partial u_{ij}}{\partial r_{ij}} \right\rangle \tag{10.94}$$

It goes without saying that the ensemble average can be replaced by the time average, as long as the system is ergodic.

While the thermodynamic quantities are relatively easy to obtain, it is much more difficult to derive the microscopic expressions for the fluxes needed in the Green–Kubo formulae. Different approaches have been developed (Irving and Kirkwood, 1967; Evans and Morriss, 1990). Irving and Kirkwood (1967) developed an approach in 1950 by comparing the ensemble average of microscopic quantities with macroscopic conservation equations. We will use a simple case, the mass conservation, to illustrate their approach.

Consider the time-averaged density at a spatial point $\mathbf{R}$ in one system of an ensemble. The ensemble-averaged density at this point is

$$\rho(t,\mathbf{R}) = \left\langle \sum_{i=1}^{N} m\delta(\mathbf{R} - \mathbf{r}_i) \right\rangle \tag{10.95}$$

where $\mathbf{r}_i$ is the particle position. The delta function with a vector argument is similar to the normal delta function. It is zero if a particle is outside some microscopic volume dV surrounding $\mathbf{R}$ and is normalized to the whole volume of the system,

$$\int_V \delta(\mathbf{R} - \mathbf{r}_i) dV = 1 \tag{10.96}$$

The time derivative of the density is

$$\frac{\partial \rho}{\partial t} = \int \sum_{i=1}^{N} m\delta(\mathbf{R} - \mathbf{r}_i) \frac{\partial f^{(N)}(t,\mathbf{r}_i,\mathbf{p}_i)}{\partial t} d\mathbf{r}_i d\mathbf{p}_i$$

$$= -\int d\mathbf{r}_i d\mathbf{p}_i \, f^{(N)}(t,\mathbf{r}_i,\mathbf{p}_i) \frac{\partial}{\partial t} \left[\sum_{i=1}^{N} m\delta(\mathbf{R} - \mathbf{r}_i(t)) \right]$$

$$= \int d\mathbf{r}_i d\mathbf{p}_i \, f^{(N)}(t,\mathbf{r}_i,\mathbf{p}_i) \left[\sum_{i=1}^{N} m \frac{\partial \delta(\mathbf{R} - \mathbf{r}_i(t))}{\partial \mathbf{r}_i} \frac{\partial \mathbf{r}_i}{\partial t} \right]$$

$$= -\int d\mathbf{r}_i d\mathbf{p}_i \, f^{(N)}(t,\mathbf{r}_i,\mathbf{p}_i) \left[\sum_{i=1}^{N} m\mathbf{v}_i \frac{\partial \delta(\mathbf{R} - \mathbf{r}_i(t))}{\partial \mathbf{r}_i} \right]$$

$$= -\frac{\partial}{\partial \mathbf{R}} \int d\mathbf{r}_i d\mathbf{p}_i \, f^{(N)}(t,\mathbf{r}_i,\mathbf{p}_i) \left[\sum_{i=1}^{N} m\mathbf{v}_i \delta(\mathbf{R} - \mathbf{r}_i(t)) \right]$$

$$= -\frac{\partial(\rho(t, \mathbf{R})\bar{\mathbf{v}}(t, \mathbf{R}))}{\partial \mathbf{R}} \tag{10.97}$$

where

$$\rho(t, \mathbf{R})\bar{\mathbf{v}}(t, \mathbf{R}) = \sum_{i=1}^{N} m\mathbf{v}_i \delta(\mathbf{R} - \mathbf{r}_i(t)) \tag{10.98}$$

defines the velocity field $\bar{\mathbf{v}}$. Equation (10.97) is simply the equation of continuity,

$$\frac{\partial \rho}{\partial t} + \nabla \bullet (\rho \bar{\mathbf{v}}) = 0 \tag{10.99}$$

and $\rho\bar{\mathbf{v}}$ is the macroscopic mass flux. From eq. (10.98), we see that the macroscopic expression for the mass flux is

$$\mathbf{J}_m(t) = \sum_{i=1}^{N} m\bar{\mathbf{v}}_i \tag{10.100}$$

which is an apparent result. We can follow similar procedures for the time derivative of the momentum flux, $\rho\bar{\mathbf{v}}$, and compare the obtained expression with the macroscopic momentum conservation equation. This procedure leads to a microscopic expression for the shear stress tensor,

$$\tau_{xy} = \frac{1}{V} \left[\sum_i m(v_{xi} - \bar{v}_x)(v_{yi} - \bar{v}_y) + \frac{1}{2} \sum_{i=1}^{N} \sum_{j \neq i=1}^{N} x_{ij} \frac{\partial u(r_{ij})}{\partial y_{ij}} \right] \tag{10.101}$$

where x_{ij} and y_{ij} are the projections of $\mathbf{r}_{ij}$ along the x- and y-directions, respectively. The above expression is valid only when the interatomic potential can be expressed as a pairwise sum, as in eq. (10.90). For a system in equilibrium, the average velocity is zero, that is, $\bar{\mathbf{v}} = 0$ (which will be assumed in all following expressions). Similar procedures also lead to a microscopic expression for the heat flux

$$\mathbf{J}_Q(t) = V \left[\sum_{i=1}^{N} \left(\mathbf{v}_i h_i + \frac{1}{2} \sum_{j=1, j \neq i}^{N} \mathbf{r}_{ij} (\mathbf{F}_{ij} \bullet \mathbf{v}_i) \right) \right] \tag{10.102}$$

where $\mathbf{F}_{ij}$ is the force interaction between the pair of particles i and j. Again, the above expression is valid for pairwise potential only and modification should be made for a three-body potential as in the Stillinger–Weber potential (Volz and Chen, 2000). Other methods of deriving microscopic expressions for the shear stress and heat flux have been developed (Hardy, 1963; Hansen and McDonald, 1986; Evans and Morriss, 1990). Among these methods, one is designed to explore the Einstein relation that relates transport coefficient to the autocorrelation function, as represented by eq. (9.52). This approach leads to the following microscopic expressions for the shear-stress tensor and heat flux (Hansen and McDonald, 1986)

$$\tau_{xy}(t) = \frac{d}{dt} m \sum_{i=1}^{N} v_{ix} y_i(t) \qquad (10.103)$$

and

$$\mathbf{J}_Q(t) = \frac{d}{dt} \sum_{i=1}^{N} r_i(t) h(r_i) \qquad (10.104)$$

Although these simple derivative-type expressions are attractive, they cannot be used directly for simulations in periodic systems that are often implemented in actual molecular dynamics simulations. Detailed discussion of the reasons can be found in Haile (1992).

10.3.6 Thermostatted Ensembles

In the linear response theory we assume the Boltzmann statistics, which implies canonical ensembles. In a typical molecular dynamic simulation, however, the equations of motion solved correspond to a microcanonical ensemble with constant energy, volume, and number of particles. Anderson (1980) first considered the simulation of a constant pressure and constant enthalpy ensemble with a fixed number of particles. He reasoned that in a constant pressure ensemble the volume should be allowed to change, and introduced a scaled system with coordinate $\mathbf{r}' = \mathbf{r}/V^{1/3}$ and a new Lagrangian for the scaled system, from which a new set of equations of motion was derived. The equations include the volume $\Pi(=V)$ and its momentum conjugate $\dot{\Pi}$ as the variables in addition to the rescaled coordinates $\mathbf{r}'$ and their momenta $\mathbf{p}'$. Anderson proved that the ensemble average of a phase variable in the rescaled microcanonical system with generalized phase space coordinates $(\mathbf{r}', \mathbf{p}', \Pi, \dot{\Pi})$ is equivalent to the average of the variable in an ensemble with constant pressure, constant enthalpy, and the same number of particles of the original system. We will not go into the details of Anderson's derivation since our interest here is in the canonical ensemble with a constant temperature.

Anderson himself was not able to construct a similar approach for the canonical ensemble (constant N, T, and V). Instead, he suggested a stochastic process in which the particles in the simulation domain collide with a thermal reservoir at the desired temperature. If we denote the colliding frequency as ν, then during each time interval Δt of the numerical integration of the equations of motion the fraction of the particles going through the collision is $\nu \Delta t$. These particles can be chosen by associating a random number to each particle at each time step. If this random number is less than $\nu \Delta t$, the particle goes through the collision with the thermal reservoir and assumes a new velocity based on the Boltzmann distribution. Anderson also recommended that the best rate of collision with the thermal bath is of the order of

$$\nu = \frac{2akV^{1/3}}{3\kappa_B N^{2/3}} \qquad (10.105)$$

where a is a constant of the order of one, and k is the thermal conductivity. Anderson proved that the ensemble thus simulated was identical to a canonical ensemble. This method of imposing a constant temperature on the simulated system is called the Anderson thermostat.

Anderson's success in treating systems of constant pressure, constant enthalpy, and constant number of particles on the basis of a new Lagrangian inspired others to develop a similar method for the canonical ensemble. Nosé (1984) succeeded in doing so by constructing a new system with rescaled velocity and time,

$$\mathbf{v}'_i \equiv \mathbf{v}_i \text{ and } dt' = s \, dt \qquad (10.106)$$

and treating s as a new dimensional coordinate. For the scaled system, Nosé introduced a new Hamiltonian of the form

$$H_N(\mathbf{r}, s, \mathbf{p}', p_s) = \sum_{i=1}^{N} \frac{\mathbf{p}'_i \cdot \mathbf{p}'_i}{2ms^2} + u(\mathbf{r}) + \frac{p_s^2}{2M_N} + (N_f + 1)\kappa_B T \ln s \qquad (10.107)$$

where $\mathbf{p}'_i = m\mathbf{v}'_i$ is the scaled momentum of particle i, p_s is the momentum conjugating to s, T is the system temperature, and N_f is the number of the degree of freedom in the momentum of the original physical system. Nosé showed that this Hamiltonian is a conserved quantity. The Hamilton equations of motion for the newly constructed system are then

$$\frac{\partial \mathbf{p}'_i}{\partial t'} = -\frac{\partial H_N}{\partial \mathbf{r}_i} = -\frac{\partial u(\mathbf{r})}{\partial \mathbf{r}_i} = \mathbf{F}_i \qquad (10.108)$$

$$\frac{\partial \mathbf{r}_i}{\partial t'} = \frac{\partial H_N}{\partial \mathbf{p}'_i} = \frac{\mathbf{p}'_i}{ms^2} \qquad (10.109)$$

$$\frac{\partial p_s}{\partial t'} = -\frac{\partial H_N}{\partial s} = \sum_{i=1}^{N} \frac{\mathbf{p}'_i \cdot \mathbf{p}'_i}{ms^3} - (N_f + 1)\kappa_B T \frac{1}{s} \qquad (10.110)$$

$$\frac{\partial s}{\partial t'} = \frac{\partial H_N}{\partial p_s} = \frac{p_s}{M_N} \qquad (10.111)$$

Eliminating $\mathbf{p}'$ and p_s, the equations of motion can be written as

$$\frac{d^2 \mathbf{r}_i}{dt'^2} = -\frac{1}{ms^2} \frac{\partial u}{\partial \mathbf{r}_i} - \frac{2}{s} \frac{\partial s}{\partial t'} \frac{\partial \mathbf{r}_i}{\partial t'} \qquad (10.112)$$

$$M_N \frac{d^2 s}{dt'^2} = -\sum_{i=1}^{N} m_i s \frac{d\mathbf{r}_i}{dt'} \cdot \frac{d\mathbf{r}_i}{dt'} - \frac{(N_f + 1)\kappa_B T}{s} \qquad (10.113)$$

The above procedure shows the advantage of Hamilton's approach in constructing the equations of motion over Newton's approach. Although eqs. (10.112) and (10.113) are in the form of Newton's equations of motion, the "forces" are generalized. Generalized forces are difficult to construct if one starts from Newton's equations of motion.

Nosé (1984) showed that the microcanonical ensemble average of a phase variable in the scaled new system $A'(\mathbf{r},\mathbf{p}')$ corresponds to the canonical ensemble average of $A(\mathbf{r},\mathbf{p})$ in the original physical system. To do so, he first evaluated the partition function of the scaled new system, Ω, which is a microcanonical ensemble and thus with its partition function equal to the number of states,

$$\begin{aligned}
\Omega &= C \int d\mathbf{r} \int d\mathbf{p}' \int ds \int dp_s \delta(E - H_N) \\
&= C \int d\mathbf{r} \int d\mathbf{p} \int ds \int dp_s s^{N_f} \\
&\quad \times \delta\left[E - \left(\sum_i^N \frac{\mathbf{p}'_i \cdot \mathbf{p}'_i}{2m_i s^2} + u(\mathbf{r}) + \frac{p_s^2}{2M_N} + (N_f + 1)\kappa_B T \ln s\right)\right] \\
&= C \int d\mathbf{r} \int d\mathbf{p} \int ds \int dp_s \frac{s^{N_f+1}}{(N_f + 1)\kappa_B T} \\
&\quad \times \delta\left\{s - \exp\left[-\left(H(\mathbf{r},\mathbf{p}) + \frac{p_s^2}{2M_N} - E\right)\Big/(N_f + 1)\kappa_B T\right]\right\} \\
&= C \int d\mathbf{r} \int d\mathbf{p} \int dp_s \frac{1}{(N_f + 1)\kappa_B T} \exp\left[-\left(H(\mathbf{r},\mathbf{p}) + \frac{p_s^2}{2M_N} - E\right)\Big/\kappa_B T\right] \\
&= \frac{C}{(N_f + 1)} \left(\frac{2\pi M_N}{\kappa_B T}\right)^{1/2} \exp\left(\frac{E}{\kappa_B T}\right) \int d\mathbf{r} \int d\mathbf{p} \exp\left[H(\mathbf{r},\mathbf{p})/\kappa_B T\right]
\end{aligned}$$

(10.114)

where C is a normalization factor and $H(\mathbf{r},\mathbf{p})$ is the Hamiltonian of the original system. In going from the second to the third equality, we have used the following property of a delta function,

$$\delta[g(s)] = \delta(s - s_0)/g'(s) \tag{10.115}$$

The last double integral in eq. (10.114) is the core of the canonical distribution. Following a similar procedure, it can be readily shown that the microcanonical ensemble average of a phase function $A'(\mathbf{r},\mathbf{p}')$ is equal to the canonical ensemble average of $A(\mathbf{r},\mathbf{p})$ $[= A'(\mathbf{r},\mathbf{p}')]$ in the original physical system,

$$\begin{aligned}
\langle A'(\mathbf{r},\mathbf{p}')\rangle_{NVE} &= C \int d\mathbf{r} \int d\mathbf{p}' \int ds \int dp_s A'(\mathbf{r},\mathbf{p}')\delta(E - H_N)/\Omega \\
&= \int d\mathbf{r} \int d\mathbf{p} A(\mathbf{r},\mathbf{p}) \exp\left[-H(\mathbf{r},\mathbf{p})/\kappa_B T\right] \Big/ \int d\mathbf{r} \int d\mathbf{p} \exp\left[-H(\mathbf{r},\mathbf{p})/\kappa_B T\right] \\
&= \langle A(\mathbf{r},\mathbf{p})\rangle_{NVT}
\end{aligned}$$

(10.116)

which proves that a numerical simulation of the new equations of motion is equivalent to one in which the original system is maintained at a constant temperature. The proof also shows that the final results are independent of the choice of M_N. In an actual molecular dynamics simulation, M_N represents the strength of the coupling between the system and a thermal reservoir. A small M_N corresponds to a low inertia of the thermal reservoir and a rapid temperature fluctuation when the system temperature changes, while a large M_N leads to a slow, ringing type of response to a sudden temperature change (Frenkel and Smit, 1996).

Although we have proved the equivalence of the microcanonical ensemble average of the scaled new system and the canonical average of the physical system, the time rescaling by s, as in eq. (10.106), makes the time averaging in the physical system a little trickier. Equation (10.116) implies that we must use the rescaled time to calculate the physical system time correlation functions. Because s is a function of t', and this function can be random as eq. (10.111) dictates, the real time t also fluctuates in the Nosé thermostat. If we use the real time to calculate the time average, we find

$$\overline{A} = \lim_{\tau \to \infty} \frac{1}{\tau} \int_0^\tau A[\mathbf{r}(t), \mathbf{p}(t)] dt$$

$$= \lim_{\tau \to \infty} \frac{\tau'}{\tau} \frac{1}{\tau'} \int_0^{\tau'} A'[\mathbf{r}(t'), \mathbf{p}'(t')] \frac{dt'}{s}$$

$$= \lim_{\tau' \to \infty} \frac{1}{\tau'} \int_0^{\tau'} A'[\mathbf{r}(t'), \mathbf{p}'(t')] \frac{dt'}{s} \bigg/ \left[\lim_{\tau' \to \infty} \frac{1}{\tau'} \int_0^{\tau'} \frac{dt'}{s} \right]$$

$$= \overline{(A'/s)}/\overline{(1/s)} = \langle A'/s \rangle_{NVE} / \langle 1/s \rangle_{NVE} \qquad (10.117)$$

which clearly shows that the time averages in these two different time scales are not equal. This problem was solved by Hoover (1985). He showed that, by reverting to the physical time scale $dt = sdt'$, s can be eliminated from eqs. (10.112) and (10.113) and thus time rescaling is not necessary. The corresponding equations of motion are

$$\frac{\partial \mathbf{p}_i}{\partial t} = \mathbf{F}_i - \zeta \mathbf{p}_i \qquad (10.118)$$

$$\frac{\partial \mathbf{r}_i}{\partial t} = \frac{\mathbf{p}_i}{m_i} \qquad (10.119)$$

$$\frac{\partial \zeta}{\partial t} = \frac{1}{M_N} \left[\sum_{i=1}^N \frac{\mathbf{p}_i \bullet \mathbf{p}_i}{m} - N_f \kappa_B T \right] \qquad (10.120)$$

where ζ is effectively a friction factor. Under Hoover's treatment, the phase variables are $(\mathbf{r}, \mathbf{p}, \zeta)$. Note that there is one fewer variable than in Nosé's set of phase variables $(\mathbf{r}, \mathbf{p}, s, p_s)$. The Nosé and Hoover approaches to establish a canonical ensemble equivalent to microcanonical simulation are together called the Nosé–Hoover thermostat. Both Nosé (1984) and Hoover (1985) had combined their approaches with that of

Anderson's (1980) to construct a new Hamiltonian and its corresponding equations of motion for an ensemble with constant temperature, pressure, and numbers of particles.

In molecular dynamics simulation, another technique, velocity rescaling in every time integration step to keep the system kinetic energy a constant that corresponds to the desired temperature, is often used. This technique, however, does not sample any known statistical ensemble.

10.4 Solving the Equations of Motion

After establishing the equations of motion and the interatomic potential, the next step of a molecular dynamics simulation is to integrate the equations of motion, subject to appropriate initial and boundary conditions.

10.4.1 Numerical Integration of the Equations of Motion

Newton's equations of motion are second order in time and are nonlinear, arising from the nonlinear interaction force among atoms. An explicit time integration scheme is usually used, taking the potentials evaluated at previous time steps. In a typical molecular dynamics simulation, most of the computation time is spent evaluating the interatomic force. Thus a simulation algorithm should minimize the force computations. It is well known that explicit integration schemes are typically conditionally stable, and a stable algorithm usually requires small time steps. Because of the nonlinear force term in the equations of motion, a simple criterion for numerical stability cannot be obtained, though an analysis by linearizing the potential leads to a criterion that can be used to guide the selection of the time steps. In addition to numerical stability, the time step selection should also consider other characteristic times of the atoms, such as the vibrational period of the atoms in a crystal. The time steps should be much smaller the shortest period to capture the phonon picture in a crystalline solid. Another major consideration in the numerical solution is the accuracy of the integration. This latter consideration sets apart several methods that are widely used to carry out numerical integration of the equations of motion: the Verlet method, the leapfrog method, and the predictor–corrector method.

Both the Verlet method and the leapfrog method are based on the Taylor expansion of the particle coordinate variable $r_i(t)$. The Verlet method (Verlet, 1967) calculates the position at time step $(t + \Delta t)$ from

$$\mathbf{r}_i(t + \Delta t) = 2\mathbf{r}_i(t) - \mathbf{r}_i(t - \Delta t) + \Delta t^2 \frac{d^2 \mathbf{r}_i(t)}{dt^2} + O(\Delta t^4)$$

$$= 2\mathbf{r}_i(t) - \mathbf{r}_i(t - \Delta t) + \Delta t^2 \left(\frac{\mathbf{F}_i}{m_i}\right) + O(\Delta t^4) \qquad (10.121)$$

This method has a local truncation error that varies as Δt^4 and hence is third order in position. The velocity at time t is calculated from

$$\mathbf{v}_i(t) = \dot{\mathbf{r}}_i(t) = \frac{\mathbf{r}_i(t + \Delta t) - \mathbf{r}_i(t - \Delta t)}{2\Delta t} + O(\Delta t^2) \qquad (10.122)$$

and is accurate to the first order. The lower accuracy in velocity does not affect the solution of the equations of motion since the position integration does not involve velocity directly. The velocity computation accuracy, however, does impact the calculation of other statistical properties, such as total energy and transport properties, which can in turn affect the time steps required in the numerical simulation. The velocity Verlet algorithm is algebraically equivalent to eq. (10.121), but it calculates the velocity from (Swope et al., 1982)

$$\mathbf{v}_i(t + \Delta t) = \mathbf{v}_i(t) + \frac{\mathbf{F}_i(t + \Delta t) + \mathbf{F}_i(t)}{2m} + O(\Delta t^4) \qquad (10.122a)$$

which has the same order of accuracy as the Verlet algorithm.

Both the Verlet and the velocity Verlet schemes are excellent in preserving the energy of the simulated systems, which makes them particularly attractive. This energy preserving characteristic is in a sense more important than formal order of accuracy, since there is no way to accurately treat the phase-space trajectory of a complicated N-body system.

The leapfrog method is based on the following Taylor expansion,

$$\dot{\mathbf{r}}_i\left(t + \frac{\Delta t}{2}\right) = \dot{\mathbf{r}}\left(t - \frac{\Delta t}{2}\right) + \Delta t \bullet \frac{d^2 \mathbf{r}_i(t)}{dt^2} + O(\Delta t^3)$$

$$= \dot{\mathbf{r}}\left(t - \frac{\Delta t}{2}\right) + \Delta t \bullet \left(\frac{\mathbf{F}_i(t)}{m_i}\right) + O(\Delta t^3) \qquad (10.123)$$

$$\mathbf{r}_i(t + \Delta t) = \mathbf{r}_i(t) + \Delta t \bullet \dot{\mathbf{r}}(t + \Delta t/2) + O(\Delta t^3) \qquad (10.124)$$

and thus both position and velocity are of the same 2nd order in accuracy. However, in the leapfrog scheme, the velocity and position are not evaluated at the same time. This can be ratified by also computing the velocity at time t, from

$$\dot{\mathbf{r}}_i(t) = \dot{\mathbf{r}}(t \mp \Delta t/2) \pm \frac{\Delta t}{2} \frac{d^2 \mathbf{r}_i(t)}{dt^2} + O(\Delta t^2) \qquad (10.125)$$

This velocity, at the same time as the position, however, only possesses first-order accuracy.

The predictor–corrector algorithms, as their name implies, are composed of predictions and corrections to the predictions. From the position $\mathbf{r}_i(t)$ and velocity $\mathbf{v}_i(t)$ at current and previous time steps, the position $\mathbf{r}_i(t + \Delta t)$ and velocity $\mathbf{v}_i(t + \Delta t)$ at the end of the next step are first predicted by some form of Taylor expansion. Using this new position, the potential is reevaluated, and similarly with other quantities involved in the prediction step. Finally, the values at time step $(t + \Delta t)$ are corrected using combinations of the predicted and previous values of the position and velocity. There are different ways to combine the predicted and previous values, leading to different orders of accuracy. Textbooks on molecular dynamics should be consulted for more detailed descriptions (Haile, 1992; Rapaport, 1995). Compared to the Verlet and velocity methods, the predictor–corrector method does not have comparable energy conservation characteristics and is less popular.

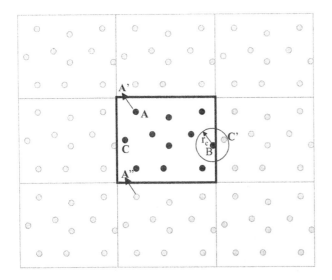

Figure 10.1 Periodic boundary condition used in molecular dynamics simulations.

10.4.2 Initial Conditions

Because there exists no way to determine exactly the position and velocity of each atom in the system, a molecular dynamics simulation usually starts with assumed position and velocity distributions and runs the program for a period of time until the system is randomized and approaches the desired statistical distribution, usually the Boltzmann statistical distribution for a system at equilibrium. During the calculation, the total momentum of the system should be checked and ensured to be zero for systems with no bulk motion.

10.4.3 Periodic Boundary Condition

The number of atoms that can be simulated by molecular dynamics is usually very small compared to real systems. Typically, a molecular dynamics system contains only a few hundred or a few thousand atoms. Large-scale simulation using high-performance computers can deal with several millions of atoms. This impressive number is still small when converted into real crystals. For example, a cubic silicon crystal of 1000 Å has ~50 million atoms inside. The small number of atoms in a typical molecular dynamics simulation means that the results may be affected by surface effects because a large proportion of atoms is on the surface. Even the "internal atoms" are affected by the surface because the force from boundary atoms can extend into the inner region of the system. Thus molecular dynamics is an ideal tool to investigate surface effects in nanosystems. To simulate large systems, the periodic boundary condition is often used to overcome the artificial surface effects caused by the small number of atoms in the system.

The periodic boundary condition is equivalent to filling the whole space with identical copies of the simulated region, as shown in figure 10.1. This periodic boundary condition affects the numerical scheme and creates artifacts that must be addressed in the analysis of the results. In a numerical scheme, under the periodic boundary condition, the atoms that leave the simulation region through a particular boundary immediately reenter the region through the opposite boundary. For example, atom A in figure 10.1 crosses into the next cell at A' at $t + \Delta t$, and this is equivalent to reappearing at A'' in the primary cell. Another effect of the periodic boundary condition is that an atom lying within a distance r_c of a boundary interacts with atoms in the adjacent copies of the system, or equivalently, with atoms near the opposite face. For example, assuming the nearest neighbors of atom B include C', we should then consider the interaction of B and C in the primary cell. In general, the force acting on particle i is

$$\mathbf{F}_i = \sum_\alpha \sum_j F[\mathbf{r}_{ij} - (n_x L_x + n_y L_y + n_z L_z)] \quad (10.126)$$

where r_{ij} is the distance between two particles in the cell and L_x, L_y, and L_z are the lengths of the simulation cell in each direction, n_x, n_y, and n_z are integers, and $(n_x L_x + n_y L_y + n_z L_z)$ is a translation vector, which effectively moves the atom at C to C', within the interaction range of the potential. Understandably, only the nearest neighbors of the copies to the original cell are important to the force correction in most cases, due to the rapid decay of interatomic potentials.

The use of periodic boundary conditions removes unwanted surface effects at the expense of introducing artificial periodicity into the simulated systems. We mentioned in section 10.3.4 that this periodic boundary condition makes the derivative-type flux expressions obtained from the Einstein relations inaccurate. For equilibrium properties—particularly thermodynamics—and local structures, available evidence suggests that the effects of periodic boundary conditions are small (Pratt and Haan, 1981a, 1981b). For transport properties, the effect is more pronounced, and these effects must be dealt with more carefully. For example, in phonon simulation (Volz and Chen, 2000), periodicity means that the longest phonon wavelength is the primary unit cell length, thus low-frequency phonons cannot be excited. The forced periodicity causes artificial correlation among atoms, which is important in the calculation of the autocorrelation functions.

10.5 Molecular Dynamics Simulation of Thermal Transport

There are two prevailing ways of simulating thermal transport. One is the equilibrium method and the other is based on simulating nonequilibrium systems. The equilibrium method is only suitable for transport properties, whereas the nonequilibrium method can be used to study actual transport processes.

10.5.1 Equilibrium Molecular Dynamics Simulation

The equilibrium method first obtains the history of individual particles in an equilibrium system, from which the transport properties are extracted on the basis of the linear response theory we discussed in section 10.3. For example, the

Green–Kubo formula for the thermal conductivity of an isotropic solid, according to eq. (10.86), is

$$k(T) = \frac{V}{3\kappa_B T^2} \int_0^\infty \langle \mathbf{J}_Q(t) \bullet \mathbf{J}_Q(0) \rangle dt$$

$$= \frac{V}{3\kappa_B T^2} \int_0^\infty \overline{\mathbf{J}_Q(t) \bullet \mathbf{J}_Q(0)} \, dt \tag{10.127}$$

where the instantaneous heat flux of the system per unit volume can be calculated from eq. (10.102). To evaluate the second integral in eq. (10.127), the autocorrelation function must first be computed as a function of the time delay. The autocorrelation function is often evaluated after solving the equations of motion, during which the basic information needed to calculate the desired properties, such as the position, velocity, and force, is stored at a predetermined time interval Δt_s. Such a time interval is typically longer than the time step Δt used in integrating the equations of motion, because storing the data at every Δt would require a large amount of storage space. In addition, we also need to decide when to start storing the data and when the stored data are useful, by judging whether the system is already in a truly equilibrium state. As explained before, it takes time for the system to randomize the initial conditions to reach a true equilibrium state. Only data after such a state is achieved are useful. To compute the time average of a correlation function, we resort to its definition in eq. (10.37). We assume that the time step of the stored molecular dynamics simulation results is constant, Δt_s. The time span of the useful data (after the system reaches equilibrium) is $M_t \Delta t_s$, where M_t is the total number of useful time steps in the stored results. According to eq. (10.37), the time correlation calculation should sample all possible time origins in the data set. If there are M sets of time origins for a given time delay t, the latter must be a multiple of Δt_s, that is, $t = j \Delta t_s$, and the autocorrelation function can then be calculated from

$$\overline{\mathbf{J}_Q(t) \bullet \mathbf{J}_Q(0)} = \frac{1}{M} \sum_{k=1}^M \mathbf{J}_Q(t_k) \bullet \mathbf{J}_Q(t_k + t) \tag{10.128}$$

Taking $t_1 = 0$ as the first useful time origin, that is, when the system is truly at an equilibrium state, the available time origin M is dependent on the length of the time delay t,

$$M = M_t - \frac{t}{\Delta t_s} = M_t - j \tag{10.129}$$

where j is an integer less than M_t.

Numerical simulations of transport properties were first performed on liquids (Alder and Wainwright, 1958, 1967; Levesque et al., 1973). Alder et al. (1970) simulated the thermal conductivity and viscosity of hard-sphere fluids and found, for example, that the thermal conductivity results are in close agreement with the Enskog theory (section 9.1.2) at all densities, but that the shear viscosity differs at higher densities (figure 9.4).

The larger shear stresses from molecular dynamics simulations than those of the Enskog results were attributed to the presence of a slowly decaying, positive tail in the shear stress autocorrelation function, $\sim t^{-3/2}$ for time in the range from 10 to 30 mean collision time (Erpenbeck and Wood, 1981).

Molecular dynamics simulation has been used to investigate the thermal conductivity of various solids. Crystal solids with van der Waals potentials were first studied by Levesque et al. (1973). Ladd et al. (1986) computed the thermal conductivity of fcc crystals with an interatomic potential proportional to the 12th power of the interatomic separation. They also compared the results from the direct Green–Kubo formula with those obtained from the phonon transport picture, and found good agreement between the two sets. The equilibrium method was also applied to simulate the thermal conductivity of amorphous materials (Lee et al., 1991; Michalski, 1992; Feldman et al., 1993) and complex crystals such as clathrate (Dong et al., 2001). The simulation of low thermal conductivity materials, such as amorphous materials and inert crystals that can be described by the van der Waals potential, is relatively easy because the interatomic potentials are simple. More recently, molecular dynamics simulations of the thermal conductivity of highly conductive crystalline materials were reported, including Si (Volz and Chen, 2000), β-SiC (Li et al., 1998), and diamond (Che et al., 2000). For the simulation of high thermal conductivity materials, it is found that direct integration of the static Green–Kubo formula from numerical simulation usually does not lead to good results. Both Li et al. (1998) and Che et al. (2000) used exponential fitting of the autocorrelation function. The latter group considered two exponential decays, one representing the optical phonons and the other the acoustic phonons, and found that optical phonons contribute less than 0.1% to the diamond thermal conductivity. Volz and Chen (2000) argued that the long-wave and low frequency modes are excluded due to the periodic boundary condition in a molecular dynamics simulation, and thus opted to obtain the thermal conductivity from the spectral analysis of the autocorrelation function. Their method is equivalent to the assumption of an exponentially decaying autocorrelation function, and leads to the following expression for the frequency-dependent thermal conductivity,

$$k(\omega) = \frac{V}{3\kappa_B T^2} \int_0^\infty \langle \mathbf{J}_Q(t) \bullet \mathbf{J}_Q(0) \rangle e^{-i\omega t} dt$$

$$= \frac{V}{3\kappa_B T^2} \int_0^\infty \langle \mathbf{J}_Q(0) \bullet \mathbf{J}_Q(0) \rangle e^{-t/\tau} e^{-i\omega t} dt$$

$$= \frac{V \langle |\mathbf{J}_Q(0)|^2 \rangle}{3\kappa_B T^2 (i\omega + 1/\tau)} \tag{10.130}$$

$$|k(\omega)| = \frac{k(0)}{\sqrt{1 + (\omega\tau)^2}} \tag{10.131}$$

where τ is the heat flux autocorrelation function decay time and ω is the frequency of excitation. Thus, by fitting the frequency-dependent thermal conductivity obtained from direct Fourier transformation of the heat flux autocorrelation function with eq. (10.131), the thermal conductivity at the zero frequency limit can be obtained. Figure 10.2(a) shows an example of the such a fit, and figure 10.2(b) the simulated

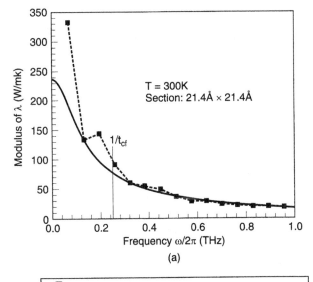

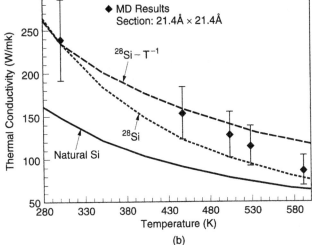

Figure 10.2 (a) Fit of the modulus of the spectral thermal conductivity by eq. (10.131). (b) Static thermal conductivity of silicon crystals obtained as a function of temperature; dots are molecular dynamics simulation results, lines are experimental results for natural and isotope-enriched silicon (Volz and Chen, 2000).

temperature-dependent thermal conductivity of bulk silicon crystals. Also marked in figure 10.2(a) is the minimum frequency allowed in the simulation domain, below which the frequency-dependent data cannot be trusted because of the artificial autocorrelation introduced by the periodic boundary conditions.

We also note that, in addition to the above-described simulation based on purely the autocorrelation function of the equilibrium system, there have also been efforts to solve the perturbed equations of motion (Ciccotti et al., 1978; Evans, 1982; Gillan and Dixon, 1983). Ciccotti et al. (1978) developed a method to simulate two equations; one is the original equations of motion for an equilibrium system, and the other is the equations of motion for a perturbed system. The autocorrelation of the heat flux vector arising from the perturbation is computed and used to derive the thermal conductivity of the simulated system. The perturbed heat flux depends on the wavevector of the perturbation and thus

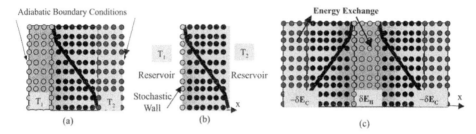

Figure 10.3 Different nonequilibrium molecular dynamics methods for simulating heat conduction: (a) constant temperatures are imposed in two regions of the simulation domain; (b) stochastic wall method; (c) heat flux method.

the thermal conductivity must be extrapolated to zero wavevector. Ciccotti et al. (1978) encountered some difficulty in such an extrapolation, which was addressed by Gillan and Dixon (1983). The latter authors showed that the revised scheme is equivalent to the results of the Green–Kubo method for a Lennard–Jones crystal. Evans (1982) developed a similar approach, however, without relying on direct computation of the perturbed heat flux. He argued that the Green–Kubo formula is not compatible with the periodic boundary condition because the equations of motion are discontinuous under this condition (Evans and Morriss, 1990). Evans called his method the homogeneous nonequilibrium molecular dynamics. There has not been any effort, however, to quantify the difference between the direct Green–Kubo formula-based simulation results and the homogeneous nonequilibrium molecular dynamics simulations, although Evans's paper (1982) shows that his method leads to excellent agreement with experimental data for argon crystals.

10.5.2 Nonequilibrium Molecular Dynamics Simulations

Nonequilibrium molecular dynamics methods are widely used to investigate fluid flow (Koplik and Banavar, 1995) and heat transfer processes (Chou et al., 1999; Maruyama, 2002). Due to the limitation on computational power, most nonequilibrium molecular dynamics simulations have so far been performed for one-dimensional heat conduction. In the lateral direction, periodic boundary conditions are often used. For heat transfer, two different approaches are taken to create the nonequilibrium transport conditions: impose a temperature difference to calculate the heat flux ($\Delta T \rightarrow Q$), or impose a heat flux to calculate the resulting temperature distributions ($Q \rightarrow \Delta T$). Within each of these categories, the actual implementation methods vary. We will discuss some of these methods.

Early works focused on the $\Delta T \rightarrow Q$ approach (Hoover and Ashurst, 1975; Levesque et al., 1973; Ciccotti and Tenenbaum, 1980; Tenenbaum et al., 1982). The key issue is how to impose the hot and cold reservoirs. Figures 10.3(a–c) illustrate various methods that have been used to impose hot and cold walls. In the hot and cold reservoir method [figure 10.3(a)], part of the simulation domain is designated as hot and the other part as cold. The average temperatures in the reservoirs are monitored and maintained as predetermined temperatures. Ashurst (1974) and Hoover and Ashurst (1975) designed methods to maintain the hot and cold reservoir temperatures. In the stochastic wall method [Figure 10.3(b)] (Lebowitz and Spohn, 1978; Ciccotti and Tenenbaum, 1980; Tenenbaum et al., 1982), no atoms are included in the hot and cold reservoirs. Instead, the temperature difference is established between the two ends of the simulation domain by the following method. Once an atom moves across the simulation domain during a simulation step and enters the virtual reservoir, its velocity components are reassigned by sampling the following Boltzmann statistics,

$$f(v_y) = \left(\frac{m}{2\pi \kappa_B T}\right)^{1/2} \exp\left(-\frac{mv_y^2}{2\kappa_B T}\right) \text{ (similarly for } v_z) \quad (10.132)$$

$$f(v_x) = \frac{m}{\kappa_B T} v_x \exp\left(-\frac{mv_x^2}{2\kappa_B T}\right) \text{ (} v_x \text{ pointing to simulation domain)} \quad (10.133)$$

and its position is reassigned to $\mathbf{R} + \mathbf{v}(t)\Delta t$, where $\mathbf{R}$ is the computed position in the virtual reservoir. If this position falls back into the simulation domain, the subsequent motion of the atom follows the same normal equations of motion as other particles. If this newly assigned position is still outside the simulation domain, it is moved onto the stochastic wall by translating along the x-direction from the newly calculated position.

One concern with specifying the temperatures of hot and cold sides is that a large temperature difference is needed to establish a converging temperature profile. To overcome the problem, a constant heat flux method was developed (Kotake and Wakuri, 1994; Ikeshoji and Hafskjold, 1994). The heat flux is added into (or extracted from) the hot (or cold) regions by rescaling the velocity in the region,

$$\mathbf{v}_i = (1+\alpha)\mathbf{v}'_i + \boldsymbol{\beta} \quad (10.134)$$

$$\boldsymbol{\beta} = -\frac{\alpha \sum m_i \mathbf{v}'_i}{\sum m_i} \quad (10.135)$$

where $\mathbf{v}'_i$ and $\mathbf{v}_i$ are the velocities before and after the energy addition, respectively; α is a scaling factor and $\boldsymbol{\beta}$ maintains the momentum before and after the energy addition at zero so that the reservoirs remain stationary, and the summation is over all the particles in each region. Energy conservation requires that the added energy ΔU (or extracted energy, by setting ΔU negative) is

$$\Delta U = \frac{1}{2} \sum m_i (\mathbf{v}_i^2 - \mathbf{v}_i'^2) \quad (10.136)$$

Equations (10.134–10.136) uniquely determine α and $\boldsymbol{\beta}$ for a given ΔU. In figure 10.3(c), two symmetric regions are simulated, a treatment that would also allow a periodic boundary condition to be applied for the large nonequilibrium unit cell. The corresponding heat flux along the x-direction is

$$J_x = \frac{\Delta U}{2\Delta t L_y L_z} \quad (10.137)$$

where L_y and L_z are the lengths in the y- and z-direction, respectively. In addition to the method of imposing a heat flux described above, another method is to swap the atoms in the cold and hot sides randomly (Osman and Srivastava, 2001).

After the proper methods of imposing the hot and cold walls have been chosen, the trajectories of the atoms in the whole simulation domain are obtained from solving the equations of motion. The local thermodynamic and transport properties such as temperature, pressure, heat flux, and so on are calculated on the basis of similar microscopic expressions as given in section 10.3.4. However, these properties are no longer averaged over the whole simulation region, but only over all atoms in a small segment, such as between x and $x + \Delta x$ for heat conduction along the x-direction. The choice of Δx depends on the statistical fluctuation. Often, for heat conduction problems, it was found that $\Delta x \approx 1$ atomic plane is reasonable (Tenenbaum et al., 1982; Hafskjold and Ratkje, 1995). To reduce statistical fluctuations, such properties are often averaged over some time interval of the steady-state results.

The nonequilibrium molecular dynamics methods are relatively easy to implement and are usually faster than the equilibrium method because the latter requires the calculation of the autocorrelation function, which can take a long time to decay. However, nonequilibrium molecular dynamics simulations suffer from more drawbacks, as we discuss below.

First, the statistical foundation of nonequilibrium molecular dynamics is not as soundly established as that of equilibrium molecular dynamics simulations. For the constant reservoir temperature method shown in figure 10.3(a), there have been efforts to simulate the reservoirs using the Nosé–Hoover thermostat (Poetzsch and Bottger, 1994), but most simulations often use a simple velocity rescaling scheme to maintain the reservoirs obeying a Boltzmann statistical distribution. Moreland (2004) compared the velocity rescaling scheme and the Anderson thermostat in the simulation of the thermal conductivity of carbon nanotubes diamond nanowires and observed differences in the results obtained for the two schemes. For the segment in between the reservoirs Newton's equations of motion are used, but it is not clear what kind of canonical ensemble such a simulation corresponds to, nor it is clear that the system is at local equilibrium. Hafskjold and Ratkje (1995) considered the criteria of local thermal equilibrium for heat and mass transport in a binary mixture and studied the Onsager reciprocity relations by molecular dynamics simulations (Onsager, 1931a, 1931b). They concluded that local equilibrium can be reached within a few atomic layers away from the hot/cold walls. Their conclusion should not be extended to solids, however, in which phonons may have long mean free paths.

On the practical side, non-homogeneous boundary conditions impose limits on the particle mean free path and the maximum wavelength. The simulation domain must be longer than the mean free path to reduce the effects of the reservoirs. The technique is thus best suitable for systems where the mean free path is small, such as liquids and amorphous materials. In addition, the temperature difference and gradient must be artificially large to reduce the computational error. It is thus difficult to determine the reference temperature of the system.

Most nonequilibrium molecular dynamics simulations show some rapid, and sometimes even discontinuous, change of temperature distributions near the hot and cold walls. The major reason for this phenomenon is because the velocity distributions of the particles near the walls are not similar to those in the internal region of the simulation domain. This is similar to photon (or phonon) transport between two parallel plates as shown in figures 7.4 and 7.5. We assume that the particle distribution coming into the simulation domain obeys the Boltzmann statistics. Inside the simulation domain, if the transport is by diffusion, the particle distribution is described by eq. (6.51). The transition from the Boltzmann distribution at the boundary to this diffusion distribution leads to the rapid change in the temperature distribution. If the simulation domain is not long enough, the internal region may never reach equilibrium as required for eq. (6.51) to be valid. However, the temperature distribution can still be quite linear in the central region. This is also the case of figure 7.4, which shows that the temperature distribution (or internal energy for photons) can be quite linear even when the mean free path is much longer than the separation of the two plates. In many molecular dynamics simulations, it is often taken that local equilibrium is reached when the central section shows a linear temperature profile. The slope of this temperature profile is used to calculate the thermal conductivity of the "bulk" material. This treatment is dangerous since we know that the solution of the Boltzmann equation for phonon transport in thin films does not lead to the same thermal conductivity as for bulk. So far, however, most nonequilibrium molecular dynamics simulations in thin films target low thermal conductivity crystals, such as argon crystals, for which this issue is less severe.

10.5.3 Molecular Dynamics Simulation of Nanoscale Heat Transfer

Nanostructures fall into the interesting regime that is reachable by direct molecular dynamics simulation. Both equilibrium and nonequilibrium molecular dynamics simulations have been used to study nanoscale heat transfer phenomena.

One interesting and challenging example is the thermal conductivity of carbon nanotubes. Various molecular dynamics simulation methods have been attempted on this problem because of the difficulties involved in measuring the thermal conductivity of a single carbon nanotube (Kim et al., 2001). Equilibrium molecular dynamics simulations have been used to simulate the thermal conductivity of carbon nanotubes and reported a thermal conductivity of 2980 W m^{-1} K^{-1} for a (10,10) carbon nanotube (Che et al., 2000), based on a cross-sectional area of 2.1×10^{-19} m^2 (assuming a hollow cylinder with a 1 Å thick wall). Berber et al. (2000) employed Evans's homogeneous nonequilibrium molecular dynamics simulation for carbon nanotubes and reported a high thermal conductivity of 6600 W m^{-1} K^{-1} for a (10,10) single-walled carbon nanotube, based on a cross-sectional area of $\sim 8 \times 10^{-19}$ m^2 (assuming a hollow cylinder with a 3.4 Å thick wall). Osman and Srivastava (2001) carried out a nonequilibrium molecular dynamics simulation of (5,5), (10,10), and (15,5) carbon nanotubes and reported a thermal conductivity of $\sim$2800 W m^{-1} K^{-1} for the (10,10) nanotube based on a cross-sectional area of 8×10^{-19} m^2 (using the same area as in Berber et al.'s simulation). Maruyama (2002) performed a nonequilibrium molecular dynamics simulation and found that the thermal conductivity increases linearly with length for tubes up to 0.5 μm and that the value obtained for (10,10) nanotubes is less than 500 W m^{-1} K^{-1}, based again on $A_c = 8 \times 10^{-19}$ m^2. Moreland et al. (2004) simulated longer (10,10) nanotubes and observed convergence of thermal conductivity as a function of length, but the converging value is $\sim$830 W m^{-1} K^{-1} for a cross-sectional area of 14.5×10^{-19} m^2. The large variations in the reported values based on various methods suggest the difficulties in simulating materials of high thermal conductivity.

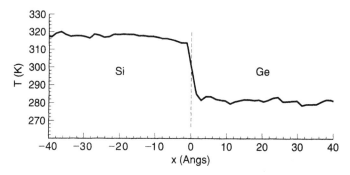

Figure 10.4 Temperature distribution in the vicinity of a Si-Ge interface (courtesy of J.B. Freund).

On the other hand, for solid nanostructures such as thin films and nanowires, the thermal conductivity is often reduced because of the size effects discussed in chapter 7. In a sense, all nonequilibrium molecular dynamics simulations previously reported, although targeting bulk materials, are actually for nanoscale thin films because the thickness values of these simulated domains are generally in the nanometer range. Thermal conductivity of thin films and superlattices has been simulated with various nonequilibrium methods (Lukes and Tien, 2000; Abramson et al., 2000; Daly et al., 2002) and also with the equilibrium method (Volz et al., 2000). Volz and Chen (1999) simulated the thermal conductivity of silicon nanowires using a nonequilibrium method and observed convergence of the thermal conductivity in relatively short nanowires and a large reduction of the thermal conductivity value compared to that of bulk silicon.

Molecular dynamics is ideally suited for interfacial problems, provided the interface region can be decoupled from bulk. How to implement such a decoupling is still an open question. There have been interesting studies using molecular dynamics to simulate, for example, thermal boundary resistance phenomena (Maiti et al., 1997). Figure 10.4 shows the temperature drop at a Si–Ge interface (Freund et al., 2002). However, the thermal boundary resistance values may suffer from the same size dependence problem as the simulation of the thermal conductivity of thin films, because the distributions of phonons coming toward the interface affect these values (Chen, 1998).

Short time scale heat conduction processes also fit well into nonequilbrium molecular dynamics simulations. Volz et al. (1996) investigated the validity of the Fourier law, using molecular dynamics simulation for solid argon under transient heating. Molecular dynamics simulations are also used to study laser–material interaction processes (Kotake and Kuroki, 1993; Wang and Xu, 2002, 2003).

Liquid–vapor interface is another topic that can be well addressed by molecular dynamics simulations. Equilibrium molecular dynamics simulations have been extensively used to simulate liquid–vapor interface and surface tension (Chapela et al., 1977; Maruyama, 2000; Yang et al., 2001; Sinha et al., 2003).

10.6 Summary of Chapter 10

This chapter introduces the molecular dynamics methods, particularly emphasizing their statistical foundation and their use in simulation of thermophysical properties. Molecular dynamics simulations seek the solution of the equations of motion governing the trajectory of all the molecules in the system and analyze the trajectories on the basis of statistical principles to extract desired information.

There are different ways of establishing the equations of motion. Newton's equations of motion are the most familiar but they have limitations. Two alternative methods to establish the equations of motion are introduced: the Lagrange equations of motion and the Hamilton equations of motion. After discussing the equations of motion in section 10.1, we further discuss the interatomic potential, which determines the fidelity of the simulation to true systems. The simplest potential is the van der Waals potential. Other potentials for covalently bonded crystals, for metals, and for water are also introduced. Since most of the computational time in a molecular dynamics simulation is spent on evaluating the interatomic potential, the complexity and computational efficiency of the potential affects a molecular dynamics simulation significantly.

The statistical foundations needed to analyze a molecular dynamics simulation are discussed in section 10.3, particularly for equilibrium systems. A molecular dynamics simulation traces the time history of a system, while statistical analysis as introduced in chapter 4 is based on ensemble averaging. The ergodicity of most practical systems ensures that the ensemble average is equivalent to the time average. Linear response theory is a fundamental tool for extracting transport properties from simulation of equilibrium systems. The Green–Kubo formula relates the autocorrelation of the flux (heat flux, charge flux, momentum flux) fluctuation of an equilibrium system to transport properties. The linear response theory is applicable to a wide range of problems and is not just limited to the analysis of molecular dynamics simulation results. For example, the Kramers–Kronig relation and the fluctuation–dissipation theorem are often encountered in different disciplines. The microscopic expressions for thermodynamic and transport properties, needed in the Green–Kubo formula and for other analyses are introduced. Finally, the equations of motion in different ensembles are discussed. Although a molecular dynamics simulation usually mimics a microcanonical ensemble, other types of ensemble can be simulated. The Nosé–Hoover scheme for simulating canonical ensembles is given as an example.

In section 10.4, numerical integration schemes of the equations of motion are introduced. One important technique used is the periodic boundary condition, which effectively increases the size of the simulation domain. The periodic boundary condition, however, also introduces new complications that need to be addressed in a simulation (such as potential calculation), and creates artifacts that can affect the simulation results.

In section 10.5, we discuss the simulation of thermophysical properties and thermal transport processes. Equilibrium molecular dynamics is often used to extract the thermal conductivity of bulk materials, but artifacts due to periodic boundary conditions need to be properly addressed. In addition, various nonequilibrium molecular dynamics simulation methods have been employed to simulate thermal transport. Most efforts using nonequilibrium molecular dynamics simulations have been geared toward extracting the thermal conductivity of bulk materials. For this purpose, the nonequilibrium molecular dynamics simulations are trickier than the equilibrium methods because the boundaries have much stronger effects on the simulation results.

Molecular dynamics simulations are ideal for nanostructures. Both equilibrium and nonequilibrium molecular dynamics approaches are being used, although the nonequilibrium method does find a natural fit. A number of recent studies of heat transfer in nanostructures are reviewed.

10.7 Nomenclature for Chapter 10

a	potential truncation radius, m	H	Hamiltonian, J
a_{ij}	function in Tersoff potential, eq. (10.30)	j_Q	local heat flux, W m^{-2}
		J	flux of current, A m^{-2}
A	variables of the phase space (**r**,**p**)	J_Q	heat flux, W m^{-2}
		k	thermal conductivity, W m^{-1} K^{-1}
A_{sw}	parameter in Stillinger–Weber potential	**k**	wavevector, m^{-1}
		K	kinetic energy, J
A_t	parameter in Tersoff potential, eV	L	Lagrangian, J
		m	mass of each particle, kg
b_{ij}	function in Tersoff potential, eq. (10.27)	M	total number of available time origins for calculating autocorrelation function
b_t	parameter in Tersoff potential		
B	phase variables of the phase space (**r**,**p**)	n	number of the degree of freedom in coordinates
B_{sw}	parameter in Stillinger–Weber potential	n_t	parameter in Tersoff potential
		N	number of particles
B_t	parameter in Tersoff potential, eV	N_f	degree of freedom in the momentum space in original system
c_t	parameter in Tersoff potential, Å		
C	normalization factor	**p**	momentum of the system
d_t	parameter in Tersoff potential	p_i	generalized momentum component
D_t	parameter in Tersoff potential, Å	**p**$_i$	generalized momentum of particle i
e	electron charge, C	p_s	momentum conjugate to s in Nosé rescaled system
E	total energy, J		
E	electric field, V m^{-1}	P	pressure, N m^{-2}
f_A	function in Tersoff potential, eq. (10.24)	Q	effective charge, C
		r	generalized particle coordinate
f_c	cut-off radius function, eq. (10.26)	$\dot{\mathbf{r}}$	generalized velocity
f_R	function in Tersoff potential, eq. (10.25)	r_i	component of the coordinate or particles in the system
$f^{(N)}$	N-particle distribution function	**r**$_i$	vector coordinates of one particle
F_i	force acting on particle i	r_{ij}	distance between particles i and j, m
F$_{ij}$	force on particle i by particle j	**r**$_{ij}$	vector connecting particles i and j
$F(t)$	external disturbance	**R**	spatial coordinate
g	function in Tersoff potential, eq. (10.29)	R_t	parameter in Tersoff potential, Å
h	local energy density, J m^{-3}		
h_t	parameter in Tersoff potential	s	new variable in rescaled system of Nosé thermostat

S	motion integral defined by eq. (10.5)	σ	length parameter in Lennard–Jones potential, m; electrical conductivity, $\Omega^{-1}\,m^{-1}$
t	time, s		
T	temperature, K		
u_i	potential on particle i	τ	time interval or heat flux autocorrelation function decay time, s
u_2	two-body potential, J		
U	potential energy, J		
$\mathbf{v}$	particle velocity, m s^{-1}	τ_{xy}	shear stress tensor, N m^{-2}
V	volume, m^3	ϕ_{BA}	after-effect function
Z	number of charges per particle	χ_{BA}	susceptibility
		ω	angular frequency, rad.Hz
α	scaling factor in heat flux method	Ω	partition function

Subscripts

α_t	parameter in Tersoff potential	0	equilibrium
β	defined by eq. (10.135)	AA	autocorrelation of variable A
Γ	Liouville operator		
δT	local temperature disturbance, K	AB	correlation between A and B
Δt	time interval, s	m	mass
Δt_s	time interval used for computing autocorrelation function	N	Nosé thermostat
		OO	between oxygen atoms in water potential
ε	parameter in Lennard–Jones potential	Q	heat
		sw	Stillinger–Weber potential
ζ	friction factor in Hoover thermostat	t	Tersoff potential
η_{ij}	function in Tersoff potential, eq. (10.31)	Superscripts	
$\lambda_1, \lambda_2, \lambda_3$	parameters in Tersoff potential, J	$'$	real part, perturbation, or rescaled phase-space coordinates, or before energy addition in heat flux boundary conditions
μ	chemical potential, J; dynamic viscosity, N s m^{-2}		
		$''$	imaginary part
ν	rate of collision in Anderson thermostat, s^{-1}	$\bullet$	time derivative
		$-$	time average
Π	volume in Anderson rescaled system for constant pressure simulation	$\sim$	Fourier transform
		$*$	complex conjugate
		$=$	second-order tensor
ρ	background charge density, C m^{-3}; local mass density, kg m^{-3}	Symbol	
		$\langle \rangle$	ensemble average

10.8 References

Abramson, A.R., Tien, C.-L., and Majumdar, A., 2002, "Interface and Strain Effects on Thermal Conductivity of Heterostructures: A Molecular Dynamics Study," *Journal of Heat Transfer*, vol. 124, pp. 963–970.

Alder, B.J., and Wainwright, T.E., 1958, "Molecular Dynamics by Electronic Computers," in *Proceedings of the International Symposium on Statistical Mechanical Theory of Transport Processes*, ed. Progigine, I., Interscience, Wiley, New York, pp. 97–131.

Alder, B.J., and Wainwright, T.E., 1967, "Velocity Autocorrelations for Hard Spheres," *Physical Review Letters*, vol. 18, pp. 988–990.

Alder, B.J., Gass, D.M., and Wainwright, T.E., 1970, "Studies in Molecular Dynamics. VIII. The Transport Coefficients for a Hard-Sphere Fluid," *Journal of Chemical Physics*, vol. 56, pp. 3813–3826.

Allen, P.B., and Feldman, J.L., 1993, "Thermal Conductivity of Disordered Harmonic Solids," *Physical Review B*, vol. 48, pp. 12581–12588.

Allen, M.P., and Tildesley, D.J., 1987, *Computer Simulation of Liquids*, Oxford University Press, Oxford.

Anderson, H.C., 1980, "Molecular Dynamic Simulations at Constant Pressure and/or Temperature," *Journal of Chemical Physics*, vol. 72, pp. 2384–2393.

Ashurst, W.T., 1974, "Determination of Thermal Conductivity Coefficient via Non-Equilibrium Molecular Dynamics," in *Advances in Thermal Conductivity*, ed. Reisberg, R.L., and Sauer, H.J., Jr, University of Missouri-Rolla, MO, pp. 89–98.

Balamane, H., Halicioglu, H., and Tiller, W.A., 1992, "Comparative Study of Silicon Empirical Interatomic Potentials," *Physical Review B*, vol. 46, pp. 2250–2279.

Berber, S., Kwon, Y.-K., and Tomanek, D., 2000, "Unusually High Thermal Conductivity of Carbon Nanotubes," *Physical Review Letters*, vol. 84, pp. 4613–4616.

Biswas, R., and Hamann, D.R., 1985, "Interatomic Potentials for Silicon Structural Energies," *Physical Review Letters*, vol. 55, pp. 2001–2004.

Callen, H.B., and Welton, T.A., 1951, "Irreversibility and Generalized Noise," *Physical Review*, vol. 83, pp. 34–40.

Car, R., and Parrinello, M., 1985, "Unified Approach for Molecular Dynamics and Density-Function Theory," *Physical Review Letters*, vol. 55, pp. 2471–2474.

Chapela, G.A., Saville, G., Thompson, S.M., and Rowlinson, J.S., 1977, "Computer Simulation of a Gas–Liquid Interface. I," *Journal of the Chemical Society Faraday Transactions II*, vol. 73, pp. 1133–1144.

Che, J., Cagin, T., Deng, W., and Goddard, W.A., III, 2000, "Thermal Conductivity of Diamond and Related Materials for Molecular Dynamics Simulations," *Journal of Chemical Physics*, vol. 113, pp. 6888–6900.

Chen, G., 1998, "Thermal Conductivity and Ballistic Phonon Transport in the Cross-Plane Direction of Superlattices," *Physical Review B.*, vol. 57, pp. 14958–14973.

Chou, F.-C., Lukes, J.R., Liang, X.-G., Takahashi, K., and Tien, C.L., 1999, "Molecular Dynamics in Microscale Thermophysical Engineering," *Annual Review of Heat Transfer*, vol. 10, pp. 141–176.

Ciccotti, G., and Tenenbaum, A., 1980, "Canonical Ensemble and Nonequilibrium States by Molecular Dynamics," *Journal of Statistical Physics*, vol. 23, pp. 767–772.

Ciccotti, G., Jacucci, G., and McDonald, K.R., 1978, "Thermal Response to a Weak External Field," *Journal of Physics C: Solid State Physics*, vol. 11, pp. L509–L513.

Ciccotti, G., Frenkel, D., and McDonald, K.R., 1987, *Simulation of Liquids and Solids: Molecular Dynamics and Monte Carlo Methods in Statistical Mechanics*, North-Holland, Amsterdam.

Cook, S.J., and Clancy, P., 1993, "Comparison of Semi-Empirical Potential Functions for Silicon and Germanium," *Physical Review B*, vol. 47, pp. 7686–7699.

Daly, B.C., Maris, H.J., Imamura, K., and Tamura, S., 2002, "Molecular Dynamics Calculation of the Thermal Conductivity of Superlattices," *Physical Review B*, vol. 66, pp. 024301/1–7.

Daw, M.S., and Baskes, M.I., 1983, "Semiempirical, Quantum Mechanical Calculation of Hydrogen Embrittlement in Metals," *Physical Review Letters*, vol. 50, pp. 1285–1288.

Daw, M.S., and Baskes, M.I., 1984, "Embedded-Atom Method: Derivation and Application to Impurities, Surfaces, and Other Defects in Metals," *Physical Review B*, vol. 29, pp. 6443–6453.

Ding, K., and Andersen, H.C., 1986, "Molecular Dynamics Simulation of Amorphous Germanium," *Physical Review B*, vol. 34, pp. 6987–6991.

Dong, J., Sankey, O.F., and Myles, C.W., 2001, "Theoretical Study of the Lattice Thermal Conductivity in Ge Framework Semiconductors," *Physical Review Letters*, vol. 86, pp. 2361–2364.

Erpenbeck, J.J., and Wood, W.W., 1981, "Molecular Dynamics Calculations of Shear Viscosity Time-Correlation Functions for Hard Spheres," *Journal of Statistical Physics*, vol. 24, pp. 455–468.

Evans, D.J., 1982, "Homogeneous NEMD Algorithm for Thermal Conductivity—Application of Non-Canonical Linear Response Theory," *Physics Letters*, vol. 91A, pp. 457–460.

Evans, D.J., and Morriss, G.P., 1990, *Statistical Mechanics of Nonequilibrium Liquids*, Academic Press, London.

Feldman, J.L., Kluge, M.D., Allen, P.B., and Wooten, F., 1993, "Thermal Conductivity and Localization in Glasses: Numerical Study of a Model of Amorphous Silicon", *Physical Review B*, vol. 48, pp. 12589–12602.

Frenkel, D., and Smit, B., 1996, *Understanding Molecular Simulation*, Academic Press, London.

Freund, J.B. et al., 2002, private discussion.

Gillan, M.J., and Dixon, M., 1983, "The Calculation of Thermal Conductivities by Perturbed Molecular Dynamics Simulation," *Journal of Physics* C, vol. 16, p. 869ff.

Green, M.S., 1952, 1954, "Markoff Random Processes and the Statistical Mechanics of Time-Dependent Phenomena," *Journal of Chemical Physics*, vol. 20, pp. 1281–1295; vol. 22, pp. 398–413.

Grefett, J.-J., Carminati, R., Joulain, K., Mulet, J.-P., Malnguy S., and Chen, Y., 2002, "Coherent Emission of Light by Thermal Sources," *Nature*, vol. 416, pp. 61–64.

Guillot, B., 2002, "A Reappraisal of what we have Learnt during Three Decades of Computer Simulations on Water," *Journal of Molecular Liquids*, vol. 101, pp. 219–260.

Hafskjold, B., and Ratkje, S.K., 1995, "Criteria for Local Equilibrium in a System with Transport of Heat and Mass," *Journal of Statistical Mechanics*, vol. 78, pp. 463–494.

Haile, J.M., 1992, *Molecular Dynamics Simulation*, Wiley, New York.

Halicioglu, T., Pamuk, H.O., and Erkoc, S., 1998, "Interatomic Potential with Multi-Body Interaction," *Physica Status Solidi B*, vol. 149, pp. 81–92.

Hansen, J.P., and McDonald, I.R., 1986, *Theory of Simple Liquids*, Academic Press, London, chapter 7.

Hardy, R.J., 1963, "Energy-Flux Operator for a Lattice," *Physical Review*, vol. 132, pp. 168–177.

Hoover, W.G., 1985, "Canonical Dynamics: Equilibrium Phase-Space Distribution," *Physical Review A*, vol. 31, pp. 1695–1697.

Hoover, W.G., and Ashurst, W.T., 1975, "Nonequilibrium Molecular Dynamics," in *Theoretical Chemistry: Advances and Perspectives*, ed. Eyring H., and Henderson, D., Academic Press, New York, vol. 1, pp. 1–51.

Ikeshoji, T., and Hafskjold, B., 1994, "Non-equilibrium Molecular Dynamics Calculation of Heat Conduction in Liquid and Through Liquid–Gas Interface," *Molecular Physics*, vol. 81, pp. 251–261.

Irving, J.H., and Kirkwood, J.G., 1967, "The Statistical Mechanical Theory of Transport Processes. IV. The Equations of Hydrodynamics," in *John Gamble Kirkwood Collected Works, Selected Topics in Statistical Mechanics*, ed. Zwanzig, R.W., Gordon and Breach, New York, pp. 51–75.

Jorgensen, W.L., Chandrasekhar, J., Madura, J.D., Impey, R.W., and Klein, M.L., 1983, "Comparison of Simple Potential Functions for Simulating Liquid Water," *Journal of Chemical Physics*, vol. 79, pp. 926–935.

Kim, P., Shi, L., Majumdar, A., and McEuen, P.L., 2001, "Thermal Transport Measurements of Individual Multiwalled Nanotubes," *Physical Review Letters*, vol. 87, pp. 215502/1–4.

Kohler, F., 1972, *Liquid State*. Verlag Chemie, Weinheim, chapters 8 and 9.

Koplik, J., and Banavar, J.R., 1995, "Continuum Deductions from Molecular Hydrodynamics," *Annual Review of Fluid Mechanics*, vol. 27, pp. 257–292.

Kotake, S., and Kuroki, M., 1993, "Molecular Dynamics Study of Solid Melting and Vaporization by Laser Irradiation," *International Journal of Heat and Mass Transfer*, vol. 36, pp. 2061–2067.

Kotake, S., and Wakuri, S., 1994, "Molecular Dynamics Study of Heat Conduction in Solid Materials," *JSME International Journal B*, vol. 37, pp. 103–108.

Kubo, R., 1957, "Statistical-Mechanical Theory of Irreversible Processes," *Journal of the Physical Society of Japan*, vol. 12, pp. 570–586.

Kubo, R., Yokota, M., and Nakajima, S., 1957, "Statistical-Mechanical Theory of Irreversible Processes, II. Response to Thermal Disturbance," *Journal of the Physical Society of Japan*, vol. 12, pp. 1203–1211.

Kubo, R., Toda, M., and Hashitsume, N., 1998, *Statistical Physics II*, 2nd ed., Springer, Berlin.

Ladd, A.J.C, Morgan, B., and Hoover, W.G., 1986, "Lattice Thermal Conductivity: A Comparison of Molecular Dynamics and Anharmonic Lattice Dynamics," *Physical Review B*, vol. 34, pp. 5058–5064.

Lebowitz, J.L., and Spohn, H., 1978, "Transport Properties of the Lorentz Gas: Fourier's Law," *Journal of Statistical Physics*, vol. 19, pp. 633–654.

Lee, I.-H., and Chang, K.J., 1994, "Atomic and Electronic Structure of Amorphous Si from First-Principle Molecular-Dynamics Simulations," *Physical Review B*, vol. 50, pp. 18083–18089.

Lee, Y.H., Biswas, R., Soukoulis, C.M., Wang, C.Z., Chan, C.T., and Ho, K.M., 1991, "Molecular-Dynamics Simulation of Thermal Conductivity in Amorphous Silicon," *Physical Review B*, vol. 43, pp. 6573–6580.

Levesque, D., Verlet, L., and Kürkijarvi, J., 1973, "Computer 'Experiments' on Classical Fluids. IV. Transport Properties and Time-Correlation Functions of the Lennard–Jones Liquid Near Triple Point, *Physics Review A*, vol. 7, pp. 1690–1700.

Li, J., Porter, L., and Yip, S., 1998, "Atomistic Modeling of Finite-Temperature Properties of Crystalline β-SiC. II. Thermal Conductivity and Effects of Point Defects," *Journal of Nuclear Material*, vol. 255, pp. 139–152.

Lukes, J.R., and Tien, C.L., 2000, "Molecular Dynamics Study of Solid Thin Film Thermal Conductivity" *Journal of Heat Transfer*, vol. 122, pp. 536–543.

Luttinger, J.M., 1964, "Thermal of Thermal Transport Coefficient," *Physical Review*, vol. 135, A1505–A1514.

Maiti, A., Mahan, G.D., and Pantelides, S.T., 1997, "Dynamical Simulations of Nonequilibrium Processes–Heat Flow and the Kapitza Resistance across Grain Boundaries," *Solid-State Communications*, vol. 102, pp. 517–521.

Maruyama, S., 2000, "Molecular Dynamics Method for Microscale Heat Transfer," *Advances in Numerical Heat Transfer*, ed. Minkowycz, W.J., and Sparrow, E.M., Taylor & Francis, New York, vol. 2, pp. 189–226.

Maruyama, S., 2002, "A Molecular Dynamics Simulation of Heat Conduction in Finite Length SWNTs," *Physica-B*, vol. 323, pp. 193–195.

Michalski, J., 1992, "Thermal Conductivity of Amorphous Solids above the Plateau: Molecular-Dynamics Study," *Physical Review B*, vol. 45, pp. 7054–7065.

Moreland, J.F., Freund, J.B., and Chen, G., 2004, "The Disparate Thermal Conductivity of Carbon Nanotubes and Diamond Nanowires Studied by Atomistic Simulations," *Microscale Thermophysical Engineering*, vol. 8, pp. 61–69.

Mori, H., 1958, "Statistical–Mechanical Theory of Transport in Fluids," *Physical Review*, vol. 112, pp. 1829–1842.

Nosé, S., 1984, "A Molecular Dynamics Method for Simulations in the Canonical Ensemble," *Molecular Physics*, vol. 52, pp. 255–268.

Onsager, L., 1931a, "Reciprocal Relations in Irreversible Processes I," *Physical Review*, vol. 37, pp. 405–426.

Onsager, L., 1931b, "Reciprocal Relations in Irreversible Processes II," *Physical Review*, vol. 38, pp. 2265–2279.

Osman, M.A., and Srivastava, D., 2001, "Temperature Dependence of Thermal Conductivity of Single-Wall Carbon Nanotubes," *Nanotechnology*, vol. 12, pp. 21–24.

Poetzsch, R.H., and Bottger, H., 1994, "Interplay of Disorder and Anharmonicity in Heat Conduction: Molecular-Dynamics Study," *Physical Review B*, vol. 50, pp. 15757–15763.

Polder, D., and Van Hove, M., 1971, "Theory of Radiative Heat Transfer between Closely Spaced Bodies," *Phys. Rev. B*, vol. 4, pp. 3303–3314.

Pratt, L.R., and Haan, S.W., 1981a, "Effects of Periodic Boundary Conditions on Equilibrium Properties of Computer Simulated Fluids, Theory," *Journal of Chemical Physics*, vol. 74, pp. 1864–1872.

Pratt, L.R., and Haan, S.W., 1981b, "Effects of Periodic Boundary Conditions on Equilibrium Properties of Computer Simulated Fluids. II. Application to Simple Liquids," *Journal of Chemical Physics*, vol. 74, pp. 1873–1876.

Rapaport, D.C., 1995, *The Art of Molecular Dynamics Simulation*, Cambridge University Press, Cambridge, UK.

Rytov, S.M., Kravtsov, Y. A., and Tatarski. V.I., 1987, *Principles of Statistical Radiophysics*, vol. 3, Springer-Verlag, Berlin.

Sinha, S., Dhir, V.K., Shi, B., Freund, J., and Darve, E., 2003, "Surface Tension Evaluation in Lennard–Jones Fluid System with Untruncated Potentials," *Proceedings of 2003 ASME Summer Heat Transfer Conference*, HT2003-47164, Las Vegas, NV, July 21–28.

Stillinger, F.H., and Rahman, A., 1974, "Improved Simulation of Liquid Water by Molecular Dynamics," *Journal of Chemical Physics*, vol. 60, pp. 1545–1557.

Stillinger, F.H., and Weber, T., 1985, "Computer Simulation of Local Order in Condensed Phases of Silicon," *Physical Review B*, vol. 31, pp. 5262–5271.

Swope, W.C., Andersen, H.C., Berens, P.H., and Wilson, K.R., 1982, "A Computer Simulation Method for the Calculation of Equilibrium Constants for the Formation of Physical Clusters of Molecules: Application to Small Water Clusters," *Journal of Chemical Physics*, vol. 76, pp. 637–649.

Tenenbaum, A., Ciccotti, G., and Gallico, R., 1982, "Stationary Nonequilibrium States by Molecular Dynamics. Fourier's Law," *Physical Review A*, vol. 25, pp. 2778–2787.

Tersoff, J., 1988a, "New Empirical Approach for the Structure and Energy of Covalent Systems," *Physical Review B*, vol. 37, pp. 6991–7000.

Tersoff, J., 1988b, "Empirical Interatomic Potential for Carbon, with Application to Amorphous Carbon," *Physical Review Letters*, vol. 61, pp. 2879–2882.

Tersoff, J., 1989, "Modeling Solid-State Chemistry: Interatomic Potential for Multicomponent Systems," *Physical Review B*, vol. 39, pp. 5566–5568.

Toda, M., Kubo, R., and Saito, N., 1995, *Statistical Physics I*, 2nd ed., Springer, Berlin.

Verlet, L., 1967, "Computer 'Experiments' on Classical Fluids. I. Thermodynamical Properties of Lennard–Jones Molecules," *Physical Review*, vol. 159, pp. 98–103.

Volz, S.G., 2001, "Thermal Insulating Behavior in Crystals at High Frequencies," *Physical Review Letters*, vol. 87, pp. 074301/1–4.

Volz, S.G., and Chen, G., 1999, "Molecular Dynamics Simulation of Thermal Conductivity of Silicon Nanowires," *Applied Physics Letters*, vol. 75, pp. 2056–2058.

Volz, S.G., and Chen, G., 2000, "Molecular Dynamics Simulation of Thermal Conductivity of Silicon Crystals," *Physical Review B*, vol. 61, pp. 2651–2656.

Volz, S., Saulnier, J.B., Lallemand, M., Perrin, B., Depondt, P., and Mareschal, M., 1996, "Transient Fourier Law Deviation by Molecular Dynamics in Solid Argon," *Physical Review B*, vol. 54, pp. 340–347.

Volz, S.G., Saulnier, J.B., Chen, G., and Beauchamp, P., 2000, "Computation of Thermal Conductivity of Si/Ge Superlattices by Molecular Dynamics Techniques," *Microelectronics Journal*, vol. 31, pp. 815–819.

Wang, X., and Xu, X., 2002, "Molecular Dynamics Simulation of Heat Transfer and Phase Change during Laser Material Interaction," *Journal of Heat Transfer*, vol. 124, pp. 264–274.

Wang, X., and Xu, X., 2003, "Molecular Dynamics Simulation of Thermal and Thermomechanical Phenomena in Picosecond Laser Material Interaction," *International Journal of Heat and Mass Transfer*, vol. 46, pp. 45–53.

Yang, C., Chen, M., and Guo, Z.Y., 2001, "Molecular Dynamics Simulation of the Specific Heat of Undercooled Fe–Ni Melts," *International Journal of Thermophysics*, vol. 22, pp. 1303–1309.

Zwanzig, R., 1965, "Time-Correlation Functions and Transport Coefficients in Statistical Mechanics," *Annual Review of Physical Chemistry*, vol. 16, pp. 67–102.

10.9 Exercises

10.1 *Equations of motion for a constant pressure system.* To simulate a constant pressure system, Anderson (1980) designed a scaled system with a Lagrangian of the form

$$L(\mathbf{r}', \dot{\mathbf{r}}', \Pi, \dot{\Pi}) = \frac{1}{2}m\Pi^{3/2}\sum_{i}^{N}\dot{\mathbf{r}}_i \bullet \dot{\mathbf{r}}_i$$

$$-\frac{1}{2}\sum_{i=1}^{N}\sum_{j=1}^{N}u(\Pi \mathbf{r}'_{ij}) + \frac{1}{2}M\dot{\Pi}^2 - \alpha\Pi$$

where α is the system pressure, and $\mathbf{r}'(t)$, $\dot{\mathbf{r}}'(t)$ and $\Pi(t)$ are related to the original system particle position $\mathbf{r}(t)$, momentum $\mathbf{p}(t)$, and volume $V(t)$ (volume fluctuates in a constant pressure ensemble) as follows,

$$\Pi(t) = V(t), \mathbf{r}_i'(t) = V(t)^{-1/3}\mathbf{r}_i(t), m\dot{\mathbf{r}}'(t) = V(t)^{-1/3}\mathbf{p}(t)$$

Answer the following questions:
  (a) What is the generalized momentum conjugate to Π?
  (b) What is the Hamiltonian of the scaled system?
  (c) Derive the Hamiltonian equations of motion for the scaled system.
  (d) Derive the Lagrange equations of motion for the scaled system.

10.2 *Linear response theory for particle mobility.* If particles in a system are acted upon by an external force $\mathbf{F}(t)$, the Hamiltonian of the system is then

$$H = H_0 + \sum_i \mathbf{r}_i \bullet \mathbf{F}$$

where H_0 is the Hamiltonian of the unperturbed system. The particle mobility μ is defined in relation to the average velocity by

$$\langle \mathbf{v} \rangle = \mu \mathbf{F}$$

Derive the following Green–Kubo formula for the particle mobility,

$$\mu = \frac{V}{3\kappa_B T} \int_0^\infty \langle \mathbf{v}(t) \bullet \mathbf{v}(0) \rangle \, dt$$

where $\mathbf{v}$ is the average instantaneous velocity of all the particles in the system.

10.3 *Kramer–Kronig relation for thermal conductivity.* Given a frequency dependent thermal conductivity of the form

$$k(\omega) = \frac{k(0)}{i\omega + 1/\tau}$$

show that the real and the imaginary parts of $k(\omega)$ obeys the Kramer–Kronig relations.

10.4 *Microscopic expression for temperature.* We can derive the microscopic expression for temperature from the thermodynamic definition of temperature in a microcanonical system,

$$\frac{1}{T} = \left(\frac{\partial S}{\partial E}\right)_{NV}$$

where E is the system energy, comprising a kinetic and a potential part, $E = K + U$. The entropy of a microcanonical system is, according to the Boltzmann principle,

$$S = \kappa_B \ln \Omega$$

where Ω is the number of microcanonical states, which can be expressed as (Haile, 1992)

$$\Omega = \frac{1}{(2\pi\hbar)^{3N} N!} \int d\mathbf{r} d\mathbf{p} \theta(E - H)$$
$$= \left(\frac{m}{\hbar}\right)^{3/2} \frac{1}{N!\Gamma(3N/2 + 1)} \int d\mathbf{r} \, (E - U)^{3/2} \theta(E - H)$$

where $\hbar$ is the Planck constant divided by 2π, H is the system Hamiltonian, θ is a step function and its derivative is a delta function, and $\Gamma(N)$ is the gamma function. From here, show that temperature can be expressed as an average of the system kinetic energy

$$\frac{3}{2}\kappa_B T = N \langle K \rangle$$

10.5 *Effects of Boundary Conditions on Nonequilibrium Molecular Dynamics Simulation.* To appreciate the potential effects of hot and cold walls on the molecular dynamics simulation result, we consider phonon heat conduction across a thin film as shown in figure P10.5. Solve the phonon Boltzmann equation numerically for the following two boundary conditions, using the gray body approximation and varying the phonon Knudsen number from 0.01 to 100.

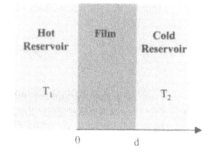

Figure P10.5
Figure for problem 10.5

(a) The incident phonons from the two reservoirs are isotropic.

(b) The arithmetic average of the intensities of incident phonon from the reservoir and phonons coming toward the boundary from the opposite direction inside the film is isotropic. Iteration is needed for this latter case.

(c) Examine the temperature distribution and temperature gradient as a function of location for each case. Show that the effective thermal conductivity obtained from case (a) is lower than from case (b) when the phonon Knudsen number is large.

(d) Consider various ways of imposing hot and cold walls in nonequilibrium molecular dynamics simulation and elaborate on what might be the distributions of the phonons entering into the transport region from the reservoirs.

10.6 *Equilibrium Molecular Dynamics Simulation of Thermal Conductivity of Nanowires.* Develop an equilibrium molecular dynamics simulation algorithm for the thermal conductivity of silicon nanowires.

10.7 *Nonequilibrium Molecular Dynamics Simulation of Thermal Conductivity of Nanowires.* Develop a nonequilibrium molecular dynamics simulation algorithm for the thermal conductivity of silicon nanowires.

Appendix A

Homogeneous Semiconductors

Semiconductors have a moderate band gap. For example, silicon has a band gap of 1.1 eV and GaAs has a band gap of 1.42 eV at room temperature. Silicon is widely used in electronic devices while GaAs in semiconductor lasers (Sze, 1981). Silicon is an indirect gap semiconductor with six conduction bands along the [100] direction, as shown in figures 3.17(b) and 3.18(b). The conduction band can be described by the following elliptical dispersion relation,

$$E - E_c = \frac{\hbar^2}{2}\left(\frac{(k_x - k_{x0})^2}{m_L^*} + \frac{k_y^2}{m_t^*} + \frac{k_z^2}{m_t^*}\right) \quad (A1)$$

where k_{x0} is the location of the conduction band minimum along the [100] direction, $m_L^* (= 0.98\,m)$ is the effective mass along the long-axis direction of the ellipsoid and is called the longitudinal effective mass, $m_t^* (= 0.19\,m)$ is the effective mass in the other two directions, and m is the free electron mass. The valence bands of silicon are approximately spherical, centered at $\mathbf{k} = 0$, with the dispersion given by

$$E - E_v = \frac{\hbar^2}{2m^*}(k_x^2 + k_y^2 + k_z^2) \quad (A2)$$

Figure 3.17(b) shows that there are actually two valence bands. The broader one has a larger effective mass ($m_h^* = 0.49\,m$) and is called the heavy-hole band. The narrower one is called the light-hole band with an effective mass $m_l^* = 0.16\,m$.

The number densities of electrons and holes can be calculated following the strategies laid out in section 4.2.2, and are given by eq. (4.64),

$$n_0 = N_c \exp\left(-\frac{E_c - E_f}{\kappa_B T}\right) \tag{A3}$$

$$p_0 = N_v \exp\left(-\frac{E_f - E_v}{\kappa_B T}\right) \tag{A4}$$

where we have used the Boltzmann distribution as an approximation to the Fermi–Dirac distribution and have replaced the chemical potential μ by E_f. We will call E_f the Fermi level, as is customarily done in electrical engineering. The condition for the Boltzmann distribution approximation to be valid is that $(E_c - E_f) \geq 3\kappa_B T$ and $(E_f - E_v) \geq 3\kappa_B T$. This condition is usually satisfied if the band gap is reasonably large (compared to $\kappa_B T$) and doping levels (impurities) are not high. Under such conditions, E_f falls into the band gap and the semiconductor is called *nondegenerate*. N_c and N_v in eqs. (A3) and (A4) are given by

$$N_c = 2K_c \left(\frac{2\pi m_c \kappa_B T}{h^2}\right)^{3/2} \tag{A5}$$

$$N_p = 2K_v \left(\frac{2\pi m_v \kappa_B T}{h^2}\right)^{3/2} \tag{A6}$$

where K_c and K_v are the numbers of identical bands along different crystallographic directions for the conduction and the valence bands, respectively; and m_c and m_v are the effective masses of the conduction and valence bands. For the conduction band of silicon, as shown in figures 3.17(b) and 3.18(b), $K_c = 6$ because of the six identical ellipsoids of the conduction bands, and m_c and m_v are the effective masses of the conduction and valence bands if these bands are spherical. The conduction band of Si is elliptical and its equivalent effective mass $m_c = (m_L^* m_t^{*2})^{1/3}$ (see problem 3.9). For holes, K_v equals 1, with an average effective mass of the two hole bands $m_v = (m_h^{3/2} + m_\ell^{3/2})^{2/3}$.

From eqs. (A3) and (A4), we have

$$n_0 p_0 = N_c N_v \exp\left(-\frac{E_c - E_v}{\kappa_B T}\right) = N_c N_v \exp\left(-\frac{E_G}{\kappa_B T}\right) = n_i^2 \tag{A7}$$

where n_i as defined by the last equation is called the intrinsic carrier concentration. If the semiconductor is very pure (intrinsic), the numbers of electrons and holes are equal. Equation (A7) leads to

$$n_0 = p_0 = n_i = \left[N_c N_v \exp\left(-\frac{E_G}{\kappa_B T}\right)\right]^{1/2} \tag{A8}$$

and the Fermi level

$$E_f = \frac{E_G}{2} + \frac{\kappa_B T}{2} \ln\left(\frac{N_v}{N_c}\right) \tag{A9}$$

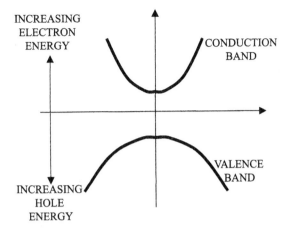

Figure A1 Typical conduction and valence bands of a direct gap semiconductor.

When impurities are added into semiconductors, the energy levels of the impurities may fall into the band gap, depending on the electronic configurations of the impurity atoms. Phosphorus, for example, has five electrons. As an impurity inside a silicon crystal, its energy level is close to the conduction band [figures A1, A2(b)]. An electron from phosphorus can easily be excited from the phosphorus atomic orbital into the conduction band orbital of the silicon crystal. The remaining four electrons form stable bonds with the rest of the electrons in the neighboring silicon atoms [figure A2(a)]. Such impurities that donate electrons to the conduction bands are called donors [figure A2(b)]. Similarly, some impurities have energy levels close to the valence band, such as boron atoms in a silicon crystal. These impurities accept electrons from the valence band, leaving mobile holes in the valence band. Such impurities are called acceptors. There are also impurities having energy levels deep inside the bandgap, such as gold inside silicon. They do not donate or accept electrons but can scatter electrons and holes and serve as recombination centers for electron and hole recombination. These deep level impurities are usually not desirable. Semiconductors with more donors are called n-type and those with more acceptors are called p-type. For doped semiconductors, eqs. (A3) and (A4) are still applicable. However, the electrons and holes are no longer equal in number as in an intrinsic semiconductor. For an n-type semiconductor, if the donor energy level is close to the conduction band edge (within $\kappa_B T$), one can assume that all the donors are ionized. If it is further assumed that the intrinsic carrier concentration n_i is much smaller than the dopant (impurity) concentration, as is usually the case in intentionally doped semiconductors with a moderate band gap (Si, GaAs, etc.), the electron concentration n_0 in eq. (A3) can be set to the donor dopant concentration, $n_0 = N_D$, which leads to

$$E_f = E_c + \kappa_B T \ln\left(\frac{N_D}{N_c}\right) \qquad (A10)$$

The hole concentration in such an n-type doped semiconductor can be calculated from eq. (A7), whose validity for doped semiconductors is ensured by eqs. (A3) and (A4). Similar operations can be carried out for p-type semiconductors. If not all the donors and/or acceptors are ionized, or if the n-type and p-type dopant concentrations are close

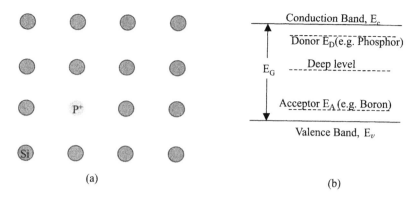

Figure A2 (a) Phosphorus inside a silicon crystal serves as a donor. An electron is released from a phosphorus atom into the conduction band. (b) Acceptors (such as boron in silicon) have energy levels close to the valence band and receive electrons from the valence band, becoming negatively charged and leaving a mobile hole in the valence band. Some impurities (such as gold in silicon) have energy levels deep inside the band gap (deep levels) and usually do not contribute electrons or holes, but they can scatter electrons and holes and serve as recombination centers.

to each other in the semiconductor, the Fermi level, and thus the electron and hole concentrations, can be calculated from

$$n + N_A^- = p + N_D^+ \tag{A11}$$

where N_A^- is the number of acceptors that are occupied by electrons and N_D^+ is the number of donors that have already donated their electrons. Equation (A11) states that there are equal numbers of positive and negative charges in the crystal; this is called the charge neutrality condition. The negative charges are free electrons in the conduction band, n, and the acceptors that have already accepted an electron (thus creating a hole in the valence band), N_A^-. The distributions of the ionized acceptors and donors are given by

$$N_D^+ = N_D \left[1 - \frac{1}{1 + \frac{1}{g_D} \exp\left(\frac{E_D - E_f}{\kappa_B T}\right)} \right] \tag{A12}$$

$$N_A^- = \frac{N_A}{1 + g_A \exp\left(\frac{E_A - E_f}{\kappa_B T}\right)} \tag{A13}$$

where E_A and E_D are the energy levels of the acceptor and the donor, respectively, and g is a factor that accounts for the degeneracy of the ground state of the impurity level, with $g_D = 2$ and $g_A = 4$.

Reference

Sze, S.M., 1981, *Physics of Semiconductor Devices*, John Wiley, New York.

Appendix B

Semiconductor p–n Junctions

As shown in figure 8.7, when an n-type semiconductor and a p-type semiconductor are brought into physical contact, electrons in the n-type region diffuse into the p-type region, leaving positively charged ions (donors) behind. Similarly, holes diffuse into the n-type region, leaving negatively charged ions (acceptors) behind. At the interface, the positively charged ions in the n-side and the negatively charged ions in the p-side establish an electrostatic potential barrier that resists further diffusion to establish an equilibrium state for the whole structure. This is reflected in the band diagram shown in figure 8.7(b). The region around the interface where negatively and positively charged ions are no longer neutral is called the space-charge region. The concentration of free electrons or holes in this space-charge region is very low compared to the number of electrons and holes in the bulk material. The built-in potential over the space-charge region can be found from the requirements that the Fermi levels are equal at equilibrium and that, far away from the space-charge region, the free carrier concentrations must be the same as in homogeneous semiconductors. If we assume that both the donors in the n-type region and the acceptors in the p-type region are fully ionized, the electron concentration in the bulk n-type region, n_{n0}, that is, away from the space-charge region, is given by eq. (A3),

$$n_{n0} = N_D = N_c \exp\left(-\frac{E_{c1} - E_f}{\kappa_B T}\right) \tag{B1}$$

where E_{c1} is the level of the conduction band in the bulk n-type. The electron concentration in the bulk p-type region, n_{p0}, according to eqs. (A4) and (A7), is

$$n_{p0} = \frac{n_i^2}{p_{p0}} = N_c \exp\left(-\frac{E_{c2} - E_f}{\kappa_B T}\right) \tag{B2}$$

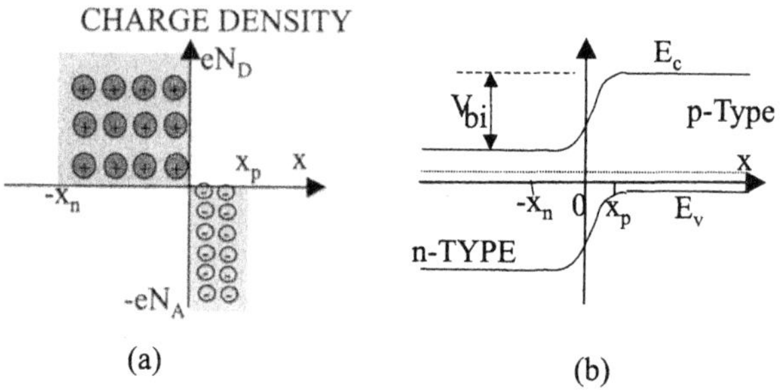

Figure B1 (a) Space charge distribution around an abrupt p–n junction. (b) Profile of conduction and valence bands.

where p_{p0} is the hole concentration in the bulk p-type region and E_{c2} is the level of the conduction band in the bulk p-type region. Eliminating E_f in eqs. (B1) and (B2), we obtain the built-in potential

$$V_{bi} = \frac{E_{c2} - E_{c1}}{e} = \frac{\kappa_B T}{e} \ln\left(\frac{n_{n0}}{n_{p0}}\right)$$

$$= \frac{\kappa_B T}{e} \ln\left(\frac{n_{n0}}{n_i^2/p_{p0}}\right) = \frac{\kappa_B T}{e} \ln\left(\frac{n_{n0} p_{p0}}{n_i^2}\right) \quad (B3)$$

where e is the positive unit charge.

Next, we would like to determine the potential profile in figure 8.7, which is redrawn in figure B1. The governing equation is the Poisson equation, that is, eq. (5.13) in the Maxwell equations. For the one-dimensional coordinate system as shown in figure B1(a), it can be written as

$$\varepsilon_0 \varepsilon_r \frac{d\mathscr{E}}{dx} = e[p + N_D^+ - n - N_A^-] \quad (B4)$$

where $\mathscr{E}$ is the electric field, ε_0 is the electric permittivity in vacuum, and ε_r the dielectric constant. Since the band edge is parallel to the electrostatic potential, we can write eq. (B4) as

$$\varepsilon_0 \varepsilon_r \frac{d^2 E_c}{dx^2} = e^2[p + N_D^+ - n - N_A^-] \quad (B5)$$

To solve the above equation, we can use the expressions for n, p, N_D^+, and N_A^- given in the previous section, with the boundary conditions that E_c must be flat as x approaches minus and plus infinity, that is, in the bulk regions. Because the expressions for these charges involve exponentials, the solutions are not easy to obtain. In chapter 9, we dealt with similar situations for the electric double layer in liquid. Here, we take advantage of the fact that in the space-charge region n and p are negligible. We further assume that the space charges have uniform profiles on each side, as shown in figure B1(a). In the n-type region, the space-charge region is uniformly distributed with donor ions and its width is $-x_n$. In the p-type region, the space-charge region is uniformly distributed with acceptor ions and its width is x_p. The values of x_n and x_p depend on

N_D and N_A and remain to be determined. Under these assumptions, eq. (B5) can be written as

$$\varepsilon_0 \varepsilon_r \frac{d^2 E_c}{dx^2} = e^2 N_D \quad (-x_n \leq x < 0) \tag{B6}$$

and

$$\varepsilon_0 \varepsilon_r \frac{d^2 E_c}{dx^2} = -e^2 N_A \quad (0 \leq x \leq x_p) \tag{B7}$$

The above two equations can be solved with the following boundary conditions,

$$x = -x_n \quad dE_c/dx = 0 \tag{B8}$$

$$x = 0 \quad E_c(x = 0_-) = E_c(x = 0_+) \tag{B9}$$

$$\left. \frac{dE_c}{dx} \right|_{x=0_-} = \left. \frac{dE_c}{dx} \right|_{x=0_+} \tag{B10}$$

$$x = x_p \quad dE_c/dx = 0 \tag{B11}$$

The solutions for the above equations are

$$E_c(x) = \frac{e^2 N_D}{\varepsilon_0 \varepsilon_r} \left(x_n + \frac{x}{2} \right) x + C \quad (-x_n \leq x < 0) \tag{B12}$$

$$E_c(x) = \frac{e^2 N_A}{\varepsilon_0 \varepsilon_r} \left(x_p - \frac{x}{2} \right) x + C \quad (0 \leq x \leq x_p) \tag{B13}$$

$$N_D x_n = N_A x_p \tag{B14}$$

$$w = x_n + x_p = \sqrt{\frac{2\varepsilon_0 \varepsilon_r (N_D + N_A) V_{bi}}{e N_D N_A}} \tag{B15}$$

where we have used the fact that $E_c(x_p) - E_c(-x_n) = eV_{bi}$. C in the above equations is an arbitrary constant, depending on the reference point for E_c, and w is the width of the space-charge region. Equations (B14) and (B15) also determine the width of the space-charge region on each side of the semiconductor. The space-charge region for the more heavily doped side is narrower.

Index

absorption, heat generation distribution due to, 224
absorption bands, 64
absorption coefficients, 166, 241, 358
   of gallium arsenide, 360–361
   of nonpolar crystals, 359
   of polar crystals, 359
   of silicon, 360–361
absorption lines, 61
absorptivity, 177
   of thin film, 189
acceptors, 97, 98, 507, 508
accessible quantum states. See quantum states, accessible
accommodation coefficients, 303, 322
acoustic branches, 104–105
acoustic impedance, 178
acoustic mismatch model, 182
acoustic Poynting vector, 169
acoustic waves
   plane, 167–169
      interface reflection and refraction of, 178–180
      longitudinal, 168
      transverse, 168–169
   propagation in thin films, 192–193
after-effect function, 465, 469–470

Ampère law, 162
Anderson thermostat, 480, 492
angle of incidence, 171
angular frequency, 44
anharmonicity, 123
anomalous dispersion, 206, 207
antireflection coatings, 190
atomic displacement vector, 350
atomic force microscope, 198
atoms, inert, 68
Auger recombination, 364
autocorrelation functions, 416, 464, 475, 487, 488–489
Avogadro's number/constant, 138, 513

ballistic–diffusive equations, 260, 331
   for phonon transport, 333–336, 337
   for thermal radiation, 331–333
ballistic transport, 199
band edge, 98
bandgap, 99
   photonic, 117
barrier height, 323
basis, of crystal, 78
Bethe theory, 324
BGK approximation/model, 236, 309–310, 313

biomolecules, 437
blackbody radiation, 13–15, 46–47
  coherence length for, 209
  emissive power for, 14–15, 46–47, 147
  in multilayer structures, 211–212, 226
  in small cavity, 158
Bloch theorem, 92, 96
Boltzmann, Ludwig, 126
Boltzmann constant, 126, 513
Boltzmann distribution function, 23, 137
Boltzmann equation, 231–233, 366–367
  derivation, 232
  diffusion approximation, 317–330
  electron, 367
  generation rate term in, 366
  for heterojunction structures, 307
  linearized, 243
  modified differential approximation. *See* ballistic–diffusive equations
  phonon, 269, 368
  range of validity, 232
  scattering term in, 353, 366, 409
  transient, under relaxation time approximation, 333
  under relaxation time approximation, 237, 242–243, 284
    for Poiseuille flow between two parallel plates, 309
    for rarified gas heat conduction, 303
    for two-dimensional problem, 284
Boltzmann factor, 129–130
Boltzmann principle, 126
Boltzmann transport equation. *See* Boltzmann equation
Born, Max, 49
Born–Mayer potential, 84
Born–von Karman periodic boundary condition, 95–96
Bose–Einstein distribution function, 25, 136–137
bosons, 31, 104, 136
Bragg condition for diffraction, 90
Bragg reflector, 190–191
  randomness in thickness of, 212
  transmissivity of, 211–213
Bravais lattice, 79
  types, 80
Brewster angle, 176
Brillouin zone, first, 88
Broglie, Louis de, 48
Brown, Robert, 412

Brownian motion, 411–416
bubbles
  equilibrium temperature of, 441
  surface tension of, 443
  vapor pressure in, 440

calories, 18
canonical partition function, 129, 405
  for diatomic gas, 139
  of rotational modes of hydrogen molecule, 133–134
capillary rise, of liquid in tube, 450
carbon dioxide (CO_2) molecules
  vibrational normal modes, 62
  vibrational-rotational absorption bands, 65
carbon nanotubes, 8, 56, 114
  multiwalled, 114
  single-walled, 114
  specific heat of, 149–150
carrier conservation equation, 357
carrier excitation, 358–363
Cattaneo equation, 259, 475
causality, 465
CDs, rewritable, 6
charge–dipole interaction, 417–419
charge neutrality condition, 508
chemical potential, 126, 135, 141–142, 249–250
  distinction from Fermi level, 249
  in doped semiconductors, 142–144
Christoffel equation, 168
closed system, energy conservation in, 16
coagulation, 427
C–O bonds, spring constant of, 74
coherence
  of electromagnetic waves, 208–213
  of electron waves, 213–214
  of phonons, 214–216
coherence length, 32, 208, 214
  for blackbody radiation, 209
  phase, 214
collision integral, 366
colloid, formation of, 426–427
complex permittivity, 164
complex refractive index, 164
condensation, of vapor in porous medium, 440
conduction, heat, 9–11
  across thin films, 299, 321
  along thin films, 201, 288–292, 313

INDEX 515

in gas, 18–20
hyperbolic one-step model, 370
hyperbolic two-step model, 370
in metals, 21
molecular dynamics simulations
  of, 494
parabolic one-step model, 370
parabolic two-step model, 370
in rarefied gas, 28–29
  between two parallel plates, 302–307,
    321–322
in semiconductors, 21
in superlattices, 300–302, 321
universal quantum thermal conductance,
  201–203
in wire geometries, 291
conduction bands, 97, 98
  of silicon, 505, 506
  See also electron band structures
conductivity
  electrical, 162, 252
  Green–Kubo formula for, 472–473
  thermal. *See* thermal conductivity
conservative equations, 16
constant energy surfaces, 100
constant heat flux method,
  490, 491–492
constant reservoir temperature method,
  490–491
constitutive equations, 16
continuity, equation of, 478
convection, heat, 11–13
  forced, 11
  natural, 11
  in nonplanar structures, 314–315
  no-slip boundary condition, 12
conventional unit cell, 79
correlation function, 464, 469–470, 487
corresponding states, law of, 449
Couette flow, rarified, 345
Coulomb electrostatic force, 65
Coulomb potential, 417
covalent bonding, 86
critical angle, 176, 180
critical point, 449
Crowell–Sze theory, 324, 326
crystal momentum, 96, 207
  in Boltzmann equation, 232
crystal planes, 80–81
  determination, 81
  Miller indices of, 80–81

crystals
  bonding potential in, 84–87
  copper, electron band structures,
    98, 99
  covalent, 86
  defects, 83
    dislocations, 83
    line, 83
    planar, 83
    point, 83
  density, 120
  diamond, 82, 86
  diffraction in, 89–91
  direction index in, 81
  electron energy states in, 91–100
    energy bands in real crystals, 98–100
    one-dimensional periodic potential,
      91–97
  gallium arsenide (GaAs), 82, 85
    electron band structures, 98–99
  germanium, 82, 86
  gold, Fermi level in, 108–109
  graphite, 82, 114
  hexagonal close-packed (hcp) structure, 82
  interatomic interactions in, 20–21
  ionic, 86
  lead, phonon dispersion for, 105, 106
  melting points of, 441–442
  metallic bonding, 86
  molecular, 84
  real, 81–83
    electron energy bands in, 98–100
  silicon (Si), 82, 86
    density, 82–83
    electron band structures, 98–99, 505
    phonon dispersion for, 105, 106
  sodium chloride (NaCl), 86
  three-dimensional, phonons in, 105
  three-dimensional photonic, 191
  vapor pressures of, 441
  zinc blend structure, 82
  See also lattices
curvature effects, 438–443
  on equilibrium phase transition
    temperature, 441
  on surface tension, 442–443
  on vapor pressure of droplets,
    439–441

dark current, 391–392
Debye approximation, 102–103, 109–110, 122

516 INDEX

Debye cutoff vector, 109
Debye frequency, 109, 144
Debye–Hückel theory, 423
Debye length, 423
Debye temperature, 144, 145
Debye velocity, 144
deformation potential, 350
degeneracy, 64
 lifting of, 68
 for one-dimensional electron model, 105–106
 in three-dimensional crystals, 106
 See also density of states
degenerate states, 58
Deissler boundary conditions, 320
density profile
 of liquids between two walls, 427
 near liquid–vapor interface, 429
density of states, 105–111
 definitions, 107
 differential, 111, 233
 electron, 107–108
 phonon, 109–110, 122
 photon, 110
 size effects on, 122
detailed balance, principle of, 182
deviation function, 242, 284
diamagnetic material, 162
diamond, thermal conductivity, 10
dielectric constant/function, 163, 164
dielectric materials
 heat conduction in, 20
 in thermovoltaic generators, 403
differential wave equation, 101
diffuse mismatch model, 183–184
diffusion length, 380
diffusion–transmission boundary conditions, 318–319
 for electron transport across interface, 322–326
 for heat conduction in thin films, 321–322
 for rarified gas flow, 326–330
 for thermal radiation between parallel plates, 320
diffusivity, 252–253, 413
dilute gas limit, 132
 criterion for validity, 133
dipole, 161
 charge–dipole interaction, 417–419
dipole moment, 162
Dirac, Paul, 48

discrete ordinates method, 316
disjoining pressure, 425, 450
dislocation density, 83
dispersion, anomalous, 206, 207
dispersion force. *See* van der Waals potential
dispersion relations, 22–23, 95
 definition, 45
displacement, 167
displacement velocity, 167
DLVO theory, 427
donors, 97, 98, 507, 508
dopants (impurities), 97, 98
 acceptors, 97, 98, 507, 508
 donors, 97, 98, 507, 508
 types, 81, 82
drift–diffusion equation, 252, 267, 379
drift–diffusion relations, validity, 260–262
drift velocity, 253, 266, 413
driving force, random, 414
droplets
 curvature effects on vapor pressure of, 439–441
 equilibrium temperature of, 441
 formation of, 432–433
 surface tension of, 442–443
Dupré equation, 431
dynamic susceptibility. *See* susceptibility, dynamic

edge dislocations, 83
effective diameter, of atom or molecule, 27
effective mass, 100
eigenvalues, energy, 51
Einstein, Albert
 on Brownian motion, 411–414
 on particle characteristics of photons, 47
Einstein model, 109–110
Einstein relation, 253, 413, 471, 479
 derivation from Langevin equation, 415
Einstein summation convention, 262
elastic stiffness tensor, 167–168
electrical conductivity. *See* conductivity, electrical
electric displacement from, 162
electric double layer potential, 421–426
electric double layers, 421–427, 437–438
 Debye–Hückel theory of, 423
electric field energy density, 165
electric field vector, 161
electric susceptibility, 162

electrochemical potential, 250, 368
electrode workfunction, 46
electrohydrodynamic equation, 266–268
electrokinetic flows, 436–438
electroluminescence, 364
electromagnetic radiation, wave–particle
    duality of, 46–48
electromagnetic waves,
    plane, 161–167
  coherence of, 208–213
  electric field of, 164
  energy flow associated, 165
  interface reflection and refraction
    of, 171–177
    boundary conditions, 172
  magnetic field of, 164
  propagation in thin films, 186–191
    field-tracing method, 186
    resultant wave method, 186
    transfer matrix method, 186, 190
  tunneling of, 195–196
electron(s)
  delocalization of, 87
  lifetime of, 366
electron affinity, 112, 323
electron band structures
  copper, 98, 99
  gallium arsenide (GaAs), 98–99
  silicon, 98–99, 505
electron–electron scattering, 240
electronic thickness, 287
electron Knudsen number, 287
electron minigaps, 191
electron–phonon coupling factor
  in metals, 357, 402
  in semiconductors, 356–357, 401
electron–phonon interactions, nonequilibrium.
    See electron–phonon scattering;
    electron–phonon transport, coupled
electron–phonon scattering, 213, 240,
    349–358
  classical collision model, 401
  in metals, 355, 356, 357
  phonon scattering in, 357–358
  in semiconductors, 355–357
electron–phonon transport, coupled
  without recombination, 367–373
    cold and hot phonons in energy
      conversion devices, 373
    hot electron effects in short pulse laser
      heating of metals, 370–371

    hot electron and hot phonon
      effects in semiconductor devices,
      370–373, 402
electron reflection, 74
electron rest mass, 513
electron scattering, 213, 240
  elastic, 213
  inelastic, 213
  See also electron–electron scattering;
    electron–phonon scattering
electron spin, 58–59
electron transmission, in nanowire, 89
electron transport
  across heterojunctions, 307–308,
    322–326
  along thin films, 285–288, 313
  in isothermal conductors, 249–253
    metals, 251–252
    semiconductors, 252–253
  with no current flow, 253
  See also thermoelectric effects
electron-volt (eV), 41, 513
electron waves
  coherence of, 213–214
  plane, 161
    interface reflection and refraction of,
      169–171
    propagation in layered media, 193–194
    tunneling of, 74–75, 195, 196
electro-osmosis, 436, 437–438
electrophoresis, 436–437
electrostatic potential, 249, 250
  flux of, 376
embedded-atom method, 461
emissivity, 177
  spectral, 15
energy
  internal. *See* internal energy
  total, 69–70
  translational, 69–70, 76
energy accommodation coefficient, 303, 322
energy carriers, 18–22, 31–32
  allowable energy levels, 22–23
  basic characteristics, 31
  simple kinetic theory for motion, 25–27
  statistical distribution, 23–25
energy conservation, in closed system, 16
energy conservation equation,
    265, 267, 374
  for electrons, 368
  for phonons, 369

energy conversion devices
  cold and hot phonons in, 373
  See also solar cells; thermoelectric devices; thermovoltaic power conversion
energy coupling length, 373
energy exchange, in semiconductors with recombination, 373–386
  current flow in p–n junction, 377–381
  energy conversion in p–n junction, 376–384
  energy source formulation, 373–376
  energy sources in p–n junction, 378, 381–384
  radiation heating of semiconductors, 384–386
energy flux, 255
  of electrons, 376
  of holes, 376
energy loss
  in metals, 356
  in semiconductors, 357
energy operator, 50
energy propagation velocity, 206
energy quantization, 42
energy source term, 374
ensemble averages, 125, 463, 481, 482
ensembles, 124–130, 228–229, 463
  canonical, 127–130, 463, 479–480
    construction, 128
  definition, 124, 125
  grand canonical, 130, 463
  microcanonical, 125–127, 463, 479, 481
  thermostatted, 479–483
Enskog equation, 409–410, 487–488
  scattering term in, 410
enthalpy, 265
entropy, 126
  of mixing, 156
  of one-phonon state, 158, 271
equation of radiative transfer, 241–242, 292
equilibrium
  definition, 17
  local, 17, 492–493
    meaning of, 260–261
equipartition theorem, 24, 132, 415
ergodic hypothesis, 124
Euler equation, 375
evanescent waves, 194–195
  acoustic, 180
  electromagnetic, 176–177, 195
  electron, 170, 194
  penetration depth of, 194, 195
  See also tunneling
expectation, 50
extensive variables, 374
extinction coefficient, 164, 242, 292–293, 361

Faraday law, 162
fast transport phenomena, 30–32
Fermi-Dirac distribution function, 25, 135, 136
Fermi level, 97, 135, 141, 506, 508
  distinction from chemical potential, 249
  in gold crystal, 108–109
fermions, 31, 104, 136
Fermi's golden rule, 235, 350
Fermi temperature, 144
Fermi velocity, 213–214
Feynman, Richard, 4
Fick's law of diffusion, 41, 413
films, thin. *See* thin films
flocculation, 427
flow line, 228
fluctuation–dissipation theorem, 413, 470–471
fluctuation force, 428
fluid flow
  in microchannels, 433–434
  in nanochannels, 434–435
Fourier law, 9–10, 244–245, 475
  derivation from kinetic theory, 26–27
Fourier series, for periodic function, 87
Fourier transformation, temporal, 465
Fredholm integral equations, 294–295
  numerical method for solving, 297–299
frequency, 44
Fresnel, Augustin Jean, 46
Fresnel coefficients, 174, 466
  with absorbing medium, 174
friction coefficient, for Brownian particles, 414
  calculation from Langevin equation, 415–416
Fuchs–Sondheimer solutions, 288
fundamental postulate, in statistical mechanics, 124–125

gas flow, rarified
  between parallel plates, 308–313
  for nonplanar geometries, 314–315
  regimes, 315
Gaussian quadrature formula, 316
Gauss–Legendre integration method, 298

gel-electrophoresis, 437
generalized coordinates, 454
generalized momentum, 455
generalized particle coordinate vector, 454
generalized particle velocity, 454
generation, 266, 365
  See also heat generation
generation rate, 366
geometrical optics, 226
Gibbs–Duhem relation, 375, 439
Gibbs equation, 439
Gibbs factor, 130
Gibbs paradox, 156
Gibbs surface, 429
Girifalco–Good–Fowkes equation, 431
gradient operator, 50
grand canonical partition function, 130
grand canonical potential, 130
Green–Kubo formula, 470, 471
  for electrical conductivity, 472–473
  for particle mobility, 503
  for thermal conductivity, 473, 475, 487
group velocity, 204, 206–207
  in Boltzmann equation, 232
  definition, 206
  electron, 352

Hamaker constant, 419, 420
  negative, 421
  for various materials, 431
Hamilton equations of motion, 453–454, 455–456, 480–481
Hamiltonian, 50, 455, 480
  expectation value of, 73
Hamiltonian mechanics, 456
Hamilton's principle, 454
hard-drive devices, 308
harmonic oscillators, 59–62
  energy levels, 22, 59, 60
  probability distributions, 61
  wavefunctions, 60–61
heat carriers. *See* energy carriers
heat conduction. *See* conduction, heat
heat diffusion equation, 17
heat dissipation, 266
heat flux, 255, 266, 268
  at constant plane, 243–244
  between two points, 198, 243
  electron, 368, 369, 376
  microscopic expressions for, 478–479

phonon, 369
  for radiative transfer between two parallel plates, 293–295, 297, 394
heat flux autocorrelation function, 488–489
heat generation, 364, 374, 383
  due to photon absorption in semiconductors, 384–386
heat transfer
  classical definition, 9
  macroscopic theory, 9–18
    energy balance, 16–17
    local equilibrium, 17, 260–261, 492–493
    scaling trends under, 17–18, 19
  in microchannels, 434
  in nanochannels, 435
  radiative. *See* radiation, thermal
  See also conduction, heat; convection, heat; heat flux; radiation, thermal
heat transfer coefficient, 11
  for common configurations, 12
Heaviside function, 335
Heisenberg, Werner, 48
Heisenberg uncertainty principle, 51–52
  for free electron, 53
  for particle in one-dimensional potential well, 55
Helmholtz free energy, 129, 375
  for hard-sphere fluid, 407
Helmholtz layer, 421
Hermite polynomial, 60
heterojunctions
  current flow across, 307–308, 322–326
  double, electrical conductivity of, 345–346
  internal cooling of, 384
holes, 98
  energy flux of, 376
Hooke law, for isotropic medium without damping, 169
"hot electron" effects, 268
Huygens, Christian, 46
hydration force, 428
hydrogen atoms
  absorption lines, 69
  degeneracy, 67
  electronic energy levels, 64–70
  interatomic distance between, 73
  photon frequency and wavelength of emissions, 69
  spring constant in, 73
hyperbolic heat conduction equation, 258–260

ideal gas constant, 513
ideal gas law, 132
impurities. *See* dopants
incoherent transport, 209
insulators, 97–98
integral exponential function of nth order, 287
integrated circuit, fabrication process, 7–8
intensity, 233, 241–242
  phonon, 301
intensity distribution, of semiconductor laser, 177, 366
intensive variables, 374
interaction potential, between two close surfaces, 426
interactions, types of, 418
interatomic potential, 20, 427, 458–462
  for covalent bonding systems, 459–461
  for metals, 461
  for water, 461–462
interfaces
  diffuse, 286, 288, 303
  liquid–liquid, 428
  liquid–polymer, 428
  liquid–vapor, 428, 494
  partially diffuse and partially specular, 287–288, 303
  polymer–polymer, 428
  solid–liquid, 427, 428
  solid–vapor, 428
  specular, 286, 287, 303
interface specularity parameter, 216, 288, 291
interference matrix
  for acoustic waves
    through multilayer structure, 192–193
    through thin film, 192
  for electromagnetic waves
    through multilayer structure, 188
    through thin film, 187, 188
interference phenomenon, in thin films, 190, 224
intermolecular potentials, 417–419
internal energy
  of electrons in crystals, 143
  of gases, 138–141
    diatomic, 138–141
    monatomic, 138
  of phonons, 144, 145
  of photons, 146–147
  size effects on, 148–150
intrinsic carrier concentration, 506

Jeffreys equation, 259
Joule heating, 165

Kapitza resistance, 180
Kelvin equation, 439, 440
Kelvin relations, 256, 257
kernel, 294
kinetic coefficients, 257
kinetic energy, microscopic expression of, 476
Knudsen layer, 329
Knudsen minimum, 308, 313, 329–330
Knudsen number, 261
  electron, 287
  for molecules, 306–307, 315
  phonon, 288
  photon, 293
Kramers–Kronig relations, 466, 503
Kronecker delta function, 351
Kronig–Penney model, 91–97
  use in superlattices, 115
Krook equation, 237

Lagrange's equations of motion, 453–455
Lagrangian, 454
Lagrangian mechanics, 456
Lame constants, 168
Landauer formalism, 198–203
  for electron thermal conduction, 226
  for phonon heat conduction, 225
Langevin equation, 414
Langmuir equation, 426
Laplace equation, 432, 438
Laplace operator, 50
lasers, 363
  quantum well, 56
  semiconductor, 5–6, 56, 99, 176, 365
    heat generation distribution, 177
    intensity distribution, 177, 366
    short-pulse heating of metals by, 370–371
    vertical-cavity surface-emitting, 190
latent heat, 441
lattice(s)
  Bravais, 79
    body-centered cubic (bcc), 80
    face-centered cubic (fcc), 80, 81–82, 85, 88
    simple cubic (sc), 80
    types, 80
  crystal systems, 80
    triclinic, 79, 80
  description in real space, 78–81

INDEX 521

one-dimensional, 87, 91–97
  first Brillouin zone of, 95
  Wigner–Seitz unit cell of, 95
reciprocal, 87–90
  symbols for directions of, 89
two-dimensional, 78
lattice constants, 79
  of face-centered cubic (fcc) crystal, 85
lattice displacement, 351
lattice vibration, 100–105
  energy quantization in, 103–104
  normal modes, 102
  one-dimensional lattice chains
    diatomic, 103, 104, 105, 121
    monatomic, 100–103
    polyatomic, 104–105
  phonons in three-dimensional crystals, 105
  See also acoustic waves
lattice waves, 20–21
  damped, 215–216
  See also acoustic waves; lattice vibration
leapfrog method, 483, 484
left-hand rule, 167
Legendre transformation, 455
Lennard–Jones potential, 84, 406, 435, 458–459
  for argon crystal, 121
  parameters for noble-gas crystals, 85
  Tolman length for, 443
  truncation of, 459
  for water, 462
level degeneracy, 59
lifetime, of electrons, 366
Lifshitz theory, 420
light
  speed of, 513
  wave–particle duality of, 46–48
linear response theory, 464, 466–473
  calculation of dynamic susceptibility under, 466–471
  for particle mobility, 502–503
linear response to internal thermal disturbance, 473–476
Liouville equation, 229–230, 408, 464, 467
  linearized, 464
Liouville operator, 467
liquid membrane, stretching of, 428, 429
liquids
  bulk, 405–416
    kinetic theories of, 408–411
    See also Brownian motion

local energy density, 269
localization, wave, 212–213
local momentum density, 269
London force. *See* van der Waals potential
Lorentz number, 253, 513
luminescence, 364
lumped heat capacitance, 41
Lyman series, 69

Madelung constant, 86
magnetic disk drive, 6
magnetic field energy density, 165
magnetic field vector, 161
magnetic induction, 162
many-body problem, 453
mass conservation equation, 263
mass flux, 478
mass–spring system, 60
material waves, 48–49
Matthiessen rule, 237
Maxwell, James Clerk, 46–47
Maxwell distribution, 24
  displaced, 247
Maxwell equations, 15, 162–163
  applicability, 163
mean free path
  electron, 213–214
  for gases, 27–28
    definition, 27
    derivation, 27–28
  phonon, 41, 245
    estimation of, 246–247, 289–291
melting points
  of gold nanoparticles, 442
  of small crystals, 441–442
  of substances in pores of inert materials, 442
metallic bonding, 86
metals, 97–98
  alkali, 97
  electron transport in, 251–252
  electron–phonon scattering in, 355, 356, 357
  energy loss in, 356
  heat conduction in, 21
  noble, 97
microchannels
  fluid flow in, 433–434
  heat transfer in, 434

microscopes, 197–198
  atomic force, 197
  scanning tunneling, 197–198
Mie scattering theory, 241
Miller indices of crystal plane, 80–81
minority carrier devices, 381
mobility, 252–253, 401
molecular-beam-epitaxy (MBE) technique, 55
molecular chaos assumption, 232, 409
molecular dynamics simulation
  solving equations of motion, 483–486
  statistical foundation for, 462–483
    linear response to internal thermal disturbance, 473–476
    microscopic expressions of thermodynamic and transport properties, 476–479
    thermostatted ensembles, 479–483
    time average versus ensemble average, 462–464
    See also linear response theory
  of thermal transport, 486–494
    equilibrium method, 486–490, 494
    homogeneous nonequilibrium method, 493
    nanoscale heat transfer, 493–494
    nonequilibrium methods, 490–493, 494, 503–504
  velocity rescaling, 483, 492
  See also interatomic potential; motion, equations of
molecular partition functions, 130–134
molecular reflection, 303
  diffuse, 303
  specular, 303
molecular vibration frequency, 61
molecules, scattering of, 242
moment of inertia, 63
momentum accommodation coefficient, 303
momentum conservation equation, 264–265, 267
momentum flux, 478
momentum operator, 50
momentum relaxation time, 253, 358
momentum scattering, 267
Monte Carlo simulations, 316–317
MOSFET devices, 5, 370–373
motion, equations of, 453–458
  of ball thrown into air, 456–458
  for constant pressure system, 502
  equivalence of, 456–458

  Hamilton's, 453–454, 455–456, 480–481
  Lagrange's, 453–455
  Newton's, 453, 454, 455, 481
  solving, 483–486
    initial conditions, 485
    numerical integration, 483–484
    periodic boundary condition, 485–486
multidimensional transport problems, 316–317
  discrete ordinates method, 316
  Monte Carlo simulations, 316–317
  spherical harmonics method, 316

nanochannels
  fluid flow in, 434–435
  heat transfer in, 435
nanoscale heat source, transient heat transfer near, 336, 337
nanowires. *See* quantum wires
Navier–Stokes equations, 13, 263–266
Nernst equation, 422
neutron transport, 282
Newton, Isaac, 46
Newtonian mechanics, 456
Newton's equations of motion, 453, 454, 455, 481
Newton shear stress law, 13, 41, 247–249
  generalized, 266
Newton's law of cooling, 11
nonmagnetic material, 162
normal modes, 102
Nosé–Hoover thermostat, 482, 492
Nosé thermostat, 481–482
no-slip boundary conditions, 12, 434
N-particle distribution function, 229
numerical integration schemes, 297–298
Nusselt number, 11

Ohm law, 162, 251–252, 472
one-particle distribution function, 230–231
Onsager reciprocity relations, 257–258, 492
optical branches, 104–105
optical constants
  complex, 164
  of representative materials, 361, 362
optical fiber, 176
optical interference filters, 115, 116–117
optical thickness, 293

orbitals, 66–68
   1s-, 66–67
   2p-, 67
   2s-, 67
osmotic pressure, 412

pair potential. *See* two-body potential
paramagnetic material, 162
partial coherence theory, 210–211
particle, wavelength of, 48
particle conservation equation, 52
particle density, definition, 229
particle flux (current), 52, 73, 161, 255
particle regime, 32
partition functions
   canonical, 129, 405
      for diatomic gas, 139
      of rotational modes of hydrogen molecule, 133–134
   electronic, for diatomic gas, 138
   grand canonical, 130
   molecular, 130–134
   rotational, for diatomic gas, 138
   vibrational, for diatomic gas, 138
Pauli exclusion principle, 59
Peltier coefficient, 256, 376, 382
periodic functions, Fourier series expansion of, 87
periodic table, 67–68
   electron configuration of first 30 elements, 68
permeability, 162
   of free space, 513
permittivity, 162
   complex, 164
   of free space, 513
perturbation potential, 350
phase-breaking length, 32
phase coherence length, 214
phase space, 228
phase transition, size effects on, 438–443
phase velocity, 204, 207
   definition, 45
phonon(s), 20–21, 103–104
   average wavelength, 182
   coherence of, 214–216
   naming of concept, 103
   propagation in thin films, 191–193
   in three-dimensional crystals, 105
   tunneling of, 196–197
phonon entropy density, 271

phonon generation source, 333
phonon heat conduction
   across cylindrical thin shells, 346, 347
   across spherical thin shells, 346, 347
   across thin films, 299, 321
   along thin films, 201, 288–292, 313
   ballistic–diffusive equation, 333–336, 337
   in nonplanar structures, 315
   temperature definition in, 336
   in wire geometries, 291
phonon hydrodynamic equations, 268–274
   0th-order, 271
phonon–impurity scattering, 239
phonon interference filters, 191
phonon Knudsen number, 288
phonon–phonon scattering, 215, 239
phonon Poiseuille flow, 273, 313
phonon-polaritons, 200–201
phonon scattering, 215, 237–240, 291
   boundary, 239
   diffuse, 291
   normal process, 238, 239, 268–269
   specular, 291
   three-phonon scattering, 237–239
   umklapp process, 238, 239, 268–269, 272–273
   See also electron–phonon scattering; phonon–impurity scattering; phonon–phonon scattering
phonon transport. *See* phonon heat conduction
phosphorus, in silicon crystal, 507
photoluminescence, 364
photon(s), 22
   absorption of. *See* photon absorption
   concept of, 47
   emission of. *See* photon emission
   See also electromagnetic waves
photon absorption, 358–363
   by electrons in crystals, 359–363
      in metals, 360, 361
      in semiconductors, 360–363
   by gas molecules, 359
   by phonons, 359
   energy condition for, 61
   in semiconductors, 384–386
photon emission
   energy condition for, 61
   wavelength, 73
photonic crystals, 7, 117
photon Knudsen number, 293

photon scanning tunneling
  microscope, 198
photon scattering, 240–242
  elastic, 241–242
  inelastic, 240–241
photovoltaic cells, 391–395
Planck, Max, 47
Planck blackbody radiation law, 147
Planck constant, 47, 513
  divided by 2π, 48, 513
Planck–Einstein relations, 48, 73
Planck's law, 14
plane of incidence, 171
plane waves, 160–169
  complex representation, 160–161
  definition, 160
  interface reflection and refraction
    in, 169–185
  scalar, 160
  superposition of, 204–206
  vector, 160
  See also acoustic waves;
    electromagnetic waves;
    electron waves
plasmons, 201
p–n junctions, 376–384, 509–510
  band structure, 377
  built-in potential, 510
  energy conversion in, 376–384
    current flow, 377–381
    energy sources, 378, 381–384
    under forward bias, 378
Poiseuille, Jean Louis Marie, 433
Poiseuille flow
  phonon, 273, 313
  rarified gas, 308–313
Poiseuille law, 433
Poisson–Boltzmann equation,
  323–324, 423
Poisson bracket, 467
Poisson equation, 307, 323, 383, 510
position operator, 50
potassium, electron configuration, 68
potential wells. *See* quantum (potential) wells
Poynting vector, 165
  acoustic, 169
  time-averaged, 165–166
Prandtl number, 11
predictor–corrector method, 484–485
pressure, microscopic expression of, 477
primitive lattice vectors, 78

primitive unit cell, 79
  Wigner–Seitz, 78, 79
  in reciprocal space, 88
principal quantum number, 66
principle of detailed balance, 182
principle of equal probability, 125
proton mass, 513

quantum dots, 56, 114
  electron density of states in, 122
  electron energy states in, 75
  electron specific heat in, 157
  phonon specific heat in, 157
quantum efficiency, 364
quantum number
  definition, 54
  principal, 66
quantum (potential) wells, 111–114
  electron density of states in, 112–113
  formation, 111–112
  one-dimensional with finite height, energy
    levels, 75
  one-dimensional with infinite height,
    energy levels, 53–55
    wave function, 76
  phonon density of states in, 113–114
  phonon dispersion in, 113–114
  power factor of, 280
quantum states, accessible, 124
quantum well lasers, 56
quantum wires (nanowires), 8, 114
  electron density of states in, 122
  electron transmission in, 89
  growth of, 440–441
  square
    electron energy levels in, 56–58
    thermal conductance of, 201–203
quasi-Fermi levels, 379–380

radial distribution function, 405
  of crystalline solids, 405, 406
  of gases, 405, 406
  of liquids, 405, 406
radiation, thermal
  ballistic–diffusive equations for, 331–333
  between concentric cylinders, 314, 320–321
  between concentric spheres, 314, 320–321
  between two parallel plates, 292–299
    diffusion–transmission boundary
      condition for, 320
  in macrostructures, 7, 13–16

in microstructures, 7
in nanostructures, 7, 199
  between small gaps, 199–201
  thin films, 201
radiative transfer, equation of, 241–242, 292
radiosity, 293
random driving force, 414
ray tracing, 209
recombination, 266, 363–366
  Auger, 364
  nonradiative, 364, 365
  radiative, 364, 365
recombination length, 385
recombination rate, 365
  net, 365
  surface, 366
recombination velocity, surface, 366
reflection
  total internal
    of acoustic waves, 180
    of electromagnetic waves, 176
reflection coefficient
  for acoustic waves through thin film, 192
  for electromagnetic waves through thin film, 188
  for electron wave, 170
  for horizontally polarized shear wave, 178
  for transverse electric wave, 174
  for transverse magnetic wave, 174
  for vertically polarized shear wave, 179
reflectivity, 74–75
  of acoustic wave, 178–180
    through multilayers, 193
  for air/glass interface, 175
  for air/silicon interface, 175
  for dielectric materials, 175
  of electromagnetic wave, 174–175
    through multilayers, 190
    through thin film, 189–190
  of electron wave, 170
  for gold, 175
refractive index
  complex, 164
  of silver, 206–207
refrigeration
  thermoelectric, 6, 280, 387–389
    coefficient of performance, 388
    figure of merit, 388
    maximum temperature difference, 389
relaxation, of excited state due to photon absorption, 363–365

relaxation time, 31, 236–237, 354
  electron–electron, 365
  electron–phonon, 240, 365
    in metals, 355
    in semiconductors, 355
  energy, 253, 358
  for molecules, 242
  momentum, 253, 358
  phonon, 41, 245
  single-mode, 357–358
  total, 237
relaxation time approximation, 236
response function. *See* susceptibility, dynamic
Reynolds number, 11
Rice–Allnatt equations, 409
Richardson constant, 325, 327
right-hand rule, 166–167
rigid rotor, 63–64
  energy levels, 63
Rosseland diffusion approximation, 279, 319
rotational constant, 63
rotational temperature, 134

saturation current, 381
saturation pressure, 449
scanning tunneling electron microscope (STM), 197–198
scattering, 233–242, 291
  elastic, 349
  of electrons. *See* electron scattering
  incoming, 235, 241
  inelastic, 349
  of molecules, 242
  momentum, 267
  outgoing, 235, 241
  of phonons. *See* phonon scattering
  of photons. *See* photon scattering
  in semiconductors, 240
  small angle, 358
scattering integral, 234–237
scattering matrix, 235, 351
scattering phase function, 241
Schottky barrier, 322
Schottky theory, 324
Schrödinger, Erwin, 48
Schrödinger equation, 49–52, 452
  for free particles, 52–53
  for harmonic oscillator, 60
  for hydrogen atom, 65
  for particle in one-dimensional potential well, 53–55

Schrödinger equation (cont.)
    for rigid rotor, 63
    steady-state, 51
screw dislocations, 83
second sound, 260, 272
Seebeck coefficient, 254
    along thin film, effects of interface
        scattering on, 344–345
    of metal, 281
    of nondegenerate semiconductor, 257–258
    of nondegenerate silicon, 280–281
    of quantum well, 280
Seebeck voltage, 255
selection rule, 61
semiconductors, 97–98
    band structure near minima, 99–100
    direct gap, 99
    electron mobility in, 252–253, 401
    electron transport in, 252–253
    energy exchange in. See energy exchange
    energy loss in, 357
    extrinsic, 117–118
    heat conduction in, 21
    homogeneous, 505–508
    hot electron and hot phonon effects
        in, 370–373, 402
    indirect gap, 99
    intrinsic, 97, 506
    nondegenerate, 506
        Seebeck coefficient of, 257–258
    n-type, 97, 98, 507
        chemical potential in, 142–144
    p–n junctions. See p–n junctions
    p-type, 97, 98, 507
    radiation heating of, 384–386
    scattering in, 240
shear stress, 248, 266
    microscopic expressions for, 478, 479
    See also Newton shear stress law
shear stress autocorrelation function, 488
shear viscosity, Green–Kubo expression
    for, 476
Shockley equation, 381
signal velocity, 206
silver, refractive index of, 206–207
single particle potential, 458
size effects
    classical, 28–29
    on density of states, 122
    on electron and phonon conduction parallel
        to boundaries, 283–292
    on internal energy, 148–150
    on phase transition, 438–443
    quantum, 29–30
    on specific heat, 148–150
    on transport in nonplanar structures,
        313–316
    on transport perpendicular to boundaries,
        292–307
    See also ballistic–diffusive equations;
        diffusion–transmission boundary
        conditions; Poiseuille flow
skin depth, 166
slip boundary conditions, 327–328
Snell law, 173
    with absorbing medium, 174
    for horizontally polarized shear wave, 178
    for transverse wave polarized in plane of
        incidence, 179
solar cells, 391–393, 395
    efficiency, 393
    electrolytic, 395
solar radiation, 14, 361–363
solid angle, 111
solids, atomic vibrations in, 62
solvation force, 427
source function, 285
space-charge regions, 324, 377, 509, 510–511
specific heat
    of carbon nanotubes, 149–150
    of electrons in crystals, 141–144
    of gases, 138–141
        diatomic, 138–141
        monatomic, 138
    of gold, 147–148
    of helium, 24–25
    of hydrogen, 140–141
    of phonons, 144–146
        Debye model, 144–145
        Einstein model, 145–146
    of photons, 146–147
    size effects on, 148–150
    of titanium dioxide nanotubes, 149
    volumetric, 374
specific thermal boundary resistance, 182
spectral emissive power, 14
specularity parameter, 216, 288, 291
speed of light, 513
spherical harmonics functions, 63
spherical harmonics method, 316
spin down, 58
spin quantum numbers, 58

spin up, 58
statistical mechanics, fundamental postulate in, 124–125
statistical thermodynamics, 123
See also ensembles; internal energy; specific heat
Stefan–Boltzmann constant, 15, 513
Stefan–Boltzmann law, 15
steric force, 428
Stern layer, 421
stiffness tensor, 167–168
Stillinger–Weber potential, 86, 459, 461
  for silicon, 459, 460
Stirling approximation, 132
stochastic wall method, 490, 491
Stokes law, 412
stop bands
  phonon, 193
  photon, 117, 190–191
strain tensor, 167
stress tensor, 167
  calculation of components of, 169
superlattices, 55, 115
  energy states in, 115–116
  heat conduction in, 300–302, 321
  Si/Ge, 115
  thermal conductivity of, 215, 289, 290, 302, 321, 494
  unit cells of, 116
surface effects, investigation of, 485–486
surface forces
  due to molecular structures, 427–428
  for various materials, 431
surface impedance, 187
surface states, 428
surface of tension, 429
surface tension, 428–433
  of bubbles, 443
  curvature effects on, 442–443
  of droplets, 442–443
surface waves, 394
  tunneling of, 199–201, 394
susceptibility
  dynamic, 465–466
    calculation under linear response theory, 466–471
    frequency and wavevector dependent, 471
  electric, 162
system energy, microscopic expression of, 476

temperature
  classical definition, 9
  equivalent equilibrium phonon, 184–185
  equivalent equilibrium photon, 296
  microscopic expression of, 476–477, 503
  for monoatomic gas, 24
temperature slip, 295
temperature tensor, 267
Tersoff potential, 460–461
  for silicon, 461
thermal boundary resistance, 180–185, 494
  analogy for photons, 225
  at higher temperature, 183
  at imaginary interface, 184
  at low temperature, 182–183, 185
  specific, 182
thermal conductivity, 10, 27
  of amorphous materials, 488
  of carbon nanotubes, 493
  of common materials, 10
  of crystalline materials, 488
  of diamond, 488
  effective, for fast transport processes, 475
  of electrons, 253, 372
  of gallium arsenide, 246
  of gases, 28, 248–249, 280
  of gas slab, 322
  Green–Kubo formula for, 473, 475, 487
  of liquids, 41, 410–411
  of metallic films, 288
  of phonons, 245–246, 279, 372
    reduction of, 389–390
  of silicon crystals, 488, 489
  of silicon nanowires, 494
  steady-state, 475
  of superlattices, 215, 289, 290, 302, 321, 494
  of thin films, 29, 321, 494
    3ω measurement method, 39–40
    AC calorimetry measurement method, 38
    laser pulse measurement method, 40–41
    membrane measurement method, 38–39
    steady-state measurement method, 38–40
thermal de Broglie wavelength, 131
thermal equilibrium, 9
thermal length, 209, 214
thermal radiation. *See* radiation, thermal
thermal reservoir, 127

thermal velocity, 214
thermal waves, in dielectrics, 260
thermionic emission, 307, 322–326
thermocouple, 255
thermodynamic potential, 126
thermodynamics, first law of, 16, 255, 374
thermoelectric devices, 6–7, 373, 386–391
   cooling/refrigeration, 6, 280, 387–389
      coefficient of performance, 388
      figure of merit, 388
      maximum temperature difference, 389
   nanostructure benefits, 390–391
   power generation, 387, 389–390
      maximum efficiency, 389
thermoelectric effects, 254–258
thermovoltaic power conversion, 391–395
   basic principles, 391–393
   dielectric-coupled generator, 403
   efficiency, 393
   nanostructure benefits, 394–395
thin films
   absorptivity of, 189
   colors of, 224
   electrical conduction along, 285–288, 313
   phonon heat conduction across, 299, 321
   phonon heat conduction along, 201, 288–292, 313
   wave propagation in, 185–194
Thomson coefficient, 257
Thomson effect, 256–257
Thouless length, 214
three-body potential, 459–460
time average, 463
titanium dioxide nanotubes, specific heat, 149
Tolman length, 443
Tolman theory, 442–443
trajectory, 456
transfer matrix method, 186, 190
transition probability, 358
transition rate, 234
translational energy, 69–70, 76
transmission coefficient
   for acoustic waves through thin film, 192
   for electromagnetic waves through thin film, 188
   for electron wave, 170
   for horizontally polarized shear wave, 178
   for transverse electric wave, 174
   for transverse magnetic wave, 174
   for vertically polarized shear wave, 179
transmission electron microscope, 73
transmissivity, 74–75, 170
   of acoustic wave
      through multilayers, 193
      through Si/Ge-like superlattice, 193
   of electromagnetic wave, 174–175
      through multilayers, 190
      through thin film, 189–190, 209–210
   of electron wave, 170
tunneling, 195–198
   of electromagnetic waves, 195–196
   of electron waves, 74–75, 195, 196
   of phonons, 196–197
   of surface waves, 199–201, 394
tunneling diode, 197
two-body potential, 458–459

units, conversions of, 512
unit step function, 335
universal gas constant, 138
universal quantum thermal conductance, 201–203

vacuum permittivity, 162
valence bands, 97, 98
   heavy-hole, 505
   light-hole, 505
   silicon, 99, 505
valence electrons, 69
van der Waals equation of states, 407
   isotherms predicted by, 407
van der Waals potential, 84, 419, 430, 431
   between one atom and surface, 420–421
   between two nanowires, 449
van Hove singularity, 110
velocity autocorrelation function, 416
velocity rescaling, 483, 492
velocity slip, for rarified gas flow, 326–330
velocity Verlet method, 484
Verlet method, 483–484
vibrational temperature, 138
virial theorem, 477
viscosity
   dynamic (absolute), 13, 248–249
   kinematic, 13
   of liquids, 410–411, 413, 433–434
      solution, 414
      solvent, 414

of nanofluids, 449
shear, Green–Kubo expression for, 476
viscosity tensor, 167

wafers
  semiconductor, 81, 82
    primary flat, 81, 82
    secondary flat, 81, 82
water molecules, vibrational normal modes, 62
water potentials, 461–462
wavefunction, 49
waveguides, 176
wave localization, 212–213
wavenumber, 44
wave packets, 205, 206
  momentum of, 206
  traveling and interference of, 208–209
wave regime, 32
waves
  basic characteristics, 44–46
  energy in, 46
  horizontally polarized shear (SH), 178
    reflection coefficient for, 178
    transmission coefficient for, 178
  inhomogeneous, 224
  longitudinal, definition, 44
  longitudinally polarized shear (L), 179
  material, 48–49
  standing, definition, 45
  transverse, definition, 44
  transverse electric (TE-polarized), 171
    reflection coefficient for, 174
    transmission coefficient for, 174
  transverse magnetic (TM-polarized), 171
    reflection coefficient for, 174
    transmission coefficient for, 174
  traveling, definition, 45
  vertically polarized shear (SV), 179
    reflection coefficient for, 179
    transmission coefficient for, 179
  See also acoustic waves; electromagnetic waves; electron waves; evanescent waves; plane waves; thermal waves
Wiedemann–Franz law, 253
Wien displacement law, 14
Wigner function, 233
Wigner–Seitz cell, 78, 79
  in reciprocal space, 88
work of adhesion, 430, 431
work of cohesion, 430
work function, 323

X-ray diffraction, 89, 90–91, 121

Young, Thomas, 46
Young equation, 432–433

zeta potential, 436–437, 438

Units and Their Conversions

Measures for small quantities
milli 10^{-3}; micro 10^{-6}; nano 10^{-9}; pico 10^{-12}
femto 10^{-15}; atto 10^{-18}; yactto 10^{-21}

Length
1 m = 10^3 mm = 10^6 μm = 10^9 nm = 10^{10} Å

Time
1 s = 10^3 ms = 10^6 μs = 10^9 ns = 10^{12} ps = 10^{15} fs

Force
1 N = 10^3 mN = 10^6 μN = 10^9 nN = 10^{12} pN = 10^{15} fN = 10^{18} aN

Energy
1 J = 10^3 mJ = 10^5 Dyne = 10^6 μJ = 10^9 nJ = 10^{12} pJ = 10^{15} fJ = 10^{18} aJ = 10^{21} yJ

How much is 1 eV?
1 eV = 1.6×10^{-19} C × 1 V = 1.6×10^{-19} J
1 eV $\leftrightarrow \nu = E/h = 1.6 \times 10^{-19}$ J$/(6.6 \times 10^{-34}$ J s$) = 2.4 \times 10^{14}$ Hz
1 eV $\leftrightarrow \lambda = c/\nu = ch/E = 3 \times 10^8$ m s$^{-1} \times (6.6 \times 10^{-34}$ J s$)/(1.6 \times 10^{-19}$ J$) = 1.24$ μm (photon)
1 eV $\leftrightarrow \eta = 1/\lambda = 8.06 \times 10^5$ m$^{-1} = 8.06 \times 10^3$ cm^{-1} (photon)
1 eV $\leftrightarrow 38 \kappa_B T$ (for $T = 300$ K)

How much is $\kappa_B T$ for $T = 300$ K?
$\kappa_B T$ (for $T = 300$ K) = 1.38×10^{-23} J K$^{-1} \times 300$ K = 4.14×10^{-21} J = 26 meV

Physical Constants

Physical constant	Symbol	Value	Units
Speed of light	c	2.997×10^8	m s^{-1}
Planck constant,	h	6.6262×10^{-34}	J s
Planck constant divided by 2π	$\hbar$	1.0546×10^{-34}	J s
Avogadro's number	N_A	6.0222×10^{23}	mol^{-1}
Electron rest mass	m	9.1096×10^{-31}	kg
Proton mass	m	1.67×10^{-27}	kg
Proton mass/electron mass ratio		1836.1	
One electron volt	1 eV	1.6022×10^{-19}	J
Boltzmann constant	κ_B	1.38×10^{-23}	J K^{-1}
Permittivity of free space	ε_0	8.8×10^{-12}	F m^{-1}
Permeability of free space	μ_0	$4\pi \times 10^{-7}$	s^2 F^{-1} m^{-1}
Stefan–Boltzmann constant	σ	5.67×10^{-8}	W m^{-2} K^{-4}
Ideal gas constant	R	8.314	J K^{-1} mol^{-1}
Lorentz number	L	$\sim 2.45 \times 10^{-8}$	W Ω K^{-2}
Universal ideal gas constant	R_u	8.314	J K^{-1}mol^{-1}

CPSIA information can be obtained at www.ICGtesting.com
Printed in the USA
BVOW11*1752150915

418021BV00005B/20/P